Richard L.Gregory

Mind in Science

A History of Explanations in
Psychology and Physics

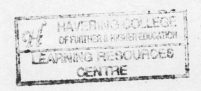

Penguin Books

to
Ma, Margaret, Mark, Romilly and Freja

PENGUIN BOOKS

Published by the Penguin Group
27 Wrights Lane, London W8 5TZ, England
Viking Penguin Inc., 40 West 23rd Street, New York, New York 10010, USA
Penguin Books Australia Ltd, Ringwood, Victoria, Australia
Penguin Books Canada Ltd, 2801 John Street, Markham, Ontario, Canada L3R 1B4
Penguin Books (NZ) Ltd, 182–190 Wairau Road, Auckland 10, New Zealand

Penguin Books Ltd, Registered Offices: Harmondsworth, Middlesex, England

First published in Great Britain by George Weidenfeld and Nicolson Ltd 1981
Published in Peregrine Books 1984
Reprinted 1988

Copyright © Richard L. Gregory, 1981
All rights reserved

Reproduced, printed and bound in Great Britain by
Hazell Watson & Viney Limited
Member of BPCC plc
Aylesbury Bucks

Peregrine Books

Mind in Science

Richard Gregory read Moral Sciences at Downing College,
Cambridge. His appointments have included Lecturer in
Psychology at Cambridge, where he designed the Special
Senses Laboratory for work on vision and hearing, and for
simulation of perceptual problems to be anticipated in
astronauts; Professor of Bionics at the Department of
Machine Intelligence and Perception, Edinburgh
University; and four visiting professorships in the United
States. He is at present Professor of Neuropsychology and
Director of the Brain and Perception Laboratory,
University of Bristol. Among his many publications are
The Intelligent Eye (1970); *Concepts and Mechanisms of
Perception* (1974); and *Eye and Brain* (third revised edition
1977). He is also the editor of the *Oxford Companion to the
Mind* (1983).

Contents

ECT Energy and information Carnot's Principle *Information, the Second Law and Maxwell's Demon* Mechanical and intelligent Demons Measuring information Measuring meaning Biological atoms: Are cells intelligent Demons? The *Origin of Species:* The origin of the Universe The origin of life by Natural means The evolutions of men and women Is man beyond the evolutionary sequence?

Thanks

Thanks are due to far more people than I can mention. An especial debt is due to Professor Oliver Zangwill, FRS, Head of the Department of Psychology at Cambridge, for most helpful comments on a draft of the manuscript. Dr Bruce Goatly wrote detailed comments, and pointed out so many mistakes that he has saved me from embarrassments, and the reader from being misled, in several places. Any mistakes that remain I have probably added since his invaluable work on this book.

John Curtis, Dr Robert Baldock and Paula Iley deserve my thanks for all they have done by way of encouragement and the work of seeing it through the press. This has been possible through the faith and support, over longer than I can fully justify, of Lord Weidenfeld.

Many people have helped with the typing of various drafts, including Libby Parsons and Sheila Somers.

I am grateful to the publishers concerned for permission to quote from R. Hanson, *Patterns of Discovery*, Cambridge University Press; G. Kirk and J. Raven, *Pre-Socratic Philosophers*, Cambridge University Press; D. de Solla Price, *Gears from the Greeks*, American Philosophical Society; and L. Wittgenstein, *Philosophical Investigations*, Basil Blackwell.

Pretext

Our central theme and aim is to relate historical and also present-day accounts of how we perceive and learn, and understand, with how science works to produce its kind of knowledge. From the beginning of recorded thought there has been if not a marriage at least conjugal relations between philosophy, psychology and physics – to form a triangle which if not eternal is as old as historical records and looks like continuing as long as we do. This 'three body' problem of their interactions, by ever-changing attractions and repulsions through recorded time is the drama of this book.

This, or rather a much better book, should have been written jointly with the American philosopher of science, the late Norwood Russell Hanson. He will not be forgotten by any of us who had the pleasure of his friendship. His originality and remarkable energy live on in his books and in our minds. Russ Hanson was an American of the heroic mould. He came to Oxford shortly after World War II and then to Cambridge, where I knew him. The record of his achievements is dazzling: scholar, athlete, ace pilot in the U.S. Air Force, with great ability as a musician – playing organ and trumpet. As scholar, he combined philosophy with physics. At Cambridge he was extremely active in building up the teaching of History and Philosophy of Science, and he rose to administrative heights in the University. After but a few years he left Cambridge, to become Professor of Philosophy at Indiana, where again he impressed his personality indelibly. Born in 1924, Russ Hanson died in 1966 in a flying accident in his Grumman Bearcat in which he hoped to break the world speed record for piston-engined planes.

Collaboration for our planned book, which was never started, was delayed by Russ Hanson's return to America, to Indiana, and was finally made impossible by the flying accident that ended his life. From the too few discussions we had I know too well what I missed in stimulation and challenge which the collaboration would have brought. The result must sadly lack Russ Hanson's mind. One may hope that something of his inspiration as scholar and friend will appear through these pages.

A major theme has turned out to be the slow change from ancient explanations of the natural world, seen as controlled by Minds of gods, with the gradual exorcism of Mind from the Universe through control of our world of technology and by the Natural Sciences which are inspired by prevailing and ancient technologies. This by now, at least within science, almost complete exorcism of Mind from the Universe of matter makes psychology a very curious science; for psychologists still generally accept Mind as controlling *some* matter: intelligent living organisms – especially ourselves. What makes organisms so special that they, alone of all matter, are controlled by Mind? Or is this, our usual view of Mind, an absurd hangover from mythological pre-scientific thinking, which was dominated by animistic projection of ourselves into the Universe? This is a claim of extreme Behaviourist psychologists – that there is no Mind as generally conceived associated even with brains. But what, then, is the *observer*: Can an observer be Mindless? And what is the questioner, the philosopher, the scientist, or the inventor who puts his ideas into practice to create in the world things that were never there before except perhaps in dreams? Surely human creativity is not illusion!

The part played by intelligent observation, as it is studied and described by psychologists and physiologists, is hardly at all allowed in physics beyond frankly inadequate statements such as: 'The eye is an optical instrument – it sees like a camera.' Psychology may be low in the pecking order of the sciences but at least it has convincingly shown this account misleadingly inadequate. Even though still largely mysterious it is clear that questions of how we see are incomparably more interesting than how a camera produces pictures; and it is absolutely worth understanding perception, and learning and intelligence, in the rich detail demanded as a matter of course for questions in physics. Nevertheless, accounts or theories of Mind invariably rely upon physical analogies, especially to engineering devices and principles, such as tensions of springs; pressures of fluids in hydraulics; telephone switching circuits and now analogue and especially digital computers. Can physics serve as a mirror in which we may see ourselves? We will not find out without looking, in many directions and in considerable detail, at concepts of the physical sciences. This we attempt to do; so we look at ancient and modern notions of force, energy, electricity and magnetism, entropy and information, and notions of the origins of things, including life – though many questions will of course remain unanswered.

I can only hope that this exploration of Mind in Science may help the reader to think for himself or herself, as it has made me think and read books and papers that I would not otherwise have met or bothered with. It has taken me to Sumerian cuneiform tablets, to Aristotle, and to the pre-Socratic philosophers whose germinal thoughts sadly exist only in tantalizing fragments. It has made me question what words such as 'Mind' and 'Mechanism' do or should mean; and to wonder whether philosophers, through the ages, have failed to appreciate vital contributions of technology to abstract thinking, and to all understanding in science. The recent description, by the historian of science Derek de Solla Price, of a Greek gear-wheel computer which simulated planetary cycles must, alone, change one's view of the Greeks – suggesting dependence upon models and processes of technology for philosophy and science in their day and in ours.

What are my qualifications to write on such themes as these? I wish they were adequate. I read philosophy (Moral Sciences) at Cambridge, where I met as teachers the minds of Bertrand Russell, Richard Braithwaite, John Wisdom, A. C. Ewing (for whom I wrote an essay every week for two years), and other great men. The weekly discussions of the Moral Sciences Club, which met in Braithwaite's rooms in King's, is still vivid in my mind. Then I read Psychology in the Department of a man whom I most deeply respect: Professor Sir Frederic Bartlett. He was strong, kind and wise. He is the grandfather of my mind.

For seventeen years I was a lecturer in his Department, which Professor Oliver Zangwill took over from Bartlett. Oliver Zangwill taught me that the history of psychology is interesting; and that intuitive 'feel' has a guiding power which does not necessarily by its seduction lead one astray. He generously supported my attempts to build up a research group whose philosophy in some ways was not his. As well as studying the human senses, we tried to invent and develop with more or less success novel instruments; and we became involved in space research, with support from the U.S. Air Force, to investigate anticipated problems in docking and moon-landing. For this we built a space simulator: an electric railway running down the central corridor of the laboratory, with electronically generated displays, and a large swing for investigating effects of acceleration on how men might mis-see in space. These were great for visitors!

Cambridge is a tolerant place. I was an experimental psychologist, but with freedom to move around in other domains, even astronomy. My father was an astronomer. Perhaps early memories of great glass eyes following the stars, and the insubstantial flame-like images of incredibly distant objects, made one think about what it is to see: to ask how vision shoots arrows through space and time – though what one sees may differ dramatically from what one knows is out there. How could my brain comprehend what was there far out in space way beyond touch? And if this is hard to explain how do we begin to comprehend what lies within another's head? What causes the flash of recognition or of amusement in another's eyes; or the questions he, hook-like, snatches from sensed reality to see things in his way? The glass lens of the telescope was not an eye after all: it was merely another window for the curious intelligence of the eye's Mind.

I have spent over twenty years studying illusions: illusions of vision and of some of the other senses. This may seem bizarre – why illusions? Why study error when truth is the aim of the experimenter? Illusions appealed to me because they are measurable discrepancies between what we see as true and what we believe to be true. This allows science, with its methods, to look into the heart of epistemological questions. It gradually became clear that errors of perception are more subtle and suggestive than accounts then available in text books of psychology. I began to realize that some well-known illusions might be generated by *procedures* employed by the intelligent visual system for deriving perceptions from hardly adequate sensory signals – when the normally useful procedures are not appropriate to the situation or the problem to be solved. Other errors or illusions could be due to disturbances of physiological mechanisms or of neural signals. So two very different kinds of causes could produce dramatic illusions; and by discovering whether inappropriateness of procedure or upset of signalling mechanisms is responsible in each case we could see significance in illusory phenomena. The emphasis upon procedures, when they are appropriate or inappropriate, seemed to me then and still does seem the road towards generalizing neural function beyond brains to computers – and so for sorting out what is logically necessary from contingent neural properties – for similar procedures can run on physically very different computers.

Since this book is a somewhat personal exploration rather than a conventional text it is fairly freely structured. It comes to only a few firm conclusions as it aims to allow and help the reader to form his or her own views and opinions on essentially controversial issues. I do, though, present some conclusions and some opinions, which the reader may or may not wish to follow. Inevitably we meet some theological issues, for we are concerned with the nature of Mind and its relation – our relation – with the Universe. Some may find the account arid or bleak as perhaps few of the traditional family relationships with the Universe remain following the mathematical accounts of physics: especially of astronomy, and biology following Darwin. But, surely, the very questioning of our relation to the Universe and how we know each other makes us special; and questions remain in plenty. In any case, whatever the Universe we live in is – it inspires awe and is wonderful, as is our existence.

Themes and Conclusions

At least from the beginning of records tools of technology have been seen as challenging the gods, by doing rather better than magic in many situations; and technologies have provided answers, and set suggestive questions which have led to quite new possibilities and new knowledge. Babylonian pickaxes proved more powerful than magic for controlling flood and drought, and this practical challenge to ancient beliefs remains today to challenge how we may prefer to see the world and Mind.

We still see tools and machines, and also procedures that clearly work, as confirmations and as tests of theories. This applies to theories of nature and of Mind. Many examples for medicine are given by Jonathan Miller in his book, *The Body in Question*. We discuss machines, in some detail, as suggesting and illuminating meanings for mental as well as physical 'function', 'motivation', 'autonomy', and 'minder', and so on. Such words apply to Mind and machines, which reflects ancient ideas of Mind suggested by technologies and theories of matter. Philosophies of Mind seem to have depended ever since the early Greeks on theories of matter and the function of devices. Conversely, and almost as important, technological shortcomings or gaps seem to have engendered long-lasting philosophical confusions, where analogies from somewhat practical experience were not available for providing conceptual models.

It is suggested that the sophistication and the effects of ancient technologies have generally been underestimated by the philosophers of the time, and also by later classical scholars, who are not generally interested in such matters as tools, toys, or processes of design and manufacture. Exceptions however include the present-day classicists and historians: R. S. Brumbaugh; G. E. R. Lloyd; G. S. Kirk – whose invaluable book written with J. E. Raven, *The Presocratic Philosophers*, I have, with their kind permission, used extensively – D. de Solla Price who has already been mentioned for his remarkable work on the First Century BC Antikythera computing mechanism, which is illustrated here with his permission. A particular and deep debt is owed to these distinguished scholars.

I hope to have made their work a little more accessible, and that these issues will be more generally considered, as they do seem central to appreciating the sources of our knowledge and understanding by invention and discovery. This view should also help to raise, justly, the academic status of technology – technology being, as we shall see it, often beautiful solutions of ancient and modern questions: to be no less than embodied intelligence – solutions on tap for our use or misuse. Here I confess to departing from Arthur Koestler's view, in his brilliantly written *The Sleepwalkers* and *The Ghost in the Machine*. Although no doubt ideas do sometimes come as though from a mysterious source, in sleep, it seems to me that Galileo, Newton, and the other great architects of our understanding were very much awake. And they were generally aware of their heritage of science and technology from past discoveries, and were dependent upon them. More particularly, I see our creativity as brain-based and not as coming from Mind quite separate from matter. This allows the possibility of machines, in practice computers, being creatively intelligent. Programs indeed have a kind of immortality and existence of their own, though they cannot operate entirely free of matter.

Current views of intelligence testing and rating along a single (I.Q.) dimension are strongly criticized for forcing the rich variety of human potentiality and accomplishments into a too narrow mould, which limits and distorts our views of ourselves and each other, and generates serious social problems. I take the view that recent developments in Artificial Intelligence, far from being de-humanizing, increase understanding of sensitive issues of our very different individual abilities, and kinds of creative potentiality; and this should be socially rewarding if, as I suggest, they dispel damaging and dangerous myths of our time such as accepting that human abilities and potentials can be measured along a single continuum. A I is the experimental philosophy that allows us to ask, with at least hope of answering, questions which until recently were rejected as philosophically *meaningless*, because no conceivable experiments could be brought to bear on them. Can anything that enriches meaning be bad?

Intelligence is described as present in *potential* form, as available in existing solutions to problems that have already been solved; in the design of tools, which are part or complete solutions, or of techniques or theories – whether the designs be those of man-made science or technology, or of solutions discovered by processes in organic evolution by Darwinian Natural Selection. Intelligence we see also in *kinetic* form, in the activities of applying data (including the selection of inherited answers) to solve current problems. On this view intelligence is by no means limited to brain activity. We see Evolution by Natural Selection as a form of kinetic intelligence which is not brain-based; and Evolution has of course solved many problems beyond our present understanding.

How structures can become more complex, though generally things tend to chaos, by trading information for energy, is discussed here; though for a more sophisticated account of these difficult issues the reader is referred to Ilya Prigogine's recent book, *From Being to Becoming*.

Looking back through the history of science at Mind seen as controlling and being controlled by matter, including Newton's view of Mind in his absolute space which is God's consciousness, we discuss mind-full alchemy, which grew into our mind-less chemistry; and ancient mind-full astrology which fathered and lived for millennia with mind-less astronomy, in which empowered by tools of mathematics we are god-like observers, of vast vistas of space and time. This takes us to current accounts of how Mind is seen as related to matter: or whether ultimately Mind is identical with matter in certain forms or activities. Here we come to consider whether *we* are machines, capable of learning and knowing. This makes us consider procedures of logic – Deductive, from the general to the particular, and Inductive from the particular to the general – in order to see whether there are Laws of Thought, which might perhaps be somewhat similar to physical laws. From here we go on to the structure of scientific knowledge itself, to ask: Does how science works to generate knowledge provide a paradigm for how organisms, including ourselves, gain knowledge by biologically endowed perception and learning? In short, does the intelligent brain work by rules like the procedures of scientific inquiry? Given this: Can the methods of science tell us about the nature of Mind?

Perceptions are considered as being *predictive hypotheses* – essentially similar to hypotheses in science. This leads to a rather detailed account of bases of perception, though not so much of its characteristic phenomena as I have discussed these in previous books. Here we become involved in particularly controversial issues, especially the range of application of Sir Karl Popper's extremely influential notion, that knowledge can only be gained by refuting hypotheses. Looking at perception and learning in this light, it seems to me very difficult indeed to believe that we do in fact see and learn only by positing and refuting hypotheses. This issue of the nature of induction and how knowledge can be gained is central in the philosophy of science, and it should be central in psychology also. The status of induction is seen here as an *empirical*, rather than a *logical* issue. The suggested conclusion is that although Sir Karl Popper's emphasis upon formulating hypotheses so that they can be refuted is admirable advice for any scientist, it is not however the case that all, or indeed most, knowledge comes by refuting hypotheses. It is suggested that the apparent assymmetry between refuting and confirming for gaining knowledge is due to the generally greater surprise – and so greater information in the technical sense – gained from refuting established knowledge than confirming by further examples what is already accepted. I support, rather, Francis Bacon's original account of Inductive procedures; both as the initial statement of a sound working model of science and also as the basis of how animals and humans learn through experience.

Having looked at ancient science and philosophy – and the book is not strictly structured chronologically as we discuss many themes – we move rather rapidly to the seventeenth century, where we meet the architects of modern science, and the origins of present-day psychology and how it was born, not without trauma, from philosophy and the physical sciences.

Many theories in physics, even apart from physiology, are quite directly relevant for considering Mind. For example, theories of the nature of light may be seen as attempts to link the world to perception, and so to Mind. The perennial difficulties in understanding the nature of light are possibly so great because light is not like matter that can be touched and handled. These very difficulties are clues to seeing how we understand, and why Mind presents such difficulties. Here we look also at further links to Mind, including the transmission of signals in nerves, and at brain function, though we do not expound neurophysiology as there are many far better accounts than I could attempt. We discuss such issues as how functions are localized in the brain and whether, and if so why, phrenology is misleading; why it is so difficult to establish the physical basis of memory, and finally whether Mind is like programs run on man-made digital computers.

The new science and engineering of Artificial Intelligence is the Experimental Philosophy which at last brings effective techniques to bear on some of the most ancient unsolved questions of Mind. I was captured by the dream of A I – to leave Cambridge and join Donald Michie and Christopher Longuet-Higgens in the adventure of setting up with them the Department of Machine Intelligence and Perception at the University of Edinburgh. I had the grand title of Professor of Bionics. This was in the generation following the key theoretical insights, of Alan Turing (whom I met several times when a student) in England, and in America especially the mathematician John von Neumann, and, interfacing into psychology, Allen Newell and Herbert Simon. We built (though not until after I left Edinburgh) the Robot Freddy, which was one of the first to recognize objects from any point of view, and to assemble models from randomly scattered parts. Freddy's hardware was designed and built by Steven Salter, who worked with me on several projects including an 'intelligent' telescope camera, for minimizing the annoying effects of atmospheric turbulence, which we tested in New Mexico and Arizona, and later at the Royal Greenwich observatory. Principally responsible for Freddy's software were: Robin Popplestone ('Pop'), A. P. Ambler, H. G. Barrow, C. M. Brown and R. M. Burstall. We enjoyed active and exciting collaboration with the corresponding Departments at M I T and at Stanford, especially with the outstanding leaders in the field, Marvin Minsky and John McCarthy.

The philosophy of A I is discussed here, though hardly adequately in its details. Considering the questions which are so difficult even to formulate, let alone answer, of consciousness and computers, though I regard the philosophy and experiments of Artificial Intelligence as extremely important, I think we should be cautious over equating consciousness with computer states, or processes, at least as these are presently understood. Even though A I does at last offer a powerful technology for providing not only analogies of Mind, but even perhaps for experimenting with Mind in quite new forms, we cannot say what if anything in a computer is critically associated with consciousness – or indeed whether consciousness has any 'engineering' causal significance. It is too early to claim that Artificial Intelligence holds keys to these questions; or guess what doors of present mystery keys to consciousness may finally open, perhaps finally to make Mind at least no more mysterious than Matter. Here, with I trust due caution, I suggest a hint, based on a tentative account of meaning, for how we might think about consciousness, and how perhaps Mind is related to brain.

We end by discussing the reliability of knowledge. The more one looks, the more it seems to me *absolute* knowledge fades to melt into hardly formulated assumptions. It becomes, however, hardly rational to question things which are universally accepted as true; for in the limit there are no tools of logic or experiment powerful enough to support such doubts. We should not be entirely discouraged by abandoning the traditional philosophical quest for certainties in favour of accepting ultimate relativity of knowledge, which historically is far more acceptable in science than in philosophy. At the least, appreciating that knowledge rests on assumptions we cannot strictly justify leads to the great virtue of tolerance: even to entertaining, as host, evidence and arguments that threaten most fundamentally, by challenging beliefs that seem to be written indelibly with the special flourish of one's signature into the book of one's own Mind.

Part 1 Forging science from myth

What the hammer? What the chain?
In what furnace was thy brain?
What the anvil? What dread grasp
Dare its deadly terrors clasp?

When the stars threw down their spears,
And watered heaven with their tears,
Did he smile his work to see?
Did he who made the lamb make thee?

William Blake (1757–1827) [1]

1 Mind in myth

Our fundamental ways of thinking about things are discoveries of exceedingly remote ancestors, which have been able to preserve themselves throughout the experience of all subsequent time.

William James (1842–1910) [2]

We shall start at the beginning: with ideas and technology before the dawn of science with the Greeks, who first formulated the explicit ways of reasoning, observing and proving which became science. They showed that devices can serve us, as we were supposed to serve the Gods – and that much of what we call 'mental' can be mechanized. They started the long saga of exorcizing Mind from Matter through the suggestive power of models, toys and useful machines, and through mechanistic explanations. This did not reach its flowering until the seventeenth century, when it became possible to ask: 'Are *we* machines?' I hope that we may find out.

To discover how man saw the world before the social and psychological effects of civilization is unfortunately hardly possible, for only by language can such ideas be expressed, and perhaps conceived; but there is no written language before quite highly developed civilization. Archaeology provides indirect evidence of beliefs and rituals, but one has only to consider how difficult it would be to appreciate how and what we think from our artefacts, without access to language, to appreciate the extreme difficulty of understanding the Mind of prehistoric man. There are, however, written records two to three thousand years older than the Old Testament, and no doubt many of these do reflect still more ancient ideas and modes of thinking. We have the remarkable, only recently available, writings of the early races of Babylon – especially the Sumerian cuneiform (wedge-shaped) writing on clay tablets. It is highly fortunate that, of all the kinds of writing, marks impressed on clay are the least destructible, and some half a million tablets have been found and can be read. In fact the problems of translation, though only recently solved, turned out to be far less than for the much better known Egyptian hieroglyphics.

We do not find structured philosophical arguments in the Sumerian or the Egyptian literature. Most of the records are of transactions rather than ideas of any kind; but the narrative poems and hymns are certainly highly suggestive, and in some cases they are explicit statements of deep beliefs from the dawn of civilization.

The five or six thousand years of written records may seem a long time, but compared with the emergence of man as an identifiably separate species it is short indeed. Putting the age of man at two million years, the proportion of this time for which we have written records is only about 0.3%. We may assume that the first records reflect much earlier notions; however, this is in some cases doubtful, because the first writing dates from the establishment of politically organized societies and (as we know from many modern examples) what people believe is very much affected by where they live and which social groups they are identified with. It could well be that early questionings and 'mythological' answers changed radically as men came to work together with specialized skills, and with 'professional' priests, who dominated the Egyptians. So we should be careful not to place too much weight on these first written accounts as evidence of how Stone Age men thought, and of what they thought about; but they are all we have and they are extremely interesting. Recent archaeological dating methods are described by C. Renfrew (1973).

The records of mythological ideas, going back to about 3500 BC in the writings of Sumeria, and somewhat later of Egypt, are not logically articulated writings but are mainly records of transactions, laws, and accounts of wars, which are often inconsistent. The significance of myths, as indicating beliefs, is stated neatly by Frankfort *et al.* (1949): [3]

> The ancients, like the modern savages, saw man always as part of society, and society as imbedded in nature and dependent upon cosmic forces. For them nature and man did not stand in opposition, and did not therefore, have to be apprehended by different modes of cognition . . . natural phenomena were regularly conceived in terms of human experience and . . . human experience was conceived in terms of cosmic events. . . .
>
> The fundamental difference between the attitudes of modern and ancient man as regards the surrounding world is this: for modern, scientific man the phenomenal world is primarily an 'It'; for ancient – and also for primitive – man it is a 'Thou'.

They add later: [4]

> Just as there was no sharp distinction among dreams, hallucinations, and ordinary vision, there was no sharp separation between the living and the dead. The survival of the dead and their continued relationship with man were assumed as a matter of course, for the dead were involved in the indubitable reality of man's own anguish, expectation or resentment. . . .

We are concerned here mainly with origins of ideas, but we should look at ideas of origins. Accounts of Creation are central to myth, and remain as speculative in science, for we have not the faintest idea based on evidence of how it all started.

The universal account of Creation in myths is that the world and all it contains came into being by procreation. For both the Mesopotamians and the Egyptians, the sky was lifted from the earth, and earth and sky were lovers. For the Mesopotamians the rain not only irrigated but seeded the earth, to give what is later described as spontaneous generation. This notion continues into Greek mythology, as we see in Hesiod's *Theogony* of about the seventh century BC: [5]

Verily first of all did Chaos come into being, and then broad-bosomed Gaia [earth], . . . and Eros, who is fairest among immortal gods, looser of limbs, and subdues in their breasts the mind and thoughtful counsel of all gods and all men. Out of Chaos, Erebos and black Night came into being; and from Night, again, come Aither and Day, whom she conceived and bore after mingling in love with Erebos. And Earth first of all brought forth starry Ouranos [sky], equal to herself, to cover her completely round about, to be a firm seat for the blessed gods for ever. Then she brought forth tall Mountains, lovely haunts of the divine Nymphs who dwell in woody mountains. She also gave birth to the unharvested sea, seething with its swell, Pontos, without delightful love; and then having lain with Ouranos she bore deep-eddying Okeanos, and Koios and Krios and Hyperion and Iapetos. . . .

This essentially prehistoric notion of the rain fertilizing the earth is dramatically expressed by Aeschylus (the father of Greek tragedy, 525–456 BC) in the *Danaids*: 'Holy sky passionately longs to penetrate the earth, and desire takes hold of earth to achieve this union. Rain from her bedfellow sky falls and impregnates earth, and she brings forth for mortals pasturage for flocks and Demeter's livelihood.'

For the Babylonians the sky was male and the sea female; the reverse was so for the Egyptians, though the essential thought is the same. In both cultures, mechanical questions such as what held up the sky, or what held up the earth, were generally answered in terms of people-like or animal-like Gods. For the Egyptians the sky was sometimes a woman (Nut) arched across the earth, with the stars as jewels suspended from her form: or, (less romantic for us!) a vast cow standing all square on four legs. The God of the air (Shu) originally lifted the sky from the earth, and continued to separate them by supporting the arched sky-Goddess. The all-important sun-God Ra climbed into the sky when he became bored with humanity, to sit on Nut the sky-Goddess. In some versions she then changed herself into the cow standing over the world, but in other versions they both exist together. Such details do not seem to be too important: it is as though almost any account that gives confidence against the sky falling down will do, and two accounts combined (like belt and braces) may be better than one. [6]

1   The Egyptian sky-Goddess Nut, supported by Shu, the God of the air, *c*. 1000 BC. The stars were her jewels. (From the Greenfield papyrus, by courtesy of the Trustees of the British Museum.)

For the Egyptians, the form of the earth was flat, with a flat sky like a lid over it – shaped like Egypt itself, roughly rectangular. The earth was sometimes supposed to be floating on a vast ocean and sometimes encircled by water, as in the Bible: 'He set a circle upon the face of the deep' (Prov. viii, 27). The earth is sometimes supposed to be supported on pillars; but it may also rest on nothing. We find these in: 'For the pillars of the earth are the Lord's, and He hath set the world upon them' (I Sam. ii, 8), and: 'He stretched out the north over empty space and hangeth the world upon nothing' (Job xxvi, 7).

This notion of the earth itself being unsupported continues with the early Greek philosophers, who sometimes considered that it continued downwards so far that the problem of its support virtually disappeared. The notion of the moon, the sun, the stars and the planets ('wanderers' or 'vagabonds') as supported on spherical egg-like shells in the sky came later, with the Greeks.

For the Egyptians the Milky Way is the heavenly Nile, flowing through the land of the dead and round the earth, hidden to the north by mountains; and on this river was a boat carrying a flaming disk: the sun-God Ra. He was born every morning, to grow in strength until noon when he changed to another boat which carried him to the entrance to Dait, when further boats carried him through the night. The night journey becomes highly elaborate in the later Egyptian accounts, as Ra, through the twelve hours of night, brightens and gives life to twelve separate 'countries' of the hidden world during his nightly journey. The boat is sometimes attacked by a serpent: then the sun is eclipsed. Much the same happens for the moon every month: a sow attacks the moon on the fifteenth day, and then the moon grows pallid and dies, to be reborn next month.

These early accounts of the Natural world and of what is supposed to go on behind appearances, to make the world as it appears, are projections of the familiar social life, feelings of fear and strife and creative success. They are thus Animistic. The question for the philosophy of science is: How did we come to discover non-Animistic processes in Nature? To achieve this is to do more than 'project' Mind-like aspects of ourselves: it is to think 'objectively'.

How have we come to think objectively? To know this is to know the roots of science and to understand our relation to the Natural world. I think that we get clues by looking at the earliest recorded science: the science of Mesopotamia (back to about 3000 BC) and Greece (back to 600 BC). We should include also the Egyptians, though their science is generally disappointing. We may say at once that Mesopotamia was concerned with finding and applying procedures – with little, if indeed any, attempt to find underlying models of the Universe – while the Hellenistic Greeks developed, later, highly ingenious conceptual models to explain the movements of the planets, eclipses, and so on, based on many centuries of observations made by the Babylonians. The Greek models came to suggest, and required, new kinds of observations. Greek conceptual models, we may now say after recent discoveries, were closely linked to surprisingly sophisticated technology.

Only recently has the importance, and indeed the surprising sophistication, of much of ancient technology been realized, and the implications of this for appreciating the development of philosophy and science from myth has not yet been at all fully worked out. We shall attempt a preliminary skirmish, as this is an origin of Mind in Science. Meanwhile we look briefly at notions of Mind in the mythological accounts of cultures that most nearly led to Western science.

Babylonian myth

Civilization suddenly flowered in the Tigris–Euphrates valley with the cooperative building of large-scale irrigation canals, requiring tools such as the pickaxe, which was regarded by the Sumerians as specially important. The canals, dams and storage basins, which were first built at Erech before 4000 BC, allowed a large increase in population – with concentrations of population which became cities. Trade then lent necessity to writing, to counting, and to recording numbers. It must also have inspired far more discussion, with greater variety of experience and problems to state and solve by agreed action. These communities, especially at Erech, rapidly became extremely well organized, as we see from the vast ziggurats made of millions of sun-dried bricks, with a uniform architecture throughout Sumeria. Theirs was a 'male chauvinist' society. The name of not a single queen has come down to us. They had a democracy, but women were excluded with the slaves. Nevertheless special women were, as with the Greeks, accepted as particularly wise, as we find in the Gilgamesh Saga, the earliest known narrative (c. 2000 BC). Women (and idiots) were accepted as unwitting mouthpieces of the Gods but hardly as significant in their own right.[7]

What stands out in their view of the world is the lack of distinction between living and non-living. Stones, salt, grain and so on, all have human-like personalities. Kinds of flint that flake in particular ways, or are easily or with more difficulty formed or handled, were considered to have characteristic personalities. Reasons for the personalities of such objects were given in what to us are psychological terms. For example, flints flake easily because Flint had once fought the God Ninurta – and Ninurta punished Flint by making it flake. Objects are talked to just as people are talked to, to persuade or for seeking help or advice. There is deep sympathy with Nature, and this is bound up with magic.

Phenomena were seen to have their own wills. Mesopotamia (unlike Egypt) suffered, as it still does, from violent storms. The God Storm was the prime motivation for calamities of all kinds, including disastrous wars. Catastrophes were often attributed to wars between the Gods. There is a continual human-like struggle behind the scenes of the play that we see with the senses. The Universe was thus a mirror of human and animal passions.

These Animistic accounts described what we call physics in terms of what to us is psychology. They continued essentially unchanged until the later Greeks. Greek thinking, from the fourth century BC onwards, tended to exorcize Mind from the Universe; but this happened very gradually and was for the Greeks – or indeed for us – not complete. Their laws of the physical world had, as for the Sumerians, the status of human laws, but enacted by the Gods – law-makers of the Universe. These first attempts to discover Natural laws were surely an essential step to science: from observations linked first by memory and later by records; from observations selected for testing theoretical possibilities, and from artificially designed situations which simplified Nature so that Nature could be challenged and tested as in a defined game. It was, no doubt, this development of scientific method that generated views of the Natural world in which man appears alien to the Universe in which we live.

In Sumerian as in all early thinking, Gods preceded humans and animals, and the earth and sky were created by Gods who were themselves generally born of other Gods – sometimes in very strange ways. The Biblical account of creating, which is of course the best known in the Western world, derives from earlier Babylonian accounts; though unfortunately these are not known to survive except as passing references. However, new clues to this connection are available with the translations of the tablets – such as Adam's rib which became Eve. Why a rib? In the Sumerian Dilmun poem of paradise[8] the young god Enki ate eight forbidden plants in their paradise one thousand five hundred years before the Old Testament.

Receiving a curse, he fell ill in eight organs of his body, including his ribs. His sick organs were healed or renewed by eight specially created nurse-Goddesses, one for each of the sick organs. Now the word 'rib' in Sumerian is *Ti*, which also meant 'to make live'. The Goddess specially created to renew Enki's rib was (from the first meaning) Nin-ti, 'the lady of the rib'; but this also meant 'lady who makes live', and the meanings were switched as in puns. This seems to be the origin of Eve, mother of men.

If we ask, 'How was the Universe created?', we are hardly better than the Sumerians at finding an answer. Ancient accounts of origins were sexual combinations while ours – science's – are chemical, but both come unstuck when it comes to what started the whole business. There are many ancient analogies for creating from builders and craftsmen: this is explicit in Plato and implicit in the much earlier Sumerian literature. For the Sumerians, the pickaxe was a tool of the greatest importance: the God of the sky, Enlil, separated earth and sky with the pickaxe, which allowed sexual creation of animals and men by the male sky mating with the female earth. This is an account of the origin of species found in many ancient mythologies, as already mentioned.

Man was supposed by the Sumerians to have been made from clay. Abnormal and subnormal people (who are an intellectual embarrassment when man is made in the image of God) were made from a lump of clay by the chief God Enki when he was the worse for wear with drink: blame was thereby avoided and the good name of their favourite God saved.

What do the tens of thousands of cuneiform tablets tell us about Babylonian conceptions of Mind? They give no clear philosophy, or critical thinking; but they do give us in the first narrative and epic poems a wonderfully rich mythology. The best known poem is the Gilgamesh Epic. The hero Gilgamesh seeks immortality from a special weed beneath the sea, but it is snatched from him by a serpent. The notion of life, soul, or Mind as substance is indeed ancient, as revealed in such myths.

There are recorded disputations, prophesies and laws; as well as elaborate mathematical tablets, multiplication tables and astronomical observations, which were of great use to Greek astronomers. There are no tablets discussing abstract questions though the rich mythology abounds with references to Gods and dragons, and demons and magical practices.

There are, however, occasional references to what seem to be controlling forces of Nature – the *mes*. These seem to maintain, and perhaps originate things (Kramer, 1963).[9] Most is known of the *mes* from the myth 'Innana and Enki: the transfer of the arts of civilization from Eridu to Erech'. Various aspects of the state are described as controlled each by a different *me*. They seem to be laws which control Gods, men, social institutions, and perhaps the Universe. If so, the

Sumerians began to have explanations that were not entirely in terms of the anthropomorphic Gods. So far as we know, they had no account of Mind apart from the Gods; but perhaps the Sumerians began to appreciate, with the *mes*, laws of Nature.

The Babylonians were concerned with the times and positions of appearance and disappearance of heavenly bodies (which placed emphasis on observations low on the horizon which are difficult to make and are intermittent owing to cloud, which tends to obscure the horizon). The linked conceptual models of why clouds and why the stars move the way they do, which seem clearly different to us, were for them mixed up. It was indeed a long time before phenomena of meteorology and astronomy were clearly distinguished. This reflects the lack of conceptual models before the beginning of Greek science in the sixth century BC.

We might hazard the guess that man came to think 'objectively', and to create science, first through the power of explicit procedures; he then formulated conceptual models which, it now seems, stemmed from ancient technology. Procedures included ways of observing (such as by recording the positions of the sun from its shadow cast by a gnomon in Babylonian sundials), and they included ways of inferring and ways of calculating. The Babylonians were especially adept at devising ways of calculating. Logical procedures for testing arguments seem to have developed first in Hellenic Greece and especially with Aristotle's formulation of the syllogistic forms, and subject–predicate logic.

We owe our use of the number sixty, in minutes and seconds, to the Mesopotamian sexagesimal counting system, which was used for their astronomical work, though not for all their calculations. To them we owe perhaps the most important mathematical notions and early concepts – the use of places for numbers (corresponding to our 'units', 'tens', 'hundreds' and so on) and, later, the vitally important zero. Zero was at first indicated by a space, and later by a special symbol. The old Babylonian scribes had tables for multiplication, for squaring and for reciprocals. (Possibly our mathematical 'tables' comes from inscribed 'tablets', though they were not unearthed and recognized in modern times until the end of the last century.) They were arithmetical rather than algebraic, in that there was nothing like our use of x for any number. Instead, particular examples for illustrating general conclusions were presented.

Neugebauer (1957) says of these tablet texts: [10]

The texts fall into two major classes. One class formulates the problem and then proceeds to the solution, step by step, using the special numbers given at the beginning. The text often terminates with the words 'such is the procedure'. The second class contains collections of problems only, sometimes more than 200 on a single tablet of the size of a small printed page. These collections of problems are usually carefully arranged, beginning with

2 A Babylonian mathematical clay tablet, in cuneiform (wedge-shaped) writing, about eight inches high, eighteenth century BC. (By courtesy of the Trustees of the British Museum.)

very simple cases e.g., quadratic equations in the normal form, and expanding step by step to more complicated relations, but all eventually reducible to the normal form.

From many examples, it becomes obvious that it was the general procedure, not the numerical result, that was often considered important.

Many mathematical discoveries that we generally attribute to the Greeks were known to the Babylonians. Pythagoras' Theorem (or at least its result, good to five decimal places) was known to the Babylonians more than a thousand years before Pythagoras. The Greeks inherited the notation and a great deal of knowledge from the Babylonians. What they added, for the first time, was the notion of proof. Their interest in geometry may have come from Egypt, though Egyptian mathematics was crude and elementary, but the Egyptians did develop an excellent calendar. They divided the year into 360 days, with five adjustable extra days. This was so convenient for astronomical cycles that it was used for computation right up to and including Kepler (who used the base 60 for his calculations) in the seventeenth century A D.

The early use of lunar calendars (with the lunar cycle of $29\frac{1}{4}$ days, which is not a simple multiple of the solar year) inspired complex calculations from the outset of civilization. A solar calendar was essential for agriculture, but the lunar period was easy to observe, and the lunar calendar still remains (for the Christian and other Churches) important for religious festivals. The further fact that the solar day is not the same as the sidereal day (because, as we now know, the earth's going round the sun adds about a degree to its rotation needed for the sun to attain the same altitude daily, making the solar day about four minutes longer than the sidereal day) produced further complexities to tax and develop early ingenuity and arithmetic. The most interesting complication is the fact that the rate of solar motion is not constant throughout the year. This was not understood before Kepler's discovery that the earth moves in an elliptical orbit with the sun at one focus so that its distance and apparent velocity change seasonally. This variation in rate led to a concept whose importance is not always appreciated, although the concept goes back to Babylonian times: accepting mean positions, and mean angular velocities, of the sun. The Mean Sun was a most useful fiction. It was a crucial step away from direct observation – to mathematically convenient, but imaginary, objects.

Accepting average or mean *positions* for objects is ancient, but, most curiously, the closely related concept of average or mean *observations* came far later – hardly until the nineteenth century A D. (As we shall see, the reluctance to accept that many observations are better than one is to reject Realist accounts of perception: the reluctance is evident even today among noted philosophers.)

Such fictions as the Mean Sun (which we use today in our Mean Solar Time) may be adopted to make procedures for calculation simpler, or they may be postulated to fill awkward gaps in theoretical models. Modern physics is full of such fictions: atoms, sub-atomic particles and fields of force. These two motives – to simplify procedures and to fill gaps in models – were present in the earliest known science. Such fictions (which are sometimes incompatible with observations) have remained essential elements of science ever since.

Observations have only very gradually become accepted as important for scientific knowledge. It is worth pointing out how extremely misleading many reported observations were in classical times, and of course also later on. For some examples see Dreyer (1953). Ktesias, in the early fourth century BC, reported that from mountains in India the sun appeared ten times as large as in Greece; Artemidorus reported it as appearing, from Ge Gades while setting, one hundred times its normal size. People living on the Atlantic coast reported a hissing sound when the sun sank into the sea. Such 'observations' were gradually rejected as theoretical accounts became more powerful. The relation between observations and belief – and the reasons why some observations have power to change belief and others not – is surely far more complicated, and less clear, than it seems in classical Empiricism, which supposes that knowledge comes simply from observations. Observations depend on processes of perception that are far from simple and are still only partly understood.

There is surprisingly detailed evidence of early methods of measurement and construction which transformed the world – and human Mind – by challenging the power of the Gods as seen in myth.[11]

Egyptian myth

Egyptian mythical–religious ideas remained remarkably static over two thousand years, as did their customs and way of life, from the Fifth Dynasty to the Roman Ptolemaic period around the time of Christ. The same chief Gods, and names, appear on the tombs and the papyri throughout this long period, so it seems safe to give one single account of the Egyptian concepts of Mind. I shall draw heavily from *The Egyptian Book of the Dead* (*The Papyrus of Ani*) (Budge, 1895).[12]

The heart and not the brain was considered the seat of the Mind. The Mind was not, however, a single entity, but several. Budge comments: 'The whole man consisted of a natural body, a spiritual body, a heart, a double, a soul, a shadow, an intangible ethereal casing or spirit, a form, and a name.'

The physical body was called *khat*, which seems to be connected with the idea of things liable to decay. The physical body could after death turn into the spiritual body, *sahu*. In this form (provided it had attained sufficient knowledge) it could live with the Gods. The *ka* was the man's abstract individuality or personality endowed with all his characteristic attributes. This abstract personality had an absolutely independent existence. It could move freely from place to place, separating itself from, or uniting itself to, the body at will and also enjoying life with the Gods. The *ka* dwelt in the man's statue just as the *ka* of a God inhabited the statue of the God.

Corresponding closely to 'soul' was the *ba,* which means something like 'sublime', or 'noble'; for though it dwelt in the *ka* and was in some respects, like the heart, the principle of life in man, still it possessed both substance and form. It revisited the body in the tomb and re-animated it, and conversed with it; it could take upon itself any shape it pleased; and it had the power of passing into Heaven and of dwelling with the perfected souls there. It was eternal.

The shade or shadow of the man (the *khaibit*) may be compared with the Roman *umbra.* It too had an entirely independent existence and was able to separate itself from the body and move wherever it pleased. The equivalent to our 'spirit' was the *khu.* This was a shining or translucent intangible casing or covering of the body, which is frequently depicted in the form of a mummy. It is often translated as 'shining one', 'glorious' and 'intelligence'. The *khu* lived in Heaven, with the *khus* of the Gods, after death. A puzzling part, or aspect, of man and Mind is the *sekhem,* sometimes translated as 'power' or 'form'. It seems to be closely associated with the *khu,* and is perhaps a life force. Lastly we have the name. The name of a man was believed to live in Heaven. The name was part of the man.

There do not seem to be any texts where concepts, beliefs or ideas are questioned. There is, in this sense, no recorded Egyptian philosophy. When one considers the enormous effort that went into their religion this does seem surprising. Perhaps, though, this society with its almost unchanging beliefs, mythology and customs (except for one episode) retained its continuity and power for so long just because they were not examined critically by philosophy or challenged by science.

As is very well known, religious observances dominated Egyptian life, and decisions were always taken after divinations and omens of many kinds. Animals especially were considered powerful embodiments of knowledge, and the various virtues and ills; for example the Apis bull was regarded as the living image of Osiris, and was consulted as an oracle. He was offered food of all kinds: acceptance was favourable and rejection a sign that Osiris was against it. Apis bulls were kept in sacred precincts of the temples, and (as reported by the Greek traveller Pausanias, second century AD) those wishing to consult Apis burnt incense on the altar, deposited money to the right of the statue of the God, and placing their mouths near the God's ear, asked him a question. Then, covering their own ears, they would leave the temple and listen for the first words spoken by anyone outside as the words of the God. The babble of children at play could serve.

The combinations of animal heads on human bodies to represent the Gods is curious, and of uncertain origin. Possibly there may have been taboos on the killing of certain animals, or kinds of animal – so it may have originated as a conservation policy. Some were, however, probably suggested by their animal characteristics: the important God Thoth, God of wisdom and of writing (which itself was sacred as 'hieroglyphics' means 'sacred writing') was depicted as an ibis or a baboon, possibly because they sometimes look serious. There were local variations; for example, in regions where the crocodile was regarded as a menace, the ichneumon, which eats its eggs, was sacred; in other places the crocodile was sacred, as evidenced by finds of embalmed crocodiles (at Kom-Ombo, for example). However this may be, and the origins are prehistoric, animal-Gods were powerful symbols and judging from cave paintings may be as old as man.

The embalming of human bodies has a curious link with stone worship. It was a general early belief that the spirits of the dead could enter stones, and damp stones were special objects of veneration. In pre-dynastic and early dynastic Egypt, the bodies of the dead were buried in the dry hot sand beyond the agriculturally useful damp soil, irrigated by the Nile. These bodies were preserved naturally and indeed there are examples preserved to this day, five thousand years later. But when bodies were placed in sarcophagi the flesh disappeared, leaving only the bones: so the body was thought to have been absorbed, or eaten, by the stone (hence 'sarcophagus', from the Greek $\sigma \acute{\alpha} \rho \xi$ meaning 'flesh', and $\phi \alpha \gamma o s$ meaning 'eating'). This is an interesting case of a generalized observation serving as evidence for a magical or mythical view. There may well have been many such examples – for evidence is seldom, if ever, 'theory-neutral'. This is an issue that will assume for us great importance.

The fact that the Egyptians did not embalm the brain (which was retracted through a nostril and thrown away) strongly suggests that Mind was in no way associated with the brain. The Egyptians were, however, much concerned with medicine, and the professional class of doctors was highly esteemed. The Edwin Smith papyrus deals in detail with injuries of many kinds (forty-eight fractures resulting from accidents) and the Ebers papyrus (both named after American collectors) deals with surgery and anatomy and pharmacy, which was much practised. The most celebrated doctor is Herophilus. With permission from Pharaoh Ptolemy, he performed autopsies and possibly vivisection. He described the eye, the heart, the liver and the brain. He did regard the brain as the seat of the soul; but Herophilus was comparatively late, living in Alexandria in the third century BC when Egypt was open to Greek influence. [13]

There is a striking parallel across early cultures between the terrain and social systems on the one hand and on the other the cosmologies and accounts of the origin of the world. This can be seen in the cosmologies of Sumer, Egypt, Greece, China, India, and the Scandinavian Norse mythologies.

In Egyptian cosmology the Universe was shaped like the country of Egypt (with the Nile the Milky Way), and the Gods were organized in strict hierarchy, as were the people. For Sumer, the sky was male and violent, more or less raping the earth at the origin of the world, while the placid sky of Egypt was female. The Greeks had no ordered cosmology before their rationalizing philosophy (G.E.R.Lloyd, 1975), and the Gods were as individualistic and lacking in dogma (if Gods can be dogmatic!) as the Greeks themselves. China accepted the Universe as empty space (Joseph Needham, 1975); and also as covered by a dome, 80,000 *li* (usually about a third of a mile) distant from the earth, which was an upturned bowl. Of Indian cosmology, Richard F.Gombrich (1975) asks: [14]

. . . why is Indian cosmology so complicated? Just as the Indian system of social organisation, caste, has grown throughout history by aggregation and inclusion . . . so Indian cosmology – which remained largely a branch of Indian mythology – rarely abandoned a theory or idea, but allowed it to remain alongside the new ideas even if it was inconsistent with them. . . . We are about to meet a universe of Chinese boxes . . . the enormity and complexity are like those of the enormous and complex social system. . . .

The first records of Indian cosmology date from the second half of the second millenium BC. It started with just earth and sky, and then, R.F.Gombrich says,[15] 'the Chinese boxes start: sometimes the two, sky and earth, sometimes the three, earth, atmosphere and sky, are said each to consist of three strata, giving a total of six or nine'. In the *Hymn of the Cosmic Man* the Gods sacrifice a giant to create the Universe. The air comes from his navel, from his head the sky, from his feet the earth – and the four estates of society (the Brahmins and so on) were produced from the mouth, arms, thighs and feet. The unit of time must not go unmentioned: the smallest unit is the blink (about a fifth of a second). For the Jains, who were and still are particularly keen on cosmology, the Universe (R.F.Gombrich)[16] 'is measured in terms of a peculiar unit, the *rajju*, which is defined as the space covered by a God in six months if he flies at 2,057,152 *yojanas* in a "blink"', that is about ten million *yojanas* a second. The measurements are all given in the most careful detail and the complications are enormous.'

Scandinavian cosmology dates from the bronze age of perhaps 1600 BC. The basic scheme is a tree, encircled by a serpent, with much surrounding water and a ship for carrying the dead. The Cosmic Tree with the surrounding waters seems to represent meeting places, where Gods as men met in council. These are the kinds of world peopled by man-like Gods with a fair sprinkling of animals, and thunderbolts and floods to keep the action going, that set the scene for philosophy and science.

Greek myth

Greek society was very different from Egyptian society because the Greeks had no priesthood. There was far more individuality in Greece; but it has been suggested, by M.I.Finley (1970),[17] that Greek myths bound the very different city states together. The stories were of course transmitted verbally, and Homer's poems are generally regarded as written versions of an older oral tradition. The Greek written language was invented in Crete: first a form of picture writing, then 'Linear A' (in which, unlike Egyptian writing, most of the symbols represented syllables), and – in Cnossus and the mainland – 'Linear B' writing. Linear A has not been deciphered, and Linear B was only translated recently, by Michael Ventris, who realized that it represented an early form of Greek. These tablets (originally of unbaked clay, but preserved by baking in the conflagrations of the palace of Cnossus) are inventories. They have no 'philosophical' content.

In early Greece it seems that men saw themselves as living closely with the Gods and acting generally to please them: not too difficult as the moral standards of most of the Gods were not high! Finley says:[18]

By a variety of formalized actions, men sought to establish the most favourable possible relationship with the supernatural powers. That is to say, they tried to discover the will of the gods, and to placate and please them. The former required specialists, such as soothsayers, diviners and seers, but the rest of the activity was carried on by ordinary people.

With writing, the genealogy of the Gods became relatively fixed; then there was more to criticize specifically and argue about. According to Herodotus (II, 53), Homer and Hesiod 'first fixed for the Greeks the genealogy of the gods, gave the gods their titles, divided their honours and functions, and defined their images'. One might guess that it was this fixing by explicit statements, in writing, which revealed inconsistencies in beliefs – to create philosophical thinking. One might guess also that the powers of metal tools showed that the ancient ways of the Gods could be challenged and improved by tools, planned technology and critical creative thinking. I suspect that these are the primary sources of philosophy and science.

Why should philosophy and science flower in Greece rather than Egypt? Possibly because the Egyptian learned men and scribes were in the priesthood, which inhibited individual thinking as heresy. A subtle balance seems to be required for viable societies between freedom to express doubt, and encouragement to see new problems and invent new solutions to problems. Orthodoxy inhibits originality, but inspires group loyalty and allows very large-scale cooperative works (such as pyramid building) to be undertaken. Greece and Egypt were extreme cases. In their own ways they are equally fascinating, but one could be an individual in Greece. The study of original.minds grew in the society which appreciated individuals; and in which individuals could recognize themselves, and accept conflicts of opinion and belief as interesting.

As has been pointed out, there was no priesthood in Greece; and the mythology was not consistent between cities or states. The various myths seldom, however, conflicted to produce social discord, which appears very odd to us now, with our experience of the 'Mind-less' violence of religious discord that runs through Christianity and is with us today. How did the Greeks hold inconsistent views and yet live in comparative harmony? This continued until the sixth century BC, and it was then that the myths became explicitly formulated and subject to criticism. In short, the trouble began with the start of philosophy and science. Where there is organized religion there is also professional 'dogmatic' criticism. This evidently produces discord. The discord can lead to ways of settling arguments – especially by observation and experiment – without mere warring opinion expressed in fights to the death. When myths are simply accepted there is no such problem; however, when they are formulated with an eye to consistency and held as true – as inventories, transactions and technical processes are formulated with demanded truth – then there is trouble. The law courts can legislate human affairs, and practical success or failure sooner or later tests technology; but for myth, and indeed philosophy, there is no such tribunal. Then, disagreements become disputes, without means of resolution except by combat.

Early Greek writers give us both accounts of the myths and their significance for life, and also criticisms which ultimately lead to modern science. The earliest expressions of Mind in myth are those of the poet and systematist of earlier poems, Homer. Homer is now thought to have lived in the eighth century BC, rather than around 1200 BC, as previously thought.

In the *Iliad* and the *Odyssey* we see wonderfully expressed the ancient notions of a world of magic – of larger-than-life but essentially human Gods who are not too moral but disport themselves with fun and fantasy and wit. At the same time, they are not quite free to do as they like: they – even Zeus – are bound by the power of the Fates. So they bow to Natural laws of some kind.

Homer discussed Mind in many places. He distinguishes between mortal Mind and immortal soul. There is for him always a double drama: the drama of the 'wheel of life' of the immortal soul, and the drama of the human Mind, which is as much a puppet of the Fates as a master of its own fate. It is thus bound up with the will and the judgements of the Gods – and shares their ultimate dependence on the inevitable cycles of the Universe.

Of course, Homer lived somewhat before systematic philosophy, and expresses pre-philosophical (essentially magical–mythological) thinking, which we take to be the way that men thought before the impact of philosophy, and before the physics we date from Thales of the sixth century BC. The world is seen by Homer in the *Odyssey* to be 'sympathetically' related to man. There is a great deal (too much for our taste) of divine intervention in Homer. Wounds inflicted by a weapon are healed by that same weapon; men could change into other forms; trees, rivers, and the night, were personified as having in their own right powers over Gods and men. As we shall see, Homer had a quite specific account of the soul. But all this is to be found not systematically stated or reasoned as in later Greek philosophy, but for the most part in passing references and allegory. The later Greeks, on the whole, avoided allegory.

Homer was an uncertain figure to the later Greeks with no established place or time for his life though he had great influence. He was regarded as the supreme poet, with highly important things to say. There was as a consequence a tradition of looking for hidden significance, or reading meanings, into Homer as though he were a divine sage. (This 'reading of meanings' was indeed the original sense of 'etymology'.) This was not, however, in general a feature of Greek philosophy, and not at all what Socrates did; but it is a feature of mystical thinking and writing where some individuals are supposed to have special experience and knowledge that cannot be expressed directly. Greek commentators on Homer did try to draw such very special significance from him, though he does not have to be read in this way. He reads as a supreme story-teller, with striking freshness and enthusiasm though touched with melancholy. He accepts a wonderful past, of gold and ivory and superhuman heroes. He looks to the past for knowledge, while the later Greek philosophers looked to the new methods of thinking and experiment to solve problems in the future. In this sense, there is something backward-looking, and with this melancholic, about Homer; while the Empiricism of, for example, Aristotle is imbued with optimism in unlimited progress for knowledge and society. This switch from looking to the past to looking at the future for knowledge and solution to particular and general problems, is a fundamental switch from the pre-philosophy of myth to philosophy and experimental science. This is surely one of the great benefits of technology: it is inherently look-ahead – though we may fear what lies ahead. This is not to be found in Homer; he is a voice from before the iron-forged tools that severed us from our myth–magic origins. Its origins are stories told at the dawn of pre-literate Minoan civilization. Its faith was in an idealized heroic past that never existed.

The Greek poet Hesiod (*c*. eighth century BC) describes in the *Theogony* how the Universe was born from chaos, and gives a systematic account of the Gods. This very systematization spelled the end of myth; it lies half-way between mythical and scientific thinking, both in its date and its content. Hesiod, as Homer, describes the Gods as having human form and essentially human powers and weaknesses. Though the ancient myths continued as stories, and as references for human behaviour, they were eroded by doubts with the development of formulated philosophy and observational and theoretical science. Only dogmatic theology could hold its own against questions and explicit attempts to find answers. With this the charm went out of myth.

The early Greek philosopher-scientist Anaxagoras (500–428 BC), who was of the Ionic school but taught in Athens for thirty years, was indicted for saying that the sun was a white-hot stone, and not a deity. The conflict between myth held as true and science had begun. Anaxagoras was an effective scientist: he appreciated the reason for eclipses, and was the first to realize that the moon (on which he thought people lived) shines by the sun's reflected light. He looked for and found mechanisms in nature. Mind had its place – for controlling Nature's mechanisms. It was Anaxagoras who first distinguished clearly between Matter and Mind, saying: 'Mind rules the world and has brought order out of confusion.' He thought of Matter as particles, though not atoms as in the Atomic Theory but rather as *seeds* present in various proportions, as the properties of things. Mind is, however, pure. I shall quote him in full: [19]

All other things have a portion of everything, but Mind is infinite and self-ruled, and is mixed with nothing but is all alone by itself. For if it was not by itself, but was mixed with anything else, it would have a share of all things if it were mixed with any; for in everything there is a portion of everything, as I said earlier; and the things that were mingled with it would hinder it so that it could control nothing in the same way as it does now being alone by itself. For it is the finest of all things and the purest, it has all knowledge about everything and the greatest power; and Mind controls all things, both the greater and the smaller, that have life. Mind controlled also the whole rotation, so that it began to rotate in the beginning. And it began to rotate first from a small area, but it now rotates over a wider and will rotate over a wider area still. And the things that are mingled and separated and divided off, all are known by Mind . . . Mind arranged them all, including this rotation in which are now rotating the stars, the sun and moon, the air and the aether that are being separated off. And this rotation caused the separating off. And the dense is separated off from the rare, the hot from the cold, the bright from the dark and the dry from the moist. But there are many portions of many things, and nothing is altogether separated off nor divided one from the other except Mind.

So Mind starts Creation; but as the 'centrifuge' separating the elements grows, the world becomes more mechanical, leaving Mind almost independent of Matter. Anaxagoras is thus a Dualist: the first Mind–Matter dualist. Mind affects Matter by the purity of the Mind-substance. At first everything contained Mind; but the cosmic centrifuge left Matter free of Mind except for living things. Mind is the same for Anaxagoras in all living creatures: what distinguishes man is our hands. Mind is the source of all motion; but Mind is not structured, as mechanisms are structured. For Anaxagoras the world of objects is built from 'seeds', the infinitely divisible particles coalescing to form the corporeal objects that we experience. All

objects contain some of the parts of other objects, so objects are different according to the proportions of the primary elements, cold, dark and light, and so on –
excepting only Mind, which is not a mixture but is 'pure'.

This kind of account is unsatisfactory in that Mind, being supposed structure-
less, can do nothing except perhaps serve as prime mover. Once motion has
started, physical processes take over – though Mind may be invoked where no
physical processes can be imagined. We still live with this problem: brains can
have mechanisms, but can Mind be a mechanism?

Pythagoras (*fl.* sixth century BC) is the first individually known mathematician,
but he was also a mystic. Pythagoras was born at Samos and founded his School
of Crotona, in Magna Graecia. This was a 'way-of-life' community, which developed mathematical, astronomical and mystical–religious ideas. None of the original writings survive, so attribution of specific discoveries is difficult. He is known
mainly from Aristotle's accounts. He is described neatly by Bertrand Russell in
the *History of Western Philosophy* (1946) as 'intellectually one of the most important men who ever lived, both when he was wise and when he was unwise . . . the
influence of mathematics in his philosophy partly owing to him, has, ever since
his time, been both profound and unfortunate'.[20] Pythagoras was evidently an
extraordinary mixture of a man, combining a doctrine of transmigration of souls
with the eating of beans as an ultimate sin. Some of the Rules of the Pythagorean
order, in which women were accepted equally with men, have prehistoric origins:

Abstain from beans.

Do not eat the heart.

Do not look in a mirror beside a light.

When you rise from the bed-clothes roll them together and smooth out the impress of your
body.

This is one side of the strange Pythagorean mixture of ancient taboo, ecstatic
mysticism and mathematical rigour.

Pythagoras' Theorem (though it was in some form known to the Babylonians)
at once set up the notion of Ideal Forms as underlying logical thought. It also
introduced incommensurables.

A principal worry for the Pythagoreans was that simple ratios – 'harmonies' –
were not found for many geometrical figures which seemed especially 'good', or
perfect, forms. This applies to the diameter and circumference of the circle, and
to the sides and diagonal of the square. Pythagoras' Theorem was especially
important in showing that there was a 'neat' solution which made the diagonal of
a square commensurable with its sides, after squaring the numbers. This showed

that number relations could describe with absolute precision, though measurements erred. This later supported the notion of Ideal Forms. Lastly, the proof held for the squares of right-angular triangles of any size – including triangles too large to measure. It was thus applied to heavenly bodies, for discovering their distances when the power of mathematics reached out into space beyond human observation. The Universe indeed looked as though constructed according to properties of numbers, a notion not dead in modern cosmology.[21]

The Egyptians, who were not sophisticated mathematically, knew that a right-angled triangle could be constructed with rods of unit lengths 3, 4 and 5. What Pythagoras (or at least his school) did was to prove this. To do so, they used a construction as an aid to thinking. Procedures and tools seem necessary for scientific, rather than mythological, thinking. Pythagoras' construction led to the general proof that any right-angled triangle can be exactly described by $a^2 + b^2 = c^2$, where a and b are any numbers. (So, of course, $3^2 + 4^2 = 5^2$, or $9 + 16 = 25$, which is a particularly convenient case.) With their proof the Greeks could abandon the trial-and-error methods they had inherited and could design *any* right-angled triangle. This could be done with the formulae $a = v^2 - u^2$; $b = 2uv$; and $c = u^2 + v^2$; for any positive integers u and v, where u and v have no common factor and are not both odd. This was the most spectacular example of mathematical form corresponding in a deep way with a physical object's structure.

There were, however, equally spectacular examples of *failures*,to find elegant mathematical accounts of what 'ought' to be, in a mathematically designed Universe. By far the most celebrated example of such a converse case is that annoying fact that the diameter of a circle never divides without residue into its circumference. Circles and spheres were for the Greeks the prime examples of perfect figures: but this worry that pi (as we call it, though the Greeks did not) was an incommensurable ratio, seems to go back at least to 3000 BC, as evidenced by the egg-shaped 'circles' of standing stones that may have been designed as commensurable circle-like figures.

We now think of circles as a special case of a conic section (an ellipse of zero eccentricity) so that the concern over pi as this curious number extending for ever, 3.141592654 . . . , has largely gone. It must have been particularly galling that 'squaring the circle' did not work, while 'triangulating the square' did, though the circle is the 'perfect' figure. If mathematics is such a fickle mistress, is anything to be trusted?

Curiously enough a bother later appeared over what is essentially an extension of Pythagoras' Theorem – the celebrated Last Theorem of Fermat.

The French mathematician Pierre de Fermat (1601–65) asked what would happen if we raised the integers to powers higher than 2. Can integers a, b, c, be found for which $a^3 + b^3 = c^3$? Or, in general, when the positive exponent is greater than 2, can the equation $a^n + b^n = c^n$ be solved? The story of this is so curious that it is worth mentioning here, before we look at the impact of alternative geometries on the great mistake of *a priori* truth from numbers.

Fermat was working on the number theory of a Greek, the Greek mathematician Diophantus (*fl*. third century AD), who wrote the earliest known treatise on algebra. Fermat wrote marginal notes of his comments and proofs as he went along on a copy of this work, stating that $a^n + b^n = c^n$ is not solvable for *any* integers greater than 2. He claimed to have a proof – but that it was too long to write in the margin of the copy of Diophantus' book that he was using. It is still not known whether Fermat did indeed have a legitimate proof. It has not been proved since, except for certain numbers, and for all positive integers up to 619. No counter-example has been found. This search for generality – which is sometimes successful and sometimes not – is the appeal of mathematics and of science. Larger generalities often destroy what had seemed to be primary characteristics of the world as they destroy and change our assumptions.

I mention the fate of extensions of Pythagoras' Theorem to show by example the kind of difficulties in the Greek view that properties of numbers are properties of things, or, as Pythagoras put it, that 'Everything is number.' It seems, rather, that it is very difficult to invent number systems to fit reality neatly. There was, however, for Greeks and for much later thinkers one strong reason – which turned out to be ill-founded – for believing that properties of numbers are properties of things, and that essence of reality is changeless as numbers are changeless: this is the power of geometry. It was evident to the Greeks that geometrical arguments could give precise and logically necessary accounts of the shapes of objects in space. Deduction was thus better than measurement. It was thought that geometrical arguments from undeniable axioms could give new and certain truths about properties of objects with perfect precision, while measurements were always subject to error. This notion justifies metaphysical inferences to characteristics of the world right up to our own time.

Geometrical axioms, such as that parallel lines never meet, were universally accepted as unquestionably true, right through the history of mathematics and philosophy, until the middle of the nineteenth century. It is the most drastically misleading and longest-lasting error in thought. The error was at last revealed with the discovery of the alternative – non-Euclidean – geometries. This mistaken view of geometry had for over two thousand years profound and disastrous effects on the philosophy of Matter and for how we should think of Mind.

The ecstasy of mysticism, for the Pythagoreans, was not separate from the rigour of mathematics, but became a way of life which initiated and has inspired science ever since. Mathematical physics is remarkably similar in spirit!

For Pythagoras, numbers are reality. Relations between numbers mirror essential structures of the Universe; hence the phrase 'All is number', attributed to him. The magic power of numbers is applied to man as well as Nature. Thus his view of Ethics is summed up by Burnet in his *Early Greek Philosophy:* [22]

We are strangers in this world, and the body is the tomb of the soul, and yet we must not seek to escape by self-murder; for we are the chattels of God who is our herdsman, and without His command we have no right to make our escape. In this life there are three kinds of men, just as there are three sorts of people who come to the Olympic Games. The lowest class is made up of those who come to buy and sell, and next above them are those who come to compete. Best of all, however, are those who come simply to look on. The greatest purification of all is, therefore, disinterested science, and it is the man who devotes himself to that, the true philosopher, who has most effectually released himself from the 'wheel of birth'.

Pythagoras thought of number as being in shapes. His discoveries relating to the length of vibrating strings to musical intervals promised a profound link between numbers and physics, and also between physics and sensation. He looked for other 'harmonic ratios' by analogy with music, but harmonics were not always to be found where they should be. This incommensurability of the circle, and of the planetary cycles, started the search for symmetries and neat equations which continues today.

Empedocles (*c.* 430 BC), who lived in Sicily, was a politician with democratic leanings, a soothsayer and a poet. With him we begin to see mechanical analogies applied to the mysteries of Mind. He was however, a miracle-worker and claimed to be a God: he is supposed to have demonstrated this by jumping into the crater of Mount Etna – which rejected his sandals though accepting him!

It was Empedocles who moved away from single-substance accounts, to suppose that the world is made of earth, air, fire and water. He made physical discoveries, especially of air, which he recognized from the action of the clepsydra, a small vessel with a perforated bottom and a thumb hole on the hollow handle, used for taking water or wine out of storage vessels. ('*Klepsydra*' meant 'water-stealer'; it later came to mean 'water-clock' from the similar shape of the water-clocks in the law courts.)

When a girl, playing with a water-clock [water-stealer]* of shining brass, puts the orifice of the pipe upon her comely hand, and dips the water-clock into the yielding mass of silvery water, the stream does not then flow into the vessel, but the bulk of the air inside,

---

*Bertrand Russell is surely not correct to use 'water-clock' in this translated passage from Empedocles, in *History of Western Philosophy*, p. 72. Clearly it should read 'water-stealer', which in fact included wine-stealing by devices for stealing fluid from jars.

pressing upon the close-packed perforations, keeps it out till she uncovers the compressed stream; but then air escapes and an equal volume of water runs in.

This is interesting in showing that the common claim that the Greeks never did experiments is not true. Surely this is an elegant experimental observation.

Empedocles held that the senses work by elements in the body meeting elements outside, including love and strife. Theophrastus, who himself wrote on the senses, describes Empedocles' theory thus: [23] 'For with earth do we see earth, with water water, with air bright air, with fire consuming fire; with Love do we see Love, Strife with dread Strife.' Sight and hearing are explained in terms of how *we* understand smell, by particles entering channels of the senses, 'and when these effluences are the right size to fit into the pores of the sense organs then the required meeting takes place and perception arises'. The elements are supposed to be blended in the blood, and thought and knowledge are in the heart and the blood. Empedocles wrote on this:[24] '. . . [The heart] dwelling in the sea of blood which surges back and forth, where especially is what is called thought by men; the blood around men's hearts is their thought.' Theophrastus comments on this: [25]

And he has the same theory about wisdom and ignorance. Wisdom is of like by like, ignorance of unlike by unlike, wisdom being either identical with or closely akin to perception . . . he added at the end that 'out of these things are all things fitted together and constructed, and by these do they think and feel pleasure or pain'. So it is especially with the blood that they think; for in the blood above all other parts the elements are blended.

Consciousness is given by a certain blend of the six elements in the blood: air, earth, fire, water, love, strife. Empedocles held that the soul goes through a 'wheel of birth' lasting thirty thousand seasons of alternating dominance of love and strife. The highest incarnations occur at the fourth cycle, finally producing 'prophets, bards, doctors, and princes'. Empedocles believed that he had, in an earlier cycle, been a bush.

He distinguished between the life-soul and individual consciousness. Homer made a similar distinction, though, as pointed out by Kirk and Raven (1960): [26]

. . . when separated from the body, the surviving soul in Homer is a mere shadow, which can only be restored to conscious life by drinking blood; to Empedocles, on the other hand, it is of divine race and has fallen for the very reason that it has tasted blood.

Plato made this same distinction between immortal soul and the mortal soul which perceives. Perhaps this reflects the dawning distinction between what we know with the senses and what we discover or infer (or know intuitively or as direct knowledge) of hidden orders of reality. It may also reflect conscious and unconscious.

Thales (*fl.* 580 BC) is regarded as the first true scientist, and the first 'pre-Socratic' philosopher. He was born at Miletus, and founded the Miletus, Ionian, School. He is believed to have travelled to Egypt and Babylon, and to have been expert in land surveying, navigation and astronomy. He is sometimes credited with inventing geometry as a Deductive system, from Egyptian practical know-how, and to have predicted the solar eclipse of 585 BC. He must have been lucky, as he could not have had an adequate theory, and the cycles as known to the Babylonians did not allow prediction to a given place on earth. Otherwise, nothing is known of his life. He is not known to have written a book, and no writing of his survives. Aristotle, who refers to him, is not sure of his sources. So Thales is indeed a shadowy figure. Apart from saying that the earth floats on water, and that water is the material cause of all things, Aristotle attributes to Thales the sayings, 'All things are full of Gods,' and 'The magnet (lodestone) is alive; for it has the power of moving iron.'[27] This implies, as Burnet (1908) points out,[28] that Thales must have thought other kinds of objects *not* alive. He presumably attributed life to anything capable of producing movement in non-obvious mechanical ways. This is the germ of our word 'motivation', which still has almost mythical or mystical connotations.

It has been clearly pointed out by Robert Brumbaugh in *The Philosophers of Greece* (1964) and *Ancient Greek Gadgets* (1966) that, from the evidence of comments by later writers, Thales was a great inventor and engineer, and was a primary figure in introducing mechanical models into speculative thinking, using actual or imaginary models based on the technology of his day for explanations by analogy, which remain the basis of science. This was developed by his successor Anaximander, with his notion of semi-circular tubes of fire as a model of the sun to explain eclipses and so on. Anaximander produced the first map of the world and of the stars. Maps are of course models in this sense. Many examples of models (such as Hippodamus' tuned metal disks having radii of concordant ratios representing and producing musical intervals) are described by Brumbaugh (1966) together with toys and gadgets, some of which may well have been far less trivial in the history of human understanding. These remain as fascinating for us as they must have been for Greek children, for whom they brought Heaven down to earth.

Hippocrates (fifth century BC) the first doctor of antiquity and the accepted founder of medicine, lived at the peak of Athenian life, being contemporary with Socrates, with the sculptor Phidias and with the dramatist Aristophanes. He was the first historical figure to combat medical superstition, though he accepted the early notion of vital force, as the *thymos* in breath (or pneuma) giving life, thought and feeling. He came to recognize the brain as the seat of mental life. This he inferred from effects of accidental brain damage, and from his remarkable studies of epilepsy. He also regarded the liver as important (black and yellow bile and phlegm), stating the theory of humours which dominated medical and psychological thinking about personality for two thousand years.

The ravings of lunacy, and the sudden 'possessions' of epilepsy, however, were attributed to Gods or spirits entering the person. Hippocrates questioned the 'sacredness' of epilepsy. In *The Sacred Disease* he wrote:[29]

I am about to discuss the disease called 'sacred'. It is not in my opinion more sacred than other diseases, but has a natural cause, and its supposed divine origin is due to men's inexperience, and to their wonder at its peculiar character. . . . They really disprove its divinity by the facile method of healing which they adopt, consisting as it does of purifications and incantations. . . .

My own view is that those who first attributed a sacred character to this malady were like the magicians, purifiers, charlatans and quacks of our own day . . . being at a loss, and having no treatment which would help, they concealed and sheltered themselves behind superstition, and called this illness sacred, in order that their ignorance might not be manifest. They added a plausible story, and established a method of treatment that secured their own position. . . . These observances they impose because of the divine origin of the disease, claiming superior knowledge and alleging other causes so that should the patient recover, the reputation for cleverness may be theirs; but should he die, they may have a sure fund of excuses, with the defence that they are not at all to blame, but the gods.

For if they profess to know how to bring down the moon, to eclipse the sun, to make storm and sunshine, rain and drought, the sea impassable and the earth barren, and all such wonders, whether it be by rites or by some cunning . . . I am sure that they are impious, and cannot believe that the gods exist or have any strength. . . . For if a man by magic and sacrifice will bring the moon down, eclipse the sun, and cause storm and sunshine, I shall not believe that any of these things are divine, but human, seeing that the power of godhead is overcome and enslaved by the cunning of man.

The fact is that the cause of this affection, as of the more serious diseases generally, is the brain. . . . The brain of man, like that of all the animals, is double, being parted down the middle by a thin membrane. For this reason pain is not always felt in the same part of the head, but sometimes on one side, sometimes on the other, and occasionally all over.

Then follows an anatomical account of 'veins' (including the vagus nerve) supposed to take the breath to various parts of the body, and, when blocked or compressed, to produce paralysis, or numbness, or the symptoms of epilepsy. It is thus related to the winds – which are divine. And he says that 'This disease attacks the phlegmatic, but not the bilious. . . .' There is then an account of how phlegm descending into the veins makes the patient speechless, the eyes rolling, the intelligence failing, for:

. . . the air that goes into the lungs and the veins is of use, when it enters the cavities and [of?] the brain, thus causing intelligence and movement of the limbs, so that when the veins are cut off from the air by the phlegm and admit none of it, the patient is rendered speechless and senseless.

Some winds are good, others bad: 'At the changes of the winds for these reasons do I hold that patients are attacked, most often when the south wind blows, then the north wind, and then the others.'

Later he says:

In these ways I hold that the brain is the most powerful organ of the human body, for when it is healthy it is an interpreter to us of the phenomena caused by the air, as it is air that gives it intelligence. Eyes, ears, tongue, hands and feet act in accordance with the discernment of the brain; in fact the whole body participates in intelligences in proportion to its participation in air. To consciousness the brain is the messenger. For when a man draws breath into himself, the air first reaches the brain, and so is dispersed through the rest of the body, though it leaves the brain its quintessence, and all that it has of intelligence and sense. . . .

Wherefore I assert that the brain is the interpreter of consciousness. The diaphragm has a name due merely to chance and custom, not to reality and nature, and I do not know what power the diaphragm has for thought and intelligence.

This reference to the diaphragm is to sensations from the diaphragm: '. . . if a man be unexpectedly over-joyed or grieved, the diaphragm jumps and causes him to start'. This is interesting, as Hippocrates preferred the brain as the seat of Mind, against the direct evidence that sensations in emotional states come from the diaphragm and the heart, but never the brain. It should be noted that the Greek for 'diaphragm' and 'sensation' is the same word. Hippocrates goes on to say that the 'Sacred Disease' is just physical, and should be curable by physical means – when a doctor 'would not need to resort to purification and magic spells'.

This passage is interesting in many ways. It gives a good idea of how Hippocrates thought, of how he tried to make use of evidence available to him from his experience of physical processes – rather than from 'good' and 'bad' which still are somewhat associated with health and illness, as though they are reward and punishment. He attempted to assign causes according to his physical notions. These made him reject the heart and diaphragm, but accept the brain, as the seat of consciousness and thinking, though the former *'feel'* different according to emotional states and stress. No doubt specific mental losses associated with obvious brain damage through injury led him in this direction. This was the direction that led to the dramatic errors of phrenology and to some of the most important current research and understanding of Mind through brain function. Interpreting the evidence of brain damage remains extremely difficult, in the absence of adequate general concepts of how the brain functions. (Without this knowledge it is indeed logically impossible to assign functions to brain regions; this is a point that is still sometimes forgotten.)

The notion of the four 'elements' of earth, air, fire and water were later combined with the four 'humours' of which the body was supposed to be composed: blood (*sanguis*), phlegm (*pituita*), yellow bile (*chole*), and black bile (*melanchole*). These were later arranged by Aristotle as opposites. The Hippocratic notions continue in psychological language – 'melancholy' and 'sanguine'; and our word 'temperament' stems from the significance that Hippocrates gave to the temperature of the brain.

So gradually myth became philosophy and science. As knowledge became effective, its successes and failures served as spurs and checks on shared views of the world – giving, as John Ziman puts it, the Public Knowledge of Science.[30] Perhaps myths are essentially private. What kind of a science then is psychology? This was a most important question for Greek philosophers, and especially for Plato and Aristotle, as now for us.

Plato (*c.* 427–347 BC) is generally regarded as the greatest (or at least the greatest Rationalist) philosopher. Born probably in Athens, of an aristocratic family, not much is known of his early life, except that he served in the Peloponnesian War. He became a disciple of Socrates, who appears in many of the *Dialogues*. Plato attended Socrates' trial by the Democrats (399 BC) described so movingly in the *Apology*, the *Crito* and especially the *Phaedo*. Plato founded his own school, the Academy, in Athens in 388. His Ideal State is described in the *Republic*, and his cosmology in the *Timaeus*. His great disciple was, of course, Aristotle. Plato's basic ideas are so well known that I shall not summarize them here but we shall meet him on several occasions later. Suffice it to say that he saw abstract forms or structures as the keys to understanding and describing the world; appearance was of secondary importance. Mathematics was thus the queen of the sciences. He was, however, highly concerned with questions of moral conduct and the place of man in the scheme of things. He was concerned also with theories of meaning and language; or at least we can easily read this into his notions of symbols reflecting and so having meaning by reference to Ideal Forms, which are never attained on earth.

Aristotle (384–322 BC) was born at Stageira, a Greek colony on the peninsula of Chalcidice. When he was eighteen years of age he moved to Athens, and three years later became Plato's pupil at the Academy, where he stayed for twenty years. Later, after Plato's death, he founded the Lyceum in Athens. He is the most influential philosopher-scientist: so much so that progress was seriously held up for two thousand years, especially in logic, which made no significant advance until the first years of the present century. But of course he is not to be blamed for that; and he seldom seems dogmatic.

Aristotle gave far more weight to observation than did Plato. Aristotle founded Deductive logic, by formulating syllogistic arguments, and also (which is sometimes forgotten) he promoted Inductive procedures for science. He was a Natural Scientist with an astonishing sweep of knowledge, including physics and biology. He wrote extensively on the Mind (especially in *De Anima*). For Aristotle, the 'soul' is the principle of life, and possibly has a kind of existence apart from the body, but it is potentialities that may be realized by bodily functions and behaviour. In this he holds what we could call a Functionalistic view. Perception, he argues, is given by receiving forms of objects according to how far they are like the structures of the sensory channels. The sense organs have the potential for recognizing the object, and *become that object* with perception. He accepts that perception may be indirectly mediated, as by light, and he gives weight to processes of judgement in perception. He set out the forms of the syllogisms which became the basis and set the limits of all Deductive logic for two thousand years. He also knew of several sensory illusions, explaining these with physical analogies. Aristotle wrote extensively on dreams, on illness and much else (see chapter 14). He is the principal source of 'pre-Socratic' ideas. His notions of mechanics were seriously incorrect, but dominated science until Galileo. Aristotle's notions of mechanics seem intuitively right before they are questioned by critical experiments, and children today seem to tend to think as he did. We should know more of this.

These are some of the names that have come down to us with their actual words, or at least accounts of their individual thoughts, which changed the beliefs of their contemporaries. The later writers questioned and looked for evidence in the world, including the results, as in medicine, of carrying out systematic procedures. We may talk of this, and later science, as 'exorcizing myth'; but there are always mysteries hidden beneath current understanding – and most mysterious of all remains understanding itself. How can certain lumps of matter – us – at all understand the world, and ourselves? Ghosts remain where we have still not learned how to question, and so how to understand in ways that can be shared.

Modern accounts of myth

Of course, myths are not dead. The problem has always been, and still is, to separate mythological beliefs from true knowledge; and still for many people, what are accepted as myths are yet regarded as true. Not indeed merely true, but even far more important than the truths of science, economics, and other 'facts of life', which may be distasteful. Here perhaps lies the key: we do not always seek or want truth – at least 'objective' truth: we want rather to live in another world. It may be deeply foolish to write this off as living in a fool's paradise. Better this than no paradise at all! There are many people who believe that myths are age-old truths, deeper and more significant than science. And much of psychoanalysis is based on just this – especially of course the ideas of C.G.Jung, who accepts much of alchemy, astrology, and so on as essentially true and important, at least for understanding Mind.

Then there are modern myths, such as flying saucers; and indeed science fiction is a kind of mythology, a mythology which may have ancient roots and which is quite often prescient in predicting achievements of later technology. Possibly myths are in some cases dreams of better (or at least different) worlds, which may indeed come to pass just because they have been in the dreams of men. Perhaps the dreams of some myths do direct technology, as well as protecting us from the proverbial harshness of reality.

Can one go further, to challenge science, by asking: is science myth? Are the accounts of the cosmologists – the worlds of Newton, Einstein, Bohr and Heisenberg – myths, as the arched Goddess of the sky with jewels for stars was an Egyptian myth? After all, the accounts of science do change: indeed they change far more than do the myths of antiquity! How can we trust the ever-changing assertions of science as true? Here I suppose that we may accept that science is homing in on truth, and, indeed, few scientists suppose that we have now arrived at a fixed account, though there have been periods, especially in the nineteenth century, when this was believed. If science changes while homing in, it is rational to support science rather than myths, because truth is approaching; and much may now be true, though limited. Here the Pragmatism of the evident practical usefulness of science-inspired technology is undoubtedly effective evidence, and has been so right through civilization from the first pickaxes which set up barriers against the floods and so defied the Gods.

It might be argued that science and myth, though conflicting, may yet exist together. This is maintained by much, and perhaps all, of contemporary theology. And one might even suggest that many individuals do not need one consistent belief system: one has met scientists who clearly wear two caps, one for their religion and the other for their science. Yet another solution is to accept that there are *no* solutions, and to live either by not questioning, or by living with questions in place of answers – which may in any case be not much more than merely vacuum-filling. While individually satisfactory, the weakness here is that neither questions nor rejection of questioning lead to consistent behaviour or socially effective agreements. Intellectually, however, it seems only reasonable to accept questions as highly valuable and exciting; and indeed much of philosophy is, rightly, devoted to asking good questions. Some answers are, however, necessary for survival.

Can we understand myth? There is an astonishing variety of theories of myth. The range is from myths as primitive science – for example by Durkheim – to the Polish anthropologist Bronislaw Malinowski (1884–1942), who held that myths are sacred tales accompanying rituals and not to be considered apart from rituals;

to Levi-Strauss, who largely dismisses the content of the myth and looks for its structure (by analogy with language structure) with different levels of meaning, rather as a poem can have different levels of meaning. This is similar to Freud's 'latent' and 'manifest' content of dreams (see chapter 14). Perhaps these (and other) theories of myth are not mutually exclusive.[31]

As Sir Frederic Bartlett (1932) showed very clearly there is a very strong tendency to fill in gaps in stories when recalled, and to 'rationalize' inconsistencies. Features of stories, and pictures, from alien cultures are recalled with systematic changes or shifts towards the familiar culture. Gap-filling is basic in memory and perception. It seems that vacuums are as abhorrent to Mind as to Nature! Perceptual and memory gap-filling may generate myth – perhaps much as it generates the far more structured hypotheses of science.

Myth makes us turn to children's thinking. Do children think mythologically? Do they, for example, hardly distinguish the world from themselves? Are they unable to see things from other points of view? Jean Piaget (1896–1980) asked how, and when, children come to appreciate the world, as existing apart from the particular viewpoint from which, at any given moment, they are seeing it.

In Piaget's best known experiment on what he calls 'decentration' (1956) he set up three model mountains, which are distinguished by colour on one, snow on another and a house on the third.[32] The experimenter produces a doll, which he places at one side of the mountains, so the doll would have a different view from the child, who is asked what the doll would see. Because a verbal description would be difficult, even for an adult, the child may be given pictures of the model taken from various positions, and asked which of these is the doll's view of the mountains.[33] There are other variants of the experiment, but in all children under about nine years of age do markedly less well than adults. Children of six or seven choose the picture corresponding to their own viewpoints. This has been held by Piaget to show that children up to this age cannot visualize or conceive the world apart from their own viewpoint. It is, however, argued by Margaret Donaldson (1978), following extensive experiments, that the children do not understand the instructions; and when the situation is arranged so that they can understand what is required of them, they perform much better than Piaget claimed. This problem of communication is clearly extremely serious for investigating or appreciating children's or primitive thinking. This leaves two alternatives for understanding what is going on: thinking and beliefs may be inferred from behaviour, which is the ethological approach; or communication may be improved, which is the major aim of philosophy. The ethological approach is beginning to be used effectively for mapping the child's development before language is available; and philosophy has had a major part to play in refining language and clarifying concepts. Logic and mathematics may also be viewed in this light. Both behaviour and language are links to Mind, but the links are tenuous – so that myths and indeed Minds remain mysterious.

Margaret Donaldson's point about the difficulty of communication may also be applied to the adult. Surely it could be that our knowledge of science serves to help us see from other points of view and communicate these views.

Remarkably interesting experiments have recently been carried out on rotation of 'mental images', by Roger Shepard and J. Metzler (1971). They find that fairly simple shapes (made of three-dimensional Ls) can be visualized and mentally rotated. If a pair of such figures is presented, subjects can decide whether they are the same object by mentally rotating one figure to see if it matches the other. The time taken to rotate mentally increases exactly linearly with the required angle of rotation, and that it is quite slow and constant for each individual. Shepard tends to think that there is a kind of rotating brain analogue; but it might be successive computing of a digital kind. The issue is important for considering the physical basis of Mind – how the brain carries or represents perceptions – and it is suggestive for considering the relation between, as we might say, perceiving and conceiving. ('Mental rotation' may require successive memory comparisons and so be slow.)

There are clear differences between the situation of adults in the ancient 'mythological' world, and present-day children. The adults of pre-philosophical and pre-scientific times were as intelligent as ourselves, in their potentials for understanding, but they did not have the cultural information available to our children, or the contact with technology which gives very different experience – because the physical world *is* very different in a technological world. The child's nervous system must develop for understanding to grow, and he must acquire certain skills and knowledge for other skills and knowledge to be possible or appreciated.[34] As Jerome Bruner puts it (1952), the child must be ready for the next step.[35] Now it is interesting to speculate that this second kind of readiness may apply in the social development of philosophy and technology much as it applies innately to the developing child. Possibly we are not yet ready to understand Mind. By this, I do not mean that we are morally inadequate (though this could be a religious view), but rather that we may not as yet have discovered, or created, the necessary concepts or the technology necessary for understanding. Possibly the new concepts and technology of computers capable of making decisions are a critically important step for understanding Mind. Even this notion, though, has ancient origins, as we shall see (chapter 12).

There is lack of agreement as to how mythological thinking became 'scientific' and critical. It is easy to point to a few individuals (especially Homer and Hesiod) as creating this most dramatic change, as they lived just before Greek philosophy and the dawn of science with Thales; but this does seem naïve. The process was gradual, and is not yet complete. We may guess that this development is tied closely to technology – which is still developing.

Whereas technology is concerned with the environment, myths and much of philosophy have been and still are more concerned with Mind. Technology very likely freed us from myth – so far as we are free of myth – by showing that tools could be more powerful than Gods, when used with intelligent human initiative. But since machines for intelligent decision-making have never been part of technology until now, with very recent computers, intelligence and psychology have remained in the realm of myth. At least this is an arguable position which in various forms we shall meet and consider in our journey through these pages.

2 Tools that challenged myth

It is a moot point whether the human hand created the human brain, or the brain created the hand. Certainly the connection is intimate and reciprocal.

Alfred North Whitehead (1861–1947)[1]

Tools are products and extensions of the limbs, the senses and Mind. Their importance can hardly be exaggerated.

Tools are dependent on the properties of the materials of which they are fashioned, and upon their forms. There are subtle interrelations between properties and forms of substances upon function, which is clear to anyone – and we may suppose to early man – who makes and uses tools. Their effective use requires, and so implies an understanding of, strategies, both individual and socially organized. It is here, rather than in dexterity, that man is the tool-user, and even more the tool-maker.

Kenneth Oakley (1961) describes tools as extensions of limbs:[2]

The evolution of new bodily equipment in response to a change of environment required millions of years, but relying on extra-bodily equipment of his own making, which could be quickly discarded or changed as circumstances dictated, man became the most adaptable of all creatures. Making fire, constructing dwellings and wearing clothes followed from the use of tools, and these cultural activities have enabled man not only to meet changes of environment, but to extend his range into every climatic zone.

Comparing man in this regard with the apes, Oakley says:

One may sum up by saying that apes of the present day are capable of perceiving the solution of a visible problem, and occasionally of improvising a tool to meet a given situation; but to conceive the idea of shaping a stone or stick for use in an imagined future eventuality is beyond the mental capacity of any known apes. . . . Systematic making of tools implies a marked capacity for conceptual thought.

This conclusion is borne out by the lack of major anatomical differences between man and apes. In the opinion of the great anatomist Sir Wilfred Le Gros

Clark, the differentiation of man from ape will finally rest not on any anatomical basis, but on the human ability to speak and make tools. On the similarity of the brains and body and limbs of ape and man, Sir Wilfred Le Gros Clark says in *Man-Apes or Ape Man?* (1967):[3]

Again, the brain of the large apes is astonishingly like the human brain – smaller of course, but constructed on the same basic pattern with a similar, if simpler, pattern of convolutions of the grey matter of the cerebral cortex. . . . Some of the limb bones of a chimpanzee may be quite difficult to distinguish from human limb bones. . . . The foot skeleton and the muscles associated with it, in spite of the divergent big toe, show many striking similarities to the hominid foot skeleton . . . the big toe in man could hardly have arisen as a product of evolution unless it had been derived from a large and powerful big toe very similar to that of anthropoid apes . . . it is sometimes difficult to tell from an isolated molar tooth whether it is that of a chimpanzee or of some type of man.

It is important for philosophy to realize how recent is our appreciation of the age of man. Philosophers in the past were drastically misled. Returning to Oakley:[4]

Until about 100 years ago the possibility that man had existed for more than a few thousand years had barely been considered. According to Archbishop Usher's chronology which was still current the first man had been created in 4004 BC (on March 23rd according to a 'Kalendar' of St John's College, Oxford). With the discovery of remains of mammoth and other extinct animals in the ancient river deposits (Diluvium) of various parts of Europe, the belief gained ground that there had been several periods of creation, and the diluvial animals, or animals of the previous age, had been wiped out by a universal deluge of greater magnitude than the historic Noachian flood.

The first known crude flint tools are found in the Olduvai Gorge, Tanzania, and were made (Richard Leakey, 1977)[5] 'by knocking a couple of flakes off a tennisball sized pebble. The result is a chopper, the central piece of equipment in the so-called Oldowan industry. Crude scraper and possibly hammerstones also make up part of the tool kit.' This tool-cult disappeared from the Olduvai Gorge a million years ago, to be replaced by a second tool-culture, the Acheulian, who made tear-shaped hand axes with sharpened edges. They had also '. . . well designed choppers, chisels, scrapers, cleavers, awls, anvils, and hammerstones, as well as several variations on these. In no sense was this tool kit the result of random stone knapping.'

Flint is a most suitable material for forming sharp edges which may function as knives or axes for cutting softer substances. Tools of suitable form and substance can dominate animals and plants which are absurdly vulnerable to the sharp edges of stone tools. This lesson is perhaps two million years old and at one stroke placed man, when tool-maker, beyond the reach and power of other animals, to make him a kind of God. With the use of metals, including alloys having properties never found in Nature, possibilities became limitless, and imagination has grown with the power of tools, up to the modern discoveries of electricity, radio and space flight. These all depend on blending the properties of substances with forms, to make tools, machines and instruments for attacking and probing Nature. These interrelations are appreciated with more or less general concepts which may generate new technologies and may inspire philosophy.

The first useful metal was copper: indeed it was so useful that the Neolithic world, limited to scraping and hammering with simple hand tools, was gradually transformed into a technology in which operations entirely beyond human strength and precision were regularly performed. With metals, technology started to take over and effectively transform the bodies and Minds of men.[6]

The origin of the word 'metal' is interesting: it is somehow related to the Greek μεταλλᾶν, to seek after, explore. The Latin is *metallum,* a quarry, mine. The six metals of antiquity – gold, silver, copper, lead, tin, iron – were at first found as nuggets in surface layers, and later quarried or mined. Although gold and silver were held in special regard as tokens of wealth and for jewellery, copper (generally and at first unintentionally alloyed with lead or tin) was far more important than gold or silver. It was strong and malleable; it could in its pure soft state accrete sand as a useful abrasive for drilling and cutting stone, and in its hard forms (alloyed with tin to form bronze) it could form cutting edges, bearings, hinges, or even (later) gear wheels. Copper was, in its turn, gradually replaced by iron from about 1200 BC. The first iron to be used was from meteorites. These include quantities of nickel which makes the metal steel-like. This iron was originally used for jewellery. The Sumerian word for iron means 'Heaven metal', and the Egyptian word means 'black copper from Heaven'. The general use of iron as superior in strength, and cutting edge, to bronze must date after processing by repeated hammering and heating with the addition of carbon. This 'steeling' of wrought iron was perhaps the invention of the Chalybes (*c.* 1400–1200 BC), who were subjects of the Hittite kings and who had a monopoly of the new metal which made the swords of heroes. (The Golden Fleece, by the way, has its origin in technology, from trickling gold-bearing streams through oily fleeces, the gold sticking to the wool while the unwanted jetsam floated away.)

The great advantage of metal over stone tools was the variety of shapes that could be produced in metal, while flints could only be flaked according to their internal structure. The shapes of metal could be imposed purely by the intention and imagination of the craftsman. Further, moulds allowed many identically shaped metal parts to be made, so that a good design could be easily and accurately reproduced. The original did not have to be made in metal, but could be quite easily cut from wood or clay. No doubt this emphasized the distinction between substance and form. Moulds were usually made of clay, with piece moulds for complex shapes. The *cire perdue*, or 'lost wax', process was established to allow raised patterns to be reproduced in piece moulds. The process also allowed complex shapes to be made from a single mould.

Statues were formed from sheet metal as early as the copper bulls from El-Ubaid, before 3000 BC, and the copper statues of Pepi I and his son (*c*. 2300 BC) at Cairo, which seem to be the earliest metal statues. They were made of copper sheets, hammered with frequent annealings, to form the shapes of an inner wooden framework to which they were nailed. Chasing, and *repoussé* (hammering from the inside with forming tools), were employed for fine detail. Techniques for soldering gold and silver were developed in Ur as early as 2500 BC. Metal was used not only for jewellery, statues and weapons, but also for straps for joining wood in carts, chariots, ships and other large machines. Such composite construction remained important until the nineteenth century and is still sometimes found.

By the time of Hellenistic Greece, techniques of fine metalwork were well advanced, and indeed in some ways have never been bettered, so perhaps we should not be too surprised to find a flowering of invention in the construction of ingenious metal mechanisms. These included pumps, taxi meters with reduction gearing driven by a carriage wheel to indicate the distance travelled, fire pumps with piston and cylinder, and water-clocks with gearing to indicate the time with the pointer of a moving figure or a moving drum. There were also heavy lifting devices for building with massive stone blocks, and quite massive machines for use in theatres. The fact that these existed implies an elaborate and widespread technology in classical Greece, though it receives scant attention from the philosophers.[7]

Tools having moving parts, such as bow drills, date from the earliest Egyptian dynasties (say 3000 BC) and maybe earlier in Babylon. These were made of wood, with gradually increasing use of copper and iron.

It now appears that there were some two million years of slow but distinct progress in tool-making, before the explosion of technology with the development of tools and other machines having moving parts. The change in conceptual understanding is then so rapid that we are forced to attribute the way we think very largely to our tools and what has been created by them. This is partly by human intention, but very largely what is created is what is made easy to create by available resources, techniques and tools.

Buildings such as temples, and later cathedrals, were ostentatiously showing off the limits of the prevailing technology, as well as being peaks of artistic achievement. It is surely this union of art with technical daring which makes them so exciting. It is continuation of this same technical daring, perhaps almost without intention, that has landed man on the moon.

The immense importance of technology in moulding how we think has implications still only dimly appreciated, and we cannot hope to add much here. It does, however, imply that any psychology based only on our biological origins is going to be inadequate: what is amazing about man is how far he has escaped his origins. This is through the use of tools, and the effect on us of the technology that the tools have created. These considerations are at least as important as our biological background for understanding man and how the first mythological, philosophical and scientific ideas were conceived and developed. For example, only recently mechanical clocks have profoundly affected the way that we see time as a steady flow divided into arbitrary intervals, from the ancient notion of time cycling with the stars, and mechanical time dictates our lives and turns industrial society into a vast machine almost independent of the heavens, which is a quite new idea.

It is obvious that ways of life, and many freedoms and tyrannies, are dictated by technologies; but I wish to explore the notion that the most abstract notions of philosophy and theories of Mind stem directly from technology. Technological innovations typically come before conceptual bases by which they are understood; as understanding grows, principles can be described with increasing generalization to allow deeper analogies. It is these, I suggest, that are scientific explanations and the basis of philosophies including (most strangely) accounts of Mind. We shall look, though briefly, at some of the steps of technology by which concepts of science seem to have been reached. This is, at the same time, in large part the story of the development of human Mind.

Perhaps, much as children learn to think and understand by active exploration with hands and senses and later by making things with tools, so Science is Mind matched to aspects of reality by active use of tools. This is at least a notion that we can explore, by looking, if briefly, at some of the tools and steps of technology, and how they seem to have inspired concepts of Matter. The story is one of exorcizing Mind from Matter, until now we are left with Mind only in brains – so we are stuck with the problem that Mind is a special case, and so especially difficult for Science.

It seems safe to say that the tools of technology usurped the Gods by allowing man, with their aid, to work miracles if not always more dramatic at least more reliable than the claimed miracles for the Gods. Above all, technology led to machines capable of more and more autonomous functions. We have come to prefer for Science mechanistic explanations, perhaps finally to see Mind as the functions of brain mechanisms – and perhaps mechanisms of technology. We may look now at tools that have usurped the Gods, severed man from his past, and so changed us that we are unique in Nature.

The giant Prometheus was credited with bringing fire from Heaven for use by man. Fire is the first of a long sequence of forces to be encouraged and tamed, forces that are dangerous, though useful when used intelligently. Since it was for thousands of years very difficult to make fire on demand, fires were carefully tended; this required continuous care and preparation, and shortage of fuel necessitated planning and conservation. The maintenance and sharing of fires may well have had importance in developing social groups. There remains one people with no means of making fire: the Andamanes. Neolithic man, and perhaps late Palaeolithic man, made fire by the percussive method – banging a flint nodule against iron pyrites. The hot sparks were used to ignite tinder. (This method was used in England as late as 1827, when friction matches were invented.) Fire was also made by heating wood with friction, with a leather thong or a fire-drill. This is interesting, as it is also a tool in its own right for making holes in wood and even stone. Animal oils as well as wood were used in late Palaeolithic times for heating. Oil lamps are found at the 'Grimes Graves' Neolithic flint mines, in East Anglia; and lamps as old as 3000 BC have been found at Ur. Fire must have seemed the divine sun come to earth. It is hardly surprising that it has long-standing associations with Creation.

The special properties of materials used for making things, such as wood, metals including alloys, and water, earth and fire, took on philosophical significance through their uses and powers in technology. This underlies the long-standing notions of forms and essences of things contained in their substances. This notion is found in the concept of 'protoplasm', which from the earliest times and until very recently was supposed to imbue organisms with vitality, with life.[8]

There was, however, also an early invention that may have suggested that structure can give function independently of substance – leading perhaps to the alternative view that substances have their properties by virtue of their atomic structures rather than by the essences of their substance. I refer to the invention of building with bricks. Bricks, made from standard moulds, date from Mesopotamia of the fourth millennium BC. They were made in clay moulds and baked in the sun. Fire-baked bricks came later. Fire, indeed, was seen sometimes to change the essence of substances, and sometimes to harden, purify and even to make things immutable. These two principles – substances having hidden properties which emerge, and properties and functions given by structures – run right through technology and philosophy. They remain, also, the two principal ways of thinking about how Mind is related to brain.

We may see the gradual development of chemistry from alchemy, and of astronomy from astrology (which both ran in parallel for most of their histories) as a slow process of exorcizing Mind from Matter, through subtle lessons from technology. For chemistry, the lessons or suggestions were from purifying and alloying of metals; for astronomy they were from measurements, recorded over many centuries for predicting climatic cycles, and for Egypt the vital annual flooding of the Nile delta, for agriculture. These revealed repeating machine-like cycles, hidden from the senses with their limited memory-span, by specially contrived observations and records kept over many generations and many centuries. They revealed, also, properties of things that could not be seen or guessed at. The procedures of technology, and later experiments, were no doubt confounded with religious ritual, and even now observatories and laboratories are special places – temples – demanding qualities of dedication and honesty beyond normal living.

So science and philosophy depended on the use of many kinds of tools. Let us look at some examples.

Hand tools
The pickaxe
The pickaxe was the first important tool, following flint choppers and scrapers. It made possible large-scale changes to the environment – which had to be conceived, planned and carried out in organized groups. The pickaxe was invented, and was profoundly important in Mesopotamia, for building dams and canals, for storing water, and for controlling the frequent floods. The prominence of the pickaxe was acknowledged by the Sumerians: the God of the sky, Enkil, was also God of the pickaxe. There was a special symbol in the pre-cuneiform written language for the pickaxe which transformed how they lived and thought: the first powerful weapon to challenge Nature, and combat the droughts and floods – the will of the Gods.

The pickaxe works by storing kinetic energy from human muscle-power and concentrating it within a small interval of time onto a small region – to literally shattering effect. Thus the hardest rock can be subdued, and by planned designs vast works undertaken, so that floods are controlled and water stored, to change the environment essentially and create from the desert gardens of paradise.

The lever
Levers made possible superhuman feats, which later were often attributed to the Devil. Organized work and careful planning were called for, with rules for which methods to apply in given conditions. Understanding levers requires a kind of dynamic geometry, which became formalized as mechanics. For thousands of years the lever was far more significant than the wheel, which may be regarded as a lever fuctioning continuously through 360° of rotation. The laws of leverage apply, of course, to meshed wheels of various diameters. The construction and understanding of mechanisms depended on these insights, and mechanisms brought to life the relations of geometry. Perhaps the concepts of force, energy and power, as well as inertia, friction matching and so on, were distinguished by

considering levers. They allowed vast constructions such as the pyramids and Stonehenge to be built: thus levers and pickaxes challenged the Gods. Their challenge has grown, with the increasing power of technology to make dreams – and sometimes nightmares – come true.

The wheel

The wheel is a genuine invention and not a copy of something found in Nature. The key part of the invention is the bearing. For load-carrying this has to be well made, of suitable materials, and it needs to be lubricated to reduce friction – which is a non-obvious concept. Wheels allow continuous motion, which is important not only for transport and the potter's wheel, but also for representing – and it was used for this surprisingly early – movements of the stars.

The potter's wheel and also wheels with axles for transport date back to about 4000 BC (V. Gordon Childe, 1954).[9] There were also several early partial, or non-continuous, rotary devices, such as bow drills. It is, of course, continuous rotary motion that organisms do not have: it was the invention of the wheel, continuously rotating, that was a dramatic departure from bodily capabilities.

The oldest surviving potter's wheel is dated 3250 ± 250 BC, found at Ur. Sir Leonard Woolley (1930) describes his find as 'a thick disk of clay with its pivot hole smeared with bitumen and a series of small holes near its circumference'. The early use of the potter's wheel is easily established from the precise form, and parallel lines or grooves on even small fragments of wheel-made pots. There is not such clear early evidence for wheels used in transport: the evidence here is tomb paintings, engravings and drawings on pots or friezes. There are clear pictures of wheeled carts dated 2500 BC from Mesopotamia, and simple pictographs indicating wheeled vehicles date from 3500 BC. The earliest remains of wheels are from a hearse in a royal tomb from Kish, Mesopotamia, dated third millennium BC. In models dated 3000 BC, the rims of wheels are studded with nails; continuous rims date from about 2000 BC. All this is Mesopotamian. Most curiously, although the neighbouring Egyptians had boring tools, capable of fashioning large stone vases, they do not seem to have had the wheel until much later.

As we shall see it has recently been established that the Greeks of the Hellenistic period had not only potter's and chariot wheels but also accurately made bronze gear wheels employed in elaborate mechanisms – including what we must call computers.

Cooking utensils

It is somewhat odd that man prefers, to a large extent, cooked food to food in its natural state, and this can apply to domestic animals. After so many millions of years of adaptation to crudities, why should we so often prefer cooked food? Have the pot and fire bricks produced physiological adaptation in man, in almost historic times? Perhaps this is so, but of course there is more to it. Cooking increases tenderness, to make a greater variety of food acceptable. It has the immense advantage of reducing disease in communities, and no doubt the social needs of cooking led to rituals and mores that tended to stabilize societies. According to the anthropologist Mary Douglas, customs are largely structured around cooking.[10]

With cooking went the search for special commodities such as salt. There were in prehistoric times salt-routes, covering Europe and extending into Asia. Cakes of salt were used as money. It was produced by evaporation from lakes, and from salt springs; because these are only located in a few regions, salt was an essential trade commodity for many areas.

The preservation of food was undertaken from prehistoric times, but with no conceptual notion of the causes of decomposition. Salting, drying, smoking and pickling in brine were all used. These must have been discovered purely empirically, as also were the use and preparation of many kinds of spices and so on, that are useless without special and sometimes highly unlikely preparation. These are indeed some of the most remarkable and least understood inventions of man – or, more likely, woman.

Prime movers

The ancients were woefully short of sources of power other than men, animals, water and windmills. What they lacked were autonomous power sources other than organisms. No doubt for this reason, prime movers, initiators, were deeply mysterious – and seemed uniquely special to life.[11] This situation did not essentially change until the invention of practical steam engines in the eighteenth century, which within a very few years transformed the industrial world.

Virtually the only autonomous power source available to the ancients was the falling weight, string and pulley. This was used for giving controlled power (mainly for driving toys and automata) by the weight being placed on corn in a funnel. As the corn ran through the hole at the bottom of the funnel the weight slowly fell, drawing its string after it. The more elaborate mechanical theatres were placed on cliff tops in order to gain sufficient fall for the weight; but then steady control was difficult to attain, as the corn trick could not be used.

It is difficult to believe that Aristotle would have emphasized Animism as he did if he had been familiar with the prime movers of modern technology. Here we have a case where *absence* of technology seems to have produced mystery – or, rather, technology was not available to challenge early mystical ideas.

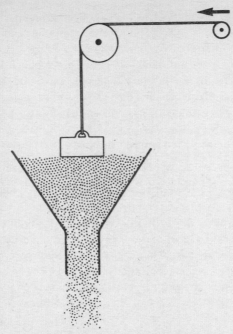

3   One of the few kinds of Greek prime movers, or motors: a funnel filled with corn or sand which, as it flowed, allowed a weight to fall slowly and give a steady pull to a string, which could drive an automaton.

Mind tools[12]

By 'tools of Mind' I mean aids to measuring, calculating and thinking. We should also include the power of speech and pictures and writing for communicating and storing knowledge and ideas. It is a question, though, whether speech is a human invention in the sense that, for example, the Babylonian pickaxe, or the wheel, were invented by men. There is evidence of a biological propensity for language which does not apply to pickaxes or wheels. According to anthropologists who have studied language in relation to culture and concepts, the vocabulary and grammatical structure of language is deeply important to how we think. Writing, however (though it was attributed to the Gods by the Egyptians), is clearly a human invention.

Language

The development of language in children raises special problems, if only because learning a language would appear to be a far more difficult task than any other accomplishment of the child – or indeed of the adult. The work of Jean Piaget has suggested that children are surprisingly weak at logical thinking, being far below adult standards up to at least the age of eight when language is virtually fully developed. How can children acquire language, though they fail at what are, to adults, very simple tests of logical thinking? Noam Chomsky's view, that there is an inherited propensity for language, might be an answer. If so, all languages would have to be essentially similar in spite of their apparent great differences: hence the interest in this respect of Chomsky's 'Deep Structure' of languages. Chomsky (1965) gave his position in these words: [13]

It seems plain that language acquisition is based on the child's discovery of what from a formal point of view is a deep and abstract theory – a generative grammar of his language – many of the concepts and principles of which are only remotely related to experience by long and intricate chains of unconscious quasi-inferential steps.

Doubt is cast on Piaget's strictures on children's logical abilities, but nevertheless it does seem extremely remarkable that a skill that so taxes the adult to learn is acquired readily by almost all children. To the skill of language we might add the skill of perceiving, if, indeed perception requires 'Unconscious Inference' as Helmholtz supposed (see chapter 12).

Many attempts have been made to relate the ways that different people think, and what they believe, to their language. Seen in this way, language is both tool and straightjacket. An anthropologist expresses this significance of language (Edward Sapir, 1931): [14]

Language is not merely a more or less systematic inventory of the various items of experience which seem relevant to the individual, as is so often naively assumed, but is also a self-contained, creative symbolic organisation, which not only refers to experience largely acquired without its help but actually defines experience for us by reason of its formal completeness and because of our unconscious projection of its implicit expectations into the field of experience.

Sapir suggested in the 1930s that this could be seen by comparing the Indo-European languages (which are all basically similar) with very different languages such as the native languages of Africa and America. This idea was developed by Benjamin Lee Whorf, who developed what has become known as the 'Whorfian Hypothesis': that language moulds thinking. It seems reasonable to assume that thinking is generally helped (and sometimes hindered) by the structure of language, which itself adapts, though slowly, to thought and action.

Whorf includes in his effects of language, effects on perception. This is well summarized by Henle (1966): [15] 'Whorf notices that Eskimo languages have a wide variety of words for different kinds of snow where we only have one. Aztec is even poorer than we are in this respect, using the same word stem for cold, ice, and snow.' Henle says, a little later: [16]

It would seem then to be consistent with what we know of mental set on other grounds to assume that the world appears different to a person using one vocabulary than it would to a person using another. The use of language would call attention to different aspects of the environment in the one case than it would in the other. Numerous illustrations of this sort may be given. The Navaho, for example, possess colour terms corresponding roughly to our 'white', 'red', and 'yellow' but none which are equivalent to our 'black', 'grey', 'brown', 'blue', and 'green'. They have two terms corresponding to 'black', one denoting the black of darkness, the other the black of such objects as coal. Our 'grey' and 'brown', however, correspond to a single term in their language and likewise our 'blue' and 'green'. As far as vocabulary is concerned they divide the spectrum into segments different from ours.

Henle then points out (I am sure correctly) that the physiological colour vision of the Navahos is not different from ours. The point is that they make different distinctions, and language developed for particular distinctions tends to create these distinctions in perception.[17]

Henle summarizes the differences between English and Navaho:[18]

English stops with what from the Navaho point of view is a very vague statement – 'I drop it'. The Navaho must specify four particulars which the English leaves either unsettled or to inference from context:

1. The form must make clear whether 'it' is definite or just 'something'
2. The verb stem used will vary depending upon whether the object is round, or long, or fluid, or animate, etc., etc.
3. Whether the act is in progress, or just about to start, or just about to stop, or habitually carried on, or repeatedly carried on, must be rigorously specified. . . .
4. The extent to which the agent controls the fall must be indicated. . . .

It is sometimes considered that these language differences are sufficiently great to make the accepted philosophy, or metaphysics, of cultures different according to their language structure and distinctions that are easy to make with their vocabularies. This is a highly technical question, on which I cannot usefully comment. It is clearly extremely difficult to establish the direction of the 'causal arrow', pointing to language affecting thought, or thought affecting language. All we can say for sure is that there are these differences between existing cultures and language, and we may assume that there were just as large differences between the first languages and ours – while their thinking was also different. It may be assumed that there is a two-way causal arrow, such that language affects thought and thought affects language. This is indeed clear in special languages, such as mathematics or symbolic logic: it is far easier to think in certain ways with these languages, which have been specially developed to be useful.

It is worth pointing out that the philosopher's use of linguistic analysis aims, with some success, both to reveal hidden meanings in various language forms, and also to help the philosopher from being misled or limited in his thinking by distinctions or implied associations in common language. These can be extremely general. For example, we tend to think of the world as made of substances having properties. This corresponds to the subject–predicate logic of Aristotle, and the subject–predicate form of all Indo-European languages. Bertrand Russell (1946) comments on this while discussing Aristotle: [19] ' "Substance", in a word, is a metaphysical mistake, due to transference to the world-structure of the structure of sentences composed of a subject and a predicate.' According to Whorf, this may have started from the significance of a container and things contained. This is different in Hopi. Henle quotes Whorf: [20]

'Our language patterns often require us to name a physical thing by a binomial that splits the reference into a formless item plus a form.' Hopi nouns, in contrast, always have an individual sense, even though the boundaries of some objects are vague or indefinite. There is no contrast between individual and mass nouns, hence no reference to container or body-type, and no analogy to provide the dichotomy of form and matter.

This should at least indicate the power of language structure to direct (and sometimes mislead) thought – and it may justify linguistic analysis in philosophy.

It is surely true of all tools, that by making some things easier they direct activity and thinking from things that are more difficult; but what is easy and what is difficult are partly set by the available tools, and so we are carried along by a sequence of largely arbitrary and sometimes unfortunate features of our technology, including our language. Human intelligence is very largely Artificial Intelligence, and even our hopes and fears (and our moral commitments, for they are set by possibilities of achievement) are largely set by existing technology.

Writing

It has recently been suggested that the first written signs originated as three-dimensional shapes made of clay, used as tokens of transactions. These were sealed in clay bags, or *bullae,* which sometimes had the forms of the tokens that they contained incised on their surface, as flat figures. These were presumably accessible records of the transactions – recorded by the tokens, representing numbers, animals, crops and manufactured objects such as beads sealed inside the *bullae.* Later the tokens were abandoned, the transactions being recorded only by their pictures. The suggestion is that the first writing was a two-dimensional representation of these three-dimensional barter-tokens. The tokens were sometimes impressed on the soft clay of the *bullae,* to leave imprints, as seals on wax. [21]

The importance of writing, for societies and for individual thought, can hardly be overestimated, and its importance was clear to the earliest civilizations in which it developed – as we may judge by the seniority accorded the Gods of writing. The

Babylonian God of writing was Nebo, who was also the God of destiny: the Egyptian God of writing and wisdom was Thoth, who recorded the virtues and sins of the soul (*ka*) judged by Osiris for acceptance to the world of the dead. The Egyptian term 'hieroglyphics' means 'sacred writing'; writing was largely secret, as it was known only to the priests and to royal scribes.

According to David Diringer (1962): [22]

What can be said with certainty is that there is no evidence to prove that any *complete system* of writing was employed before the middle of the fourth millennium BC. Representational cave-paintings and carvings on small objects have been found from as early a period as the Upper Palaeolithic (some 20,000 years or more BC), as well as circles and other symbols, full of variety and distinction. Some of these were apparently used as

Original pictograph	Pictograph in position of later cuneiform	Early cuneiform	Classic Assyrian	Meaning
				Heaven God
				Earth
				Man
				Pudenda Woman
				Mountain
				Woman Slave-girl
				Head Mouth To speak
				Food
				To eat
				Water In To drink
				To go To stand
				Bird
				Fish
				Ox
				Cow
				Barley Grain Sun Day To plough To till

4 The development of cuneiform symbols, from the first pictographs to Classic Assyrian.(From David Diringer, *Writing*, Thames and Hudson Ltd, 1962.)

property marks . . . but they are not in any way complete (that is, established and systematic) forms of writing, nor can any connection be traced between them and the ancient systems we know.

All of the earliest writing consisted of picture symbols, which became connected in conventional ways to express ideas. They became associated with sounds of the spoken language, as the sound of the name (generally of a common object) denoted by the symbol became the name of the symbol. For proper names, such as the names of kings and countries, symbols were chosen by a process of punning, from some (usually entirely irrelevant) object having the same or a similar sound for its name. Its picture symbol would then become, or be incorporated into, the way the name was written. (To make up an example we would understand: 'Babylon' might be represented as a picture of a 'baby', shown by itself 'alone'. This would be near enough in sound-values to give written signs for 'Babylon'.)

Abstract concepts cannot be at all easily indicated by pictures. (For example, to show that the baby was alone, in the example above, might be quite difficult.) The Egyptians developed a system of determinatives, as we call them, to indicate the general class, or level of abstraction, in difficult cases. The symbol for 'abstract' was a rolled papyrus scroll, tied with a ribbon and bow. This effectively meant 'It cannot be written.' A picture of the sun (a circle with a dot in the middle), when shown next to this determinative, meant 'day', or 'time', which had to be guessed as it could not be pictured explicitly. The determinatives also served to limit the number of picture symbols, which at one time got out of hand; the priests then did some pruning, and deliberately put more order into the symbols.

The alphabet

The alphabet has been called the greatest invention of man. According to Diringer[23] it

almost certainly had its origin at a single point in history, and in a specific if hitherto uncertain, place in the Near East, probably in Palestine or Syria. It was, historically, the last major form of writing to appear, and it is the most highly developed, the most convenient, and the most easily adaptable system of writing ever invented.

It may be that writing developed from individual marks to indicate ownership, and no doubt these were magic talismans. The first effective writing was detailed representational pictures, which could be combined to give simple messages; then ideographic scripts which allowed ideas to be expressed. These started as pictures, but became simplified and easier to write. For example, just a head (and still later perhaps a single curved line) might represent a person, or a God. These symbols could be combined to represent a flow of ideas, for symbols were developed for abstractions and for logical operators. Our minus sign is the Egyptian sign which was originally a picture of outstretched arms, like a policeman stopping traffic. At first it had hands and fingers on the ends; these were dropped, to leave the simple horizontal line that we still use.

Ideographic writing became phonetic, as we have said, and this in practice required cumbersome syllabaries so that the names of the symbols could be looked up. The early Babylonians and also the Egyptians had highly organized schools for the uninitiated, in which future scholars were taught to write and learned the names of the symbols, so that they could not only write but also read in their spoken language; thus language ceased to be either sacred or secret.

It seems likely that writing had profound effects on society and attitudes to the nature of things, and may indeed have inspired the Platonic notion of fixed Ideal Forms. The point is that once something is written – be it law or contract or promise – it stands apart from individuals and becomes an objective statement. This could have been a crucial step away from Animism and towards notions of shared objective social standards and truth. At the same time the words of the Gods were for better or worse fixed, unchangeable; so societies changed at their peril as social change challenged the Gods.

Counting

There are two kinds of counting. We may count things, or we may count divisions on scales. These are extremely different. The known Universe is classified, for various purposes, into what we regard as similar, and as different, kinds of things. We may count the number of people living in Cambridge; or we may divide people into the number of men and of women living in Cambridge. For some purposes men and women form the same class – people – and for other purposes we count them as different. Counting things implies classification of what are to be included together. Classifications may be on grounds of appearance, or they may be on theoretical grounds. Distinctions for forming classes are generally observable. Theoretical and not perceptually discernible distinctions come much later.

The first kind of counting, then, is of individuals, or at least what are accepted perceptually as individual objects, or features or events. The other kind of counting, of divisions of measuring scales, is always arbitrary. For length we may use feet, metres or stadia, or whatever; these units or divisions of length are chosen for general convenience. We should expect situations for which they are *not* convenient. This arises in serious form, when we consider what scale to use in the sense of linear or logarithmic or other such scales. We usually define physical reality – the Natural world – as being in its broad features linear in space and time. The choice of measuring scales can affect how things are seen or described.

The origins of counting are prehistoric, but early development of ways of manipulating numbers is preserved on Babylonian clay tablets and in later papyri of the Egyptians and the Greeks.[24] There is a tablet discovered near Babylon – the Senkereb Tablet now in the British Museum – which is dated to 2300–1600 BC and was clearly a ready-reckoner with numbers (to the base sixty) arranged in columns and rows written in the cuneiform script. The Babylonians invented the 'zero' (originally a space, and later given a special symbol) and they invented number places – corresponding to our columns of 'units', 'tens', 'hundreds', and so on. Curiously, these important inventions were not used by the Egyptians or the Romans, whose ways of representing and handling numbers were crude and cumbersome. To perform long division with the Roman number system is extremely difficult. It is also difficult or impossible for most of us without any aid: so the power and importance of notation is clear.

If we ask why the Romans had such a clumsy notation, we can only guess at an answer, which may be that for limited practical problems counters on ruled boards were adequate. The forms of the Roman numbers (I, II, III, IV, V, VI, VII, VIII, IX, X and so on) strongly suggest tally marks, made on wooden sticks and split in half as proof of transactions.

For purposes other than recording transactions or astronomical data permanent records were unnecessary, so counters and abaci would be adequate. The Babylonians were exceptional: they kept continuous running records, with calculations and predictions of astronomical observations. Counters would be far too easily moved by accident, while marks impressed on clay have a permanence which has preserved their writings to this day.

Ways of representing numbers have affected how people have thought about the relation of mathematics to the Natural world. This is clear in the patterns of dots used in Greek mathematics, which affected Pythagorean notions of number and reality. 'Triangular' numbers became very important, as also did 'square' and 'cube' numbers, from which we derive the terms 'squaring' and 'cubing'. Pythagoras (or certainly his school) wrote 'triangular' numbers as below (Boyer, 1968) for 3, 6, 10, 15 or, in general:

$$N = 1 + 2 + 3 + \ldots + n = \tfrac{1}{2} n(n + 1).$$

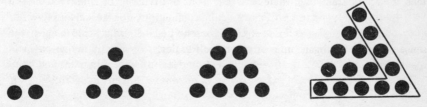

5   The outlined section indicates the gnomon.

Square numbers were formed from $1 + 3 + 5 + 7 + \ldots + (2n - 1)$ where each odd number was regarded as a pattern of dots resembling a gnomon – the Babylonian shadow-clock or sundial arranged around the two sides of the preceding square pattern of dots. (The word '*gnomon*' meant 'square', the stick casting the sun's shadow in a dial, and also 'knowing' or 'knowledge': from which derives 'cognitive'.) For 'square' numbers they look like this:

6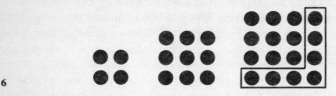

And for even 'oblong' numbers $2 + 4 + 6 \ldots + 2n = n(n + 1)$:

7

Units for measuring

Measurement, other than for example 'judging by eye', involves adopting and referring to some external standard. The senses are then used only for comparison. Counting is essential, either of the number of standard lengths or weights, or of divisions on a scale. The problem of course is to choose suitable and repeatable standards. Where these can be copied the problem is not too great (though cumulative errors can easily occur), but copying of standards over a wide area is not practicable where transport is poor. The practical problem is ancient: for upon standards of length and weight depend fairness of barter and trade, and transfer of land and other property. No doubt fulfilling these requirements served early science well.

Units of length and weight varied greatly between states and cities, and according to the uses to which they were put. Thus units for grain, for gold and even again for silver, were often different. Units for length were generally derived from the human body: thus the Greek Olympic cubit (25 digits = 18.23 inches = 463 mm) was derived from finger widths. The Egyptians had standards of length and weight stamped with the royal insignia from 3000 BC. The Egyptian royal cubit (20.63 ± 0.2 inches = 524 ± 5 mm) was divided into 7 palms and 28 digits. This curious use of 28 may reflect the 28 days of the lunar cycle.

The foot was generally accepted as two-thirds of a cubit, divided into twelve parts to give inches. The yard is the distance from the nose to the tip of the middle finger. The span (a fathom) is the distance from tip to tip of the outspread arms. The mile is one thousand double paces. Standards based on ears of corn were also used: thus an edict of AD 1324 under Edward II rejected all limb measurements, and adopted the length of three barley corns to make an inch.

The difficulty of the variations in size in humans and barley corns, and so on, was largely overcome by adopting the (very early) statistical procedure of taking the average of several examples. Thus in AD 1150 King David of Scotland declared the inch to be 'the thowmys of iii men: that is to say, a mekill man and a man of messurabil stature and of a lytell man. The thoums are to be messurit at the rut of the nayll.' The foot was established around then as the average of several (men's?) feet from the church congregation. The development and use of standards is a fascinating subject that has not been fully explored. It is salutary that the Universe is measured by commercial standards. When Napoleon defined the metre as a proportion of the size of the earth he got it wrong!

The development of sophisticated units for measuring forces and many other characteristics of the world is modern, and highly theory-laden. The current standards of length in terms of numbers of wavelengths of light of a given frequency depends of course on techniques only recently available. It is particularly interesting that with such shifts of standards from essentially arbitrary units to constants of Nature, what was empirical can become necessary – with an ever-increasing danger of fundamental tautologies. Perhaps this is the price of precision.

Abacus to computer

The notion of representing numbers with tokens (or *jetons*) placed on boards, or wires, to represent states in conceptual number space, is prehistoric. The antiquity of the abacus is indicated by the etymology of 'calculate' = Latin *calculus,* a pebble; and the diminutive *calx,* a stone. Cicero (106–43 BC) speaks of *'ad calculos vocare aliquid,'* meaning 'subject to strict reckoning'. The word 'digital' refers, of course, to the finger; and the fingers can be used as a kind of abacus. The abacus is still in everyday use in Russia. It is remarkable that so simple a device is so effective: this surely shows that the neural processes by which we do 'mental arithmetic' are very different from the abacus.

Computers are commonly classified as analogue and digital. For the computer engineer, an important distinction is that analogue computers are, generally, continuous mechanisms, while digital computers work in steps. This requires very different mechanisms, or circuits. This is not, however, the fundamental distinction: it seems far more important to appreciate that analogue devices represent input–output transfer functions (which may be logarithmic, exponential or whatever), while digital devices represent the steps of logical processes, for solving problems according to formal procedures. They thus have a deep theoretical, and not only an engineering, distinction.

Analogue computers represent more or less abstract quantities by physical parameters such as lengths, angles and voltages. Physical or electrical constraints are used to represent more or less abstract variables. The constraints may be set by using some physical principle (such as Hooke's Law for a spring), or, more usually, specially cut cams, or their electrical equivalents. Digital computers work by discrete states representing numbers, logical states and program instructions. They are machines that change according to the problem to be solved, to embody the manner of solution according to the instructions of the program. This is a subtle matter: a matter surprisingly close to the classical 'Mind–brain' problem.

Computers are in essence autonomous abaci – working without continuous human intervention. The first metal-wheeled calculating machine (adding and subtracting only) was built by Pascal in 1642, the year of Newton's birth. The concepts and first attempts to build the modern digital computer are due to one man, the Cambridge mathematician Charles Babbage (1792–1871).[25] Charles Babbage has been described as 'a pioneer who left no trail behind him', and indeed it was nearly a century before his design philosophy became embodied into successful computers. This waited upon the development of high-speed and reliable electronic switches: these became the bricks on which the ever-changing logical structure of the machine depends. It is hardly fanciful to suppose that the logical bricks, or bits of information, are the equivalent of Sumerian clay bricks – which allowed physical structures to be created and described, and which possibly launched atomic accounts of matter. Conceivably the interplay of informationally atomic bits, realized by computer switches, will prove to be atoms of Mind: the basic elements of psychology.

Although the use in an abacus of pebbles or beads on a board or on strings is prehistoric,[26] its use continued in England until the time of George II, in the exchequer – 'exchequer' meaning the cloth-covered table divided into squares, on which accounts were kept with counters. We have only to look at the current immense popularity of pocket calculators to see that arithmetic is alien to the unaided human brain – we use mechanical aids for counting and accounting because the brain is ill-adapted for these tasks. There is surely an important lesson here (see chapter 10).

The astrolabe

Calculators as simple as the abacus work by representing what are usually considered mental things – such as numbers and operations of a calculation. There are ancient tools which even go a step further, and simulate cognitive models of

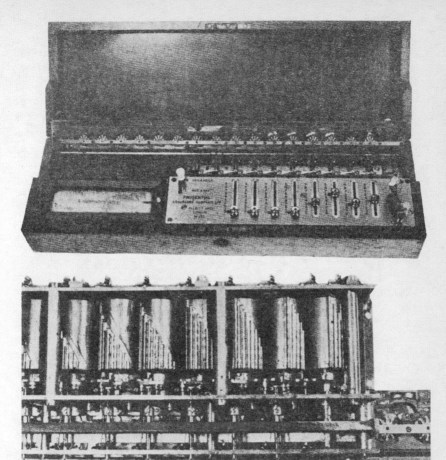

8   A calculator incorporating Leibniz's cylinders, capable of addition, subtraction, multiplication and division; first half of the nineteenth century.

reality. An early and important example is the astrolabe, which represents the heavens as a flat 'stereographic' representation engraved on brass plates.[27]

The astrolabe was used for navigation, and for telling the time at night throughout the Eastern world, and was used by the Greeks in various forms (at first in the form of globes) from the beginning of recorded science. We see in the astrolabe a device working from an internal model of the Universe – from which it gave accurate results from only crude measurements. This is highly significant for considering the importance of conceptual models for making optimum use of

9 An astrolabe. This elaborate astronomical computer, used for navigation and for telling the time at night, remained essentially unchanged from as early as the first century. They incorporate engraved plates of the paths of stars as seen from various latitudes: computation is thus based on analogue models of the heavens. This example is Persian, from the eighteenth century.

limited information. Indeed, this notion is a vital key to how the brain succeeds so well in controlling behaviour though information from the senses is intermittent and unreliable. Observations that can be made with the rete of the astrolabe are similarly intermittent and unreliable, but become useful when augmented by its internal model of star paths.

The astrolabe is fascinating as an early analogue computer, in which positions of bright stars are engraved as paths of motion on the brass plates, and read from longitude markers representing the Holy Cities. The necessary longitude parameters are fed in by selecting the nearest Holy City marker, for reading the instrument. Even the fact that it is flat is interesting, for the star paths are represented by stereographic projection, from the crystal spheres on which they are conceived to lie in space to the flat plates of the astrolabe – though linear perspective had not been invented. The situations are, however, different: what was solved for the astrolabe was how to transform shapes from one surface (a sphere) to another surface (a plane), which is not the same problem as representing objects lying at various distances in three-dimensional space to a plane. This problem was not fully solved and understood before the application of projective geometry, at the time of the French Revolution (see p. 226).

Far more elaborate and ambitious than the astrolabe is a Greek mechanism for recording movements of the heavenly bodies with geared wheel-work – the Antikythera mechanism.

Simulators of the heavens: ancient clock-computers
It is often said that the Greeks did not have a mechanical technology. This is generally attributed to their dependence on slaves, and the related low status of things mechanical. There is, however, compelling recent evidence that this view of the Greeks is not correct. They had elaborate wheel-work for simulating the heavens. There are several references in Greek literature to mechanical models of the Universe, but until recently they have been difficult to interpret as they lack detail. This situation is now transformed, with the discovery of a mechanism of the first century BC, found in a sunken Greek ship. Recently developed methods have made visible its internal mechanism of bronze gears, where it is impossible to open it up. It is fully described by Derek de Solla Price in *Gears from the Greeks* (1974).

This is the first great discovery of underwater archaeology. The mechanism was found in 1900 by Greek sponge fishermen in an ancient wreck off the shore of the island of Kythera. It is dated, from the evidence of amphorae and many other objects including bronze statues, to about 80 BC. It consisted of an elaborate chain of meshed gear wheels, made of bronze, to represent the Ptolemaic system with the earth as centre of the revolving sun and planets. There is evidence that such devices go back at least to Archimedes (c. 287–212 BC) and were probably well known, as they were exhibited in public places. A building contemporary with the Antikythera mechanism still stands in the centre of Athens which very likely housed a similar mechanism. This is the Tower of the Winds. The inside is gutted, but it still has elaborate sundials on the faces of its octagonal tower. Given its complexity and pride of place, it seems likely that there were earlier, simpler versions, in Athens and perhaps elsewhere. There are several references in the

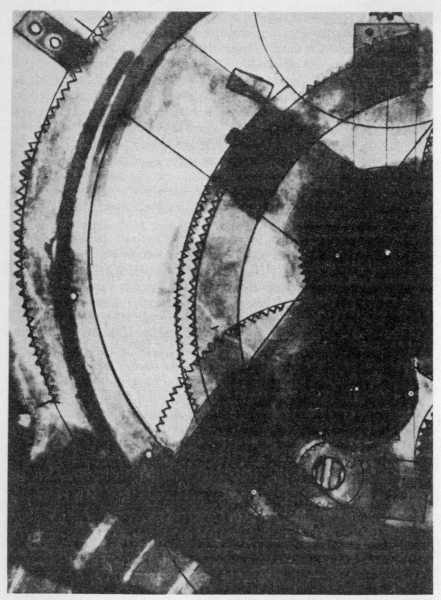

10 The Antikythera mechanism, a Greek geared astronomical computer found in 1900 on a Greek ship sunk *c*. 80 BC. This is now in the National Museum of Greece in Athens. (Copyright © Professor Derek de S. Price, Yale University.)

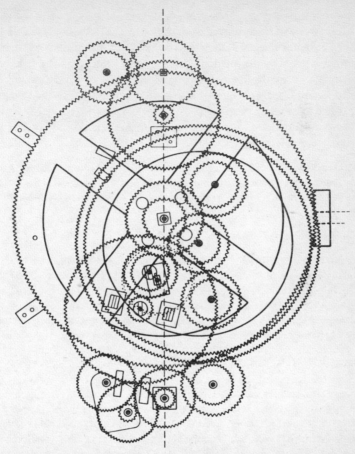

11 A diagram of the wheel-work of the Antikythera mechanism. It proved possible to count the teeth, from neutron photographs revealing structures in the bronze mass, which was sunk in the sea for two thousand years. A working model of this ancient computer mechanism has recently been built. (Copyright © Professor Derek de S. Price, Yale University.)

literature of Greece and Rome to globes with moving sun, moon, planets and stars. Until the discovery and recent full description of the Antikythera mechanism these have been difficult to interpret as they lack technical detail. It now seems that they give accurate accounts of a very long tradition of mechanisms, simulating the Universe as understood by the Greeks, and very likely affecting how the Universe appeared to them.

The Archimedean globe models were almost certainly turned by hand, and not by a mechanical clock. So time was given by the sun and stars and could be speeded up, slowed down, or reversed or stopped, at the will of the human handle-turner. Now consider the Universe controlled by the Gods: could the Gods change time, by changing the rate of their cosmic handle? If time is the flow of all events, then time is given by the rotation of the cosmos, whatever the rate may be. It is meaningless to say that the heavens are speeded up or slowed down, if time is defined in terms of these celestial events. It is possible, it seems to me, that the Greeks had ideas along these lines.

Cicero mentions two 'globes' attributed to Archimedes, the first of which he found not particularly admirable: [28]

> For I remember an incident in the life of Gaius Sulpicius Gallus . . . a most learned man as you know . . . he ordered the celestial globe to be brought out which the grandfather of Marcellus had carried off from Syracuse [in 212 BC] . . . when I actually saw it I did not particularly admire it.

But of the second he says:

> For that other celestial globe, also constructed by Archimedes, which the same Marcellus placed in the Temple of Virtue, is more beautiful as well as more widely known among the people. But when Gallus began to give a very learned explanation of the device, I concluded that the famous Sicilian had been endowed with greater genius than one would imagine it possible for a human being to possess. For Gallus told us that the other kind of celestial globe, which was solid and contained no hollow space, was a very early invention, the first one of that kind having been constructed by Thales of Miletus, and later marked [inscribed] by Eudoxus of Cnidus (a disciple of Plato, it was claimed) with the constellations and stars which are fixed in the sky. He also said that many years later Aratus, borrowing this whole arrangement and plan from Eudoxus, had described it in verse. . . . But this newer kind of globe, he said, on which were delineated the motions of the sun and moon and of those five stars which are called wanderers [planets] . . . contained more than could be shown on the solid globe, and the invention of Archimedes deserves special admiration because he had thought out a way to represent accurately by a single device for turning the globe those various and divergent movements with their different rates of speed. And when Gallus moved the globe, it was actually true that the moon was always as many revolutions behind the sun on the bronze contrivance as would agree with the number of days it was behind it in the sky. Thus the same eclipse of the sun happened on the globe as would actually happen. . . .

In *Tusculan Disputations* Cicero writes:

> For when Archimedes fastened on a globe the movements of moon, sun and five wandering stars, he, just like Plato's God who built the world in the *Timaeus*, made one revolution of the sphere control several movements utterly unlike in slowness and speed. Now if in this world of ours phenomena cannot take place without the act of God, neither could Archimedes have reproduced the same movements upon a globe without divine genius.

It is surely significant that Plato speaks of the Father, or the Creator, or the Living Being, making the Universe as a 'moving image of eternity': [29]

The nature of the Living Being was eternal, and it was not possible to bestow this attribute fully on the created universe; but he determined to make a moving image of eternity, and so when he ordered the heavens he made in that which we call time an eternal moving image of the eternity which remains for ever at one. For before the heavens came into being there were no days or nights or months or years, but he devised and brought them into being at the same time that the heavens were put together; for they are all parts of time, just as past and future are also forms of it, which we wrongly attribute, without thinking, to the Eternal Being.

Here we see Plato preserving the essentials of reality behind appearance, as timeless. A little later, he calls the observable Universe a model, with a copy:

For the model exists eternally and the copy correspondingly has been and is and will be throughout the whole extent of time. As a result of this plan and purpose of God for the birth of time, the sun and moon and the five planets as they are called came into being to define and preserve the measures of time. And when he had made a physical body for each of them, God set the seven of them in the seven orbits of the circle of the Different. The Moon he set in the orbit nearest the Earth, the Sun in the next and the morning star and the one called sacred to Hermes in orbits which they complete in the same time as the Sun . . . when the beings jointly needed for the production of time had been given their appropriate motion and had become living creatures with their bodies bound by the ties of soul, they started to move with the motion of the Different, which traverses that of the Same obliquely and is subject to it, some in larger circles, some in smaller, those with the smaller circles moving faster, those with the larger moving more slowly. . . . And in the second of the orbits from the earth God lit a light, which we now call the Sun, to provide a clear measure of the relative speeds of the eight revolutions, to shine throughout the whole heaven, and to enable the appropriate living creatures to gain a knowledge of number from the uniform movements of the same.

The significance of this modelling of the heavens was not lost on early poets and thinkers.[30] The Roman poet of the fourth century AD, Claudian,[31] wrote of Archimedes' globe: [32]

When Jove looked down and saw the heavens figured in a sphere of glass he laughed and said to the other Gods: 'Has the power of mortal effort gone so far? Is my handiwork now mimicked in a fragile globe? An old man of Syracuse has imitated on earth the laws of the heavens, the order of nature, and the ordinances of the gods. Some hidden influence within the sphere directs the various courses of the stars and actuates the lifelike mass with definite motions. A false zodiac runs through a year of its own, and a toy moon waxes and wanes month by month. Now bold invention rejoices to make its own heaven revolve and sets the stars in motion by human wit. . . . Here the feeble hand of man has proved Nature's rival.'

Not only were the heavens mimicked by mechanisms: so also was life, especially as described in Hero's *Automaton Theatre*. But automata had only levers and springs: meshed gears (including differential gears) were reserved for the heavens, for which they were ideally suited. Perhaps it was these models that led to the Ptolemaic Epicyclic account of the Universe. Was time, for the Greeks, the rate at which the handle of the copy Universe is turned? If so, what (or who) turns the handle of reality? What initiates events? For the Greeks this required soul, or Mind. For them, Mind was separate from the mechanism of the Universe – rather as Mind is generally regarded by us as somehow separate though controlling the brain and our behaviour.

I suggest that notions of Mind have been from the dawn of thought, and still are, bound up with notions of mechanism, and so with the current technology. There is also a striving for elegant solutions, which is hard to define but clearly apparent in the history of science and philosophy. It is just as apparent in the history of mechanical invention. In the design of the Antikythera mechanism we see a struggle and an elegant solution of the ancient problem of perfect ratios, or harmonies. To quote Derek de Solla Price (1975): [33]

In a device of this sort, complications will arise quite naturally both from the astronomical necessity and from the number-theoretic relations of the designed gear ratios. For example with the 235/19 ratio . . . though one may wish to use gear wheels having numbers of teeth that are easy to mark out by geometrical construction, this is not possible. The numerator has prime factors of 5 and 47, and the denominator already is a prime, so that the ratio can be achieved only by some such gear chain as for example $60 - 12 + 47 - 19$ where there will always remain wheels having the awkward numbers of 19 and 47 teeth or some multiple thereof. From astronomical necessity undoubtedly the most telling is the relation between the siderial motions of the Sun and Moon and the Moon's synodic cycle of phases; exactly that which impressed Cicero. For the Metonic cycle, for example, of 19 years and 235 synodic months, it is absolutely necessary that the Moon make $235 + 19 = 254$ siderial revolutions in a complete cycle. Either one can simply build such a consistent set of numbers into the system, for example by adding a ratio of 254/19 . . . or one can take the elegant but mechanically difficult route of providing a mechanism that starts with two of the rotation rates and generates a third by mechanical summing and differencing. This latter route seems to have been that which motivated the inclusion of a differential gear turntable in the Antikythera mechanism.

Plato, as we have seen, mentions in the *Timaeus* 'that they started to move with the motion of the Different, which traverses that of the Same obliquely and is subject to it. . .' . Could this be a reference to the intermediary differential gearing found necessary in the Antikythera mechanism, or some earlier mechanical solution for the problem of lack of harmony of the heavens represented by mechanism? It is a tempting thought. It seems very likely that these sophisticated mechanisms

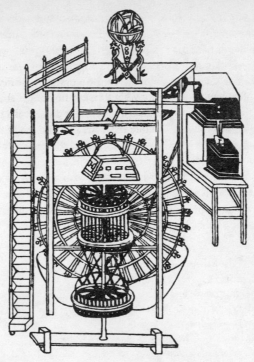

12 A tenth-century Chinese clock, which has an escapement intermediate between the clepsydra of the Greeks and Egyptians, and the mechanical escapements of the fifteenth century, which developed into the uniquely high-precision, low-friction mechanisms of seventeenth- and eighteenth-century clocks in the West. (From Joseph Needham, *Clerks and Craftsmen in China and the West*, Cambridge University Press, 1970.)

made far more important contributions to Greek philosophy than the philosophers realized, or at least admitted. One suspects that this is just as true for current technology in philosophy today!

Early mechanisms for simulating the heavens were not restricted to Western science. China had a similar tradition: a remarkable astronomical clock was made by Su Sun in AD 1088.[34] This was late in a long tradition. It was a tower thirty feet in height with a large water-wheel which powered the clock and a rotating star globe depicting the sky in 'real time', as well as elaborate manikins which popped out of doors with cards telling the hour and which rang bells and banged gongs. It had a remarkable escapement mechanism, which is a missing link between the steady water flow of clepsydra and the mechanical clocks of the Renaissance which, as Price has shown, were primarily for showing the motions of the heavens and only secondarily for telling time. Accurate clocks, and with them a conception of steadily flowing time, have been going for less than three hundred years.

Modern computer simulation is a direct development of the Greek Antikythera mechanism and its still earlier forbears. These mechanisms were, however, 'dedicated' to a particular set of computations, following a single conceptual model. They led to the elaborate astronomical clocks of the Middle Ages, following the remarkable clock (or rather astrarium, for telling the time was its least important function) of Giovanni di Dondi, built in 1348–64. Although the original is lost, exact copies from Dondi's detailed description with drawings have been constructed recently, and may be seen at the Science Museum, South Kensington, London, and at the Smithsonian Institution in Washington. These clocks profoundly affected accounts of the heavens, as we may see from pictures such as 'Two Winged Angels Operating the Revolving Mechanisms of the Sky' and from such passages as this, from Dante's *Paradiso* (Canto x) written between 1316 and 1321:[35]

> As clock, that calleth up the spouse of God
> To win her Bridegroom's love at Matin's hour,
> Each part of other fitly drawn and urged,
> Sends out a tinkling sound, of note so sweet,
> Affection springs in well-disposed breast;
> Thus saw I move the glorious wheel; thus heard
> Voice answering voice, so musical and soft,
> It can be known but where day endless shines.

What is new, essentially, about modern computers is their ability to represent very different conceptual models and perform very different functions, according to program instructions. These they may even select for themselves. Programming follows from the punched card instructions of eighteenth-century mechanical looms, and more directly from Charles Babbage's analytical engine of the 1830s. At last, with programs, we have explicit formulations of procedures, which are tested for success or failure by reliable machines carrying out instructions – and even instructing themselves. This is essentially new and at last offers promise of adequate mechanical simulators for understanding Mind. Just how much of Mind can be simulated remains to be seen. At this the ancients were singularly unsuccessful.

Simulators of life: automata

The Egyptians and the Greeks made jointed statues, including the fascinating tomb dolls of Egypt to supply the wants of the deceased as their servants did on earth. There were statues with hidden speaking tubes in temples, and many theatre machines for simulating aspects of life by automata, as described by Hero.

Aristotle (*De Anima*, i, 3) described statues that moved their arms and walked, guarding the labyrinth. But these were magical talismans, intended to mystify. Some were pure hoaxes. In no case did any mechanism begin to reproduce or represent functions with anything like the panache of the astronomical models representing conceptual accounts of the heavens.

The ancients did, however, succeed in making static stone lifelike, in their statues. It is a remarkable fact that white, stationary stone of roughly human or animal form is so evocative of life. It would seem that we, as observers, 'project' life into statues and paintings, which are really very different from the living creatures that they depict. Possibly it is because life is so relatively easy to capture, or convey, that technology was not stretched to supply this need, as it was to achieve power over Nature and represent Natural phenomena. The difference probably has a clear biological origin: survival depends on recognizing and identifying living things, and perceptual systems are 'tuned' for this – to recognize predator and prey, infant and lover; and to predict their behaviour from subtle variations of behaviour, pose and expression. It is surely this that makes it so easy to represent a face, with an expression, with the few lines of a cartoon. It is for this reason that statues 'work' though distorted from life, while scientific accounts that are distortions are correspondingly false. One might say that art evokes, while science describes and explains. These are very different, and neither should be judged by the other.

The Greeks from the fourth century BC did, however, make elaborate marionettes, which were regarded as amazing; and even more ingenious and elaborate automatic theatres, worked by falling weight motors. These simulated specific kinds of movement – of birds, of insects, and human. It is interesting that 'neuron' is the same as the Greek word for 'string' – the strings that worked the amazing marionettes.

Only very recently have the behaviour and embryological development of animals been regarded as at all amenable to description in terms of mechanistic principles – rather than due to a special 'life force', or 'animism', supposed to permeate organisms in the very substance of their protoplasm, and sometimes to vivify art, as when Prometheus breathed the fire of life into the clay of a statue.

We now generally see life as given by functional processes carried out by matter. This forces us to look carefully at the concepts of function and machine. These are concepts that have very gradually emerged from technology, and they remain far from clear. Although mechanical explanations are favoured in science, few scientists can say what they mean by 'mechanism', or 'machine', and I doubt whether there is any general agreement.

With clocks and wheeled computers, we have moved from simple tools to mechanisms. The distinction here may not be sharp, but it seems best to call things so complex, and with interacting moving parts, mechanisms rather than tools, though of course they may be used *as* tools. To design and understand mechanisms, more or less abstract concepts are needed, and these may apply to a very wide range of things, including Nature herself. Indeed, 'mechanistic' explanations are the most favoured kinds of explanation in science, if not philosophy. There are, however, some deep puzzles about how we should think of mechanisms. The Oxford philosopher Gilbert Ryle (1949) describes Mind as 'the ghost in the machine'; but even simple machines can look ghostly: especially when new properties 'Emerge' with increasing complexity of the machine. This we shall discuss in the next chapter. Meanwhile, we may consider non-functional but aesthetically interesting features of artefacts of all kinds, including machines. Is there a functinal role for aesthetics in technology?

The aesthetics of technology

Buildings, boats, and almost all human works have for us a marked aesthetic interest. This may, of course, be belittled as 'merely subjective'; but we are universally fascinated by designs. It is very clear that artefacts are always made with styles – which are usually sufficiently distinct to mark their origins in place and time, to distinguish peoples and cultures. Why should there be aesthetic styles? Why should there be (sometimes rapidly, sometimes slowly) changing fashions or tastes? The study of aesthetics may seem less interesting and profound than the study of unchanging laws, because fashions are often ephemeral, and seem arbitrary by comparisons with Natural and indeed psychological laws. But aesthetic preferences and judgements are dominant features of any society, and aesthetic styles remain as the most enduring marks of human endeavours, from the earliest cultures. I shall hazard a guess as to why adopting aesthetic styles has survival value, to people carrying out cooperative work.

Consider making an axe, a boat or a house. Some of the features of these artefacts are essential for their function, and must as much as possible be optimized for efficient function. This would apply to the cutting edge of the axe, to the length and form of the handle, and to many – though not all – features of the boat and the house. It is clear that functionally essential features will tend to become optimized through trial-and-error and experiment, within the limits of the available materials and the time and effort that may be put into construction. But what of the non-essential design features? For these, there is freedom of design – but there must be agreement as to what design to adopt for cooperative work to be possible. It must be agreed that the windows of a house should be so large, and of such-and-such proportion – though these are not functionally critical – or the house will not get built. We may suppose that a few dominant individuals set standards for features that are not functionally critical, but must nevertheless be accepted and decided upon for completion of the artefact or project. This requires a few individuals to persuade the rest to adopt particular designs, but without any functional (or indeed any 'rational') reasons. So we have leaders, and followers, of fashion.

Fashions or aesthetic styles may become fixed by magical or religious conventions, which take precedence over any individual, however powerful. Thus we find early designs functionally appropriate for wooden buildings incorporated as basic features of later stone Greek temples – and indeed later on into the radiators of Rolls-Royce cars!

Designs associated with some special artefacts acquire such prestige that these designs are transferred to other uses for which they may handicap function. This could be true of the Greek temple car radiator; and streamlined shapes are absurd for hand tools or static objects such as buildings. This is inappropriate transfer of aesthetic form, adversely affecting function, which we see even in the design of scientific instruments.

The artist has an honoured place in almost all societies. On our account of aesthetics, he not only expresses ideas, ideals, and emotions, but also, though more mundanely, he confers standards on societies so that agreement can be obtained to make cooperative action possible. This can, however, also be attained though at high cost by autocratic rule, or through rigid inflexible conventions generally imposed by fear. We may suppose that persuasion by artist-designers will produce frequent changes of aesthetic styles compared with the more rigid and seldom-changing styles in autocracies – so from the rapidity of change of styles in prehistoric communities we might be able to deduce something of their governments.

It is interesting that freely expressed art is feared by modern autocratic governments, presumably because it is a force for unpredictable change, and non-functional uniformities beyond control. Aesthetics and the persuasions of artists may be more important for technology than either technologists or artists realize or welcome.

There is clearly more to aesthetics than this. A theory of why particular aesthetic preferences come into fashion, to dominate societies for a time, would have to go much further, and I suspect would require very different concepts: psychological concepts probing deeply into human wishes, fears and beliefs. We might indeed apply the previous arguments to suggest further that wishes, hopes, and especially beliefs should be generally agreed and shared for a community to be viable. There are no doubt deep interactive forces between society and philosophy and science, which spring from and supply social needs, while at the same time they have a wonderful life of their own.

On this account aesthetics is essentially irrational (or at least *a*rational), and this applies to aesthetics of hand and Mind. At a given time in a society there are accepted ways of thinking, and more or less strongly forbidden, or ignored, ways of thinking. It is usual to criticize religions for being 'blind to new ideas', and indeed there are excesses of zeal in maintaining the *status quo*, such as the Catholic Church's attitude to Galileo. But, given that rational grounds do not exist for many important decisions that must be made for a society to be effective, what alternative is there but to adopt *a*rational beliefs where different opinions may lead to conflicting goals? It may be most important for societies to have taboos (explicit or implicit) on questioning. Just as sharing aesthetic preferences in artefacts may be essential for cooperative work (it almost turns people into an organized machine), so also tacit beliefs and taboos may be essential for a society. At the same time, if these become stuck for too long the society may atrophy. So change, and sometimes violent change, may from time to time be essential. This could justify the revolutionary, and certainly justifies individual originality in social terms.

Technology requires social planning and cooperation, often through many generations. This requires communication, discussion, leadership, acceptance of majority opinion and a combination of respect for tradition with dreams of different futures. Only through technology can ways of life which have been set by biological need and opportunity be changed; and changed most dramatically they have been, so that man's life is quite different from that of any other animal. This extends to very marked aesthetic preferences, which distinguish each human community in space and time.[36]

3 Lessons from machines

And now I see with eye serene
The very pulse of the machine;
A being breathing thoughtful breath
A traveller betwixt life and death.

William Wordsworth (1770–1850) [1]

What is a machine?

Most curiously, it is difficult to find out just what scientists or philosophers (or indeed the common man) take 'machine' to mean, and yet mechanistic explanations are generally supposed to be the most, and even the only, acceptable kind of explanations of science. This marked preference for a mechanistic explanation – whatever this may turn out to be – is frequently extended to accounts of Mind.

An account of Mind claiming to be mechanistic may, indeed, look scientifically respectable, but what would it be like? We shall hardly find out without considering, in some detail, notions of machine and concepts of mechanism, as well as something of the history and role of technology as a source of explanatory concepts of science. These are ill-understood and controversial issues over which it would be foolish to be dogmatic; but to explore them is a principal aim of this book.

A machine is generally conceived as a functional entity consisting of clearly defined parts, such that the internal functions can be understood from knowledge of the parts and how they interact. To understand its external functions – what it does – we must know about its 'environment'.

It is often considered that a machine must necessarily be man-made, though a molecule, the solar system and the entire Universe may sometimes be called machines. Man himself may be so described – and is universally regarded as embodying essential features of many man-made mechanisms such as levers, valves and servo-controlled regulatory systems. I shall question this very general use of 'machine' as including all interacting systems – including the Universe.

It is traditionally derogatory to call *us* machines ('mere' machines indeed), which implies that we are supposed to have something over and above machines. With traditions back to Aristotle and before, this something is the soul (psyche) which for Aristotle and Plato gives spontaneous motion – motivation – supposed to be lacking in machines as they need a machine-minder to set them going and direct them. Machines are seen as free of moral culpability as they are not self-directed, though they can of course be instruments for good or ill.

Machines can be more or less appropriate and more or less efficient. These judgements require a knowledge of what they are supposed to do, or are set to do. Judgements of appropriateness or efficiency cannot be made for inanimate Natural things, without theological criteria. So, at once, machines seem non-Natural (even Unnatural) because they are designed by us to serve our purposes; they are therefore in a sense outside Nature. The relation of machines to us may look like theological accounts of the relation between us and the Gods.

Machines do, however, obey, and are limited by, Natural laws: they are designed and constructed to work within the laws of physics though they display some characteristics different from anything found in Nature. It is this that makes machines so very interesting. Perhaps, most curiously, it is by the setting of strict limits to what can happen that they have their astonishing powers: so astonishing that the Gods seemed from the beginnings of civilization to pale before the evident and generally more reliable powers of even primitive technology in the hands of man.

The word 'machine' derives from the Greek μηχανή, which referred originally to stage machinery – a kind of crane allowing a God to descend into the human action. This happens in two of Euripides'[2] plays, *Orestes* and *Hippolytus,* and in many Greek plays Gods float down to comment, predict and judge. This stage machine is the origin of '*deus ex machina*' – 'The God from the machine'.

Machines are generally thought of as man-made things to carry out tasks in a soulless way; so when a machine and a man perform the same task, the one is a soulless automaton and the other has will and intention. It is of course often feared that men are in danger of becoming like machines when they work repetitively in 'soul-destroying' work, but equally, machines can save us from just this.

We apply words such as 'appropriateness' and 'efficiency' to the machine and to the man in the work situation. These words cannot be applied to inanimate Natural things without theological reference. So it is deeply misleading to call molecules, the solar system or the Universe a machine – unless indeed they are supposed to be imbued with purpose. This was common in Greek thinking, but not in modern science which restricts purpose to living organisms.

A machine is more or less efficient in carrying out some intended purpose. Only theologians, at the present time, attach similar purpose to Natural laws or to the

Universe at large; so it seems paradoxical that modern theologians object to the Universe being described as a machine! Their objection is, no doubt, that some man-made machines continue to operate by themselves without intervention – until they go wrong. Perhaps this phrase 'go wrong', applied to machines, evokes moral connotations. However this may be, clearly we have entered a thorn-bed of problems, for already it looks as though the term 'machine' implies purpose – and yet this is just what mechanistic explanations in science are supposed to avoid! But to describe a machine as being more or less efficient or appropriate, we must have a notion that it is serving a purpose that we can recognize. If we change our notion of what this purpose is, we at once change our assessment of its efficiency: if we were to decide that it had *no* purpose we would at once be forced to doubt whether it is indeed a machine.

Consider here, for example, a clock: it is a more or less efficient machine for telling the time. Telling the time is the purpose – our purpose – by which we judge it as a more or less good mechanism. The theologian who seeks purpose for the 'machinery' of the Universe ascribes its purpose to the heavenly Clock-Maker, not to us, the users. This is controversial because the heavenly Clock-Maker's use for the 'mechanisms of the Universe' cannot be read, with any agreement, from the 'mechanisms' as we see them.

Plato was concerned to ask why the Creator (or the 'Framer') produced the order of the Universe, from the preceding chaos. He takes a frankly 'psychological' type of argument in the *Timaeus* (1, 4, 30):[3]

Timaeus: Let us therefore state the reason why the framer of this universe of change framed it at all. He was good, and what is good has no particle of envy in it; being therefore without envy he wished all things to be as like himself as possible. . . . God, therefore, wishing that all things should be good, and so far as possible nothing be imperfect, and finding the visible universe in a state not of rest but of inharmonious and disorderly motion, reduced it to order from disorder, as he judged that order was in every way better.

Now what can this order, and its contrary disorder, refer to but some analogy to a (possibly man-made) machine?

Plato goes on in the same passage to put intelligence into the ordered Universe:

It is impossible for the best to produce anything but the highest. When he considered, therefore, that in all the realm of visible nature . . . nothing without intelligence is to be found that is superior to anything with it, and that intelligence is impossible without soul, in fashioning the universe he implanted reason in soul and soul in body, and so ensured that his work should be by nature highest and best. And so . . . this world came to be . . . a living being with soul and intelligence.

We then learn, most interestingly, that for Plato the Universe was created as a working model of intelligent man: 'For god's purpose was to use as his model the highest and most completely intelligible things, and so he created a single visible living being, containing within itself all living beings of the same natural order.' For Plato the physical Universe is a machine endowed with intelligence and life. It is a moving model of a deeper and timeless reality of forms and laws. The relation is something like that of design principles, to machines that we construct: an idea which persists, perhaps, in Eddington's 'God is a mathematician.'

For us, the notion that the Universe is a machine draws meaning away from the word 'machine', for we think of machines as constructed within principles of physics to achieve some aim, some purpose, which may have been in the mind of the inventor. To consider the whole Universe as a machine is to presuppose some kind of similar intention by a 'maker' who designed and constructed it. To most, if not all, scientists, this is a misleading extension of the inventor–machine analogy, because it postulates an Inventor of *everything*.

Opinion is changing as to the influence of machines and technology on the development through human history of abstract thinking. It now seems that classical scholarship has greatly underestimated both the sophistication and range of mechanisms available to the Greeks, and their effects on science and philosophy; on concepts and thought-models for explaining the Universe. There are few explicit Greek literary references to mechanisms, and even fewer to technological processes, but there is now very strong evidence indeed that machines did have widespread importance in Athenian life – even to maintaining honesty, as jurors and many officials were appointed by random-selection devices; especially the *kleroterion,* devised in fourth-century BC Athens. The Antikythera mechanism for showing the movements of the heavens is clear evidence of sophisticated mechanical design, and the mechanisms of automated theatres rivalled it, though less successfully, for representing human life.

A passage in Plato which strongly suggests analogy from practical metalwork to his conception of the Universe, is this, in the *Timaeus* (I, 5):[4]

He [the Creator] then took the whole fabric and cut it down the middle into two strips, which he placed crosswise at their middle points to form a shape like the letter X; He then bent the ends round in a circle and fastened them to each other opposite the point at which the strips crossed, to make two circles, one inner and one outer.

This strongly suggests that Plato was looking at an armillary sphere as he was writing, and that he was thinking of how it was constructed by the craftsman who made it.[5] It is often argued that technology was beneath the attention of Greek philosophers, because it was associated with the slave class; but this seems not to hold even for Plato, the most aristocratic Greek philosopher. It is rather that for the Greeks, books were for poetry and philosophy, and crafts were handed down by tradition. Even now, language is weak for describing mechanisms and engineers make great use of sketches and drawings.

We are looking for what might be called subliminal or unconscious effects, through the history of technology, of mechanisms on the development of conceptual thinking – and hence on the human Mind, for we must guess at the roots of the iceberg from its visible tips, and only few of these remain in the literature of Greece. The first known treatise on mechanical principles, mainly of levers and wheels, is *Mechanica,* written by a pupil of Aristotle.[6] This opens with the cogent thought:

Our wonder is excited, firstly, by phenomena which occur in accordance with nature but of which we do not know the cause, and secondly by those which are produced by art despite nature for the benefit of mankind. Nature often operates contrary to human expediency; for she always follows the same course without deviation, whereas human expediency is always changing. When, therefore, we have to do something contrary to nature, the difficulty of it causes us perplexity and art has to be called to our aid. The kind of art which helps us in such perplexities we call Mechanical Skill. The words of the poet Antiphon [480–411 BC] are quite true: – 'Mastered by Nature, we o'ercome by Art'.

The *Mechanica* is concerned with practical matters, such as extracting teeth, rowing boats, and the design of axes and balances. It also exhibits in one passage an intimate knowledge of the construction and properties of materials for the design of beds:[7]

Why do they construct beds so that one dimension is double the other . . . and why do they not stretch bed-ropes diagonally? . . . The ropes are not stretched diagonally but from side to side, so that the wooden frame may be less likely to break; for wood can be cleft most easily if split thus in the natural way, and when there is a pull upon it, it is subject to a considerable strain. Further, since the ropes have to be able to bear a weight, there will be less of a strain when the weight is put upon them if they are strung crosswise rather than diagonally. Again, less rope is used up by this method.

There follows a geometrical account. Geometry is static: machines move according to geometrical restraints allowing limited 'Degrees of Freedom'.

Degrees of Freedom

Mechanisms allow us to 'do something contrary to nature', as Aristotle's student put it, though without violating laws of Nature, by directing forces with constraints such that 'Unnatural' movements occur. For Aristotle, earthly movements are Naturally rectilinear, and heavenly movements Naturally circular. For us, following Descartes, and Newton's First Law of Motion, all movements are Naturally rectilinear. This is somewhat modified by Einstein. What mechanisms taught the Greeks is that all manner of movements can be produced, including even some lifelike movements of marionettes on which they devoted a great deal of ingenuity. They found also that mechanical motion could be at a constant rate, such that the motion of the sun and the length of the day became seen as variable. This was of great importance in the law courts, for when a proportion of the day was allowed for a legal defence, there was an advantage according to the time of year that the trial was held. Clock time (given by water-clocks, clepsydra, and, for short standard intervals, sand-glasses) took over sun time in Athenian courts of the fourth century BC. So mechanical restraints – from the small holes of clepsydra limiting water flow, to the fulcra of levers and the bearings of wheels, and many more – became the keys to the new freedoms (though sometimes tyrannies) given by machines. I do not know that this thought was ever explicit, but it runs, surely, underneath all political thinking: freedom must be optimized for effective action to be possible. In a state the members do become as parts of a machine, with their own freedoms and restraints according to the parts they play. Total freedom would be chaos. To put this another way, total freedom of parts allows no directed activity of the collection of the parts, which would not be an organized system or a machine of any kind. Karl Popper puts this well by considering the range from clouds to clocks: everything may be said to lie within this range; and the parts of clocks are statistical 'clouds'.

Geometrical points have three Degrees of Freedom of motion. These are the axes of three-dimensional space. Bodies of finite size have six Degrees of Freedom, for they can rotate around the three spatial axes. What mechanisms do is to limit parts to less than their six natural Degrees of Freedom, to achieve some special motion. Pivots, slides, and so on serve to limit Degrees of Freedom. For example, a wheel has one, a cylinder in a piston also one (or two if it is allowed to rotate as well as slide), and the castors of a sofa allow two, of the first kind. We might say that mechanisms are composed of selected Degrees of Freedom as a symphony is composed of notes: a machine is composed of mechanisms as a symphony is composed of tunes – themes linked sequentially and playing one off against the other.

The actions of machines can often be better described and understood by appreciating the Degrees of Freedom allowed for the functional roles of each mechanism, rather than by detailing either the substances of which they are made or the shapes of the parts – though these are essential for realizing the design in practice. Considering the forms of the parts: the scale described is most important. The atomic structure is unimportant for an appreciation of how the parts of a clock interact, though it may be important for seeing why the bearings do not wear out, or why oil reduces friction.

Traditionally, there are six 'Simple Mechanisms' which are units of machines. These are the lever, the wedge, the wheel and axle, the pulley, the inclined plane and the screw. It is purely coincidental that there are six Degrees of Freedom of parts and six Simple Mechanisms. Further Simple Mechanisms could be added, such as the valve, the piston and cylinder, the cam and follower, and the ratchet –

all known to the Greeks. They also appreciated meshed gears and the significance of gear ratios; and that meshed wheels are, conceptually, continuous levers.

It was clear from at least the fourth century BC that these Degrees of Freedom, and these Simple Mechanisms depending on them, could, in suitable combinations, do just about anything. This gave powerful impetus to atomic ways of thinking. Concepts of function, and especially localization of function in machines – as well as whether (and if so how) the function of the whole machine can be described completely from knowledge of its parts – have always been and possibly remain somewhat mysterious. A truly mechanistic explanation may be taken to be an adequate description of the function of the whole in terms of its parts, as described at appropriate resolution or scale.

Failure of a mechanistic explanation is perhaps indicated by the appearance or Emergence of new properties when the parts are combined, such that these new properties cannot be described in terms of the parts and their interactions. Emergence, in various forms, has dominated much of philosophy, as a principle by which creation and novelty may be explained; but to philosophers of mechanistic persuasion this is anathema. Most curiously, these issues are seldom if ever clearly expressed, though they have their roots in the very beginnings of human experiment and thinking.

Mechanistic explanations are rather generally supposed to eschew intention and purpose, but to carry out intentions and purposes – so they are, in this sense at least, purposive. This is no trivial point, for it is necessary to appreciate the purposes to which a machine is put before we can judge its appropriateness or its efficiency. We must know, or assume, to what ends a machine is applied before we can begin to assign appropriateness or efficiency, and our judgement changes most markedly when we change our assumption of what its purpose or use is. (This can be striking in a laboratory, when a familiar machine is used for some odd purpose for which it was not designed. I have used hair-dryers to cool the lamp-houses of optical apparatus; and on one memorable occasion two hundred brassière straps were bought, on a laboratory order form, and coated with luminous paint to provide lines of adjustable length, for a visual experiment!) In such cases, appropriateness and efficiency are not judged by the original designer's intentions but by the current use to which they are put, and this must be understood. Perhaps almost all invention is seeing that what exists can be applied efficiently in different ways.

Such considerations as these make the usual view of machines as purposeless odd: for if they did not serve some evident purpose, they could hardly be claimed as machines. We might say that nothing can be a machine unless it carries out functions, more or less efficiently, to some discernible end goal. Notions of purpose are essentially bound up with machines.

13 A Greek oil lamp, with a servo-controlled feed for the wick. Possibly this design never actually worked; but it was at least conceived by, or before, Hero in the first century. Theoretical appreciation of servo-control was not however attained until the middle of the twentieth century; which left much of biology entirely mysterious until very recently.

This is to say that we see machines carrying out purposes more or less directly set by us – their designers and users. But can machines seek goals for themselves? This is so, at least to a limited extent, for any servo-control system. This may be as simple as a windmill aiming into the wind, with error signals from its fan-tail wheel; or a Watt's governor for controlling the steam to compensate for engine load; or a thermostatically controlled water-heating system. It is interesting that servo-control was adopted by the Greeks, in oil lamps and for controlling water flow for clepsydra (described by Hero of Alexandria; Woodcroft [1851]; Brumbaugh [1966]; Mayr [1970]). The first clear account of servo-control applying in physiology is Walter B. Cannon's *Wisdom of the Body* (1932), which, though extremely important and influential, was written without appreciation of the theoretical basis of servo-mechanisms – whether man-made or biological. It is extraordinary that these principles were not appreciated until the middle of this century, especially by Norbert Wiener (1948),[8] with his founding of Cybernetics – the term meaning 'steersman' in Greek. With the recognition and design of sophisticated servo-systems, it at last became clear that machines need not be mere repetitive devices but can be self-adaptive, as circumstances change, and may even select goals of their own choosing.

Concepts of servo-control are of course basic for physiology and behaviour, including skill (for example Kelly, 1968). It is clearly so important that it now seems a tragedy in the history of science, and something of a criticism of physics, that it has taken so long to appreciate the basic principle of dynamic stability; of

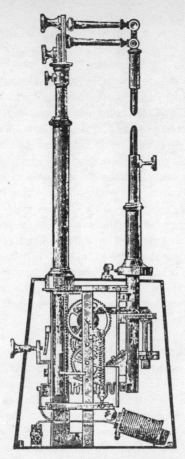

14 The servo-controlled feed for an electric arc lamp, mid-nineteenth century. The separation between carbons was maintained, as they burned away, by monitoring the current. This worked, but the theory for such devices (including Watt's governor) was not appreciated until the Second World War. This served essentially the same function as in Hero's oil lamp.

error correction, and of goal-seeking. The principle is of course that signals of discrepancies between aim and success are fed back to correct errors, so machines can strive for success. To describe such machines we must know the goal they seek and the success of their performance. This sounds very like teleology creeping into mechanistic explanation; but servo-systems are determinate, and can be described analytically. Implications of this situation to purposive explanations in psychology are well discussed by Margaret Boden (1972a, 1972b). Although servo-control systems are purposive, they are mechanisms for they can be fully described by mechanics. It was this that, sadly, had been missed until extraordinarily recently, which is almost as disastrous for philosophy as the classical view of geometry. The practical and philosophical significance of Cybernetics can hardly be exaggerated.

Among the first to consider servo-mechanism theories of brain function was W. Ross Ashby (1952). Ashby developed from an American, R.A. Wallace, in 1952, a machine called *M. Labyrinthea* which adjusted its internal connections when presented with successive stimuli, and could learn choice points on a railway track. Ashby went on to build machines having hierarchies of homoeostats, so that there were many levels of control. There was here some attempt to build principles discovered in the peripheral nervous system, and beautifully described by C.S.(Sir Charles) Sherrington, in *The Integrative Action of the Nervous System* (1905). At last machines appeared which were not simply repetitive, but were adaptive, and could learn. The great lesson from these early analogue devices, which were probably first envisaged by Kenneth Craik (1943) (see pp.371–3), was that remarkably simple machines could display lifelike behaviour which for the Greeks, and right up to the twentieth century, were supposed the prerogatives of higher forms of life. This is so though these devices could, once the principles of homoeostasis and servo-control were fully appreciated, be fully described by mechanics.

What are 'mechanistic explanations'?
What, then, are these mechanistic explanations, which are almost universally regarded as the best – or even the only – kind of explanation appropriate for science? It is usually held that these are the right sort of explanation, in fact, because they do not introduce purpose; but we find the notion of purpose essentially bound up with function, and function essential for 'machine'.[9]

One might perhaps say that although machines carry out functions (and cannot be claimed as machines if there are no discernible functions), this is not so for the causal processes of the machine; and that mechanism descriptions are causal, rather than functional, descriptions. Thus a crank or an eccentric (both delightfully psychological!) are mechanisms, which may or may not be carrying out functions. What function, if any, they carry out depends on their place in a machine, and the use that the machine serves. I think this is the best way of considering the matter: that mechanism descriptions are causal rather than functional, though 'machine' implies functions. On this account, mechanistic explanations in physics are based on mechanisms rather than machines.

This is not, however, so for biology; for we do see functions in the lens of the eye, the heart, and all manner of other mechanisms of physiology. Here, causal description is not adequate. The engineering-type accounts of machines with

recognized functions are appropriate and necessary for biology. Functional machine accounts are extended to physics in theological accounts. It remains to be
seen whether our functional biological accounts will come to seem tainted by
some hidden theology in our thinking, or whether the notion of purpose will
survive for machines and organisms.

How are functions found in machines?

Clearly, some parts of machines, and their precise shapes and positions, are highly
important for function. This must have been evident from the time of the first
simple tools. The cutting edge of an axe must be sharp, and also the head must be
heavy, to make an efficient axe. It is a conceptual problem to establish the functional significance of the sharpness of the edge and the heaviness of the rest, when
both are needed for an efficient axe. Once cutting, and the stored energy of inertial
mass, are appreciated, the functions of the sharpness and the mass are clearly
distinguished – though both 'reside' in the same lump of matter, be it of flint,
bronze, steel or whatever. The point is, and it is an important point: functions are
theoretical concepts which can only be seen in terms of purpose. This applies to
machines and to biology: thus we speak of the function of the liver or the heart as
serving the rest of the organism by carrying out their proper functions, and mal-
functions as handicapping the organism.

When we speak of the function of the entire organism, rather than its organs, or
of an entire machine rather than its parts, then we have to see the organism or
machine within a context within which it plays its part. (Thus a policeman has
functions to perform: he is a 'functionary' in his society.) No doubt the engineering
notion of restraints of Degrees of Freedom applies *mutatis mutandis* to individuals
in society, as freedoms and restraints are set by convention and by law, so that a
society works something as a machine works. This is even clearer in cases such as
the crew of a sailing ship, where each hand becomes quite literally a specialized
part of the machine; and whether ropes are pulled, or the ship steered, by man or
by machine is arbitrary. As the technology of machine intelligence increases, we
may expect the interchangeability of man and machine to extend to high levels of
decision-taking.

Consider now, though, the *location* of functions in a sailing ship. The decision-
taking captain is on the foredeck and the steersman is at the wheel. They can be
given precise spatial coordinates. Again, the halyards, the belaying pins and the
binnacle have positions in space; but is this to say that the *functions* necessary for
running the ship are spatially located? In a sense; but functions are actions or
activities while parts are spatially defined things – and functions may be shared
between spatially separated parts and also, many parts may have to be combined
to give one function.

Sometimes functions cannot be equated with *any* physical parts of machines. This point is made very effectively (if surprisingly) by Sigmund Freud, when he considers localization of consciousness in the brain in *The Interpretation of Dreams* (1900). Freud takes an explicit analogy from certain kinds of devices or machines, especially the images of optical instruments, pointing out that in some instruments the image is not located in a physical structure. He goes on to suggest that consciousness is perhaps not located in, or to be identified with, any brain structures. There are many other cases of functions separated from structures: cooking inside an oven, or before a fire, for example.

It is curious that functions are often supposed to be associated rather simply with the parts of machines, yet they are seldom if ever identified with particular parts in isolation, and are not removed, one by one, in any simple way as the parts of a machine are removed. One finds, rather, that bizarre things happen when parts are removed; or nothing may happen, except perhaps under special conditions such as extreme demands or loading. For example, spokes of a bicycle wheel can be removed, one by one, with little effect until there is a sudden collapse. Removing parts from an electrical circuit may produce output characteristics that were not present – such as whistles for a radio or complex patterns for a television.

This may seem somewhat at variance with our experience of mending machines by changing parts, and designing each part to give the functions required for the machine to work. The point is that parts contribute, often in subtle ways, to functions requiring interactions of many parts and sometimes of the entire machine. To understand how the parts combine to give functions requires deep concepts of mechanics, engineering and physics, and may require knowledge of computing or other abstract procedures.

It follows that 'mechanistic explanation' is not a simple notion, and is indeed exceedingly hard to understand. It is probably only ever adequate as a first crude account. The 'direct' experience of tools and machines can be misleading and seduce us into equating functions with discernible parts. In truth the relations between parts, and their causal interactions, and the functions that they achieve, are highly complex and subtle beyond common understanding. It is particularly difficult to say *where* functions are located. This is a most serious problem for brain research.

Recognizing functions in machines

To recognize function we must understand how the system or machine works. Thus it is necessary to appreciate that halyards and men hoist sails as parts of a highly integrated mechanism before the sail-hoisting function (of sailor, or halyard, or block or whatever) can be appreciated. It is logically impossible to localize function before functions are appreciated – so it is logically impossible to localize function without a conceptual model of how the machine – or the organism, or the society – works.

I have spent some time on this somewhat recondite matter. It is important for thinking about localization of brain function. I have suggested previously that neurology is sometimes naive on this point (Gregory, 1961). It may be that the evident separateness of machine parts has been misleading, suggesting that functions can be isolated and located as easily as parts that can be seen and handled as separate things interacting. I would not at all wish to say that neural functions cannot be established; but adequate conceptual models are surely required to identify functions – and they must be identified before functions can be defined or located.

Brain functions are sometimes located by ablating or stimulating brain regions, and then noting behavioural changes. One has only to consider, say, the sailing-ship example to see how tricky this is. If a man falls off the mast, the ship may come into the wind and circle for hours. We know that the rest of the crew are looking for him, hopefully before he drowns, but it requires a vast knowledge to understand such behaviour – there is no simple connection. Or the trouble could be a jammed sheet, which by a series of mishaps causes a mast to break. This is all purely mechanical; but to appreciate the breaking of the mast (or the resulting behaviour of the ship) from the jammed sheet block requires detailed functional knowledge of the working of ships – and ships are much simpler than brains! The question here for neurology is: can observed effects of ablation provide useful evidence for a conceptual model of brain function? Surely it can, but arguments from malfunction to normal function are fraught with difficulties even for quite simple machines.

Physics is in a most interesting position here, for it is uniquely rich in conceptual models and yet its subject matter is not what we are calling machines; and (we have suggested) it is impossible to ascribe functions in this sense to the systems that physics describes. It seems silly to say, for example, that the moon has a *function* circling the earth; though it does incidentally serve us: it gives us tides and allows the sun's corona to be seen during a total solar eclipse. To say that it has functions, as machines and their parts have functions, is to ascribe purpose to the Natural world – which generates the notion that the world was built for us, or for Gods, with intention.

What physicists do is to establish causal relations, rather than functional relations. Causes can, at least sometimes, be established by experimental procedures (as we shall discuss in detail in chapter 8), and described with physical concepts and units without reference to purpose, which is impossible for functions. It may be that the development in physics of physical units, and such 'causal' concepts as mass and force underlying experience, have exorcized functions from physics.

This is a step away from technology, which is always concerned with functions and purposes, and so indirectly with Mind.

We can, however, talk about function in biology on the Darwinian paradigm, because there are criteria for 'better than': so words like 'efficiency', 'design', 'success', and 'failure' have meaning for organisms as they do for machines, though not within physics.

It is often said that physics cannot answer 'Why?' but only 'How?' questions. 'Why?' questions can, however, be asked of machines, and of organisms, whether seen as specially created or as having evolved by Natural Selection. 'Why?' questions are acceptable in biology on either paradigm, because on either paradigm there are criteria of success and failure; though this is not so for physics. It is absurd to say that a hydrogen atom is better than a carbon atom, but there are grounds for saying that some species are better, or more successful, than others. This is so also for machines. It is an open question how far the Creator-Inventor paradigm of biology (which is theological) or Darwinian-type Evolution (which minimizes individual inventive genius) apply to technology. In either case, however, degrees of success and failure, and functions rather than causes only, can meaningfully be applied to machines and to organisms.

Emergent properties: ghosts in machines?
A machine such as a clock is made up of several generally distinct parts: the pendulum, the escape wheel, the anchor with its pallets, various gears, and spindles and bearings and so on. Where, then, is the clock? Is the clock – the time-keeping ability – something more than the parts? After all, if the parts were disassembled, lying around in a box, there would be no time-keeping and no clock but bits and pieces that could be assembled as a clock. The structural arrangements of the parts is most important for the parts to form a clock. It may seem that the clock and its time-keeping emerge from the structure of the parts. This is as though, in Gilbert Ryle's famous phrase again, there is a ghost in the machine, which lurks among the pieces that we can see, and which is incarnated when the pieces are assembled as a clock. Time-keeping might thus be thought of as an Emergent property of the pieces, which cannot be described or explained by knowledge of the pieces alone.

All machine functions (and chemical properties arising from combining atoms into molecules) have this characteristic of emerging from their parts. This is, at least, how it seems.

Is this because the world is layered, such that explanations cannot ever cross layers of complexity of structure? If so, must mechanistic explanations necessarily be inadequate for understanding complex machines, let alone chemistry, or life? These are deep questions at the heart of understanding.

We may call the layered view 'Mechanistic Pessimism'. On the other hand, 'Mechanistic Optimism' would suppose that complex machines, or whatever, showing surprising properties (Emergence) can be fully described with sufficient knowledge of their parts. (This is known as a 'Reductionist' description: 'reducing' the whole to its parts.) There are many instances of seeming layers of irreducibility melting away with conceptual advances in physics. Will the Mind–brain gap disappear, as vast divides, such as between electricity and magnetism have closed or been bridged in physics? Here we can learn a great deal by considering what these 'bridge' explanations are like.

Conceptual gaps disappear when explanations are sufficiently general. Often a quite new conceptual scheme is needed. Indeed, the appearance of 'Emergence' may well be a sign that a more general (or at least different) conceptual scheme is needed. This is not 'reducing' in the sense of 'the whole being reduced to its parts'. It is, rather, the discovery or creation of appropriate conceptual schemes or theories which may be extremely general that is important. It is the role of good theories to remove the appearance of Emergence. (So explanations in terms of Emergence are bogus.) But this is not quite reducing 'the whole to its parts'. It is indeed very far from clear what are the 'parts' – even of a machine.

It is important to note that functional descriptions are confused when machine parts are taken as the units of function. For understanding functions, it is irrelevant how the machine is assembled, or how it can conveniently be taken to pieces. Confusion between machine parts and functional units can generate mysterious Ghostly Emergence.

I think that we can dispel Ghostly Emergence from machines, and from chemistry, by saying that when parts are combined into more complicated (or 'molecular') structures, these structures have different interfaces with each other and with their local environment, and so when combined they display different properties from the isolated parts, and from the parts combined in different ways. Thus, for example, two halves of a wheel separated are obviously functionally very different from the complete wheel, just because, and with no mystery, there is not a continuous rim for the half-wheels, so they will not run on a surface. In other words, what emerges is given simply by the difference at interfaces, as the parts are changed, or added or removed. When the interfaces are different, the behaviour of the machine is correspondingly different. (And it does not matter what is inside the parts, for only the interfaces are significant.)

If the behaviour seems to be more than the sum of the parts (or if we are surprised when parts are added), this is because we have forgotten to describe the new structure, which may be extremely different when new parts are added, as it may have critically different interface characteristics.

We can see how clocks work when we know sufficient mechanical principles to appreciate their structure functionally. It does not in the least matter into what parts they can readily be taken to pieces: all that matters is certain critical features of the structure of the assembled clock. We are so used to assembling machines from parts, and 'taking them to pieces', that it is all too easy to think of the functions of machines as contained in the parts, so that functions emerge as the final critical part is put into place, and the machine comes into life. These are temptations from technology that seduce the most wary.

I have suggested that there is no evidence, apart from past failures of explanation, to believe that the world is epistemologically layered so that explanations of a broadly mechanical kind cannot be applied throughout science to give unified explanations of the complex from the simple. Even if it is not intrinsically layered in this way, it may, however, still be impossible to describe, say, all of chemistry in terms of physics, for: (1) it might be too complicated, even for any possible computer to give sufficiently detailed accounts, or (2) we would need principles for the complicated cases that are not required for the simpler cases. It is, for example, absurd to expect to understand how a clock works from descriptions of its molecular structure, even though it is nothing but a molecular structure. This is not because of a time-keeping ghost in the clock which emerges whenever the clock is wound up, but because it is shapes of its large-scale structures (by comparison with the molecular scale) that are significant for the time-keeping. For this the molecular structure does not in the least matter. (Similarly we should try to discover what is likely to be the relevant structure scale that gives the key brain functions for behaviour and consciousness.) Let us try to bring all this together.

There are three traditional ways of thinking of Emergence upon combining parts of a machine:

1   That Emergent properties may be thought to be present (if only potentially present) in the parts. So Meccano parts have, on this view, properties of cranes and bridges and trucks, and the time-keeping of clocks, embedded in them – waiting to come out when the parts are assembled as a crane, a bridge, a clock, or whatever. On this account, the parts are supposed to have something of the essence of crane-hood, bridge-hood and clock-hood, the essences being ghosts, or potential functions, which become manifest when the parts are assembled.

On this view, Mind would be in the substance, the protoplasm, of the brain. This is expressed by the brothers Čapek [10] in their conception of Robots in the play *R.U.R.* (1921), the Robots being made of an artificial protoplasm which 'had a raging thirst for life', which is the notion of Vitalism.

2   That Emergent characteristics inhabit separate complexity layers, such that complete description or explanation is not possible for phenomena in layers of high complexity from any knowledge, however complete, of lower complexity

systems. These are called 'irreducible' characteristics, for they cannot, it is supposed, be 'reduced' to simpler terms.

This kind of Emergence we may divide into two classes, which I call 'Pessimistic' and 'Optimistic'. Pessimistic Emergence supposes that explanation is in principle and for ever impossible across certain complexity layers of the world. (On this view, Mind might for ever be mysterious as explained in terms of brain structure or anything else. Perhaps it cannot, in principle, be 'reduced' to physical or mechanical, or be described in any terms that we can understand.) Optimistic Emergence supposes that these 'layers' are not absolute, but result from present difficulties that might be overcome. These may be, for example, inadequate computational power to describe the complex in terms of the simple. Later technology or conceptual insights might allow complete description of the complex in terms of the simple. (On this view, physiological accounts of brain structure and function should ultimately allow us to understand Mind in terms of mechanics of the brain. We would then say that Mind (and perhaps consciousness) has been reduced to a physiological concept. Psychology would then belong to physiology. And physiological-psychology would lose its hyphen!)

3   That Emergent features, functions, or phenomena are not ghostly essences of parts to become manifest when the parts are combined into new structures; but rather the complete structure has these properties – given by the interface characteristics of what are taken as component parts. On this view, it may not be possible to explain the function of the whole in terms of the parts from which it has been assembled, because the unassembled parts do not have the interface characteristics of the components as arranged in the complete system: machine, molecule, brain, or whatever. The parts from which it is made, or assembled, are arbitrary units for the functional description of the whole, and may be a poor start and inadequate or misleading.

On this view, the technology of machines, as they are assembled from parts, has proved an unfortunate paradigm for describing functional processes of interactive systems – including all machines and brains. To give a previous example (p. 87), a wheel made up of sections is not a wheel until the sections are joined up – simply because there is no continuous circumference until the parts are joined. Once joined, it is irrelevant what the parts of the wheel were; and given the parts, all that matters is to see that they *could* make a continuous circumference.

This is not a pessimistic view of Emergence. It holds that it is silly to expect part–whole descriptions to work when the concepts for describing the whole system are not expressed when the parts are described. But there is no reason why they should be appreciated or expressed in descriptions of parts. Considering again a clock: the concepts of the isochronous pendulum, or the action of the pallets and the escape wheel, would not be applied to the parts until they are seen as parts of a clock. It is ridiculous to describe a rod of Invar steel lying in a box as a pendulum (it could be used for all manner of things), though it is a pendulum when freely swinging. Once freely swinging it does not matter conceptually of what parts it is made, though it may matter to the constructor of the clock. So describing how clocks are *made* is not a good way of describing how they *work*.

We do, however, often learn by constructing things. This is the technological (practical) rather than the theoretical (conceptual) way of understanding. But it can be highly misleading – to generate insidious Ghostly Emergence.

On this view it is probably misleading to describe brain function by taking easily distinguishable parts of the brain as functional units. It is, however, useful to structure the description of the whole into observable units which reflect the functional components, to give the description that is simplest and most convenient and effective for a given purpose. There may be no single description that is the best description for all purposes – and no single best selection of functional units (which may not be components or parts) for describing how a machine, or the brain, works.

The reader might consider that I have abandoned mechanistic explanation with this analysis of Emergence as appearing when explanations are inappropriate or not sufficiently general or 'abstract'. If so, I think that this is a sign of how unclear we are about 'mechanism' and 'machine'. I think that it is a simple mistake to suppose that machines should be described adequately by accounts adequate for their parts.

Should we conclude that 'mechanistic' covers everything that can be made as a machine? There may be practical difficulties of no conceptual interest, so let us revise this as: 'Mechanistic' covers everything that can be *conceived* as a machine. This at once suggests the important notion advanced by the pioneer computer scientist at Manchester, Alan Turing: that anything that is conceivable can be represented by a simple machine, capable of a very long string of two alternative states. What this suggests is that we cannot conceive of what we cannot represent by a machine code. This is similar to saying that we cannot conceive of anything that we cannot express in language. This in turn is to say that anything claimed as non-mechanistic is nonsense. But this merely removes 'mechanistic' as a useful word in our vocabulary. For the moment we shall leave these difficult issues here, though they will crop up later in various forms.

If these arguments show that 'mechanistic explanations' are not what they seem, or are often taken to be, then perhaps we have made a conceptual advance. We should revise our meaning of 'mechanistic' as we come to see more clearly what it is to be a machine.

Are rules and procedures Emergent?

It is most important to appreciate that rules, procedures and causal laws (all of which may allow inference) are not additional parts of the structures of things to be described as spatial structures are described. The best discussion known to me of this point is chapter 5 of Gilbert Ryle's *The Concept of Mind* (1949): [11]

Law-statements are true or false but they do not state truths or falsehoods of the same type as those asserted by the statements of fact to which they apply or are supposed to apply. They have different jobs. . . . A law is used as, so to speak, an inference-ticket (a season ticket) which licenses its possessors to move from asserting factual statements to asserting other factual statements. It also licenses them to provide explanations of given facts and to bring about desired states of affairs by manipulating what is found existing or happening.

The point is that the laws and rules (including rules of games such as chess) are not contained in the structure of the world, or the structure of a chess set, in the way that molecular or clock structures are facts of the world. But when we come to consider Mind (or computers) this situation takes a sudden dramatic turn, for now we see material systems not simply obeying laws (Natural laws of physics, as the moon obeys Newton's Laws) but rather as carrying out rules, which may be essentially inference rules, very different from physics and its laws.

The moon does not have to solve the equations that Newton solved, to orbit the earth; it obeys them without solving them. (This is part of why we should not regard the moon as intelligent, or as a component of a machine.) But an astronomer's brain, or his computer, goes through sequences of physical states that *represent* the inference rules required for prediction and understanding, by obeying 'artificial' or designed restraints, such that more or less logical procedures or processes are carried out. This is not a mirroring of some kind of deep logical structure of the world for the inference steps carried out may follow more or less arbitrary and conventional rules, which are not laws of physics. The rules of grammar of a language, the rules of logic, arithmetic and mathematics, are not laws of physics – though vital for describing the physical world. Thus digital computers, though machines, function according to programs which are outside physics.

What we see here are procedures of inference carried out by machines within physical laws, but not to be described by physical laws – because they are not sufficient or even at all relevant for inference. Concepts quite outside physics are necessary for understanding computers, and indeed clocks. Thus logarithms are not part of the physical world, though they can be carried out by physical processes of a suitably designed machine. So the logarithmic features may appear as an Emergent property of the physics of the machine. This is a mistake, for machines do not obey, though they are ultimately limited by, physics: they obey the dictates of their human designers, who build their slides and pivots and their cams to produce results not found in Nature.

These seem to be some of the lessons of machines. We may ask now: what do machines do? Here is a rough classification of uses to which they have been put.

Kinds of machines
Hand tools
These are for cutting, breaking, moving (cleaving, sawing and drilling), joining all manner of materials including stone, soil, wood and metals, moving heavy objects and pumping water. These may be worked by hand power, or they may be power tools (see below).

Power tools
These are for grinding corn, pumping, cutting stone or metals (often with higher precision than possible by hand), especially with lathes, which are ancient; the potter's wheel is similar.

Power sources
These were limited in classical times virtually to water power and falling weights. Rate of fall was in later Greek times controlled by placing the weight in a hopper of seed which ran out through a small hole. Perhaps because of the paucity of mechanical power sources, the ancients regarded prime movers as very special – and essentially alive. (Of course power 'sources' is a misleading term: they do not create but only convert one form of energy into another, and store and direct energy to where it is needed.) None of this was clear until recently, as energy is a modern concept. Labour-saving devices were not important where slaves were plentiful, but the Greeks were fascinated by gadgets and toys.

Simulators
In particular, these are simulators of the heavens (notably the classical Antikythera mechanism) and of life: automata and automatic theatres, which date at least from the fourth century BC, were very popular in classical times.

Deceivers
These were used by the Egyptians and Greeks especially for temple miracles, in which divine influences were simulated by mechanical means.

Honesty protectors
These protected one from criminals and temptation: locks and keys, and lottery machines for random selection of jurors, and many officials, from the Athens of the fourth century BC.

Calculators

These include the abacus, prehistoric in origin; wheeled calculators from AD 1642; and wheeled computers (conceived in the 1830s by Babbage), to work from programs and be largely autonomous. (One might say that the programs convert the device into a different machine, according to needs or circumstances, while most machines are 'committed' to one or a very limited range of functions.)

Clocks [12]

Water-clocks (clepsydra) were developed for the Athenian law courts, because the use of proportional divisions of the solar day from a sundial (the gnomon) gave those pleading their cases a time advantage according to the season. Clocks marking out six of our minutes became standard in the law courts; and the public clock in the Agora gave clock time as standard for Athens in the fourth century BC; this clock was a globe, turned from a slowly sinking float in a cistern draining through a small hole, which replaced the sun, the moon and the stars as arbiter of time. The first mechanical clocks were models of the heavens. Time-keeping was secondary. Mechanical clocks were set individually from sundials (ideally with correction for the equation of time), until the railway telegraph and, later, radio time signals gave public time. This was itself set from astronomical measures until the very recent change to 'atomic time' which now *defines* regular time. So again the heavens are not the basic time-giver. One can see a series of abstract and practical concepts, and social implications here.

Weapons

These can be both for attack and defence; they are of course a major feature of technology, and exercised the minds of some of the greatest thinkers, notably Archimedes and, much later, Leonardo da Vinci. The drama of defence *v.* attack weapons, each step stimulating a corresponding counter-advance, is quite like organic evolution, having its own impetus to unforeseeable conclusions.

Toys [13]

Many, ultimately most useful, inventions seem to have started as components of mechanical toys, which have an ancient history. Machines can entertain us, and indeed mechanisms have an inherent fascination for many people and have been cultivated for their own sake, as part of our aesthetic world. Some machines are – there is no other word – beautiful. The familiar phrase 'Necessity is the mother of invention' is only half true: inventions may be conceived for fun or pure delight; and of course invention is the mother of new necessities.

I have not included here intelligent machines, for we are looking here at the roots of technology, and for hints of how machines may have inspired early concepts for understanding (and sometimes misunderstanding) Nature and Mind.

It is not even entirely agreed by authorities at the present time that there are any machines that can of their own accord show intelligent behaviour, or produce original intelligent solutions. These are questions that we shall discuss later (chapter 10).

Let us now summarize concepts of functions. We are regarding machines as those things that perform tasks, with more or less efficiency, by carrying out functions through their parts moving according to restraints that lie within the laws of physics but are not laws of physics.

We have suggested that functions are related to parts in subtle ways that are not at all obvious to untutored inspection, and that require conceptual understanding. Here the engineer is in a much stronger position than the craftsman, though practical experience of machines is important and may be necessry for conceptual understanding. It is further suggested that it is this kind of understanding that has grown into and still underlies physics and philosophy.

It should be interesting to try to break down machines conceptually into kinds of functional elements. Many functions are often combined in a single machine.

Perhaps these are some of the more significant functions found in machines:

1   To transfer, or move, substance or structures through space at prescribed rates, direction, and paths of motion. This includes amplifying (or magnifying), and attenuating (or reducing) motions or structures, including signals.

2   To transform forces, paths of motion, structures, and kinds of energy.

3   To match inputs to outputs, and forces to loads, and (with codes) information sources to channels and receivers.

4   To multiply, generally by copying and reproducing by successive addition.

5   To divide into smaller pieces; generally destructive but often necessary.

6   To store, generally statically (as in a wardrobe, or information in a book), but sometimes dynamically (as in computer delay loops, and structures in life).

7   To release energy (in the form of appropriate forces) in particular situations, for initiating functions sequentially.

8   To convert one form of energy into another, or, for atomic energy, convert mass into energy.

9   To reduce errors, or control, especially in servo-systems. (It is interesting that 'error' is meaningless without a notion of intended goal; and errors – being discrepancies between what is and what is required – are not in any normal sense physical, though they underlie the entire concept of control.)

10   To protect, or isolate, parts or the whole machine against irrelevant disturbance; so a machine is its own more or less isolated mini-Universe, whose purpose can be understood by what it offers to other mini-Universes – other machines and organisms, including us.

11   To compute, to handle, store and infer information.

12   To simulate, to copy, selected more or less general or abstract aspects of situations.

13   To represent, to simulate with a code (such as language) or a map.

Causes and functions

Finally, is it important to distinguish between functional and causal descriptions of machines? Are functional accounts useful, as causal accounts are obviously useful? What functional accounts of machines and organisms do is to select which causal processes are significant. Without a causal conceptual model one does not know how it works, and without a functional notion one does not know what it does. Causal and functional models serve as filters for looking at, investigating and describing machines and organisms. This does not quite apply, though, for physics – unless we take a theological view of the Universe. With the demise of theology in science, the Universe – apart from organisms and machines – lacks functions but retains causes. The notion of cause is by no means clear, and it is slipping out of the most general descriptions of modern physics. Both cause and function remain of first importance, however, for technology and biology.

It is time now to consider some of the explanatory concepts of science that, I suggest, developed in large part by lessons from technology and machines that have tended to exorcize the Gods from the Universe and Mind from Science.

4 Exorcisms of Mind

No exorciser harm thee!
Nor no witchcraft charm thee!
Ghost unlaid forbear thee!
Nothing ill come near thee!
Quiet consummation have;
And renowned be thy grave!

William Shakespeare (1564–1616) [1]

Alchemy to chemistry

To appreciate the pre-Natural Science concepts of matter, we must try to forget our present ideas and knowledge of atoms and elements, and their constituent particles (protons, neutrons and electrons), and we must forget about the immense energies that we now know to be necessary to split atoms into *their* constituent parts: neutrinos, quarks and so on.

The alchemists used gentle heat, from spirit lamps – seeing heat as associated with life. [2] They tried to change substances by analogy with seeds of plants changing earth into wood, leaves and flowers. It was hoped that a little gold could serve as a seed to grow more gold – if only one could get the heat and the surrounding substances and astrological conditions right. The alchemical processes become confused (as we see it) with astrology, because it was thought that the motions and patterns of the stars initiated new things. Thus it was important to carry out an experiment (as it was to carry out war or marriage) when the stars were in a favourable aspect. As a result of this philosophy, or paradigm, it was virtually impossible ever to repeat or control an experiment. Up to seven hundred distillations were required for some alchemical experiments – which was almost impossible before something went wrong such as the retort breaking – and if the experiments had to be conducted when the stars were favourable there could be no precise replication. So there were always acceptable reasons for why an experiment that was claimed would work, never worked again: the stars might not be right. There are indeed now much the same logistical difficulties for learning from history, or carrying out social or psychological experiments, for isolation is not possible for living organisms.

Alchemy is generally associated with transmuting base metals into gold, but this was not the original intention and was seldom its main aim. Alchemy was the precursor of the scientific study of matter – chemistry – and many concepts, techniques, apparatuses and words in modern chemistry derive directly from the alchemists, who go back to the later Egyptians and the Greeks. The Egyptian God of learning and writing, Thoth, is regarded as the father God of alchemy. Thoth was taken over by the Greeks, as Hermes: giving us the word 'hermetic', as in 'hermetically sealed'. The word 'alchemy' itself probably comes from a Greek word meaning infusion of plants, with the Arabic article 'al' added later.

A basic notion of alchemy is that there is an essential substance underlying the four elements of physical reality: air, earth, fire and water. This was called the *prima materia*. A principal aim was to remove the aspects of four elements, which were supposed to impose upon it, and to extract the *prima materia* – the essence of matter. The *prima materia* was early identified with mercury; but not with mercury as we find it, but rather with the 'mercury of the philosophers' which was the essence (or the soul) of mercury. What the alchemists tried to do was to remove from mercury its earthy principles. The water or liquid principle must be removed to 'fix' it. The *prima materia* then obtained had to be treated with sulphur (or sulphur and arsenic) to confer on it the desired properties that were missing – for the *prima materia* itself had no properties, as it was pure matter. The sulphur was not, however, always ordinary sulphur, but rather some principle derived from it – the active principle of the philosopher's stone, the elixir of life.

The metals, which until the seventeenth century were generally supposed to consist of mercury and sulphur, were associated with the planets. The same signs were used for the planets and their corresponding metals: gold for the sun, silver for the moon, iron for Mars, and so on. We find this in Chaucer's *Chanouns Yemannes Tale*: [3]

The bodyes sevene eek, lo! hem heere anoon:
Sol gold is, and Luna silver we threpe,
Mars iren, Mercurie qyksilver we clepe,
Saturnus leed, and Juppiter is tyn,
And Venus coper, by my fader kyn!

Gold was important not only for its monetary value but also because it was 'incorruptible'. In this it was sometimes taken to embody the immortal soul.

Hence, selling the soul to the Devil for the philosopher's stone – for the secret of transmutation of life.

Alchemy became, with the influence of the Swiss physician Paracelsus in the sixteenth century,[4] the foundation of medicine. What we see as mystical associations of matter and life (perpetuated in our still current term 'organic chemistry') perhaps led to this on the whole beneficial application of alchemical techniques into the arts and sciences of medicine. In its various forms it attracted some of the greatest scientists, including Newton and Leibniz, before it gave way to the atomic theory of matter of John Dalton at the end of the eighteenth century. It is remarkable that according to modern knowledge the constituents of atoms are not unlike the *prima materia* of the alchemists – and the elements *can* be transmuted, by radioactivity.

The alchemists were great inventors of experimental apparatus: indeed, they invented the laboratory. Much of the glassware in a modern chemical laboratory could have come straight out of a mediaeval alchemical laboratory. Although experiment owes an enormous debt to the alchemists, they failed – though the ancient astronomers without possibility of experiment but only observations remarkably succeeded – to find underlying order in Nature. Perhaps for this reason the alchemists continued to put Mind into Matter. We still do just this now, for the matter of the nervous system!

Alchemy continued, and indeed reached its peak in the seventeenth century with the scientists who opened the curtains on the dawn of modern science – especially Robert Boyle (1627–91) and Isaac Newton (1642–1727). Boyle is far less well known than Newton but his achievements were remarkable, and were largely inspired by his religious sense that God was to be found in matter and laws, and could be discovered by careful observation and experiments, by following the Inductive procedures recently set out by Francis Bacon (1561–1626) in the *Novum Organum* (1620).

As a son of the richest man in the country, the first Earl of Cork, Robert Boyle had the resources to carry out experiments with costly materials and specially built apparatus, such as furnaces and the vacuum pump which he designed with the help of Robert Hooke, and which he put to remarkable use. He showed that the Aristotelian notion that Nature 'abhors a vacuum' is incorrect, and that it is the weight of air, and its springiness, which are important. He saw life as a kind of combustion, comparing the behaviour and death of creatures in a vacuum with the loss of fire. Boyle was an alchemist with truly religious fervour, believing that some metals are noble and others base: and this terminology is with us still. Boyle succeeded, with help from the astronomer Edmund Halley, in getting an Act of Parliament of Henry IV changed to make it legal to 'multiply' gold by alchemical means, though the proceeds of this growing of gold from seeds of noble or base substances had to go to the Mint for coinage. Even the best scientists, including

15 David Teniers' *The Alchemist* (1610–90). The apparatus of the alchemist's laboratory is remarkably similar to that of the modern chemistry laboratory. (By courtesy of Stichting Johan Maurits Van Nassau, Mauritshuis, Den Haag.)

Boyle,[5] claimed positive results for transmuting base metals into gold; but these were no doubt due to gold lost from previous processes, or experiments appearing, in their retorts, as gold in recognizable form.

Metals found as veins in the rocks were seen as alive, and mines were frequently left alone for ores to grow. This was reasonable, as many metals rust away or at least tarnish in the air when taken out of the ground. The seeking of purity and strength in metals; their use for precious ornament and coins, weapons, and tools; the purity they gain through purging by fire; the dramatic changes occurring with heating and alloying – all these made metals special: endowed with lifelike, and indeed God-like, characteristics of the stars.

Perhaps most interesting for us is the alchemist's insistence that there is one underlying substance, and that all the varieties of matter are due to different organizations or arrangements of fundamental particles. This is the basis of atomic theories, and it is one of the two or three most powerful ideas in science. Structures underlying observable matter were seen as keys to the Universe and to the spiritual world. This remains as a central feature of C. G. Jung's (1944, 1964) psychology, which is still influential in some quarters.

Alchemists included magical or religious *oratory* in the experiments of their *laboratory*. It gradually became clear however that it was not the oratory, the incantations, or the moral claims of the alchemist that produced the observable results or significant discoveries; but rather the laboratory techniques, and the new conceptual models and explanations, which were quite different from the lore of the virtues of the metals. With this, alchemy gradually transmuted into chemistry.[6]

'Inorganic' and 'organic' matter

The distinction here is traditionally between 'inanimate' and 'animate'. The word 'animal' means, of course, 'animated matter': matter that is self-motivated, to move of itself. Animals and men were thus distinguished from 'inanimate' objects, which must be pushed to move. This distinction continued down through the structure of matter, investigated by the alchemists, and by inorganic and organic chemists.

Chemistry is still divided into inorganic and organic. It was originally thought that 'organic' reactions occur only in living organisms. This was believed until the German chemist Friedrich Wöhler[7] produced, in 1828, the 'organic' substance urea, by heating the 'inorganic' substance ammonium cyanate. It was previously believed that urea could only be produced by the living kidney. But Wöhler produced 'organic' urea in the laboratory, in a glass beaker, simply by gentle heat.

This was so startling that it was for a time rejected as impossible. Then many other, supposedly organic substances, including acetic acid in 1845, were synthesized. This destroyed the ancient vitalistic view that 'organic' substances are unique to life – and that life is a property of 'organic' matter having vital forces. The distinction between 'inorganic' and 'organic' chemistry was then redefined according to whether the carbon atom was involved, carbon being generally associated with life. This at one stroke destroyed Vitalism as related in this way to matter, for carbon had to be accepted as an inorganic substance like any other chemical element.

The crucial importance of structures of molecules became clear when it was discovered, by Justus Liebig[8] and Friedrich Wöhler in the 1820s, that very different molecular substances can have the same atoms, in the same proportions. These molecules were named (by Berzelius) as 'isomers'. Mechanisms for forming molecules became clear with the notion of valencies of atoms – the number of 'hooks' they have by which they can form (or not form) molecular structures.

The critically important molecules for life turn out to be large – made up of many atoms – and so they can have a large variety of structures, formed of combined atoms. So far the position is clear. Can we now go on to the final step: to say that life, and consciousness, are given *mechanistically* by the structure of complex molecules? Or are life and Mind perhaps for ever mysterious Emergent properties of complex Matter?

In terms of our previous distinction between two conceivable kinds of Emergence (see pp. 89 *et seq.*), the question becomes: Are life and Mind cases of 'Pessimistic' or 'Optimistic' Emergence? Are they outside the explanatory power of science? Are they outside the physics of the Universe?

Astrology to astronomy

The blue dome of the sky – the moon, the sun and the stars – do not look exceedingly distant. There were indeed bridges in many mythological accounts joining earth and Heaven. In the ancient Scandinavian mythology, the *Edda*,[9] rainbows were bridges for Gods and heroes to cross, for example the Bilfrost rainbow bridge which linked the Great Palace of the Gods with earth. Looking back at the first recorded mythology of the Sumerians, we find at its start the notion of the 'Cosmic Tree' which links earth and Heaven. The ziggurats were symbols of this very early tree idea, allowing men to climb to Heaven. Jacob's ladder in the Bible is just this notion.

For the Babylonians and the Greeks, no clear distinction was made between the weather and the stars – between meteorology and astronomy. The distinction could hardly be clear when the stars were seen as lying only just beyond the clouds. And the state of the weather obviously does affect us, very directly. We are affected – and the state and fate of crops and harvest are affected – by the seasons. So in its origins astrology is highly reasonable.

Predicting human fate from the skies no doubt followed the appreciation of astronomical cycles for predicting seasonal changes of weather. The most obvious astronomical cycle is, however, the lunar month, but it is not useful for predicting the seasons, without complicated conversions to the solar cycle. To the Babylonians belongs the immense credit of relating these cycles, using sophisticated arithmetical methods which they invented. They also discovered more subtle cycles, from observations recorded over many hundreds of years, which became the basis of Greek and all later astronomy. The Babylonians did not, however, develop conceptual models to explain the motions of the stars, except that the stars were fiery Gods enacting their own drama.

Astronomical cycles, and especially those associated with the seasons, showed that human destiny is determined by the stars – and so by the Gods. In the heavens lay the power of life and death; of anger through storm and the gifts of the sweet water brought by rain. For the Greeks, this was seen as a puppet play: it is hardly surprising to find the chorus in Greek plays explaining the reasons for human action in terms of the conflicts and loves of star-Gods, with man subject to the Fates from which even the Gods were not entirely free.

Astrology is astronomy reversed. For astrology, man looked at the stars to discover his nature and destiny, while with astronomy he looks out to discover the nature and future states of the Universe. For the early mythologies of the Mediterranean cultures the Universe is the observer of us. With the development of science, in Greece, this reversed to make us observers of the Universe. It is this view that we still hold, with occasional lapses in the astrology newspaper columns.

Astrology might better be called 'astromancy', by analogy with such words as 'necromancy' (the black arts) and 'chiromancy' (telling the future of individuals from lines on the palms of their hands). It goes back to the Babylonians before 3000 BC, and spread to Greece in the fourth century BC. It continued in remarkably happy partnership with astronomy until the birth of modern Natural Science, with the realization of the Copernican view that the earth is a heavenly body, not essentially different from the planets, and plainly not the habitation of Gods. Early astrology was confined to the fates of rulers and nations. Prognostications for lesser individuals came later.

The word 'horoscope' is a combination of two Greek words meaning 'time' and 'observer'. In the Babylonian world, astrology was the responsibility of priests, who were called 'inspectors' or 'observers'. Indeed, their divinations neatly bridged the heavenly and what was taken as earthly embodiments of Mind, especially the shining liver of sacrificed animals, and sometimes men: the liver, and not the brain, being accepted as the seat of the soul or the Mind. There is a vast store of Babylonian astrological predictions and comment, preserved on clay tablets that were kept in the king's libraries and written in cuneiform by the official astrological priest-observers. I shall now take a brief look through this window, to the most distant recorded past.

It is well worth looking at source material available from translations of the cuneiform tablets. The astrologers from the major cities sent reports of astronomical events and omens of many kinds to the king; these form a fascinating account of Babylonian life and beliefs, though much remains difficult to understand. I shall quote some of these from a collection of astrological tablets from the Royal Library of Nineveh, which is associated especially with the King of Assyria, Assurbanipal (668–626 BC), though much of the library of tablets is earlier.

The importance of calculation for the Babylonians is seen in two tablets:[10]

When at the Moon's appearance its right horn is long and its left is short, the king's hand will conquer a land other than this. When the Moon at its appearance is very large, an eclipse will take place. When the Moon at its appearance is very bright, the crops of the land will prosper. When the day is long according to its calculation, there will be a long reign. The thirtieth day is completed this month. In Elul an Eclipse of Elam. From Nirgal-itir.

The comment 'When the Moon at its appearance [i.e. on the horizon] is very large, an eclipse will take place' is perhaps the earliest reference to the Moon

Illusion. It pre-dates Ptolemy's account by nearly a thousand years. It is also perhaps the earliest recorded confusion between an error of observation – an illusion – and a Natural phenomenon. (The moon does change in subtended angle, but only by an amount too small to be discerned, except at rare solar eclipses that are 'annular' about once in seventy years when the moon is at its greatest distance.) Its changes of subtended angle are insignificant compared with its apparent increase in size when near the horizon, or close to mountains, giving the Moon Illusion. Why should exceptional apparent size, as in our 'harvest moon', be accepted for predicting a lunar eclipse? Lunar eclipses only occur at full moon; but, as they certainly knew, not at *every* full moon. 'Very large' could hardly refer here to the fullness of the moon. It seems more likely that some eclipses had happened to occur at optimum conditions for the Moon Illusion and this false generalization suggested the erroneous eclipse prediction. It is fascinating to see the gradual classification of phenomena for prediction.

The importance of calculations for linking the lunar with the solar calendar is clear in the second tablet: [11]

When the Moon out of its calculation delays and does not appear, there will be the invasion of a powerful city. (On the sixteenth [?] it appeared). When the Moon does not appear, the gods intend the council of the land for happiness. (On the fourteenth and fifteenth god was not seen with god.) When the Moon is not seen with the Sun on the fourteenth or fifteenth of Elul, Lions [?] will die and traffic will be hindered. When the light of the Moon and Sun is dark, the king with his land and people will repel the angry. The Moon and Sun will be eclipsed, [or] for some months god will not be seen with god.

The sun and moon are of course the Gods, who may or may not shine together. We see again the importance of calculation – related to observation and prediction – together with astrological prediction, in a further tablet: [12]

When the Moon is fiercely bright, the month will bring harm. Two consecutive months, Iyyar and Siwan, the day has 'turned back'. When a parhelion (?) stands on the right of the Sun, Ramman will inundate, (or) rains and floods will come. Mars has turned and proceeded (gone straight forward) into Scorpio. It is evil. Later on they shall finish the first part of the matter; I will finish what is going forth with the king; we shall yet see the decisions which will come and remain. From Nabu-ahi-friba.

Some of the astronomical predictions are unambiguous and specific: 'When the Moon appears on the thirtieth of Sebat, a total eclipse will happen. From the chief scribe' (no. 81). The following two are surprisingly frank: were they sent to the king? 'When Regulus approaches in front of the Moon and stands, the king will live many days; the land will not be prosperous' (no. 197). 'The omen which is unlucky for the king is good for the land; the omen which is good for the land is unlucky for the king' (no. 199). Appearances of haloes were important omens: [13]

When the Moon rides in a chariot in the month Sililiti, the rule of the king of Akkad will prosper and his hand will overcome the enemy. (The month Sililiti is Sebat.) In Sebat a halo surrounds [the Moon] in the Pleiades (Surgi). When a halo surrounds the Moon and Aldebaran . . . the king will fare well: there will be truth and justice in the land. From Nabu-ikisi of Borsippa.

Earthly events could be omens, and were sometimes used for predicting astronomical events – astronomy, astrology, and meteorology were not distinguished but were all one: [14]

When a storm comes upon the land, the crops will be increased, the market will be steady. When a storm prevails in the land, the 'reign' of the land will rule great power. When a storm bursts in Sebat, an eclipse of Kassi will take place. From Asaridu, the king's servant.

And: [15]

When it thunders on the day of the Moon's disappearance, the crops will prosper and the market will be steady. When it rains on the day of the Moon's disappearance, it will bring on the crops and the market will be steady. Long live the lord of Kings! From Asaridu.

Another tablet seems to refer to a clock of some kind: for *Abkallu sikla* means 'measure-governor'. (*'Bil-riminu-ukarrad-Marduk'* is its name.) It could have been a water-clock. Possibly it stopped owing to a block in the flow of water. Water-clocks were used to divide the day and night into *kasbu* (two hours), two *kasbu* forming a watch, and three watches making up the night: [16]

When the Moon on the fourteenth or fifteenth of Sebat is not seen with the Sun, a copious flood will come and the crops will be diminished. The *Abkallu sikla 'Bil-riminu-ukarrad-Marduk'* stopped last night: in the morning it shall be explained. O King! thou art the image of Marduk, when thou art angry to thy servants! When we draw near the king, our lord, we shall see his peace!

The fate of astrology

Astrology suggested that man follows the rules of the ordered machine of the heavens, which is occasionally disturbed by quixotic perturbations of the Gods who put spanners in the works. And so intelligence was in the heavens rather than in men – except that human intelligence is needed to read and understand the signs for discovering (rather than controlling) the future. This switch from the Universe as active observer, run by people-like Gods turning the wheels and judging our puppet performances, is the first view of reality. The sun and moon were literally regarded as eyes looking down on us. For the Egyptians, millennia before Greek philosophy, the sun was the eye of Horus, and Horus was active on his own account. Thus Horus the King fights the dog Seth – losing his eye during eclipses, and tearing off Seth's testicles in revenge. In this violent Universe Horus becomes Osiris, who rescues the eye from Seth, and is recreated as Horus. These are human-like cosmic battles explaining astronomical events. When the Greeks explained eclipses as occurring mechanically, such explanations appeared childish. When Newton showed that the mechanism required no Minder to keep it going, astrology lost its credibility. This at least is how I see it.

Astrology is, however, still not entirely dead. Many newspapers publish your 'stars'. Is there any scientifically acceptable evidence? A French statistician, Michel Gauquelin (1967), analysed case histories compiled by Krafft in his *Treatise on Astrobiology*. After showing errors in Krafft's analysis he produced his own much larger sample of 2,800 French professional men, whose birth was recorded to the nearest hour, and claimed from computer analysis that, for example, doctors tended to be born under Mars, actors under Jupiter, and scientists under Saturn – as they are supposed to be by astrological lore. The distinguished psychologist H.J.Eysenck examined the claims of a later Gauquelin study, in which more than two thousand men and women were sent personality questionnaires and their birth dates noted: when the results were put through the computer the predicted relations between personality and birth date, he found, came out very clearly. Now if these and similar claims are accepted, do we have to say that the Sumerians of Babylon were, after all, essentially correct in their astrological beliefs? We – or at least almost all scientists – would look now for some kind of naturalistic explanation. This might be in terms of the seasonal weather during pregnancy, or some such; but perhaps this could not produce such specific effects as are claimed for the stars.

There is a curious class of almost naturalistic explanations which creeps into these debates, postulating physical forces that are too weak, or too variable, to be observed reliably. For example, Colin Wilson mentions one of these (in an *Observer* article, 2 October 1977). While discussing a reported relation between the sun-spot cycle and the level of albumen in women's blood through the ovarian cycle, he writes:

Furthermore, for some odd reason, boilers can be descaled by introducing water 'activated' by floating in it a small glass buoy containing neon gas and a drop of mercury: the mercury slides about on the glass, releasing static electricity which causes the neon to glow. Presumably the water is affected by some kind of ionisation. But this reaction is not constant: it varies according to the position of the sun, moon and various other planetary factors.

It is important to point out that if indeed this were true, Natural Scientists should consider the planetary and lunar positions when designing laboratory experiments. (It would be advisable for *Nature* each week to publish the prevailing astrological positions of the sun, moon and planets, and to append the contributors' day and hour of birth to their names!) The fact that for most of us this is ridiculous shows that we are largely free of these ancient thoughts. But are any of us quite free?

Other examples of phenomena having small and difficult-to-measure causes, used for character assessment and prediction, are pendulums swinging in the hand, patterns of tea leaves, and orders of pre-shuffled cards. These are attempts to escape the mechanistic accounts of science, by selecting phenomena that (we may think) cannot be predicted and so might not be explicable in Newtonian terms. Strictly, I suppose science cannot counter such cases, except by checking the validity of the claimed predictions. This indeed applies to any claim not based on adequate specific knowledge or general theory. If pendulum swinging, tea leaves or astrology yielded consistent and otherwise surprising predictions, then they would have to be taken seriously. After all, it is by looking at surprising phenomena that science has often made advances – often to develop adequate explanation. So why are most of us reluctant to accept these alleged phenomena into the panoply of science? Is it because the evidence of prediction is not strong enough? Is it because it would do too much violence to how we see the Natural world? Is it because it would do too much violence to how we see Mind? At least it is worth asking such questions; for if they are not asked they will hardly be answered, and we miss an opportunity to discover what and how we believe.

The ancient view of the Universe as a living Mind-ful organism has been challenged by tools and concepts of technology. We see this especially in the long histories of chemistry and astronomy, with their roots in alchemy and astrology. But we are still quickened by magic and myth. With each new understanding, scientists are fascinated and inspired by new questions, new mysteries which seem magical. And the original great questions of origins and purpose remain unanswered. Are they outside science?

Mechanistic explanations are sometimes blamed for producing disenchantment. This is I think a mistake – a mistake based on not appreciating the wonder of machines. However this may be we shall go on to look at historical concepts of astronomy and chemistry – of space and matter – from their frankly mythical–magical beginnings at the dawn of recorded speculation and experiment, which is the beginning of science. Undoubtedly with its progress mentalistic explanations have receded. In this, 'the exorcism of Mind from Matter' is fair; but yet Mind remains. The exorcism is not complete. There are still ghosts in the machines that are us. Whether or not our ghosts will depart, as we learn more, lies in the future.

For now, we are forced as we have always been to live without understanding of most of what we do. We still live largely by myth and often absurd fiction. This extends to how we see the world by the senses, especially of sight. Perception and conceptual understanding are by no means one, and by no means consistent with each other. Here, perhaps, is the womb of art.

Observing with instruments

Perhaps astrology remained acceptable to astronomers for so long because both used the same kinds of measurements and observations. There were no observations that began to rule out Mind as controlling the Universe until planetary motions were accounted for by frictionless mechanism. Regular cycles had been known for thousands of years, which might and did suggest that the Universe is a machine (serving the Gods), but machines having inhumanly regular cycles are in common experience directed by human intelligence. So heavenly cycles would allow Mind, just a stage removed from the observed machine-like regularities. They would not preclude controlling Minds: Gods behind the scenes. So astronomy was not incompatible with astrology, and Gods remained as machine-minders of the Universe.

Thus while astronomy gradually displaced astrology, astrology remained active. Although astrology and astronomy went hand in hand and were hardly distinguished for thousands of years, astronomy did come into conflict with particular theologically endorsed accounts of the Universe; then there was trouble, as Galileo found to his cost and our enlightenment.

Astrology was not concerned with the scale of things: this was the preserve of astronomy, which demonstrated with unparalleled drama the power of measurement and calculation. This use of tools did indeed invade the province of the Gods. The first measurements of the size of the earth and the distances of the moon and sun were made a century after Aristotle, by Aristarchus and Eratosthenes (276–194 BC)[17], who was the librarian of the Great Library of Alexandria. He set the scale of things by measuring the shadow of the sun cast by widely separated gnomons. Such measurements as these showed how wrong appearances can be. Aristarchus of Samos (fl. 284–264 BC) went further, to measure the distance of the sun and moon, and to propose that the sun is the centre of the Universe and not the earth. The heliocentric account was not however acceptable to Aristarchus' successors until Copernicus (1473–1543). Aristarchus employed geometrical reasoning, which was correct, though not based on very appropriate observations. This is such an important example of observation and reasoning, forcing his Mind away from appearances towards deeper understanding, that it seems worth giving his hypotheses and propositions here:[18]

1   That the moon receives light from the sun.

2   That the earth is in the relation of a point and centre to the sphere in which the moon rises.

3   That, when the moon appears to us halved, the great circle which divides the dark and the bright portions of the moon is in the direction of the eye.

4   That, when the moon appears to us halved, its distance from the sun is less than a quadrant by one-thirtieth of a quadrant.

5   That the breadth of the earth's shadow is that of two moons.

6   That the moon subtends one-fifteenth part of a sign of the zodiac.

A zodiac is 30°, so on this basis the moon would subtend 2°, when it should be 0.5° diameter, a difference which could have been easily discovered by measurement; but nevertheless this is a clear case of specially selected observations – rather than the general appearance of things – being used with reason to change basic understanding. The historical fact that this had to be done again, two thousand years later, to dislodge man from the centre of the solar system, is the clearest evidence of our arrogance!

History repeated itself, indeed, for our position in our galaxy. Working with reflecting telescopes of his own manufacture, Sir William Herschel (1738–1822) inferred from star counts in various directions that we live in a disk of stars and, in 1785, he placed us at its centre. It was not until the 1920s that the American astronomer Harlow Shapley (1885–1972), counting globular clusters, and working with the sixty-inch Mount Wilson telescope, found that the solar system is situated not in the centre but near the edge of the galaxy.

These new perceptions were forced by highly and deliberately selected observations made with the aid of instruments used for measuring, and counting large samples of stars of particular types. Many tools of measurement – from rulers and balances, to standing stones used for determining celestial angles, and clocks of all kinds – externalize the basis of judgement. The observer becomes merely a null indicator. His task is merely to note differences, of the simplest kind, although the phenomena being measured may be highly complex. There are, however, tools or instruments of a very different kind: instruments for extending the senses. The most dramatic examples are telescopes and microscopes. It is interesting to learn how entirely new observations were made with microscopes, and with the first telescope to be turned to the sky – which revealed through eyes and without need of measurement that the Universe is not at all what it had seemed to be.

Fortunately Galileo described in detail how the telescope transformed his perception and our knowledge of the heavens. Galileo Galilei (1564–1642) – he is known by his Christian name – was born at Pisa, studied medicine, and became essentially the founder of modern science. It seems almost destiny that Newton was born in the year of Galileo's death. At this point we are concerned only with Galileo's first telescopic observations, around 1610. Galileo did not invent the telescope, though he was the first to understand it and the first to turn it to the sky.

16 Tycho Brahe at work with his assistants, measuring star positions. These observations were the numerical basis of Newton's *Principia*. Large quadrants were preferred for this work for some time after the invention of the telescope. (© Burndy Library, Electra Square, Norwalk, Connecticut.)

He saw Saturn as a strange triple body. He never came to see it as a ring, surrounding the globe of the planet. Galileo did, however, realize that his telescope revealed stars previously invisible, and that Saturn is not a simple spot of light; so it was clear to him that observations aided by man-made instruments reveal that Nature has surprises in store – secrets kept from unaided eye and Mind.[19]

Galileo wrote to Johann Kepler (1571–1630) of his discovery that Saturn is not a disk or a sphere – though he hid the announcement in an anagram: 'SMAISMRMILMEPOETALEVMIBANENUGTTAVIRAS.' He later gave the key to Giuliano de Medici: *'Altissimam planetam tergeminum observavi'* ('I have observed the uppermost planet threefold'). He described what he saw as two handles, 'like two servants supporting an elderly gentleman'. To Galileo's astonishment, two years later they disappeared! This was, of course, when the ring became edge-on to the earth, but Galileo did not realize that it was a ring, so could neither see nor understand its behaviour. Nearly half a century later, and after a false start, Christian Huygens came to see, in 1656, the 'handles' as the edges of a flat ring. He protected his discovery in a kind of anagram: 'AAAAAAA CCCC D EEEEE G H IIIIII LLLL MM NNNNNNNNN OOOO PP Q RR S TTTTT UUUUU', which hid the message: *'Annulo cingitur, tenui, plano, nusquam cohaerente, ad ecliptam inclinato'* ('surrounded by a thin, flat ring that nowhere touches and is inclined to the ecliptic'). So at last an adequate conceptual hypothesis was available for observing the ring.

Galileo came to see the bright moons of Jupiter as bodies circling around the globe of Jupiter. This vitally important observation was not, however, accomplished at a sitting. Here is Galileo's account of what happened:[20]

On the seventh day of January in this present year 1610, at the first hour of night, when I was viewing the heavenly bodies with a telescope, Jupiter presented itself to me; and because I had prepared a very excellent instrument for myself [which he later called his 'old discoverer', and which still exists, in the science museum in Florence] I perceived . . . that beside the planet there were three starlets, small indeed but very bright. Though I believed them to be among the host of fixed stars [he had noticed that stars invisible to the naked eye became visible with the telescope] they aroused my curiosity somewhat by appearing to lie in an exact straight line parallel to the ecliptic, and by their being more splendid than others of their size. Their arrangement with respect to Jupiter and each other was the following:

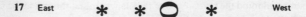

17   East   ✳   ✳   ◯   ✳   West

that is, there were two stars on the eastern side and one to the west. . . . I paid no attention to the distances between them and Jupiter, for at the outset I thought them to be fixed stars. . . . But returning to the investigation January eight – led by what, I do not know – I found a very different arrangement. The three starlets were now all to the west of Jupiter, closer together, and at equal intervals from one another. . . .

18   East      West

At this time, though I did not yet turn my attention to the way the stars had come together, I began to concern myself with the question how Jupiter could be east of these stars when on the previous day it had been west of two of them. I commenced to wonder whether Jupiter was not moving eastward at that time, contrary to the computations of the astronomers, and had got in front of them by that motion. Hence it was with great interest that I awaited the next night. But I was disappointed in my hopes, for the sky was then covered with clouds everywhere.

On the tenth of January, however, the stars appeared in this position with respect to Jupiter:

19   East   *** * O**   West

that is, there were but two of them, both easterly, the third (as I supposed) being hidden behind Jupiter. As at first, they were in the same straight line with Jupiter and were arranged precisely in the line of the zodiac. Noticing this, and knowing that there was no way in which such alterations could be attributed to Jupiter's motion, yet being certain that these were still the same stars I had observed (in fact no other was to be found along the line of the zodiac for a long way on either side of Jupiter), my perplexity was now transformed into amazement. I was sure that the apparent changes belonged not to Jupiter but to the observed stars, and I resolved to pursue this observation with greater care and attention.

And thus, on the eleventh of January, I saw the following disposition:

20   East   *** * O**   West

There were two stars, both to the east, the central one being three times as far from Jupiter as from the one farther east. The latter star was nearly double the size of the former, whereas on the night before they had appeared approximately equal. . . .

I had now decided beyond all question that there existed in the heavens three stars wandering about Jupiter as do Venus and Mercury about the sun, and this became plainer than daylight from observations on similar occasions which followed. Nor were there just three stars; four wanderers complete their revolutions about Jupiter.

Galileo continues this account with further observations and ingenious arguments for interpreting these as rotations or orbits around Jupiter:

Here we have a fine and elegant argument for quieting the doubts of those who, while accepting with tranquil mind the revolutions of the planets round the sun in the Copernican system, are mightily disturbed to have the moon alone revolve about the earth and accompany it in an annual revolution about the sun. Some have believed that this structure of the universe should be rejected as impossible. But now we have not just one planet rotating about another while both run through a great orbit around the sun; our own eyes show us four stars which wander around Jupiter as does the moon around the earth, while all together trace a grand revolution about the sun in the space of twelve years.

Writing in 1632 in *Dialogue concerning the Two Chief World Systems*, Galileo imagines looking at Jupiter from one of her moons (Dialogue Three):

. . . they are dark in themselves, and receive their light from the sun; this is obvious from their being eclipsed when they enter into the cone of Jupiter's shadow . . . but to anyone on Jupiter they would look completely lighted only when they were at the highest points of their circles. In the lowest point – that is, when between Jupiter and the sun – they would appear horned from Jupiter. In a word, they would make for Jovians the same changes of shape which the moon makes for us Terrestrials.

Having learned to see the 'starlets' as planets moving in orbits around the globe of Jupiter, he is able to bring to bear the richness of his conceptual knowledge to predict new observations, and draw analogies to other things that he has seen, and other appearances. Galileo does discuss something of this: [21]

. . . you must know that the principal activity of pure astronomers is to give reasons just for the appearances of celestial bodies, and to fit to these and to the motions of the stars such a structure and arrangement of circles that the resulting calculated motions correspond with those same appearances. They are not much worried about admitting anomalies which might in fact be troublesome in other respects.

We shall later take up questions of how knowledge and perception are related.

There is much here to ponder. How much is perception affected by, and dependent upon, knowledge or belief? It seems that there is two-way traffic between perception and knowledge: between appearance and accepted reality. Technology extends the senses and the Mind by setting up models for drawing analogies by which invisible relations, structures and laws may be discovered. At the same time it provides instruments so that the observer may be fed with quite new signals or data; or in other situations become but a null indicator; or sometimes a mere compiler or counter of classified events, or of features such as the number of hairs on a flea or the kinds of stars in a region of sky. As science becomes 'objective', so the observer disappears. And yet new observations such as Galileo's out of the blue change for ever our seeing and understanding.

5 Concepts in science

*Nothing puzzles me more than time and space; and yet
nothing troubles me less, as I never think about them.*

Charles Lamb (1775–1834)[1]

We shall consider now some of the key concepts of ancient and modern science, especially concepts that seem most relevant to Mind in Science. These will be divided, somewhat arbitrarily, into ancient and modern.

Ancient concepts
Substance
Matter – substance – seems safe, solid and real. The commonsense (implicitly accepted) account of reality is that the real world is solid substance pushing things around, and providing mechanisms that may be chemical or biochemical but hardly mental. How have we come to think of matter in this way?

Matter is supposed to exist in space and to move in time. Mind is not supposed to exist in space, though it is allowed to change, or have events, in time. We allow, though, two times: 'physical' and 'mental' time. Mental events are supposed to occur in physical time though mental time may speed up or slow down, and in memory may leap backwards or reverse. There is also 'mental' space, which in perception and imagination may more or less correspond to physical space. But we do not think of mental events as *being* spatial even when we perceive or imagine spatial relations: though we do think of mental events as occurring in physical time – so here there is a space–time asymmetry between Matter and Mind.

It may be thought that we appreciate physical objects, and the space and time in which they lie, by perception. It may also be thought that we know our own Minds by 'inner' perception. But notions of space and time have frequently changed through the history of physics. There still is no agreement as to how we should think of mental space and time, or indeed of any mental events – or even whether there are mental events.

We think of thought changing (and even of 'changing our Minds'), and yet thoughts do not lie in space, so clearly they are not changing in the sense that objects change form or move in space. We may however think of objects changing in non-spatial ways, for example changing colour or temperature. These are not at least macroscopic movements in space. It is not clear how physical and mental time are supposed to correspond. These uncertainties show that we do not have clear direct knowledge, either of physical or of mental space or time; but yet we draw analogies from the one to the other – from mental to physical and physical to mental – though concepts of neither are fixed or certain.

The essence of the notion of matter is continuity of something, though sensation is intermittent. It is just this supposition of continuity, though our perception and knowledge are discontinuous, which separates the 'outer' world of physics from the 'inner' world of experience. It might be said that it is this which distinguishes Matter from Mind. Mind is, however, sometimes supposed to have, or be, a kind of substance. This is due to the same need to postulate substance to give continuity. Freud's 'subconscious Mind' serves the same purpose: it allows motivations, fears, purposes and so on, to continue through gaps in consciousness, as matter is supposed to give continuity to physical events that otherwise would appear separate and isolated.

The ancient controversy (p. 116) between the positions of Parmenides and Heraclitus – between unchanging and changing as the basis of physical reality – is alive now in present theories of matter. It is the controversy between indestructible unchanging atoms and changing fields. Is physical reality indestructible units, or is it ever-flowing flux? To put this another way: what is the *status* of matter? In thinking about form and matter, what Heraclitus and Parmenides questioned remains central. Their conflict reflects, also, current divergent views of Mind.

Substance is especially, and no doubt was primitively, associated with sensations of touch. Perhaps it is the many illusions of vision – including seeing objects in the wrong place in mirrors, reflected brokenly in water, and the differences in how things appear to vision as the light changes – that by contrast give to touch its special quality of seeming real. Perhaps, also, it is our muscular sense of pushes and pulls as making things happen that gives touch this quality of contact with reality. There is a strong tendency to think that if something is due to pushes and pulls, transmitted by substance, then it is 'Natural' – free from supernatural forces or metaphysical complexities. Newton was indeed criticized for introducing 'action at a distance' into his scheme of universal gravitation. It looked supernatural because of the absence of substance for conveying pushes and pulls to the planets and the stars. Substance is reassuring to physicists, as it is to the common man seeking a safe Universe. The power of the nineteenth-century ether concept as a substance for conveying light through space is an excellent example of its seductive power in physics.

Mechanisms working with pushes and pulls between their parts seem to be the real processes of Nature. Everything else that may be suggested generally seems unsatisfactory, and a kind of magic. We seem now to be deeply 'programmed' to accept, and prefer, this kind of explanation – though this was not so for mythological or magical accounts. Is it possible that we became 'programmed' to prefer substance–mechanism explanations from our long history of tool-making?

Possibly some mathematicians have a similar sense of reality attached to mathematical accounts, but this is not so for the rest of us. At least for non-mathematicians, mathematics may be beautiful, and it may be useful, but its descriptions seem less real than the touch world of push–pull mechanisms, which seems solid and real and safe. Possibly the notion of matter being various forms of substance is a seeking after this same safe reality to support the all-important notion of structure in science. We are impelled to ask: structures of what? And yet, music has form and structure – but no substance!

Distinctions between matter, substance, form and structure were discussed at the start of recorded thought. According to Kirk and Raven, for the time of Heraclitus (*fl*. 500 BC)[2]

> It must be constantly remembered that no firm distinction between different modes of existence had yet been envisaged, and that what to us is obviously non-concrete and immaterial, like an arrangement, might be regarded before Plato as possessing the assumed ultimate characteristic of 'being', that is, concrete bulk. To put this another way, the arrangement would not be fully distinguished from the thing arranged, but would be felt to possess the same concreteness and reality as the thing itself.

This is still a problem for us, at any rate when we consider the ultimate nature of matter which is, finally, described purely in terms of structures which may even be merely statistical. If we then go on to consider the 'stuff of Mind' we get even more uncertain how to think. The question to consider is: what does the notion of substance (or stuff) add to the notion of structure?

There is a tradition as old as science that there are but a few unchanging substances though there is an enormous variety of properties of matter, and kinds of phenomena. The first known scientist, Thales, started this move with the ace throw, by suggesting that everything is made of just one substance: water. This was soon however by general acceptance increased to three or four substances – stabilizing at air, earth, fire and water. Two hundred years after Thales, Aristotle gives an interesting historical summary of the earlier ideas in his *Metaphysics*. (It might be as well to state here that 'metaphysics' does not mean anything like 'supernatural': it means the books of Aristotle that were placed in sequence after his *Physics*. The pejorative meaning of 'metaphysics' is recent, and unfortunate.) In this historical section of the *Metaphysics,* Aristotle distinguishes between what are now translated as 'matter', 'substance', 'attribute', and 'element' in the following passage: [3]

> Of the first philosophers, then, most thought the principles which were the nature of matter were the only principles of all things. That of which all things that are consist, the first from which they come to be, the last into which they are resolved (the substance remaining, but changing in its modifications), this they say is the element and this the principle of things, and therefore they think nothing is either generated or destroyed . . . for there must be some entity – either one or more than one – from which all other things come to be, it being conserved. . . . Thales, the founder of this type of philosophy, says the principle is water . . . getting the notion perhaps from seeing that the nutriment of all things is moist.

This notion of unchanging substance allowing change and variety of things is perhaps the key idea in Physical Science. It led to the atomic theory of the pre-Socratic Leucippus (whose works unfortunately are lost), and Democritus, of the fifth century BC, to the English chemist, two and a half thousand years later, John Dalton (1766–1844), and to the current excitement of sub-atomic particles. These last, it might be added, are too unfamiliar in their supposed properties to give the reassurance of the classically primal substances, or indeed of the hard 'billiard ball' atoms of the nineteenth century. Perhaps the classical notion of substance underlying form is however disappearing from current physics. We need not feel too embarrassed by the question, what is Mind?, for this is no more mysterious than what is Matter?, in physics. For both Mind and Matter, *substance* as underlying but without attributes is more confusing than helpful. [4]

Motion
Motion is clearly related somehow to matter and to time, but just how is it related? This ancient question has the deepest implications.

The 'flux and solder' of Heraclitus and Parmenides
I hope that the reader will not too much object to this jokey title. I defend it as the *Shorter Oxford English Dictionary* defines 'flux' as 'to flow', and 'solder' as 'to make solid or firm' – which is exactly what is meant here.

Heraclitus was born somewhat before 500 BC, of an aristocratic family, in the Greek colony in Asia Minor, Ephesus. Although he was an Ionian in time, he was separate by place and thought from what we regard as Ionian philosophy. He regarded fire as the sole substance of physical reality. He held interesting ideas as to the power and significance of opposites. He wrote before Aristotle formulated the notion of logical contradiction – so we must be careful in considering just what Heraclitus meant by 'opposites' and 'different'. It is necessary to get this reasonably clear before we can comprehend the distinction between the views of Heraclitus – who held that 'All is flux' (or change) – and Parmenides, who held that 'Nothing changes'. These are sometimes put in terms of matter and sometimes in terms of form.

Considering matter, Russell puts it neatly: [5]

Heraclitus believed fire to be the primordial element, out of which everything else had arisen. Thales . . . thought everything made of water, Anixemandes thought air was the primitive element; Heraclitus preferred fire. At last Empedocles suggested a statesman-like compromise by allowing four elements, earth, air, fire and water. The chemistry of the ancients stopped dead at this point.

The most famous of Heraclitus' aphorisms, and he has many preserved in the surviving fragments of his writings, is: 'You cannot step twice into the same river; for fresh waters are ever flowing in upon you.' Or, as Kirk and Raven no doubt more accurately if less elegantly have it: 'Upon those that step into the same rivers different waters flow. . . . It scatters and . . . gathers . . . it comes together and flows away . . . approaches and departs.' Kirk and Raven comment: [6]

According to Platonic interpretation, accepted and expanded by Aristotle, Theophrastus, and the doxographers [holders of opinion rather than reason], this river-image was cited by Heraclitus to emphasize the absolute continuity of change in every single thing: everything is in perpetual flux like a river.

A good example, related in its implications, of Heraclitus' love of opposites, is his: 'Sea is the most pure and the most polluted water; for fishes it is drinkable and salutary, but for men it is undrinkable and deleterious.' Apposite to the continuing human predicament is: 'Good and desirable things like health or rest are seen to be possible only if we recognise their opposites, sickness or weariness; so probably there would be no right without wrong.'

In summary, Heraclitus saw the Natural world as dynamic and full of strife – and he extolled the virtues of war, holding that stability is given only by the balance of opposites in a sea of flux. For him there are no fixed things: only a temporary *status quo* through equal oppositions. This is very much how we now see stable states in dynamic systems, including societies. Heraclitus was drawn to seeking notions of balance among the sea of change; surely we too seek this, in society, in human relations and in Nature.

Heraclitus' notion of the souls of men is the same as his notion of matter: 'For souls it is death to become water, for water it is death to become earth; from earth water comes-to-be, and from water, soul.' Also, 'A dry soul is the best.' On the other hand he was uncertain as to the soul's spatial extension: 'You would not find out the boundaries of the soul, even by travelling along every path: so deep a measure does it have.' For Heraclitus the soul, like matter, is made of fire and it interacts with the Natural world of matter. It comes into being from moisture (which opposes fire) and thus is born of conflict of opposites. After death, the better souls, especially the souls of those who die in war, join the celestial fire, which is the substance of the stars.

Parmenides (b. *c*. 515 BC) lived in the Greek settlement of Elea in southern Italy. He opposed Heraclitus (who, as we have seen, held that all is flux) by holding that nothing changes. Parmenides was the most significant pre-Socratic philosopher, and the one whose writings (in unlovely hexameter verse) most completely survive. The well-known paradoxes of motion and time of Zeno were born of Parmenides' questioning the possibility of change.

Parmenides' poem is in two major parts: 'Way of Truth' and 'Way of Seeming'. It is (especially the first part) a remarkable logical *tour de force:* it presents Nature as a logically closed loop that can be entered at any point and discovered by reasoning without observation. It is thus the single parent of all later metaphysics and Rationalist philosophy. His particular challenge was to question the possibility of change, and to deny the possibility of gaining knowledge from perception. We still feel the echoes of his Cartesian-like doubts. Parmenides rejects change (flux) with the first known linguistic-type argument: that words can only refer meaningfully to what is, and so we cannot speak of the past or the future, or of things ceasing or becoming, because then they are not. This is the first known philosophy based on a linguistic consideration. Its power depends on a theory of language – that words or sentences only have meaning when they refer to what exists. Plato is later influenced by this same consideration: to postulate Ideal Forms in order to allow meaning for words describing abstractions, or perfect objects, which cannot exist in the world known to the senses – such as the perfect forms described by geometry. There was no adequate alternative theory before our century, until Bertrand Russell's Theory of Logical Constructions. With this, Parmenides' argument against change might be said to collapse. The moral here is the demonstration of the interaction between very general *theories* (of language or logic or whatever) and what is allowed to *exist*. One might say that philosophical positions or theories are *filters* which sometimes disallow what seems clearly true. It is clear that things change – so how could any philosophical argument have the power to deny it? Why does not the common shared observation of change, which we cannot doubt or see as illusion, serve to show that something must be wrong with any argument denying it? This is part of the essential conflict between Rationalism and Empiricism.

Paradoxes of motion: Zeno

Zeno, of the Greek colony of Elea, in Lucano, Italy, lived in the fifth century BC, though his dates are not known. He was a favoured disciple of Parmenides, and they visited Athens where they met Socrates, at that time a young man. Plato lets Zeno speak in the dialogue, the *Parmenides,* and his paradoxes of motion are discussed in detail by Aristotle, who claims to refute them; though they were revived by Lewis Carroll (Charles Dodgson), and by Henri Bergson and Bertrand Russell. They remain of interest.

Aristotle called Zeno the inventor of dialectic, referring to Zeno's ploy of deducing two contradictory conclusions from a postulate of his adversaries. He argued not from premises generally admitted as true, or from evidence, but rather from weaknesses of argument rather than of evidence; so Zeno was a true and pure philosopher. This kind of argument can be extremely useful for shaping and refining language for expressing difficult concepts and new ideas. It is the creative, rather than the merely semantic–inertial side of philosophy.

Beneath Zeno's paradoxes of motion lies an attack on the Pythagorean doctrine that things are made of many things, or of as many kinds of particles as there are kinds of substances. Zeno is opposed to Atomism of any kind.

Zeno has four paradoxes of motion: 'Achilles and the Tortoise'; 'the Flying Arrow'; 'the Stadium'; and 'the Row of Solids'. Taking the last of these, Burnet (1908) says: [7]

> If the one had no magnitude, it would not even be. . . . But, if it is, each one must have a certain magnitude and a certain thickness, and must be a certain distance from another, and the same may be said of what is in front of it; for it, too, will have magnitude, and something will be in front of it. . . . So, if things are a many, they must be both small and great, so small as not to have any magnitude at all, and so great as to be infinite.

The argument continues to show that Atomism does not allow us to say that adding one thing to another can increase its size; or that if part is taken away, what remains will be any smaller.

Zeno's conclusion is that things must be made of an infinite number of things; which is to say that they are not composed of particles, or atoms, but are continuous. [8]

The celebrated argument of 'Achilles and the Tortoise' is set out by Aristotle, who thus preserved it:

> You cannot get to the end of the race-course. You cannot traverse an infinite number of points in a finite time. You must traverse the half of any given distance before you traverse the whole, and the half of that again before you can traverse it. This goes on *ad infinitum,* so that there are an infinite number of points in any space, and you cannot touch an infinite number one by one in a finite time.

This concerns a single moving object. The problem of the two moving objects of Achilles and the tortoise is essentially the same, though more complicated:

Achilles will never overtake the tortoise. He must first reach the place from which the tortoise started. By that time the tortoise will have got some way ahead. Achilles must then make up that, and again the tortoise will be ahead. He is always coming nearer, but he never makes up to it.

The 'Flying Arrow' argument adds the length of the object: 'The arrow in flight is at rest. For, if everything is at rest when it occupies a space equal to itself, and what is in flight at any given moment always occupies a space equal to itself, it cannot move.' What Zeno has done here is transpose points of space to points of time.

The 'Stadium' argument introduces motion of bodies in opposite directions:

Half the time may be equal to double the time. Let us suppose three rows of bodies, one of which (A) is at rest while the other two (B, C) are moving with equal velocity in opposite directions, (fig. 21(*a*)). By the time they are all in the same part of the course, B will have passed twice as many of the bodies in C as in A (fig. 21(*b*)).

21 Zeno's Stadium paradox. (From J. Burnet, *Early Greek Philosophy*, A.&C. Black (Publishers) Ltd, 1908.)

What Zeno's arguments really show is that – whether or not space and time are made up of discrete points – we must *describe* them with continuous measuring scales. To generalize: the manner of description may have to be very different from what is described. It is thus very dangerous to infer the structure of reality from language (or mathematical) structures. There is still a temptation to do this, which must be resisted to avoid being trapped by Zeno's paradoxes. But though mathematics is dangerous in this way, mathematical physicists are forced to live dangerously.

Laws of motion: Aristotle to Newton

Aristotle gave a Law of Motion which was accepted until the time of Galileo in the early seventeenth century. In the *Physics* he gives three rules, setting out the relations between distance, time, motive power and resistance (friction), in linear motion of objects pushed by an external mover.[9] Aristotle's Law is: 'the ratio of distance traversed to time elapsed varies directly with the motive power and inversely with the resistance.' If F is force or 'motive power', R is resistance or 'friction', s is the distance, t the duration and k is a constant depending on the measuring units, then Aristotle's Law of Motion may be written:

$$F/R = k \cdot s/t$$

Aristotle restricted the application of his law to cases where the motive power **F** is greater than the resistance **R,** which he conceived as an opposed force. He considered only what we would call 'average' velocities. The notion of instantaneous velocity did not appear until the fourteenth century, when the notion of resistance became closer to Galileo's and Newton's later notion of friction, as a dissipation of kinetic energy of a moving body, this being given as velocity times mass.

It seems remarkable to us now that mediaeval gunners thought of their projectiles as going up in a straight line and then suddenly falling vertically. It is surprising that Aristotle's notion (which children have to learn to reject from their initial assumptions), that heavier bodies fall faster than lighter bodies, was believed for so long. One has only to imagine a heavy and light body attached by a string and released, to question whether the string would break; or whether one object made of different masses would disintegrate as it fell before hitting the ground. The notions of acceleration were remarkably difficult to conceive, even though there were fairly clear examples open to observation. Galileo, as is well known, succeeded largely by the clever experimental trick of slowing down the phenomenon, with balls running on inclined planks and troughs. He altered the usual world in order to see its basic properties – so he was a forerunner of the makers of cyclotrons and linear accelerators for studying matter in special conditions. This is entirely different from astronomy, where observation is limited to what may be seen without modification. It is remarkable that Galileo was adept at both kinds of experiment. Galileo discovered that in free fall (or in his sloweddown conditions), bodies fall with distance increasing to the square of time, giving parabolic trajectories. This was the key to Newton's theory of universal gravitational attraction, discovered by changing the world to match the senses.

Newton's three Laws of Motion, which are the pivot of his description of the Universe, are (1) that a body continues at rest or uniform motion unless affected by force imposed on it; (2) that the chance of motion is proportional to the motive force applied, and is in the direction of that force; and (3) that to every action there is always opposed an equal reaction.

Evidently these are counter-intuitive, for they differ greatly from the Aristotelian account, which was accepted without question until the fourteenth century, and even then challenged only by exceptional men, such as the Oxford mathematician and theologian at Merton College, Thomas Bradwardine (1290–1349). The idea of instantaneous velocity, which seems to us so obvious, was a stumbling-block to this paradigm of motion which we now accept without question.

A greater stumbling-block at the entrance of this, our common sense, was steady motion without cause. It was previously assumed that for continuous motion there must be continuous force. But this only holds where there is friction which is absent for planets. Before this was realized, it was assumed that planets were pushed round by intelligent angels. Newton's theory removed cause, and with cause intelligence, from the planetary system. It is curious that for Plato, highly regular motion was a sign of intelligence, while by the seventeenth century the complex motions of the planets rejected cosmic intelligence. Possibly by then people had become blasé over repetitive motion of machines, such as water-driven pumps which had begun to replace humans, with their variability. Galileo removed this necessity for continuous applied force, or cause, not only from constant-velocity motion but also from the acceleration of falling bodies. He wrote: 'The facts establish that the velocity is not constant, and that the motion is not uniform. It is necessary then to place the identity, or if you prefer the uniformity and simplicity, not in the velocity but in the increments of velocity, that is in the acceleration.' So for Galileo the acceleration of 'Naturally' falling bodies does not require a cause. This is so also for Einstein, though not for Newton. This notion in a way persists for us in 'automaton' – 'moving of itself', without intelligence.

Animus and inertia

The Greeks lacked our concepts of inertia and frictional loss. There had to be an activating animus to initiate and maintain motion. Thus Aristotle supposed that the continuing motion of a projectile was given by puffs of air ('pneuma'), moving from in front and pushing it along from behind. Animals, of course, had animus – they were self-motivated. I have suggested that this conceptual difficulty for the Greeks may have been due to the lack of prime movers in their technology.

The order of the motions of the stars and planets was seen by the Greeks as evidence of the workings of an ordered Mind in the heavens – in the *kosmos* by contrast with the *khaos*. The Greeks extended the symbols and lore of astrology to include earthly substances: medicines, drugs, metals, colours and all animal life. All these and more were placed under the influence of planets. Considering colours, Saturn was associated with grey, Jupiter with white, Mars with red, and Venus with yellow. Mercury was supposed to vary its colour according to circumstances. As we have seen, metals were assigned: to the sun, gold; to the moon, silver; to Jupiter, electrum; to Saturn, lead; and to Venus, copper. In earlier periods Mercury (quicksilver) was associated with the planet Mercury because of its changeable character, as a solid and a liquid. In this way not only man but also animals and the entire Natural world were brought under the protection of the heavenly bodies through astrological lore. Perhaps most interesting is the allocation of the parts and organs of the human body to planetary protection. Mercury, as representing divine intelligence, was guardian to the liver, the seat of human intelligence and Mind.

What is meant now by 'Natural', or Mind-less? Here we do well to look at Aristotle's account of astronomy: especially his remarkably interesting book *On the Heavens* (*De Caelo*). Aristotle often speaks of Natural motions of bodies, and the Natural states to which bodies tend to go. Thus for him earthly objects Naturally move downwards. This is why most objects fall to the ground – and why the earth is now at the centre of the Universe, for being made of earth, the earth fell to the centre of the Universe as this was the natural place for earth. Fire naturally rises, which is one suggestion for why the sun and stars are fiery. (He later came to think that their heat is due to friction of their movement through the air.) There are for Aristotle two Natural kinds of movement: rectilinear and circular. The stars move circularly because this is their Natural motion. Earth and fire move down, and up, with Natural rectilinear motion. Not all movement is for Aristotle Natural. Objects can be forced into Unnatural states. This is expressed in *De Caelo:* [10] 'Nature is a cause of movement in the thing itself, force a cause in something else . . . .'

This continues with the theory that things continue in motion by the air pushing them along. As there are several relevant points in this passage, I shall quote it *in extenso:*

All movement is either natural or enforced, and force accelerates natural motion (e.g. that of a stone downwards), and is the sole cause of unnatural. In either case the air is employed as a kind of instrument of the action, since it is the nature of this element to be both light and heavy. Insofar as it is light, it produces the upward movement, as the result of being pushed and receiving the impulse from the original force, and insofar as it is heavy the downward. In either case the original force transmits the motion by, so to speak, impressing it on the air. That is the reason why an object set in motion by compulsion continues in motion though the mover does not follow it up. Were it not for a body of the nature of air, there could be no such thing as enforced motion. By the same action it assists the motion of anything moving naturally.

Aristotle's view of Natural and Unnatural motion is closely related to his account of the origin of things. This is a sophisticated version of generation by a 'seed'. Generation, or creation of things, is distinguished from changes in things. This is the original use of the word 'element'. The elements are eternally unchanging, but they can provide the potentiality for change:

It remains to decide what bodies are subject to generation, and why. Since, then, knowledge is always to be sought through what is primary, and the primary constituents of bodies are their elements, we must consider which of such bodies are elements and why.

. . . Let us then define the element in bodies as that into which other bodies may be analysed, which is present in them potentially or actually . . . and which cannot itself he analysed into constituents differing in kind.

Aristotle continues:

Now if this definition is correct, elemental bodies must exist. In flesh and wood, for instance, and all such substances, fire and earth are potentially present, for they may be separated out and become apparent. But flesh and wood are not present in fire, either potentially or actually; otherwise they could be separated out.

There are three sources of motion in Aristotle's philosophy: (1) Nature, which is the 'source' of motion in the thing itself which is seeking its goal or Natural state; (2) external movers, which energize by conferring the 'virtue' of the goal state; and (3) external forces, which constrain or force Unnatural movements on objects. These are remarkably like the ways we think of animal or human actions! The first is self-motivation; the second is persuasion by another, who has 'arrived' and knows better; the third is physical force or restraint which overcomes the individual's will. Basically, for Aristotle, things move and change and become according to their own nature. This is the root meaning of 'Nature': Natural Science is the study of the potentialities and behaviour of things. The Unnatural is generally seen as imposed, and generally destructive.

Conflict is in Aristotle's Natural world very evident. Potentialities are realized and change produced by conflict between contraries, especially by conflicts between the elements. We find this, indeed, embalmed in present-day language, when speaking of the weather as 'the elements' or, more strikingly, 'the raging elements'.

It is, perhaps, usually assumed that it was the success of mathematically expressed laws that finally exorcized Mind from the motions of the stars. This is, however, somewhat strange, for mathematics is traditionally an activity of Mind, and things designed to exhibit mathematical principles are accorded high praise as evidence of the intelligence of the designer. It is also often assumed that it was the *mechanical* model of the Universe developed by Kepler[11] and Newton in the seventeenth century that made Mind redundant. But is this quite right? For why, we may ask, did not Ptolemy's mechanism of epicycles exorcize Mind?

Kepler started by thinking that five kinds of regular polyhedrons govern the five then known planetary orbits (in the *Mysterium,* 1596), and he believed, perhaps for all his life, that the planets are pushed round by angels. There was, evidently, prolonged strong resistance to thinking of the Universe as a mechanism without controlling Mind. This persisted through Ptolemy's epicycle machine, and Kepler's laws of planetary motion, and his reference to the mathematics of conic sections: that they follow elliptical orbits. The decisive change followed Newton.

Mind was not exorcized from astronomy until after Kepler formulated his First and Second Laws of Planetary Motion, which were the basis of the mathematical account of the motions of heavenly bodies developed by Newton in the *Principia* of 1687. It was the *Principia* that did the trick.

So why did the particular mathematical account of the *Principia* exorcize intelligence from the heavens, though mathematical–mechanical model accounts of preceding ages did not? I shall hazard a guess. Perhaps it was that though designing mechanisms evidently required creative intelligence, and intelligence is clearly continuously needed to maintain mechanisms in working order, Newton's account showed that no external controlling forces were needed – and no maintenance – because a frictionless mechanism would continue for ever.

It was the development of nearly frictionless clockwork in the sixteenth and seventeenth centuries that may have led to the concept of self-maintaining mechanisms having no frictional losses. This, at one stroke, removed the need or reason for a 'Machine-Minder' for the heavens. On this account it was the understanding of friction, and what it would be like to have no frictional losses, that finally exorcized Mind from the Universe of the stars.

Is motion absolute or relative?

It must from the beginning of rational thought and discussion have been obvious that motion may be perceived either because objects are moving in relation to the observer, or because the observer is moving in relation to them. Are object and observer movement physically equivalent? It turns out that they are not the same when acceleration is involved. It was this that led many physicists, including Newton, to speak of absolute position and absolute space as a special reference for motion. Under the notion of accelerated motion we include any motion not in a straight line: circular motion is continuously accelerated motion.

Newton considers this, concluding that there *is* absolute space, with his bucket argument.[12] He imagines a bucket suspended by a rope and twisted so that it rotates. First there is, in the initial condition, no relative motion between the bucket and the surrounding world (the room it is in) and no relative movement between the bucket and the water. The surface of the water is flat. Then, the bucket is rotated: at first there *is* relative motion between the bucket and the room; there is *no* relative motion between the water and the room. The water is still flat. But with continuing rotation (we may say of the bucket) because of frictional drag the water rotates: then there is relative motion between bucket and room, and no relative motion between bucket and water; and there *is* relative motion between the water and the room. The water is now curved.

The generalization is that the water is flat when it is stationary with respect to the room, irrespective of its motion with respect to the bucket; and its surface is curved, concave, when it is rotating with respect to the room irrespective of its rotation with respect to the bucket. Newton concluded that choice of reference for describing rotation is not arbitrary – so there is an absolute space.

A beautiful example of the asymmetry of motion in this situation is the observation of the planet Jupiter with a telescope. It appears markedly oblate, its bulging middle being attributed to its rotation round its axis. But rotating with respect to what? To us, as observers? But then, if we were to move, however fast, in a circle round the planet, would it appear oblate? If so, if one observer moved round very fast, and another more slowly, the bulge must be greater for the first than for the second observer. This seems absurd.*

Considering accelerated motion, especially rotation, observations are consistent when described by reference to the 'fixed' stars. This is the basis of Mach's Principle, suggested by the Austrian physicist Ernst Mach (1838–1916).[13] He supposed that if the bucket that Newton used were extremely massive – comparable in mass with the total matter of the Universe – then the bucket would become the reference, in place of the 'fixed' stars. This led to the essential feature of Einstein's Theory of Relativity: that space is curved by the mass of matter. So light is bent in space because it travels in the shortest distance through mass-curved space. This is different from both Newton's and John Toland's views (see p. 141). We shall discuss this and Relativity further elsewhere (chapter 19).

There seems to have been throughout the history of science an ebbing and flowing of opinion on whether space has causal properties or is merely the receptacle of matter where all the action is. For Einstein, matter is not much more than frozen energy: space for him seems richer than matter in its properties. In particle physics, it is the spaces between the particles with their curious forces; and in chemistry it is the forms of structures that determine physical properties, together with energy transfers. In Quantum Mechanics it is probability distributions that determine properties of matter, though not of space (see chapter 19).

Perhaps we should go quite against Immanuel Kant's stricture that the space and time we experience by motion are innate categories of human perception and conception. Perhaps we should abandon the distinction between matter and space given by the senses. Why, indeed, should sensory experience be trusted to give us ultimate categories – and why should we know them intuitively?

Form

The philosopher of classical times who most effectively pursued the aim of going behind perceived reality, to what may be 'deeper' accounts, was Plato. Plato looked for ultimate unchanging Forms as lying behind the changing appearance of objects. To him the unchanging Forms were, at least 'philosophically', more real

---

* Though this kind of 'absurdity' is allowed for velocities approaching the speed of light in Einsteinian Relativity.

than objects as sensed. This started a tradition of questioning, and sometimes rejecting, what we seem to know by the senses. It also favoured moral values of timeless abstractions over sensory experience and pleasure, which is the origin of puritanism.

Aristotle, on the other hand, regarded perception as giving us essentially correct, true, knowledge of the physical world. This he held in spite of his awareness of several perceptual illusions. His philosophy was essentially a 'commonsense' philosophy, in that he was always at pains to avoid conflict with the common beliefs of his time. Plato was very different: he was not averse to saying that matter as known by the senses was unimportant; his Forms being the essential reality though unsensed – ideal circles, ideal trees, ideal women – and it was Ideal Forms that were for him described by geometry, and are the subject of science. For Plato words referred to, and had their meaning by, these Ideal Forms (a theory of meaning we call 'Nominalism'), while for Aristotle this was a chimera. For him, words are what we make them. This is a fundamental division of philosophies, which has profound psychological implications.

There are here deep implications for theories of knowledge (for epistemology). It is entirely right and proper to ask: what do we mean by 'exist'? We might wish to say that only physical objects exist; but what about laws of Nature, gravity, force, and so on? What about experience? Do experiences of pain, or of colour, exist? If not, what are we talking about when we ask whether someone is still in pain, or whether she likes the colour of a dress? However this may be, forms and structures (which may or may not be embodied in matter) had, and still have, immense importance in science – though they are not matter.

To Aristotle, Plato's Forms do not exist but may be accepted as generalizations from experience, as when we categorize different-looking objects as 'trees', or 'women', or 'circles', or whatever. For Aristotle they are created by human Mind, while for Plato they are the underlying reality of the physical world.

Reality as unchanging perfect Forms is described in Plato's late work, the cosmological phantasmagoria, the *Timaeus*. Particular objects that we experience are modelled on them, made by an intelligent agent – the Craftsman. The Craftsman does not create matter or substance, but rather (as a furniture-maker) he fashions objects from available materials, according to given designs. Plato's materials, or substances, were the same four as those of Empedocles:[14] fire, earth, air and water. But Plato took the immensely significant step of relating them to geometry. He identified each with a regular solid:

Fire   Tetrahedron (four faces)
Earth   Cube (six faces)
Air   Octahedron (eight faces)
Water   Icosahedron (twenty faces)

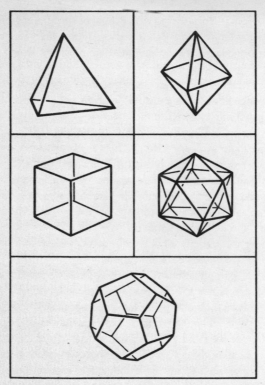

22 The five Platonic regular solids. The first four were associated with fire, air, earth, and water, and also with metals and planets; the fifth was not given particular significance.

These four figures Plato constructed from two kinds of triangle, the isosceles and right-angled triangles (fig. 22).

Sensations of touch, and of burning by heat, and so on, are explained by Plato from the shapes of the particles making up substances. Physical properties such as lightness are also explained in this way – as given by the shapes and sizes of the particles, though they are too small to be seen or otherwise sensed individually. The burning sensation of heat is given by 'the dividing and cutting effect of its particles because of the sharpness of its edges, the smallness of its particles and its speed of movement – all of which give it the force and penetration to cut into anything it encounters – can be explained when we remember the formation of its figure . . .'. It is worth pondering upon this as a type of explanation. The observed characteristics are explained by supposing them to reside in fundamental particles, though these are of such a scale that they are individually unobservable. Note that what is to be explained for an object's properties is postulated as a property of the invisible particles of its substance.

But this leaves the processes of cutting essentially unexplained – for if knives or needles cut because their particles are like knives or needles, the explanation is circular. This is avoided by carefully expressed accounts of Emergence (see chapter 3).

Atomic forms

Atomic theories do not need to say what the atoms are made of. They do not need to have an account of substance. It is sufficient to show how properties are given by formal patterns. This was not clear to Plato (whose account of forms is not in terms of arrangements of atomic particles), but is the basis of atomic accounts of matter, starting with the pre-Socratic philosophers Leucippus and Democritus.[15]

An essential feature of atomic accounts is that forms and other characteristics of matter must not be supposed present in the atoms themselves. Plato was not playing the 'atomic game' when he suggested that the sharpness, hotness and so on of objects is given by sharp or hot atoms.

The point of atomic accounts is to explain object properties by the forms or patterns of atoms, together with laws of atomic forces, rules for combination, and so on. This gets away from substance explanations. The Greek Atomists were forced to say that space between atoms exists for it would be impossible for atoms to move if they had to move from existence to non-existence – so space exists. Lucretius (c. 94–55 BC) expresses this in his great poem *On the Nature of Things:* [16]

> All nature, then, as self-sustained, consists
> Of twain of things: of bodies and of void
> In which they're set, and where they're moved around.
> For common instinct of our race declares
> That body of itself exists: . . .
> . . . Again, without
> That place and room, which we do call the inane,
> Nowhere could bodies then be set, nor go
> Hither or thither at all – as shown before.
> Naught, saving body, acts, is acted on;
> Naught but the inane can furnish room.

Such considerations led Lucretius to describe atoms in this way:

> Moreover, were there not a minimum,
> The smallest bodies would have infinities,
> Since then a half-of-half could still be halved,
> With limitless division less and less.

Then what the difference 'twixt the sum and least?
None: for however infinite the sum,
Yet even the smallest would consist the same
Of infinite parts. . . .

Or, to see this in a prose translation: [17]

And yet all things are not held close pressed on every side by the nature of body; for there is void in all things. . . . There is then a void, mere space untouchable and empty. For if there were not, by no means could things move; for that which is the office of body, to offend and hinder, would at every moment beset all things; nothing therefore, could advance, since nothing could make a start of yielding place.

Lucretius applies this argument to spaces between objects and to spaces within objects: 'If there were no empty space, these things would be denied the power of restless movement – or, rather, they could not possibly have come into existence, embedded as they would have been in motionless matter.' He concludes that 'Material things are of two kinds, atoms and compounds of atoms. The atoms themselves cannot be swamped by any force, for they are preserved indefinitely by their absolute solidity.' This is the modern view, though now the indivisible particles are 'sub-atomic'. [18] It was Ernest Rutherford who (in 1911) inferred that atoms must have small dense nuclei, from his discovery that particles from radioactive decay bounce off metal foil over a wide range of angles. This same argument, from similar data though with particles accelerated to higher energies, has now been applied to what were until very recently regarded as the indivisible particles of atoms, suggesting that protons are made up of much smaller particles ('quarks') bound together with previously unsuspected forces. So the chemical mechanism notion is now extended below the nineteenth-century atomic scale. These sub-atomic mechanisms (and the properties of their newly discovered 'parts') are, however, very odd indeed compared with macroscopic mechanisms or man-made machines. One might say that in order to preserve a mechanistic account of matter, very peculiar particles, and very peculiar force interactions for sub-atomic parts, have had to be supposed. It is a most interesting question whether such mechanistic accounts of matter might become (or are already) so bizarre that mechanistic explanations may fall into disfavour for the physics of matter. It is also a question whether conceptual models so different from mechanisms known to us via the senses *are* mechanistic. They appear bizarre just because they do not seem to be like mechanisms.

There is a class of particles in the new Gauge theory of nuclear particles called 'gluons'. These are supposed to stick the particles together – so there is a direct analogy drawn from our experience of gluey sticky substances. Is it safe to draw such analogies from the macroscopic world of our experience, to the fundamental particles supposed to underlie sensed reality?

The particles called 'charmed' particles are so named because they bring charm to the Universe, by the symmetry that they confer. They were sought to gain symmetry, and when found, after incredible effort, they were indeed charming. The term 'colour' is used for three related properties (named 'red', 'green' and 'blue') which were given colour names to help visualize the additive and subtractive properties of these particles, all three combining to give 'white', as in additive colour mixture of light as appreciated with the eyes.

What is the status of fundamental particles, and forces of matter? They are generally thought of as 'logical constructs' from observations. Modern particle theorists allow extremely remote relations between the properties of their particles, isolated in cyclotrons and linear accelerators, and 'observed' indirectly in bubble chambers. The new move is to look for the ultimate particles not in familiar objects but in extreme conditions, where 'matter' is subjected to exceptional forces.[19]

The ancient assumption that there are immutable fundamental particles of limited variety of form, creating by combinations a far greater variety of observed phenomena, lives on in the latest attempts of particle physics to describe the nature of matter. Here, surely, there lives on also something of the ancient ambiguity as to whether properties are given by various substances, or by various arrangements or forms or structures – of it matters not what.

These theories stand or fall by the degree and range of symmetry that they offer. New particles are sought where new particles would increase the symmetry of the current account. This is incredibly like Plato, linking the elements to the regular solids described by the geometricians.

It might be thought that the current generation of particle physicists are playing a sophisticated joke on Plato, by giving their particles and forces such names as 'charm' and 'colour'. This could hardly mislead though, for we are unlikely to think that the charm of women, or the colour of their hair or eyes, is the colour of the fundamental particles of which the atoms of their being are composed.

Statistical forms

It is interesting to consider forms that are maintained not by physical forces, but rather by statistical properties. An example is a candle flame. The particles of hot carbon that make it up are ever-changing; and yet it looks like a thing, an object with clearly defined edges and characteristic shape. Its edge is of course given by the cooling of the particles, which quite suddenly become too cold to emit visible light. The flame is larger in the infra-red, and the same is true of the sun. It seems to have a sharp edge; but of course the corona is tenuous, and in some ways we are inside the sun as we are bathed in its radiation.

Another statistical example is hair: each hair grows until it falls out, after about two years, so it does not get longer and longer but stops at a certain length. This is logically like the candle flame, and the edge of the sun, and the same applies to the limits of distributions of species. Concepts of statistical forms are, however, essentially modern and were not discussed by the Greeks.

Time[20]

Time, like space, is a characteristic both of the physical world, in that events occur in 'physical' time, and of psychology in that we experience sequence and extension in 'mental' time; and we remember events from the past, including situations that no longer exist in the physical world. We also of course anticipate, or predict, future events. So (as for space, and force and many other concepts) time is dual, applying (though differently) in physics and in psychology.

It is difficult to discover how the ancients thought of time. It is indeed difficult to describe how we think of time. Our distinction between 'physical' and 'psychological' time is not unlike the usual distinction made between the 'appearance' and the 'reality' of matter, of motion, of cause and of space.

Time is seen as a single-direction flow of events. For the ancients it may have been seen rather as cyclic; for they were dominated by the seasons and astronomical cycles, while we are dominated by clocks. The various forms of water-clocks (clepsydra) in antiquity were all unreliable. Sundials were essential, and remained essential until radio time signals of this present century, for setting clocks.

Sundials cannot 'gain' or 'lose' time – if time is defined as the movement of the sun across the sky, and if the dial is properly set up and correctly divided. There are, however, many ways of dividing the dial: the Egyptians did not have hours of equal length, as judged by our clock time; they divided the day into equal angular changes of the sun (actually the length of the shadow from a gnomon) through each day. Their sundials were divided into equal divisions while ours are not, for the shadow moves faster (by reference to mechanical-clock time) around midday than in the evening or morning. This gave the Egyptian clepsydra-makers a problem, for their clocks had to record each hour as a different length, as we would see it. This they did by changing the separation of the hour marks on the clepsydra bowls from which water flowed through a small hole. Further, the divisions were drawn to compensate for changes in the rate of flow of the water, as its viscosity varied with the changes of temperature (in Egypt very regular) through the day and night. This was, however, done without theoretical understanding.

We would say that days lengthen in the spring and summer, and are shorter in winter. The Egyptians defined the day as twelve of their hours of daylight and twelve of night, so the clepsydra were marked differently for each day of the year. This must have given the Egyptians a rather different sense of time from ours, if indeed technology can affect how we 'see' time.

Mechanical clocks with escapements were not known in the Western world until the thirteenth century – though as recently discovered by Needham, Ling and Price (1960) the Chinese did have an escapement clock, driven by water power but quite unlike a clepsydra, as early as the tenth century A D. This was, however, more than a millennium after our present concern with time in the classical world. Chinese technology advanced parallel to the Western classical tradition but faded away in our mediaeval period. This early parallel development of mechanism is, however, extremely interesting (see fig. 12, p. 67).

In both cases elaborate mechanical clocks were built, with their emphasis not on telling the time but on representing the heavens – as in the Antikythera mechanism of the first century BC described above. It seems that clocks as time-keepers were added much later to astronomical models (see fig. 10, p. 62; fig. 11, p. 63).

That time was seen as cyclic from the revolutions of the heavenly bodies is clear from the writings of both Plato and Aristotle. Plato's account seems to be in part an inheritance from the Pythagorean school, with its emphasis on harmonic relations. Timaeus tells how the Gods made a moving image of eternity on the principle of numbers that we call time: 'With the design the Deity created the sun, moon, and the five other stars which we call planets to fix and maintain the numbers of time.' In the *Timaeus*, Plato speaks of the souls of beings living on the cosmological bodies, together with a reference to the moving bodies as 'instruments of time', in this passage: 'He planted some of them on earth, others on the moon, and others in the different instruments of time.'

Aristotle describes the ancient 'cycling' notion of time, and also how this relates to counting time intervals, in a particularly interesting passage: [21]

. . . a uniform rotation will be the best standard [of time] since it is easiest to count.
Neither qualitative modification nor growth nor genesis has the kind of uniformity that rotation has; and so time is regarded as the rotation of the sphere, inasmuch as all other orders of motion are measured by it, and time itself is standardised by reference to it. And this is the reason of our habitual way of speaking; for we say that human affairs and those of all other things that have natural movement and become and perish seem to be in a way circular, because all these things come to pass in time and have their beginning and end as it were 'periodically'; for time itself is conceived as 'coming round'; and this again because time and such a standard rotation mutually determine each other. Hence, to call the happenings of a thing a circle is saying that there is a sort of circle of time. . . .

Plato had already described the Mind as moving in circles, following (or at least moving by analogy with) the heavenly motions. Plato talks in the *Timaeus* of the

'Souls of the World', as though the Universe is a kind of living organism. Aristotle thinks this unsound: [22]

> . . . for he [Plato] clearly means 'the soul of the world' to be some such thing as what is called mind; it is nothing like either the perceptive or desiderative faculty; for their movements are not circular. But mind is one and continuous in the same sense as the process of thinking; thinking consists of thoughts. But the unity of these is one of succession, like that of numbers. . . .

Aristotle advances several arguments against what might be called Plato's 'cycling Mind' theory: [23] 'Furthermore, that which moves not easily but only by force cannot be happy; and if the soul's movement is not part of its essence, it will be moved unnaturally.' This refers to Aristotle's view that circular and linear motions are the only Natural motions. All other motions are imposed and Unnatural. It reads oddly because we distinguish only with deep puzzlement between Natural and Unnatural. This is a good example of a 'psychological' explanation used for physics. It is discussed in detail in *De Caelo* (see also pp. 123–5). [24]

Is time absolute or relative?

Time, like space, may be conceived of as absolute or relative. The question is: is time no more than a succession of events, or does it flow on regardless, somehow behind events? This distinction goes back to the Greeks, though we seem to have few clear statements of classical concepts of time; possibly because change was seldom regarded as fundamentally important. It seems, though, as we have said, that the ancients thought of time as cyclic rather than as a linear progression. The cyclic sequences of the motions of the heavenly bodies seem to have inspired a cyclic account of time which is difficult for us to appreciate, perhaps because of the profound effect of mechanical clocks on how we think of time. The question for us is: what do clocks measure?

To take any cyclic event as the ultimate measure of time (such as the sun rising), poses the difficulty of deciding which cyclic events to select. This is one reason for accepting some kind of absolute theory of time in physics. This is implicit in Galileo's writings, and was made explicit by Newton's teacher at Cambridge, Isaac Barrow (1630–77). Barrow took the step of removing the concept of time from motion. He wrote: [25]

> Time denotes not an actual existence but a certain capacity or possibility for a continuity of existence; just as space denotes a capacity for intervening length. Time does not imply motion, as far as its absolute and intrinsic nature is concerned; not any more than it implies rest; whether things move or are still, whether we sleep or wake, Time pursues the even tenour of its way. Time implies motion to be measurable; without motion we do not perceive the passage of Time.

Barrow continues:

> . . . strictly speaking the celestial bodies are not the first and original measures of time; but rather those motions, which are observed round about us by the senses and which underlie our experiments, since we judge the regularity of the celestial motions by the help of these.

It is worth emphasizing this last point of Barrow's: that we accept the periodicities of the heavenly bodies as regular – and so as fit for measuring time – because they seem to us to be regular. This puts characteristics of the observer as primary to what he observes; though he may be, and indeed clearly is, affected by what he observes. This two-way relation could be profoundly important, and not only for considering time.

Newton spoke of time as continuous, at constant rate, independently of what happens. This may be common sense, but it has serious conceptual difficulties if time can only be perceived or measured from events. Newton's position was well described by John Locke in *Essay Concerning Human Understanding* (1690), written three years after Newton's *Principia*. Locke wrote: [26]

> We must, therefore, carefully distinguish betwixt duration itself and the measures we make use of to judge of its length. Duration, in itself, is to be considered as going on in one constant, equal, uniform course; but none of the measures of it which we make use of can be known to do so, nor can we be assured that their assigned parts or periods are equal in duration one to another. . . .

This is more sophisticated than Newton's account. As G.J.Whitrow (1975) says, [27] 'Unfortunately his [Newton's] definition of absolute time is of no practical use! We can only observe events and actual processes in nature and base our measurements on them.' Newton's 'useless' view is however the commonsense notion of time as continuing behind events which most of us hold.

The situation changed dramatically with the publication, in 1905, of Einstein's paper in which he pointed out that time depends on the notion of simultaneity. If no signals can travel faster than light then observers at different distances will each have to compensate for his distance to judge simultaneity of a commonly observed event (such as the explosion of a supernova). This, if it is possible, would allow a single objective time for all observers. But this does not hold as generally possible when they are in relative motion. This follows from Einstein's notion that physical phenomena are the same for all observers, whatever their relative linear motions. On this view time is not absolute, as believed from the Greeks to Newton, and until the beginning of the present century. We shall discuss this further in chapter 19.

The direction of time

The classical physical laws are reversible, and yet events occur in irreversible sequences. Why cannot the mechanism of Nature turn backwards? If we defined time in terms of sequences of events as they occur, or as they are experienced, then it would be meaningless to speak of reversing time. One might go on to say that we know nothing of time apart from events – so reversed time is a meaningless notion. But this is not quite the whole story, for it seems that many sequences of events cannot be (or at least never are) reversed. Here ciné films projected backwards are most revealing. Some reversed sequences cannot be distinguished from playing the film forwards, but other sequences look odd (and funny) when reversed; some look frankly impossible. Why, though, should reversed sequences of events be impossible – if the laws of Nature are time-reversible?

An example of reversibility of events without change is the continuous vibration of a sine wave. This is symmetrical in time, so reversal has no effect. But playing music backwards sounds odd, when the notes form an unfamiliar time order or when the instrument is percussive. Pianos, as their notes are struck and then die away, sound odder than organs whose notes are more constant. A bird's flight looks strange in reverse, mainly because it is flying backwards spatially. A man walking backwards looks odd, not only because he is going backwards but also because his movements are different from when he walks forwards. Things that look impossible when played backwards are, for example, such things as fragments of clay rushing together to form a pot, or drops of water coming together to form a stream entering a hose pipe. Such backward-going films make us postulate (or seem to see) new forces for controlling the bits of clay, or particles of water. The ultimately odd (gruesome) science fiction film would surely be dissection reversed, of an animal or a man. To see the organism created from a scalpel – to become alive and walk out of the dissection room with a backward smile – would make the ultimate Hammer horror film. But is even this backward sequence impossible for reality? The answer seems to be that any of these reversed sequences could occur, but they are extremely unlikely.

Time's arrow given by Entropy – the loss of organization, or loss of temperature differences – is statistical and it is subject to local small-scale reversals. Most striking: life is a systematic reversal of Entropy, and intelligence creates structures and energy differences against the supposed gradual 'death' through Entropy of the physical Universe. In this sense, life is reversed time. Indeed, we are created from bits and pieces to form the order that looks impossible in time-reversing film.

Will we ever have a physics capable of making such processes look reasonable? It is the measure of the concept of Natural Selection that increases in the complexity and order of organisms in biological time can now be understood. Natural Selection, with reversed Entropy, is supposed only to apply to living organisms.

This might be taken to imply that the processes of life can never be described or explained in terms of physical science, not because of their evident extreme complexity but for their reversal of Entropy. We may, however, consider simple physical processes that in certain conditions do give increasing orderliness with time. We cannot unscramble omelettes, but mixtures and solutions can be separated by physical means. Though a drop of ink will diffuse through water gradually, to make it evenly coloured as the particles intermingle, if left long enough without disturbance this process will reverse, to leave the ink as a layer on the top (or bottom) of the water – which is then free of ink. If there is some physical difference between the ink molecules and water molecules, and if their bonding force is not too high, time will apparently reverse for such non-living systems (see pp. 154–6). Chickens can unscramble omelettes: by eating them!

These are local reversals of Entropy, but we think of time as a kind of universal flow of almost all events. So we do not think of local reversals from the usual as time reversals, but rather as local reversals of Entropy. Although we do not allow local reversals of time, the General Theory of Relativity does suppose (against strong counter-intuition) that time can slow down or speed up. Slowing and speeding up of time is predicted for clocks in rapid motion relative to the observer. There is experimental confirmation of this from atomic clocks carried in orbiting satellites.

Einstein came to these and other conclusions of Relativity Theory by considering the observer in his descriptions of physics. He did not, however, include descriptions of what the observer is like: it could be any observer, situated in a given place and moving with a given velocity (or acceleration) with respect to what is observed. Einstein's central notion is that the physical laws of the Universe should be the same for any (perhaps ideal) observer. Since no observer can receive information from distant sources faster than light, it is not possible to give an 'objective' definition of simultaneity of distant events, without specifying distance. So time and space become a combined concept: space–time. This will be discussed more fully in chapter 19, when we consider the place of the observer in modern physics.

Concepts of space, matter and motion

Space was originally regarded as breath: a kind of animate gas that filled the Universe. At the time of Pythagoras, space was generally called '*pneuma apeiron*'. Later, it was called '*kenon*' (void). The notion of space as nothing came later – after Aristotle. It is indeed still a controversy in physics whether space is nothing or whether it has properties and so is a kind of something.

The modern notion of the Universe divides it into two kinds: objects, made of matter, and space between objects. This notion was made explicit, as we have seen, by Lucretius (see p. 129) in his poem *De Rerum Natura:* 'All nature then, as it exists, by itself, is founded in two things: there are bodies and there is a void in which these bodies are placed and through which they move about.' This was not clear to Aristotle, who had great difficulties in conceiving movement and change with his view of space. Movement and change presented insuperable problems for Aristotle because he regarded space (the 'Void', or the 'Inane') as a substance. But it is impossible for two objects to be in the same place at the same time – so how is movement or change possible? This was a difficulty generated by his concept of space. There were also logical difficulties in thinking about movement; especially the paradoxes of Zeno (see p. 119). Possibly these difficulties drew attention away from observed phenomena to unchanging realities. But these, also, are important in modern science, where 'invariances' are regarded as at least as significant as observed interactions and changes of objects.

One of the properties assigned to space was to maintain the form of objects. Another attribute of space was to make objects fall downwards. Space was generally regarded as finite: in which case it must be within a void. But if so, space itself could not be mere void. The notion of space setting limits to objects was conceived by considering a man standing at the edge of space: could he reach out his arm (or throw a spear)? This 'thought experiment' recurs indeed as late as Locke's *Essay* (1690), where it is asked whether, if God placed a man at the edge of the Universe, he could not stretch his hand beyond his body.[28]

The argument as to whether space 'exists' as some kind of substance, or whether it is mere void between objects, has continued ever since the first Greek philosophers. One might say that modern 'field' theories in physics ascribe properties to space that affect matter, while 'particle' theories look to ultimate parts (as of machines) to explain properties of matter and interactions between objects with neutral space. Particle physics considers constituents of atoms (electrons, protons, neutrons, and now quarks and so on) as parts of sub-atomic mechanisms. This is an extension of the Atomism of Democritus of classical times (fifth century BC) and of John Dalton two thousand years later (AD 1766–1844), to explain chemical reactions in terms of structured arrangements and rearrangements of atoms. As for macroscopic mechanisms, in chemistry both the structure of the parts and energy transfers between parts are crucially important for description and explanation of what is observed under various conditions.

Field theories seem less intuitively appealing. It seems odd to ascribe properties to space. And yet this is a strong tradition in Greek philosophy; and in the thinking of a supreme mechanist, René Descartes – with his 'vortices', which are kinks of space, analogous to Einstein's curved space for gravity, and to Maxwell's equations relating electricity and magnetism with light.

It is all very well to say that matter is particles interacting; but no atomic theory allows that they are in contact. There must therefore be space within matter; but this evokes the spectre of action at a distance, in the space between interacting particles. Descartes tried to avoid action at a distance, by supposing that the space between massive objects is distorted – the distortions being 'vortices'. Unfortunately this was taken by others, including Newton, to mean vortices of fluid matter. Now vortices of turbulent fluid are extremely complex and difficult to understand, and they are not at all what Descartes intended. It is now clear that he was close to Einstein's notion of distorted, or curved, space–time – which describes gravity without the concept of force. Force is not a factor because in this curved space the bodies are moving in what are for them 'straight lines'.

Straight lines are defined as the way light travels. Einstein's prediction that light should appear bent by a certain amount round the massive body of the sun – to deflect nearby star images visible in a total eclipse – was confirmed by the Eddington eclipse expedition of 1919 (see chapter 19), and this was an important prediction.

Considerations of magnetism bring out clearly the search for particles of suitable shapes to serve as parts of mechanisms of Nature. Starting from sympathies of selective attraction, in mediaeval accounts of magnetism, Descartes postulated right-handed and left-handed screw-shaped particles, corresponding to the north and south poles of magnets. Newton, in an early paper, is very close to Platonic thinking by supposing two kinds of magnetic matter to form streams, one passing through the pores of the Earth, acquiring an 'odour', making it 'sociable' to iron, while the other stream has a different 'odour' which makes it less sociable to iron, and more 'sociable' to the ether filling its pores. The two streams pass each other in the magnet by following different paths, each avoiding the other's 'odour'.

Newton held a corpuscular theory of light, but light particles do not need a medium for their propagation, though this does seem necessary for waves. So why did Newton need ether for the propagation of light as he conceived it? He needed ether to explain the bending of light by refraction. He explained refraction by supposing that ether is denser in space and rarer in the pores of bodies. Further, the sun was fuelled on ether, to maintain its heat; gravitational force, keeping the planets in their orbits, was given by the push of ether. The ether was, for Newton, a rarefied form of matter – not essentially different from the matter known to the senses – supporting the subtle mechanisms causing the phenomena that we observe. He seems also to have thought of the ether as the substance of life; rather as Galen thought of air as the 'prama' breathed to give life. Newton mixed mechanical principles with his concept of 'sociability' in chemistry and combustion. Indeed, his alchemy was not entirely separate from his physics.

The ether was invoked by Newton to allow a mechanical account of action at a distance. The causes of motions in familiar mechanisms are normally seen as mediated by matter joining the parts. The ether was postulated as a universal substance to extend this analogy of earthly mechanisms through space to the entire Universe.

Newton performed an experiment to determine whether there is indeed an all-pervading ether. For his experiment he constructed a pendulum, eleven feet in length, the bob being an empty box. He marked points of return for three swings, having calculated the weight of air in the box. He then filled the box-bob with heavy metal. This had seventy-six times the inertia – which he attributed to the ether in the pores of the heavy metal. In this first experiment he concluded that the resistance to the outside of the box – by air – was five thousand times less than of the internal resistance of the ether in the 'pores' of the metal. This suggested that the ether had far greater resistance than air. Newton repeated this experiment in a more sophisticated form – and obtained just the opposite result! He finally decided that there was no ether. From this he concluded that there was action at a distance.[29] We may note that this fundamental change of thought, or belief, was given by an experiment – an experiment which he had already found to be difficult to perform without disastrous error. His new result made him reject what we might call the strictly 'matterlistic' philosophy, in favour of force as the fundamental of the Universe and not dependent on matter for its transmission. Newton's final position was that matter is inert, sluggish with inertia, and that space is active and closely linked with God. He described space as God's sensorium.

The theological implications of Newton's Universe obeying laws in principle allowing prediction to any future time, and running by itself as a frictionless machine maintained by inertia, did not by any means pass unnoticed during his lifetime. There was hot debate in theological circles, and though Newton did not join in openly in print, he did write upon the subject in private. Perhaps most revealing is the manuscript MSS ADD 3970, fols 619r, at the Cambridge University Library. It was written just after the publication of his last book, the *Opticks* of 1704. Here is part of this manuscript:

> By what means do bodies act on one another at a distance? The ancient philosophers who held Atoms and Vacuum attributed gravity to atoms without telling us the means unless perhaps in figures: as by calling God harmony and representing him and matter by the God Pan and his Pipe, or by calling the sun the Prison of Jupiter because he keeps the Planets in their Orbs. Whence it seems to have been an open opinion that matter depends upon a Deity for its laws of motion as well as for its existence. . . . These are passive laws and to affirm that there are no other is to speak against experience. For we find in ourselves a power of moving our bodies by our thought. Life and will are active principles by which we move our bodies, and thence arise other laws of motion unknown to us.
>
> And since all matter duly formed is attended with signs of life and all things are framed with perfect wisdom and nature does nothing in vain; if there be an universal life and all space be the sensorium of a thinking being who by immediate presence perceives all things

in it, as that which thinks in us, perceives their pictures in the brain: these laws of motion arising from life or will may be of universal extent. To some such laws the ancient Philosophers seem to have alluded when they called God Harmony and signified his actuating matter harmonically by the God Pan's playing upon a Pipe and attributing musick to the spheres made distances and motions of the heavenly bodies to be harmonical and represented the Planets by the seven strings of Apollo's Harp.

So Newton ended up thinking that God runs the show, by permeating not matter but space. The opposite position was argued by John Toland (1670–1722) who was an Irish theologian; he was brought up a Catholic but became a Protestant before he was sixteen. His works, following a debate and prosecution in the House of Commons, were burned by the public hangman in Ireland.[30]

John Toland argued – counter to Newton – that motion is inherent in matter. He held that there is only one essential matter; that 'Earth, and Water, and Air and Fire, are interchangeably transformed in a perpetual Revolution; Earth becoming Water, Water Air, Air Aether, and so back again in mixtures without End or Number'; and that this is so because[31]

All the Parts of the Universe are in this constant Motion of destroying and begetting and destroying: and the greater Systems are acknowledged to have their ceaseless Movements as well as the smallest Particles, the very central Globes of the Vortexes turning about their own Axis; and every Particle in the Vortex gravitating towards the Centre.

Toland argued that bodies gravitate to the centre of the earth because of their inherent harmonious activity. This is essentially Aristotle's view. What Toland was attacking was Newton's account that matter is inert. So we might say that Toland put life (with hardly any reference to God) into matter, and Newton put life (and God) into space.

Toland was not of course a scientist; he was, though, a learned man, sharing friendship with Locke and Leibniz. But his view had scientific implications which allowed it to be attacked. It was attacked not only as being atheistical, but also because it did not begin to explain the order of things. If matter is 'innately dynamic', why should it move according to Newton's laws? Why should objects not have propensities to movement in all directions at once – and so be motionless? This is a powerful argument against self-volition for the objects of the Natural world. It seems, indeed, to distinguish inanimate from animate – non-living from living. It gives a strong reason for emphasizing interactive forces in physics – hence Newton put his *modus vivendi* into space rather than matter for physics, while emphasizing properties of matter for biology.[32]

Theologically, if the dynamism of the Universe is in individual objects, then 'inanimate' matter must be as we suppose ourselves to be – free to move much as they like by will. But then where is God? For Newton, God set up and maintains the laws of motion operating in space. So, provided there is an absolute space, there could be a place for God, or Mind, in the Universe. Newton himself did not seem to realize that his Universe did not require maintenance if it was frictionless, for he wrote that God is the 'ever-living Agent, who being in Places, is the more able by his Will to move the Bodies within His boundless uniform Sensorium'.

It was essential for Newton that there should be motion with respect to an absolute space for space to be effective in this kind of way, and to allow a place for God. Newton did consider that all motion might be relative, but abandoned this early in his life, from considerations such as his bucket argument (see p. 125).

Modern concepts
Electricity and magnetism

The word 'electricity' (coined by William Gilbert in 1600) comes from the Greek for 'amber' ($\mathring{\eta}\lambda\epsilon\kappa\tau\rho o\nu$), whose properties of attraction were known to Thales in the sixth century BC. The magnetic properties of the lodestone (magnetite) were also known to Thales. Many phenomena of electricity and magnetism are curiously lifelike, and have very often suggested analogies between Natural and Unnatural that have generally proved deeply misleading. On the other hand, electrical (and very recently also magnetic) properties of neural activity have turned out to be essential keys to understanding the physiology of the nervous system. The notion of magnetic and later electrical 'influence' is derived from flowing fluid, which may be embodied in such occult bodies as the philosopher's stone, or in such psychological notions as people having social influence. It is a word found often in the history of static electricity, as in 'influence machines'.

Both electricity and magnetisim have been used as cures for disease – sometimes with frightening effectiveness, as in electroconvulsive shock therapy (ECT) which is still used extensively in cases of depression, though with no known rationale and with a distinct possibility of irreversible brain damage.

Magnets have the property of transferring their attractive abilities – repulsive to other magnets – when rubbed upon iron or steel. This must have looked like transferring a magnetic substance by biological-looking body contact. Indeed, magnetic and sexual attraction could very easily be confused, and it is hardly surprising that people are still said to be 'electrified', and to have 'magnetic personalities', and Greek writers spoke of male and female magnets.

Magnetism was not studied as a problem for Natural Science until remarkably late: about 1600, with the highly sophisticated experimental work of the Englishman William Gilbert (1540–1603) who, after a successful career in medicine (culminating in his becoming Queen Elizabeth's personal physician), published his *De Magnete* in 1600, in which he described experiments so beautifully contrived that even Francis Bacon was impressed. He was perhaps the first to make effective use – before the *Novum Organum* (1620) – of 'Baconian methods', which are now standard scientific procedures. He discovered that substances other than amber give (as he named it) electrical attraction, and he explained the compass by conceiving the earth as a vast magnet, suggesting also that electricity and magnetism are ultimately the same force or 'fluid'. With his experiments and conclusions, magnetic and electrical forces and phenomena began to move from the realm of the 'Unnatural' to Natural Science. It is fascinating to see how such experiments (often at first misinterpreted) gradually forced quite new ways of thinking.

Particularly interesting experiments were performed in Paris by Charles du Fay (1698–1739), who reported to the Academy of Sciences in 1733 and 1734 that all bodies could be electrified, whatever their substance. This disproved the earlier notion of Jean Desaguliers that some bodies are 'electrics' and others 'non-electrics' for du Fay found that 'non-electrics' are conductors of electricity and may be electrified when insulated. He showed that the human body can be electrified (by insulating someone from the ground), and that when someone else approached the electrified person, they both experienced prickling sensations. He found also that sparks occurred (rather as though trying to join together the two people) when one was electrically influenced.

Du Fay found that a charged body attracts another, but that after contact they are mutually repelled. (The cynic might hold that this is indeed the electrical paradigm of human passion.) He found as well that some substances produce electricity, which repels some substances but attracts others: he named these 'vitreous' and 'resinous' electricity. Vitreous electricity was produced by rubbing glass, rock crystal, precious stones, hair or wool; resinous electricity was produced by rubbing amber, thread or, among others, paper. He found that each repels its own kind and attracts the other kind.

The first machine for producing static electricity was built by Otto von Guericke (1602–86) in 1660. It consisted of a rotating globe of sulphur which was touched lightly by the dry hand, when sparks would fly. The sulphur ball was cast in a glass globe, which was then broken. Actually the glass globe would have served better than the sulphur, but [33] 'it was imagined that the electricity was emitted with the sulphurious effluviam which was produced by the friction. This effluviam was

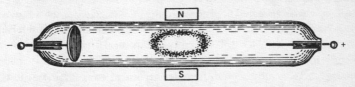

23 The influence of a magnet on the glow from electricity passed through a near-vacuum. The glow was similarly affected by movements of a nearby hand.

thought by Robert Boyle (1627–91) to be glutinous, to carry light bodies, such as small pieces of paper with it as the glutinous effluviam passed back to the electric, or rubbed amber.' Newton cast doubt on this notion by showing that the back of a charged glass plate is also charged. The 'electric effluviam' was often associated with the 'Vital Spirits' of the human body; a favourite experiment being to suspend a boy horizontally by silk cords and charge him up – when not only would his hair stand on end but he would emit sparks. The silk cords were later replaced by insulated stools to stand on. The experiment became extremely popular, especially when the spark from a finger was made to ignite gunpowder.

People paid money for the new sensation of electric shock. At first, extremely mild shocks (before the invention of the Leyden jar) were described in terms of terror and anguish, but later investigators suffered dangerous shocks as a matter of course. A particularly dramatic demonstration was given to the King of France by 180 guardsmen who, with joined hands, fell down as a man when a charge was applied to the first and last man. Here was a profound force, at once mysterious and effective.

The American statesman and scientist, Benjamin Franklin (1706–90),[34] suggested, from particularly brilliant experiments, that what had seemed two kinds of electricity are but one: 'positive' and 'negative' electricity. With his experiment of flying a kite in a thunderstorm, Franklin established that lightning – that ancient most potent weapon of the angry Gods – could be brought to earth, and is electricity. This he did in a field near Philadelphia, in June 1752, and rather surprisingly lived to tell the tale. Benjamin Franklin's invention of the lightning conductor for protecting property must surely have seemed sacrilegious: one can imagine fascinating heated clerical discussions before it was admitted as theologically safe to protect church steeples from the 'act of God' of being struck by lightning.

How did Franklin come to think that lightning might be electricity? What made him suggest the experiment? He set down in his Minutes how he came to think that this would be a worthwhile experiment (quoted in a letter to Dr Lining, in Bakewell):[35]

Nov. 7, 1749: Electric fluid agrees with lightning in these particulars: 1. giving light; 2. colour of the light; 3. crooked direction; 4. swift motion; 5. being conducted by metals; 6. crack or noise in exploding; 7. subsisting in water or ice; 8. rending bodies it passes through; 9. destroying animals; 10. melting metals; 11. firing inflammable substances; 12. sulphurous smell; 13. the electric fluid is attracted by points. We do not know whether this property is lightning, but since they agree in all the particulars in which we can compare them, is it not probable that they agree likewise in this? Let the experiment be made.

In spite of this elegantly set out Inductive argument by analogy, he did not have complete faith. For his first trial, he used his son's kite and took him along in case people thought him foolish.

Franklin's idea was rapidly developed: M. de Romans, of Nerac, succeeded with a seven-foot kite in drawing down streams of sulphur-smelling fire an inch thick and ten feet long. A celebrated casualty was Professor Richmann of St. Petersburgh, who (on 26 August 1753) killed himself with his 'electrical gnomon'. His assistant saw 'a glow of blue fire as large as his fist' dart from the rod of the gnomon to the Professor's head about a foot distant. It killed him instantly – with a red spot on the forehead and his left shoe burst open.

The notion of electricity as a fluid which we regard as false was curiously confirmed with the invention of the Leyden jar in 1745 by E.G. von Keist of Kammin, Pomerania, Germany, which was used for storing electricity from influence machines. The Leyden jar was a glass bottle partly filled with water. Thomas Kuhn has recently pointed out that the Leyden jar was invented through a false analogy – from the notion that electricity is a fluid: [36]

One of the competing schools of electricians took electricity to be a fluid, and that conception led a number of men to attempt bottling the fluid by holding a water-filled glass vial in their hands and touching the water to a conductor suspended from an active electrostatic generator. On removing the jar from the machine and touching the water (or a conductor connected to it) with his free hand, each of these investigators experienced a severe shock. . . . The initial attempts to store electrical fluid worked only because investigators held the vial in their hands while standing upon the ground. Electricians had still to learn that the jar required an outer as well as an inner conducting coating and that the fluid is not really stored in the jar at all.

The fact that the Leyden jar worked must have perpetuated this notion of electricity as fluid – to be stored, as other fluids, in jars. The emphasis on hands is also interesting: the first electrical machines for producing static electricity (invented by von Guericke in 1660) functioned by the operator's hand lightly touching a rotating ball of sulphur. The hand was replaced by a fixed cushion by Johan Heinrich Winckler of Leipzig in about 1733, when at last it became clear that this electricity was after all a Natural phenomenon. At the same time, though, 'animal magnetism' and various electrical forces were supposed to be a principle of life, or of Mind.

Kuhn seems to be referring to Professor Muschenbroeck, whose hope was 'to collect electricity within a phial to prevent its dispersion, and thereby to store up an increased quantity of the electric fluid.' [37] Interestingly, Muschenbroeck did not himself succeed, though he tried the experiment; but M. von Kliest, Dean of the Cathedral of Camin, and later M. Cuneus, did succeed – but only because they grasped the glass phial in their hands. The importance of this was not at first realized, so the experiment was not repeatable. When it was realized, the hand itself (its Vital Spirits) was – wrongly – accepted as the key to the situation. The modern explanation, which rejects the original 'commonsense' analogy that electricity fills the jar as a fluid would, is clearly expressed by Bakewell (1853):[38]

It is now ascertained that a jar when highly charged does not contain more electricity than it did before it was applied to the conductor. The effect produced by charging is not to increase the quantity, but only to disturb the natural electricity previously present in a latent state on the inside and outside of the glass. There is injected into the inside, by connection with the electric machine, an amount of positive electricity, whilst an equal amount of negative electricity is driven from the outside by the force of electrical induction; and unless the electricity on the outer surface of the glass can thus be driven off by affording it a connection with the ground, the inside cannot receive a charge.

I have given this in some detail, as Kuhn's point is particularly interesting and the full story gives a clear example of the power of experiments to change 'commonsense' analogies, while also being difficult to appreciate when they run counter to what seems obvious – that electricity is a fluid.

Electricity was throughout this period associated with the ancient 'Vital Spirits'. There were many suggestions that it would provide the key to life. One of these schools became known as Mesmerism.

Mesmerism

Friedrich Anton Mesmer (1734–1815)[39] studied medicine at Vienna. Starting with a strong interest in astrology, he turned to the idea that the stars influence men by electrical or, later, by magnetic forces. This led him to try to cure diseases by stroking his patients with magnets. He described his philosophy, which became a vogue, in his *De Planetarium Influx* (1766). His consulting apartments were dimly lit, with occasional music after long silences, and his patients sat around a mysterious chemical vat, with Mesmer wandering around dressed in the robes of a magician. His success led to an inquiry of physicians, and the Academy of Sciences, to investigate his claims, with Benjamin Franklin as a consultant. They admitted some of his claims to be true; but he was finally denounced as a charleton, and left Paris, where for many years he had been highly successful though never accepted by the medical profession.

Mesmer should not be dismissed lightly. He did elicit remarkable effects, and almost certainly many cures in his patients, and his honesty seems now unquestionable. He did not keep his techniques for his own use but put every effort into spreading his ideas and persuading others to use magnetism for curing functional diseases. When he found that wooden 'magnets' worked as well as iron or steel ones, he generalized 'magnetism' from mineral to 'animal magnetism' and suggested that this was a universal force obeying Newton's laws of force – and of optics as it could be reflected from mirrors. Unfortunately it was so universally present, in his view, that it is difficult to see how it could be specially induced or controlled: itself an interesting question for scientific method, which seems to

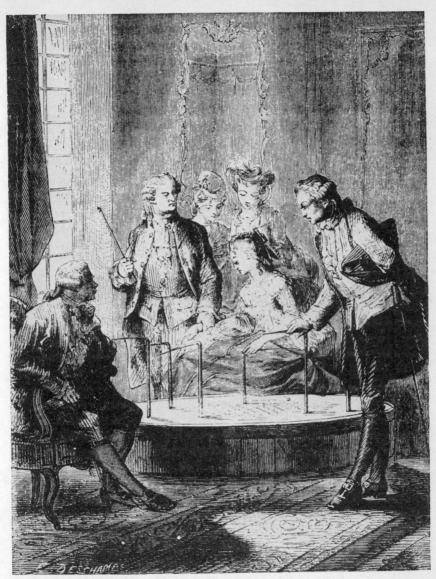

24   Anton Mesmer at work with magnetized water. (Mary Evans Picture Library.)

deserve discussion. There is no doubt that Mesmer was placing his patients in hypnotic trance states; and that he effected cures by suggestion. He was indeed the first in modern times to discover the power of hypnosis; and this led to the use of hypnosis for hysteria by Jean Martin Charcot (1825–93) at the Salpêtrière, which, in turn, very largely founded psychoanalysis. Freud was a pupil of Charcot's, and was much influenced by him, though sometimes critical.

This is a case where newly discovered phenomena of electricity, miasmic ethereal glowing lights responding to quite distant movements of the hands and showing action from a distance by magnets – suggested with subtlety and drama close connections between hidden forces of the physical world and life. And, indeed, this seemed confirmed by many of Mesmer's experiments, which were carried out entirely openly and before expert witnesses. It smacked however of the occult, and although the doctors of the time had no objection to using stage management, or the 'atmosphere' of Mesmer's consulting rooms (this was to some degree common practice), they yet jibed at his claims. Many of his arguments were specious, and perhaps he protested too much; but apart from hypnotism, which he unwittingly brought to light, he was after all right to think that subtle electrical and magnetic effects are secrets of life. It is perhaps unfortunate that hypnotism became tainted with the occultism of Mesmerism for it is still hardly respectable, in spite of genuine phenomena. Part of the trouble was the confusing doctrine of phrenology, which suggested that by stroking 'bumps' of the head with magnets desirable, or undesirable, traits of the personality might be transformed. Again, the founders of phrenology, Franz Joseph Gall (1758–1828), and to a lesser degree Johann Kasper Spurzheim (1776–1832), were bona fide scientists, Gall being an excellent anatomist with important discoveries to his credit. The evidence for phrenology was however abysmal; and yet its effect on neurology was enormous, and is not even now quite dead. Possibly Mesmer came into conflict with the Natural Scientists of his period who were beginning to rationalize magnetic and electrical phenomena; while the phrenologists carried the day (to disaster), because scientifically they were on their own. One does wonder how Freud would have fared in Mesmer's position! He might have fared badly (and he was indeed concerned to earn Ernst Mach's approval in Vienna) if he had been judged by 'hard' science. The hard scientists are quite capable of missing the point, and committing unintentional infanticide on alien ideas which may, given encouragement, grow to illuminate Mind. We might even say that physicists tend to reject Science in Mind!

ECT

Electrical stimulation of the brain by surface electrodes is perhaps the most generally used psychiatric treatment for depression, since its first use by Ugo Cerletti in 1938. The original rationale for adopting this dangerous-looking treatment – that there is a kind of hinterland between schizophrenia and epilepsy where clinical depression does not occur, and induced epilepsy might cure depression – is no longer accepted. There is no accepted theoretical basis for ECT, but in spite of some doubt there seems to be sufficient evidence that it works to justify its use. The number of deaths directly attributable to ECT is small. Typically there is, however, some short-term memory loss, which Moyra Williams (1966) has pointed out is similar to percussive brain injury; but permanent memory loss and impairment of intellectual powers may be small. It is beyond comparison with other treatments, convenient, cheap and quick; however, this has its own dangers as it may be used in cases where it is doubtfully appropriate. The situation is excellently summarized by Anthony Clare: [40]

The roots of electrical treatment lie buried deep in the history of psychiatry. That electricity can stun was known to Hippocrates and his colleagues by way of their familiarity with the electric torpedo or camp-fish found in Mediterranean waters. Sir John Pringle described how before the days of Galen the torpedo fish was applied to affected parts of the body and was particularly effective in easing persistent headaches. Earlier in 1756, John Wesley procured an apparatus that could deliver electric shocks and found the treatment so effective that he declared 'hundreds, perhaps thousands, have received unspeakable good'. A surgeon at St Thomas's Hospital, John Birch, was one of the first dedicated advocates of electricity as a therapeutic agent and in 1729 in a letter to George Adams, an instrument maker who provided Birch with many of his electricity machines, he explained how he administered the treatment to a Porter suffering from a melancholic state for almost a year. Six small shocks were passed through the brain in different directions on each of three successive days, following which the patient regained his spirits, went back to his work, and remained perfectly well for several years.

There were many other reports of electrical and other shocks apparently producing cures; but how they work (assuming that they do work) remains mysterious. Explanations range from such psychological explanations as blotting out painful memories, and providing healing effects of punishment on guilt, to such 'Natural' explanations as increasing the concentration or effectiveness of the neuro-transmitter noradrenalin in the brain. Although this treatment by electric shock has been called dehumanizing and indefensible on theoretical or empirical grounds, it is widely practised. The same objections could be raised, but seldom are for obvious reasons, against the use of anaesthetics to counter pain. Here the benefits are clear to all: we can only assume that clinical judgement is right in making extensive use of ECT, at least while alternatives to shocking the brain remain to be discovered.

Here we have come full circle, to techniques applied to Mind, via the brain, with no conceptual understanding of why they should work. Will future understanding of Mind follow the history of physics – concepts for understanding Mind arising from know-*how*, but without the know-*why* of this primitive technology?

These are cases of various kinds of force and energy applied, without understanding, to the brain. Can we understand with concepts of information?

Energy and information

Seventeenth-century physics made force a central concept. In the absence of frictional losses, descriptions of the interplay of forces served – since they conserved – to describe the dynamics of the Universe; but for machines and interactions within matter, rather than between bodies in space, the concept of energy became important, as energy is needed to maintain motion in living and non-living systems.

There are highly interesting and still largely mysterious relations between energy, structural organization and information. They intimately concern the processes of life. How do living forms gain organization through evolution of species? How can living forms attain, and maintain, more structure than their environment? How can information become structured? What are the limits to structure and knowledge?

The key concept here is the Second Law of Thermodynamics, with its related concept: Entropy. Thermodynamics is essentially a new science, which took off when it became clear that heat is not a fluid (phlogiston) but is random activity of the molecules of matter. This involves notions of order and disorder; of probabilities and randomness; of information and loss of information – all of which were entirely unknown in the ancient world. Indeed, most of these concepts are only a century old and are still perhaps in embryo form. This is a particularly difficult set of issues: I can only hope not to be misleading.

The science of thermodynamics started with the insights of the remarkable French engineer, Nicolas Léonard Sadi Carnot (1796–1832).[41] He introduced what later became known as the Second Law of Thermodynamics, in 1824. What is now called the First Law was not known at that time. It seems remarkable that Carnot succeeded in his analysis of relations between heat and mechanical energy without knowing the First Law.

The laws may be stated:

First Law: energy may be changed in form but cannot be created or destroyed.

Second Law: heat cannot by itself pass from a cold body to a warm body.

Another key notion, which we need to appreciate, is Entropy.

Entropy: loss of energy, which can no longer do useful work, through increasing randomness.

This is the ultimate pessimistic statement, for it seems to imply that the useful energy of the Universe is irreversibly declining, leading to 'heat death'. 'Entropy' can also be applied to structures; it then means that the organization of structures is continually eroded by increasing randomness. This is the next step of the pessimism of Entropy: everything deteriorates in time. There do, however, seem to be marked, if temporary, exceptions: especially living organisms.

There is also a Third Law (Nernst's Postulate).

Third Law: the Entropy of an object at absolute zero temperature is zero.

This allows Entropies, not merely relative Entropies, to be calculated.

These statements refer to macroscopic objects, and to finite intervals of time. There can be short-term fluctuations for small numbers of molecules, during which heat flow may reverse from cold to hot (lower to higher energy); and there can be short-term 'chance' gains in structural organization. These statements refer to means of populations of molecules, over finite time intervals. On average it is impossible to get useful work without destroying organization.

Carnot's Principle

Carnot's Principle states that the efficiency of an engine is proportional to the difference of the temperatures between which the engine runs, divided by the higher temperature. This is:

$$\text{Efficiency} = \frac{W}{Q_1} = \frac{T_1 - T_2}{T_1}$$

where W = work, Q_1 = heat lost, T_1 = the higher temperature 'source' and T_2 = the lower temperature 'sink'. The efficiency must be less than unity, for the absolute temperature cannot be less than zero, so T_2 must be positive. So all engines lose more heat energy than the work achieved. They therefore obey the Second Law.

Carnot realized that no work could be obtained from a heat source alone: there must be transfer of heat to a colder body. Energy can be stored, as potential energy. It may be stored by lifting a mass against a gravity force field; by compressing a spring; by creating chemical structures; and in very many other ways. Potential energy can be released as kinetic energy and kinetic energy is required to increase potential energy. So potential energy is stored kinetic energy. The form of the potential energy may be different from the form of the kinetic energy. In these cases a transducer has converted the energy from one form into the other form.

Potential energy is much more complicated than kinetic energy, as it has a far greater variety of forms. Kinetic energy is calculated simply by multiplying the mass of each part by half the square of its velocity, and taking the sum of the products,

$$E = \tfrac{1}{2} M v^2,$$

but as potential energy may be in physical distortion, or chemical bonding, etc., it cannot be analysed simply. One might perhaps say that the study of the properties of matter is the study of the forms of potential energy.

Although Entropy always increases in closed systems, and increasing Entropy reduces organization and reduces the work that can be accomplished, living organisms may seem to be exceptions. They are highly organized, and their organization is greater than the materials of which they are made. So, somehow living organisms including plants succeed in reducing their Entropy.

If we now consider the development of structure in Evolution, it seems clear that Entropy is also reduced with evolutionary development of species. How are these reversals of Entropy in life to be accounted for? The first point to make is that organisms are not closed systems. Increase in organization is always 'paid for' by energy extracted from the environment, especially from food. This is discussed most interestingly by the Austrian physicist Erwin Schrödinger (1887–1961) in his book *What is Life?* (1958):[42]

How can the events in *space and time* which take place within the spatial boundary of a living organism be accounted for by physics and chemistry? . . . The obvious inability of present-day physics and chemistry to account for such events is no reason at all for doubting that they can be accounted for by those sciences.

Schrödinger goes on to say that in principle they can be understood with statistical concepts, although:

The arrangements of the atoms in the most vital parts of an organism and the interplay of these arrangements differ in a fundamental way from all those arrangements of atoms which physicists and chemists have hitherto made the objects of their experimental and theoretical research. . . . For it is in relation to the statistical point of view that the structure of the vital parts of living organisms differs so entirely from that of any piece of matter that we physicists and chemists have ever handled physically in our laboratories or mentally at our writing desks.

The energy-trading with the environment is beautifully described by the French biochemist, the late Jacques Monod, in *Chance and Necessity* (1972),[43] when he points out that the prediction of the Second Law that 'no macroscopic system can evolve otherwise than in a downward direction' is valid 'only if we are considering the overall evolution of an *energetically isolated system*'. This argument Monod develops by considering crystals growing in a medium:

Within such a system, in one of its phases, we may see ordered structures take shape and grow without that system's overall evolution ceasing to comply with the second law. The best example of this is afforded by the crystallization of a saturated solution. . . . The local increase of order, represented by the assembling of initially unordered molecules into a perfectly defined crystal network, is 'paid for' by a transfer of thermal energy from the crystalline phase to the solution: the entropy – or disorder – of the system as a whole augments to the extent stipulated by the second law.

In other words, the organism extracts energy from the environment to increase its organization. This is established experimentally by growing bacteria (or crystals) in a medium and recording the temperature.

Information, the Second Law and Maxwell's Demon

The mathematical theory of information was put forward by Claude Shannon while working at the Bell Telephone Laboratories in 1949. This has changed how we may now think of the relation between the world of physics and the observer, as well as limits to communication – and the cost – between people, animals or machines. Since the unprecedented success of the almost observer-less accounts of the physical world in the seventeenth century, the observer has been virtually ignored; but this is changing. Much of the credit belongs to the great pioneering nineteenth-century Scottish physicist, James Clerk-Maxwell (1831–79). Maxwell's contribution to the problems of energy and information focuses on his Demon. With Shannon's Information Theory a century later, Maxwell's Demon, which challenged the Second Law, helps us to think about the limits of observation and communication in new and more rigorous ways.

James Clerk-Maxwell was a Scottish mathematical physicist who spent most of his working life at Cambridge, where he became the first Cavendish Professor of Physics. His extremely important equations relating magnetism to electricity, and to electromagnetic radiation (he predicted radio-frequency radiation some sixteen years before it was detected by Hertz), remain valid in spite of Maxwell's acceptance of the ether, which is of course now rejected by Relativity Theory. Our concern is, however, with Maxwell's Demon. Although purely fanciful, this succeeded in finally destroying the old notion of heat as a fluid (caloric), supporting the current account of heat as random molecular movement.

Temperature is now seen as the mean, or average, velocity of particles, whose distribution of velocities, in a gas, is wide. Maxwell's discussion of this (like the statistical basis of neo-Darwinism) is a concept very different from Newtonian forces. Maxwell imagined two containers filled with gas at equal temperature. The containers were joined by a trapdoor (whose friction was ignored) which could be opened or shut very rapidly, by a 'Demon' who could watch out for particularly fast or slow gas molecules as they approached the door. Maxwell raised the important question: 'Could the Demon violate the Second Law: by opening the trapdoor for fast molecules passing in one direction and opening it for slow molecules passing in the other direction?' If this could be done, at least in principle, it should result in one chamber becoming hotter, the other colder, and this would violate the Second Law. The question is: is such a molecule-sorting Demon possible? Could it indeed be done even purely mechanically: could the fast and slow molecules be separated without cognition? Maxwell conceived his

Demon in 1871, which was just thirteen years after the *Origin of Species*. Charles Darwin seems to have been the first to suggest a statistical type of model to explain systematic change or development. His theory seemed to violate the Second Law (which was understood by Carnot in 1824, but was not formulated in a way that was appreciated until Rudolf Clausius and Lord Kelvin some forty years later).

We may now consider two kinds of 'Demon'. The first kind will be simply mechanical (a 'mechanical' Demon): the second kind will be an intelligent observer of velocities of the approaching gas molecules, supposed capable of predicting which is fast and which slow before they arrive at the trapdoor – a 'cognitive' intelligent Demon.

Mechanical and intelligent Demons

Mechanical Demons are discussed by the American physicist Richard Feynman, in *The Feynman Lectures on Physics* (1972).[44] Feynman considers a paddle-wheel, set on a frictionless shaft, in a chamber in which gas molecules give random pushes in all directions to the paddles. The paddle-wheel will move backwards and forwards randomly (as we know from the motion of small particles – Brownian motion – which can indeed be observed with a light microscope). Since this motion is not unidirectional, except for small intervals of time and for small numbers of particles, Brownian movement is said not to violate the Second Law. Likewise, the randomly agitated paddle-wheel will not violate the Second Law. But now consider adding a ratchet wheel to the shaft, with a paul so that it can turn only in one direction. If the paddle-wheel now turned systematically in one direction (though jerkily), it would violate the Second Law.

The trouble is that the ratchet and paul get heated, until the paul fails to engage in the teeth of the ratchet. Further, the paul needs a force to return it to the ratchet wheel, and damping to inhibit its bouncing off the wheel before the next tooth arrives. When all these are taken into account, there is no energy left to give systematic motion. This is so even though the mechanism is entirely free of friction.

An actual example of this kind of thing is the behaviour of a diode, or a thermal junction, which when heated produces a voltage difference which can produce current and work in a circuit. Neither the diode nor the thermal junction produce unidirectional or useful current when all their parts are at the same temperature – though there is random molecular and electron motion at the junction. The general conclusion is that no such 'ratchet' can provide useful work from random thermal activity without a maintained temperature difference. One might imagine that solar cells on a space probe are exceptions; but the necessary temperature difference is here maintained by radiation from the surface of the probe, so space is the energy sink, allowing temperature gradients and so energy flow.

So far the Demon has no intelligence, no power to recognize particles or to act on predictions.[45] The question is: can the Demon gain information of the molecules, especially their approach velocities, without using as much energy as it gains by sorting the molecules with his trapdoor? In other words: can an intelligent Maxwell Demon beat the Second Law, given these limits to the transmission of information? (The problem for the Demon is acute because the gas molecules behave randomly. We might rephrase the question more generally: would it be possible to display intelligence of any kind in a random world?)

If the Demon could see individual molecules then surely it could open the door appropriately, to beat the Second Law. But can he see the molecules? The trouble is that the Demon is in the chamber; so he is at the same temperature as the gas. He therefore cannot receive energy from the molecules, so he cannot receive information: he is blind to them.

The Demon might, however, be allowed a source of light or other radiation, to see them by. The flashlight could provide information of the molecules – and this would allow him to operate the trapdoor, to produce a temperature difference. It turns out, though, that the energy (negative Entropy) of the added light is greater than the energy flow that the Demon can produce with his trapdoor.[46] The intelligent Demon, provided with information from his added light, cannot beat the Second Law.

The Demon *can* however create order – provided he is allowed to use energy. Indeed, we see this whenever we tidy a room, or correct proofs. And for life, cells create order; but at an equivalent energy cost.

To see that the Second Law is preserved in biology is extremely important, for it allows us to see living processes as within Natural Science; for the Second Law is the most general law in Natural Science. It will be similarly important to see that the increasing organization of species, by the random variation and selection of Darwinian Evolution, does not violate the Second Law – which it may *seem* to do as organization *is* clearly increasing in biological time. We should also see that individual intelligent behaviour does not violate the Second Law when creating order from chaos, as when we build houses or other artefacts. If it should turn out that increasing order through Natural Selection or by individual intelligence – Mind – violates the Second Law, biology and psychology could hardly fall within the Natural Sciences.

We have found two principles which allow life-forms to avoid heat-death (devolution of species) and, for individuals, protection from chaotic randomness. These principles are:

1   It is improbable that several mutational gene changes will reverse to an earlier structure. So Evolution is most unlikely to go into reverse.

2   Any information gain is bought at the cost of energy expenditure.

There is thus a kind of trading between energy Entropy, and information Entropy. Provided energy is available, protection from degeneration into chaos can be bought by extracting energy from the environment. Living organisms extract energy from their environment, to buy throughout their life-span protection from information Entropy. Some of the information gained may be transmitted to later generations through the genetic code and through artefacts of technology. These are ratchets. They allow the continuous creation of structure, and of information, against the background randomness of the Universe.[47]

It is tempting to believe that the laws of Nature may have developed by a kind of trial and error from randomness: somewhat analogously to Darwinian Evolution. Perhaps physics should look for ratchets capable of creating the order of Natural laws. Perhaps, indeed, animate and inanimate – living and non-living – may be distinguished by the kinds of ratchets which produced them, and continue to protect and maintain them, as islands of improbability in the primordial sea of random chaos.

One of the greatest general insights of physics is that energy can appear in many forms and be converted from one form to another. In many cases how this may be done was discovered almost, if not quite, accidentally through the know-how of developing technology. This is so for the steam engine, though perhaps less so for electricity, as the ability to control electrical energy came very late.

Such considerations as these, as well as practical design problems for communication systems, make it important to measure information.

Measuring information

Modern attempts to measure information clearly have their roots in the technological need to provide efficient communication by telegraphy, and later with the telephone (invented by Alexander Graham Bell in 1876). Suggestions for telegraphs using magnetism or electricity go right back to Roger Bacon's suggestion of 1267, that a 'certain sympathetic needle might be used for distant communication', this being, of course, the lodestone. Chains of optical telegraph towers with moving arms were well established in England in the eighteenth century; messages between London and the fleet at Portsmouth arrived within twenty minutes; and there were many suggestions for electric telegraphs following Steven Grey's discovery in 1729 that electricity could be carried on insulated wires. By the 1830s, working single-wire telegraph systems using sounders with a code began to appear. Samuel Morse (1791–1872) turned from painting to building transmitting and receiving equipment, and to inventing his code (which he first demonstrated on Saturday 2 September 1837 at Washington Square, New York, over 1,700 feet of wire). The rate of transmission of messages soon became economically crucial, especially for the transatlantic cable. (This first operated in 1858, but broke down after only four hundred messages. The first successful cable, and the recovery and repair of the earlier cable, were accomplished by Isambard Brunel's huge ship, the Great Eastern, in 1865.) At first messages could only be sent at the uneconomic rate of eight words per minute. This was not a human limit of the senders and receivers, for automatic equipment was in use. The trouble was partly the low power of the signals (which led to the developments of relays and amplification), but, more serious, it was the Inductive inertia (low band width) of the cable.

The best scientists of the time – Michael Faraday, Charles Wheatstone, William Cooke, Lord Kelvin, and as a young man whose great inventive achievements came later, Thomas Alva Edison – contributed to this incredibly important technical development. Apart from the practical importance of rapid international communication, it focused attention on information as a concept and developed notions of channels and coding that generated an insight into all kinds of communication, including language, which deeply influenced philosophy itself, as well as providing tests of models for the transmission of messages in the nervous system.

The notion of a necessary band width (frequency range) for the transmission of signals was appreciated by Lord Kelvin from his work on the transatlantic cable; but it was not formulated until 1924 by H. Nyquist in America and K. Kümpfuller in Germany, who independently stated the law that was developed to its general form by R. V. L. Hartley in 1928, that the transmission of a 'given quantity of information' is limited by the product of band width and time. Hartley went further to define information as the successive selection of signs, or words, from a given list. For this definition he rejected meaning as subjective; for it is signals, not meanings, that are transmitted.

Hartley showed that a message of N distinguishable signs (such as letters or dots and dashes) selected from a repertoire of S signs has S^N possibilities; and that the 'quantity of information' is most reasonably and usefully defined in logarithmic units, to make information measures additive. Hartley quantified information thus:

$$H = N \log_2 S$$

Dennis Gabor[48] compared the uncertainty of messages with Heisenberg's uncertainty of wave mechanics. This was developed also by Donald MacKay. The point is that – considering for example sound – the perception is simultaneously of the time and the frequency of the wave. This requires some smallest unit of 'structural information', which Gabor called a 'logon'. It bears a mathematical similarity to wave packets in Quantum Mechanics. (Amusingly, Lucretius thought of speech sounds as different-shaped atoms!)

An essential notion for quantifying information is that the less likely the symbol or event, the more information it conveys. Information rate is not defined simply in terms of the number of symbols that can be transmitted, but also in terms of the probability (or the surprise value) of their occurrence. To modify a remark of Bertrand Russell's, 'Dog bites man' does not convey much information, but 'Man bites dog' is news.

Since the simplest choice is yes or no (or, on or off for a switch) the information unit of a bit (binary choice) is useful. Norbert Wiener and Claude Shannon[49] developed the Hartley approach, by examining the statistical characteristics of signals, including the values of waves for analogue signals. They reinterpreted Hartley's Law, to define the average information of long sequences of n symbols as:

$$\mathbf{H_n} = -\mathbf{p}_i \ \log \mathbf{p}_i$$

(The minus sign makes $\mathbf{H}_n$ positive, since it involves logarithms of $\mathbf{p}_i$ which is fractional.)

It is well worth playing the dictionary game, of finding a word with the 'Twenty Questions' technique of asking: 'Is it before K?' And if it is: 'Is it before E?' and so on. Any word in a dictionary can usually be located within twelve such binary decisions, which is quite remarkable. This is surprising because we are not used to thinking in terms of powers of two.

The number of binary choices enabling selection of, say, a word from a set of words increases as follows (with the values rounded off).

Binary decisions required	Size of dictionary from which any word can be found	
1	2	
2	4	
4	16	(1.6×10^1)
8	256	(2.6×10^2)
16	65,000	(6.5×10^4)
32	4,300,000,000	(4.3×10^9)
64	19,000,000,000,000,000,000	(1.9×10^{19})
128	340,000,000,000,000,000,000, 000,000,000,000,000,000	(3.4×10^{38})
256	160,000,000,000,000,000,000, 000,000,000,000,000,000, 000,000,000,000,000,000, 000,000,000,000,000,000, 000	(1.6×10^{77})

It seems incredible that for a few hundred decisions one could in principle locate any particle in the universe. This strategy must surely have implications for how we recall memories. It is not however possible to use the amazingly efficient 'Twenty Questions' technique except for items which are ordered in some way, such as words arranged alphabetically in dictionaries, parts in stores, or possibly memories filed in some orderly way in the brain.

The greatest quantity of information that can be transmitted through a channel, with band width W over time T, in the presence of disturbing random noise, was shown by Claude Shannon to be:

$$WT \log_2 \left(1 + \frac{P}{D}\right) \text{ (bits)}$$

where P and D are mean signals, and mean noise powers. This represents a definite limit, which no channel can exceed. If, however, the *coding* of the signals is non-optimal the information rate may be very much lower.[50] It is interesting that Shannon and also Hartley produced their powerful theoretical formulations at the Bell Telephone Laboratories, somewhat after the general feel of the problems had been established by practical experience with telegraph and telephone systems. The formulations are useful for setting design limits to what is possible.

There is something odd about information, as described by Shannon's theory, which is now universally accepted as the best account; for information is quite different from anything in the Natural Sciences as it depends, not only on what has been and what is, but also on alternatives of what might be. This is so, although Information Theory gives no account of meaning. The meaning of 'meaning' remains almost as mysterious as ever, for Information Theory is limited to the statistics of signals, and transmission characteristics of channels, and not at all with understanding messages. It applies, if you like, to ignorant though intelligent Demons, who know nothing except the number of possibilities and the relative probabilities of signals. They know nothing of our world in which messages have meanings; and messages would be as meaningless to us, if we were as ignorant as Demons knowing only statistics.

Nevertheless, information is useful at least for describing the transmission of neural signals in the peripheral nervous system and the transducer characteristics of the sense organs. Also, it makes measurements of efficiency in decision-taking possible in some cases. There is an annoying difficulty here though, for to measure the amount of information we need to know not only the number of possibilities that can physically occur (for example that a door will open either inwards or outwards) but also the number of possibilities entertained by the organism, or by the decision-taking computer. Thus a door may be seen as possibly sliding, right or left, though it can physically open only inwards or outwards. If these extra possibilities are considered, the information processed is greater, though there is no change in how the door opens. This is clear from combination locks: they are effective because there are many alternatives, and it is costly in time to try out so many possibilities. It is even more costly if the safe-breaker tries out still further possibilities – including, perhaps, magic. Knowledge and theories have the power to limit possibilities, and so to reduce the number of choices to be tried in a given situation. But unfortunately we seldom if ever know just what are the restraints set by knowledge, theories or assumptions – so Information Theory can seldom be rigorously applied, outside purely engineering situations where we have full knowledge of the system.

A law was, however, established at Cambridge by Edmund Hick (1952), showing that where there are clearly defined choices in a human skill (pressing response keys to lights), the decision time t increases with the number N of the possibilities, by

$$t = K \log (N + 1)$$

The choice or decision time (the disjunctive reaction time minus the simple reaction time) is very nearly proportional to the number of bits per stimulus (where a 'bit' is a binary choice). This relation was also found in 1953 by R. Hyman, who showed that with each added possibility the choice time increases by just over a tenth of a second, and that this same increase in choice occurs no matter how the information is increased – whether by changing the relative frequencies of the alternatives or their sequential dependencies (introducing redundancy so that some tended to follow others in predictable sequences), or by increasing the number of possibilities. The maximum rate of information (bit rate) of even the most skilled human operator is surprisingly small: about 22 bits per second for an expert pianist. Speech does not exceed about 26 bits, and silent reading possibly reaches 44 bits per second. This is very low in engineering terms, where bit rates of thousands per second may be achieved in choice-making computer circuits. It is however important to note that the organism is not truly limited to the choices set by the experiment, or by the particular situation or task. Thus in Hick's experiment the subject could respond to a knock on the door: he was not deaf or blind to all except the alternative lights of the experiment. So his total bit rate is greater than measured; and it is very difficult to assess it with any accuracy because it is set by hypothetical alternatives to which he may respond, and not simply by the given alternatives of the experiment, or task situation, to which he does respond and which are recorded. No doubt attention serves to limit hypothetical possibilities, and so to speed decision-making. A totally attending individual would be a Demon![51]

An early and still important experimental treatment of Information Theory applied to continuous human performance, tracking a moving target with and without preview is a paper by Ted Crossman (1960). An essential point of the theory is that since the world and human limbs have inertia, things cannot change instantaneously, and so only limited information is required for perfect performance. All the information that can be extracted from a variable is given by sampling at twice the period of its highest frequency or band width. It turns out that the information rate for human skill of this type is not greater than 10 bits per second. It is worth noting that information measures of this kind are valuable for assessing neurological losses, for example in Parkinson's disease. Work has recently been done on this in my laboratory by Ken Flowers (1967), who found that Parkinson patients lose the power to predict their limb positions; so they can only move with 'closed-loop' servo-control, and not with the 'ballistic' predictive commands to the limbs which are essential for fast, accurate skills. I suspect that although these ideas were around twenty years ago they have not been developed in neurology as much as they might have been, and that much that remains intuitive and vague could be quantified for establishing key features of perception and behaviour, and assessing precisely with objective measures the bases of impairment through disease or injury.

In a much-quoted paper bearing the delightful title, 'The magic number seven plus or minus two', the American psychologist George Miller (1956) suggested that there is an absolute limit, of about seven bits, for the immediate span of apprehension for single stimulus dimensions. Thus we can estimate at a glance, without counting, up to about seven dots spaced randomly. But the *effective* information in a perceptual span (the 'specious present') can be greatly increased by 'chunking' into larger units.[52] This is a form of coding of data, requiring decoding, of course. The most powerful coding appears to be language. Coding can set the number of bits per chunk; but the number of chunks that can be retained in immediate memory is limited to around seven. Presumably Chinese ideogram characters are efficient chunks each conveying a lot of information; but such chunking has the disadvantage that it is inflexible and cannot readily be applied to a wide variety of situations.

Much of learning is chunking bits of information into large units which can be stored in memory.

One might say that hypotheses of science serve as chunks for conveying large amounts of information economically, and that object perception is the chunking of bits of sensory information so that we *see* objects.

Measuring meaning

The intelligent Demon is a kind of tennis-player – but not a philosopher for he knows nothing of the world and derives no meaning from his information of the velocities of the particles approaching him. The Demon is aware only of a random world of particles, so for him there are no laws of regularity to discover. His predictions can only be for individual particles. No meaning can be found in such a Universe.

It is commonly said that Information Theory has nothing to say about meaning, but this is not entirely true. The point is made very clearly by Donald MacKay, in a paper entitled 'Meaning and Mechanism' (1969). This suggests that meaning is related to conditional readiness. The meaning of a sentence (or perceived event) changes the pattern of possibilities for future action. Donald MacKay gives the following working definition of meaning, as the[53]

selective function on the range of the recipient's states of conditional readiness for goal-directed activity; so the meaning of a message to you is its selective function on the range of your states of conditional readiness.

Defined in this way, meaning is clearly a relationship between message and recipient rather than a unique property of the message alone.

If we knew the selective function of the message we could apply Information Theory to quantify meaning. This might be possible in some restricted cases. MacKay goes on, in another paper, 'What makes a question?' (1969), to suggest that states of readiness are for organisms large numbers of conditional probabilities. Asking a question is a means of changing the conditional probabilities of the questioner's states of readiness.

There is nothing in this account demanding more than computer programming, and interrogating computers. If there is a 'problem of meaning', the problem lies beyond the adequacy of this notion of selections from or changes of states of readiness. These can be described in computer terms, and in principle at least can be quantified with Information Theory.

It is sometimes said that true statements have more meaning than false statements. And of course philosophers frequently deny meaning to logical, and even to contingent, impossibilities. Logical impossibilities are internal inconsistencies of the states of readiness. Thus '$2 + 2 = 5$', or 'She is a blond brunette', do not set up new states of readiness: rather, they stop the program working by setting up internal inconsistencies. It is an interesting question how general are the contradictory program-stoppers which we call 'logical' errors. Perhaps only ($p = \bar{p}$) is self-contradictory for *any* program or logical system.

Looking now at factual or contingent errors, these are (on a correspondence account of truth) mis-matches with the world. They are dislocations of reference to things – objects, events, or generalizations – lying outside the program, the logical system, or the Mind. What then happens to meaning when there are reference mis-matches to make a statement, or a perception, false? Should we say that all meaning is lost? Here there is a wide range of opinion. Logical Positivists tend to say that false statements have meaning provided we can imagine, that is, provided there are available practical states of readiness for applying match–mis-match tests. But it is not clear how immediate or direct such tests need to be – and they are often highly indirect in science – to confer meaning. Even more fundamental: it does not seem at all clear that verification tests (or, as I would rather say, 'match–mis-match tests') can be designed to apply generally.

It does seem odd to suppose that there can be adequate and always appropriate criteria for verification of meaning, for they would have to have far more generality than the range of available theories, or the existing paradigms of knowledge.

How can we expect criteria for meaning to be far more general than our cognitive mapping of the world? Would not such a claim be too like Kant's *a priori* categories of space and time, which he supposed limited our understanding – to set *a priori* limits to empirical meaning? Attempts to set limits to meaning by very general verification criteria seem to me to let in apriorism through the back door, while it is rejected at the front door by demanding operational procedures for meaningful statements. In short: I fail to see how there can be non-cognitive filters generally appropriate for legislating on the limits of meaning.

There are many senses of the word 'meaning'. Perhaps the best account of these is still C.K. Ogden and I.A. Richards, *The Meaning of Meaning* (1944). In Chapter 8, which bears the title of the book, there is a classified list of fourteen definitions of meaning which have been or are used. These include: meaning regarded by some philosophers as an intrinsic property which cannot be analysed; the meaning of words given in dictionaries; implications, such as practical consequences; emotional reactions; and what the user of a symbol does, or ought to or believes he is referring to. Perhaps we can hardly expect the same kind of measures of meaning to apply to 'meaning' in all these senses. The first (that meaning is an unanalysable and in-principle mysterious relation) we may however ignore, for this is mere pessimism. But the others are very different from each other; for some refer to the world and others to various kinds of states of the symbol user. For our purposes we might do well to ignore all senses of 'meaning' except those which could refer to states of readiness. For example, 'What is the meaning of life?' would be deemed to have no meaning in this context, if the question produces no changes in states of readiness. But, of course, it might: such a question may drastically change a person's life – perhaps leading to his taking up religion, or drink. Somewhat similarly, poetry and music may change states of readiness, and so they lie within our criteria for 'meaning'.

This criterion is broader than verification principles for Logical Positivism, because states of readiness are not dependent upon states of the world – they are not dependent on match or mis-match with the world. This criterion of meaning is not concerned with truth. Clearly false statements can be meaningful, and if we knew enough about the cognitive structure of possibilities and conditional probabilities we should indeed be able to measure meaning for both true and false statements and for veridical and illusory perceptions (see chapters 13 and 20).

Biological atoms: cells

The ancient Sumerians learned how to build enormous buildings with arrangements of identical bricks. The analogy with the cells of living organisms is not difficult to see. As William Coleman (1971) puts it:[54]

. . . the cell, while always an architectural element of prime importance, is also the critical unit of organic function above the molecular level. The cell is thus the site of metabolism and energy exchange; it is the basis of nervous and secretory activity and therefore the foundation of harmonious, integrative, organic behaviour; the cell, as manifested in the reproductive products, ensures, finally, the very continuity of life across the generations.

Knowledge of the cell came late in the history of biology; for it depended upon the microscope. Its importance was first recognized by Matthias Jacob Schleiden (1804–81) and Theodor Schwann (1810–82), who identified cells as the key structures of life. They were principally concerned with the development of cells, which they at first thought occurred in amorphous material outside the existing cells. Cell division was discovered gradually and with it the ancient notion, that organisms were preformed in miniature in the first stage of the embryo, was rejected. This was gradually replaced in the early nineteenth century by the concept of epigenesis (development of increasing complexity from more or less homogeneous material). It is interesting that several microscopists apparently saw fully-formed miniature adults in the single egg cell before epigenesis was accepted. With this notion, it at last became clear that the sperm cell did not merely stimulate the female egg cell into growth. The essential details of division (mitosis) of the cell nucleus and the exact sharing of the chromosomes (named for their colours when stained) was not appreciated until the 1930s. The chromosomes were not related to specific features (such as wing patterns of *Drosophila*) until after the *rediscovery* of Mendel's Laws of Inheritance in 1900.

The accounts of embryological development paralleled the three views of the creation of species: that they are preformed in the egg; that they develop by a built-in design 'unfolding'; or that they develop by self-creating mechanistic processes. The latter is the current view, though much of course remains to be understood.

The parallel between unfolding by supposed vitalistic controlling forces of embryo and species development, was very close to the dictum of the Vitalist Ernst Heakel (1834–1919), which had wide appeal in the nineteenth century, that

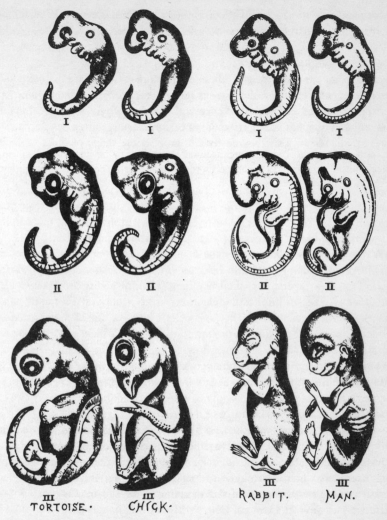

25 Embryological development in different species of vertebrates. They are hardly distinguishable at the early stages.(George John Romanes, after Ernst Haeckel, 1892; from William Coleman, *Biology in the Nineteenth Century*, Cambridge University Press, 1971.)

'Ontogeny is the brief and rapid recapitulation of phylogeny'. The notion was that the stages of development of the embryos of the higher animals mirror the sequence of species development: a view sometimes borne out by gross observation, but not by the details of embryological development. It is now regarded as a highly misleading concept.

For a recent general account of current notions of embryological development, we may turn again to Jacques Monod's *Chance and Necessity* (1972).

Are cells intelligent Demons?

Organisms are now regarded as 'chemical machines' which are maintained by a certain class of proteins, with enzymes serving as specific catalysts. Some 'regulatory' proteins act as detectors of chemical signals, recognizing and calling upon critically important molecules (including other proteins) during development and during metabolism. This they do by recognizing the shape of other molecules. Proteins are very large molecules, ranging in molecular weight from 10,000 to 1,000,000. They are made up from 100 to 10,000 amino acid residues, which are themselves limited to only about twenty different chemical types. So the enormous macroscopic diversity of organisms is given by a very small variety of microscopic structures. This is reminiscent of the great variety of models that one can make from the limited variety of parts of a Meccano set, or buildings from bricks.

While life persists there are syntheses of complex from simpler molecules: a vast increase of complexity and information in organisms flowing against the general run-down, perhaps of the Universe itself. The ability of life molecules to recognize is a key to life, to reproduction, to production of successful novelty against inertness and death. Then the cells of living organisms are cognitive Demons, for a time defying corruption, until they lose their grip, to drown and dissolve as eddies in the blind down-flowing cataract of the Second Law – to join with the Devil incarnate the Natural world of normal physics which loses a trick only to cognitive processes that can buy information with energy.

The *Origin of Species:* Charles Darwin [55]

In all mythologies men were created by the Gods, more or less in their image. Animals were separately created, together with seas, lakes and mountains.

Many writers before Charles Darwin (1809–82), and many ancient mythologies, considered transmutation of species as occurring, but generally by some inbuilt tendency to change in certain directions, and this idea is today not quite dead. On this view the Creator – God is still creating; or at least, His original plan is still unfolding. French writers seem particularly to have emphasized continuous directed change. The highly distinguished French naturalist, Chevalier Jean Baptiste Pierre Antoine de Monet Lamarck (1744–1829),[56] broke with the ancient tradition of unchanging species – to deny that there are species. His *Philosophie zoologique* (1809) was especially well known, and his *Histoire des animaux sans vertèbres* (appearing 1815–22) established him as a first-rate naturalist. This is worth pointing out, for to us his views may appear 'unscientific'. Having denied species, genera, orders, or families of plants or animals, he suggested that 'There exists throughout the Universe an astonishing activity which no cause can diminish, and everything which exists seems constantly subject to necessary change.' As Coleman (1971) puts it:[57] 'All Existence is nothing other than eternal becoming, a vast flux of evanescent qualities and unstable entities. Even classificatory categories dissolve before the insistent dynamism of nature.' This sums up also much of the thinking of the French philosopher Henri Bergson (1859–1941),[58] especially in his *Creative Evolution* (1907), and indeed this thought can be taken back to Heraclitus, who argued that change is the stuff of reality. Bergson introduced the notion of a directing life force – *'élan vitale'* – in his form of Vitalism. It is this which is essentially different from Darwin's account – which is given in terms of unguided statistical processes. For a long time suspicion of statistical concepts held back acceptance of Darwin's theory, in part because applied statistics has always been dubious – especially when applied by governments! Also, Darwin lacked any physical explanation of the random jumps of mutations, and he knew nothing of genetic inheritance. Mutations are now attributed largely to the continual bombardment by cosmic rays modifying gene structures. Most modifications are harmful, but an occasional change is beneficial to inheritance.

Whether or not individual memory from learning is transmitted genetically is a very different question. Here again Lamarck appears as a principal figure. Lamarck held that gene structure can be modified by individual experience. In Darwin's account the individual is sadly unimportant. Myths of man's importance, not surprisingly, still wield power – and indeed, if we are not important what is?

Darwin did not attempt to explain the origins of the Universe, or of matter: he took the inanimate world as given. His great achievement was to give a coherent account, with almost compelling evidence, for the creation of life-forms (species) by a trial-and-error process which did not suppose preconceived design, and so did not require a God–Creator.

Darwin saw the way in which species were distributed in different regions of the world as evidence against each species being specifically created by a God. Isolated islands and the isolated land mass of Australia were found to have their own special varieties, and even clearly unique species. It seemed too quixotic of a God to place particular species in special places. (So the argument really depends on a psychological assessment of God.) With the new knowledge of fossil sequences found in layers that gave relative dating, it became clear that life-forms changed in geological time. It was also clear that geological time was a great deal longer than had been imagined.[59] There was strong opposition to these conclusions – and of course all the arguments depended on assumptions that *could* be questioned. It could be countered that the Devil had planted fossils in the rocks, to mislead us, much as a criminal might plant misleading clues – and what answer is there to this? One can only say that it does not seem at all likely; but it must be confessed that there have been historical times when such a suggestion would have seemed more likely than Darwin's theory. It is indeed fascinating to read the contemporaries of Darwin who found his Natural Selection account implausible to the extent of being frankly inconceivable.

The notion of evolution of life-forms was by no means new in Darwin's time; indeed, his grandfather, Erasmus Darwin (1731–1802), had considered it.[60] What was original to Charles Darwin was blind forces generating successful species, and the idea that these forces are not physical forces in the usual sense and yet are acceptable to Physical Science. We would now describe these directing forces as statistical; but statistical notions came astonishingly late, being little more than vestigial in 1858, when *Origin of Species* was written. Darwin's statistical concept itself had an origin: as is well known, Darwin was inspired by a central argument of the English economist Thomas Malthus's *Essay on the Principle of Population* (1798).[61] The essential notion is that populations tend to expand faster than their means of subsistence – so the weak tend to die while the strong survive. Malthus did not question that 'Being who first arranged the system of the Universe', but he did doubt, perhaps under the influence of a depressing stage of the Industrial Revolution, that human social progress is possible. He did not see his concept as a creation principle. He saw it as remorseless laws of Nature restraining life and preventing human societies from improving: '. . . that imperious all-pervading law of nature, restrains [life] within the prescribed bounds. The race of plants, and the race of animals shrink under this great restrictive law. And the race of man cannot, by any efforts of reason, escape from it.'

Malthus did not see the struggle for survival as generating improvement. He saw it rather as producing sin and misery, from which man must die: he was a profound pessimist. Darwin (as he recalled in October 1838) was struck with the thought that under conditions of struggle for survival, 'favourable conditions would tend to be preserved, and unfavourable ones to be destroyed'. This tendency to improvement (whatever this might mean) is not in Malthus. The *Essay on Population* was a trigger, and little more to Darwin; and later to Alfred Russell Wallace (1823–1913),[62] who had the same idea, also from reading Malthus, in 1858.

Darwin was well aware that selection occurred in animal and plant breeding to improve species. He saw that the key to giving direction to a breed is 'man's power of cumulative selection; nature gives successive variations; man adds them up in various directions useful to him'. He goes on to ask whether this Unnatural selection operates in Nature: 'Can the principle of selection, which we have seen is so potent in the hands of man, apply to nature?' He sought, and found in competition as described by Malthus, Natural directive forces to replace the intentions and knowledge of human breeders. This was the decisive step for rejecting predesign and intention (and so an effective God) from the creation of species.

Charles Darwin was not content to rely on analogy or on formal argument from general principles: he took immense trouble to seek evidence for how the principles he regarded as the key worked in practice. This intense study fed back, to give evidence that the principles worked. Perhaps his most clear-cut evidence was from the diversity of plants and animals on the Galapagos Archipelago which he studied during his voyage on the *Beagle* (1831–6).

Darwin's emphasis on survival against competition as the means of improvement has clear and not altogether palatable social implications. Or should we, perhaps, define civilization as shared efforts to reduce competition? One might well think that medicine is calculated to preserve the unfit, and so in the long run to undo some of the tens of thousands of years of human species-improvement by Darwinian selection of the fittest, which is the Natural state of affairs. Although one might think that social mores serve to protect the weak, the argument is not so simple, for in a technological society individuals unable to defend themselves in primitive conditions can be important for the survival of the group. Such considerations led Darwin's friend, defender and expounder of Evolution – Thomas Henry Huxley (1825–95)[63] – to suggest that organized social groups rather than individuals are the units of Evolution. It is interesting that the opposite tack has been taken recently, dramatically described by Richard Dawkins in *The Selfish Gene* (1976), which argues that the gene is the unit of survival, the organism being merely a kind of gene-protecting robot. It is, in any case, an open question how long civilization can reverse Natural Selection without unfortunate consequences. We are almost certainly breeding ourselves in new directions, which may or may not be compatible with the kind of civilization that we plan – or that will occur in spite of our individual or political intentions.

A key problem is altruism: why do individuals risk sacrificing themselves for the benefit of other individuals, usually of their own species? How can altruistic behaviour have developed by Natural Selection? Huxley's suggestion that the group is the evolutionary unit is one possible answer, at least for gregarious species. Dawkins's very different suggestion, that genes are the evolutionary units, and that although individual genes (molecules) may be destroyed by the altruism of their owners, the similar genes of related individuals will be protected, seems to me to have severe difficulties. This is a question beyond the scope of this book and beyond my competence. It does, though, seem to make the unit not individual genes but classes of similar genes; and this seems to me logically inappropriate. Or if it is individual genes that are supposed to be protecting themselves, this seems too vitalistic for molecules. Is the unit of selection social groups, individual organisms, or the molecules of genes? Perhaps there are several selection units, much as in the competitive societies that Malthus considered.

But perhaps I am interpreting Dawkins too literally. The situation has recently been put by J. T. Bonner (1980) in these words:[64]

. . . While the immediate object of natural selection is the phenotype, the ultimate object is the gene, and in this sense genes are 'selfish'; each out for itself. But . . . one of the most effective ways genes can perpetuate themselves is by having the closely related survival machines that house them helpful and altruistic to one another. The reason for this is obvious: the genes are not selfish as individual molecules, but as a class of identical copies capable of replication in successive generations. Those genes that do make more than their share of copies are necessarily more numerous in the next generation; this is the sense in which they are selfish.

So it is not individual molecules that are selfish; rather, there are strategies by which some molecules (genes) replicate more than others, and one of these strategies is altruism between individuals bearing similar genes. This kind of thinking has been developed by W. D. Hamilton (1964), R. L. Trivers (1971), E. O. Wilson (1975) and John Maynard-Smith (1975, 1977, 1978).

Neo-Darwinism adds to Darwinian Natural Selection a theory of heredity, which is itself derived from the, at the time (and perhaps still), controversial writings of the German biologist August Weismann (1834–1914). His papers (1868–76, translated into English as *Studies in the Theory of Descent*, 1882), proposed properties of a germ plasm which are similar to the fundamental doctrine of molecular biology, that information can only genetically pass from coded DNA to messenger RNA, and not the other way round. This genetic 'diode' rejects Lamarckian inheritance of individually acquired knowledge, or adaptive behaviour. But we jump ahead, for Darwin had no knowledge of genes or mutations of genes.

The concept of evolution by mutational jumps is due to a Swiss botanist, Karl Wilhelm Nageli (1817–91). Nageli however rejected Darwin's theory, for he supposed that there is purpose in the direction of the jumps. He is heavily criticized for failing to appreciate the significance of Mendel's work. He was shown the manuscript of Mendel's paper describing his experiments on the breeding of giant and dwarf peas; his lack of interest is supposed to have prevented the work becoming known so that genetics was held up for some fifty years. Nageli's concept of mutational jumps, but without built-in directional purpose, was developed by De Vreis early in the present century.

Gregor Johan Mendel (1822–84) was an Augustinian monk. At the Abbey of St Thomas in Brunn, he carried out his plant-breeding experiments, which depended on counting the proportions of tall and dwarf peas obtained by self-pollination. He found that the varieties did not converge to a medium-height pea plant; but that the tall and dwarf characteristics were maintained, and potentially present, in each variety. This was immensely important for Darwin's theory, but unfortunately Darwin never came to hear of it.

The mutation theory was developed by the Dutch botanist Hugo De Vreis (1848–1935) who approached Mendel's discovery by seeing that something like it was needed to give the variation necessary for Natural Selection. He proposed that different characteristics might vary independently, and recombine in different ways. So was born the atomic-characteristic theory of inheritance, which later was embodied in gene and chromosome code structures – from which in turn developed modern molecular biology with the discovery by Francis Crick (b. 1916) and James Watson (b. 1928) of the structure of the long helical molecules of deoxyribonucleic acid (DNA). This, by replication, gives the physical basis of inheritance. Random changes of the DNA structure give the variation necessary for Natural Selection. The drama of this discovery is superbly presented by Watson in *The Double Helix* (1968).

Darwinian Evolution supposes that organization increases in biological time through processes of random variation and selection of the fittest to breed and survive. How is Entropy reversed by processes of Darwinian Evolution? Why does this not violate the Second Law? The problem would not arise if the Vitalists were right in thinking that life-forms are replications, or embodiments, of existing designs – that these designs exist in God's mind, or are somehow unfolded from a pre-existing design. The problems arise when we think of Evolution as the organization of species increasing by random processes, which is the basis of the neo-Darwinian theory of evolution.

We seem to be faced with a confrontation between on the one hand the Second Law and Entropy, stating that organization always *degenerates* in time, and on the other Darwinian Evolution, saying that the organization of species has been *increasing* through biological time. Since the Second Law is more general than the Darwinian principle of Evolution through random variation and selection, we might be tempted to say that this consideration refutes the Darwinian theory. (This is indeed a good example to show how difficult it is to apply the refutation notion of arriving at knowledge: for how do we know, or agree to accept, that counter-evidence or paradox is adequate for refutation? See chapter 8.)

The question now is: how can Darwinian Evolution be saved from 'refutation' by the charge that, if true, it would be violating the Second Law? The key to this lies in a statistical characteristic of mutations of genes. (The gene theory of inheritance was unknown to Darwin: he would have had no answer to this conflict with the Second Law.) Although each individual gene mutation occurs by chance, and can be reversed, it is extremely unlikely for a *combination* – representing a characteristic or a new variety – to reverse. New forms survive provided they are maintained by successful offspring. The direction of change is given mainly by which varieties breed most successfully.

This shows how apparently purposive inventive design can occur from undirected random variation, when modifications are tested and rejected or accepted by success criteria. A question for psychology is whether thinking – problem-solving and inventing – is essentially blind. Is thinking 'blind' as organic Evolution is essentially blind, and essentially unpredictable? Is there, indeed, anything equivalent to genes, in thinking? Could we, even, suppose that *concepts* are, in this sense, 'genes' – by having the property that although they can be rejected or refuted they are generally not quite reversible, and so they generally grow? The Evolution of life-forms (and technology and science) seems to move in ratchet-like steps that we call progress; though it may be difficult to say just why man is 'better' than the amoeba. This depends on the viewpoint; and from any view the amoeba may well be judged superior to man in some characteristics. For the philosopher, man's principal advantage is that we can discuss such issues while the amoeba remains silent. There are some, however, who might rate the amoeba superior for its silence.[65]

This question of 'success', and 'progress', and 'fittest' biologically and socially, has never been satisfactorily answered. As a result there is an essential circularity in Darwin's Theory of Evolution. But perhaps, ultimately, all scientific theories are circular – without necessarily making them useless or untrue.[66]

Darwin made no attempt to explain the origin of life – how life may have derived from non-living matter by Natural rather than divine processes. Neither did he consider the origin of the Universe. A complete account of life and Mind cannot ignore these questions.

The origin of the Universe

There are three main cosmologies. Each implies a different origin for Mind. These are: (1) that the Universe was created by a Divine Being who preceded matter and the physical laws; (2) that the Universe as we know it was created according to physical laws (though rapidly modified during the first few minutes by a Big Bang from a small, incredibly dense concentration of proto-matter; and (3) that the Universe has always existed, essentially in the form it has now, as there is continuous creation and corresponding loss of matter, producing a statistical steady state.

The first hypothesis implies that Mind existed before Matter, and therefore is not essentially dependent upon Matter; and *ipso facto* Mind is not dependent on brain – for the Creator would be brainless though not Mindless.

The second hypothesis gives no account of any events before the Big Bang. It may suggest that 'time' was meaningless as a word, or concept, before there were physical events; though conceivably time may be absolute, and not dependent upon events, in which case time may be said to have existed before the Big Bang origin of the Universe as we know it (as suggested by Isaac Barrow: see pp. 134–5).

The third hypothesis allows Mind always to have existed, much as it is now, in the Universe, though of course we know of Mind only on earth. The evolution of Mind would be a local phenomenon, and presumably would not be unique.

This last hypothesis, Steady State Continuous Creation and loss of matter, is to my mind by far the most elegant and intellectually satisfying. It was propounded in the early 1950s by Fred Hoyle, Hermann Bondi and Thomas Gold (see Hoyle, 1975, and Narlikar, 1977, for excellent technical accounts available to the non-specialist). The Steady State Theory can be tested by looking back in time along 'the light cone', given by the Doppler Shift which indicates the age of emitted light, assuming constant expansion of the observed Universe. It turns out that there are changes of structural features in cosmological time, so that extremely large-scale invariance, which we do not observe, would have to be supposed to save the theory. Also, the background of radiation, in the radio-frequency band, is generally interpreted as the remains of the Big Bang. An alternative might be a very large number of small-power, individually unobserved radio sources; but this is not an appealing notion. So at present this beautiful idea of a Universe essentially invariant in space and time is discounted, in favour of some kind of origin in time – a Big Bang.

The current Big Bang notion is beautifully described by Steven Weinberg (1977) in *The First Three Minutes*. This is an evolutionary account in which most of the action takes place in a few seconds of our time, with a total prior mystery.

On the notion of creation by an Intelligent Being – who may have been always present or arrived with a mental Bang – we can say very little; except that this implies Mind independent of Matter; and capable not only of fashioning Matter, as in crafts and technologies, but able also to create Matter by, and perhaps from, Mind.

It would not follow that *our* Minds can produce – create – Matter; though we can convert Matter into energy, through the relation $E = mc^2$, which is of course the basis of atomic energy. Our environment is affected in astonishing detail by the designing powers of human Minds – even to the point of footsteps of men printed on the Moon – but this is a far cry from Mind creating the Universe.

The origin of life by Natural means

From Thales onwards, the prevailing opinion has been that at least simple life-forms emerge from water; and the frequent appearance of larvae and worms and so on, of various kinds, must have perpetuated this belief with apparently the soundest evidence of seeing it happen. Aristotle was convinced of the Spontaneous Creation of life. Thus he tells us that bees can be spontaneously created (and we learn that stinging bees are not female; for Nature does not give offensive weapons to females! Aristotle concludes here that the workers are neither male nor female – as they sting, which is un-female, and they look after the young, which is un-male).[67] Aristotle considers in one passage that men may be created from earth, though usually he says that man evolved from other species.[68]

With the invention in the seventeenth century of the microscope, new life-forms were discovered – which enriched theories of Spontaneous Generation. Thus the great Dutch amateur microscopist, Anthon van Leeuwenhoek (1632–1723),[69] believed that the micro-organisms he discovered were created spontaneously in the air. This appeared to be confirmed experimentally, for even after boiling in closed vessels (an appropriate procedure) the air, or cultures, would rapidly become contaminated by air-borne organisms, and so confound the results. The brilliant work of the French chemist Louis Pasteur (1822–95)[70] finally demonstrated that there are germs in the air. He showed this, by placing a plug of gun cotton in a tube through which air was sucked into a flask. It turned out that no organisms were found in the flask – so that they must have been filtered by the gun cotton. This was confirmed by showing that when the gun cotton was dissolved in a mixture of ether and alcohol, thousands of organisms were seen in this fluid. Pasteur went on, of course, to show that the germs in the air can be causes of infection. He started with filter plugs, but finally developed his own method: S-shaped necks on the flasks, which trapped the germs though they were open to the air. It was his discovery of how difficult it is to prevent micro-organisms falling on surfaces, or entering flasks, that led to methods of sterilization, and finally to inoculation against some diseases.

Pasteur finally showed that there is no spontaneous generation of life. Of course, strictly, this could never be shown by experiment; but at least Pasteur demonstrated, with classically beautiful and simple experiments, that spontaneous generation did not occur on occasions when contamination was ruled out. It might however be added that when organisms did appear in these conditions (as still sometimes happens even in modern laboratories), it had to be assumed that some slip-up had occurred in the experiment. Since Pasteur (in spite of lingering notions of the entelechy of matter striving to become alive), it is now universally accepted that inanimate matter does not change spontaneously to living forms, even to the simplest organisms, let alone to bees or men. How, then, did life start?

The traditional account is of course that life was initially created by a Divine Intelligent Being; but this only pushes back the question to: 'How was this Being created?' And it takes the issue out of science into Divinity. Creation of human artefacts by intelligent designers and craftsmen is so familiar that it is hardly surprising if, throughout history, this notion has been projected into the creation of life by Divine Designer–Craftsmen. With Darwinian Evolution we now accept that intelligent designs can be created without an intelligent Creator, having look-ahead and purpose. This has, of course, had a profound effect on how we see our origins and our place in Nature.

The question here is: did life originate by some Divine intervention, or miracle; or did it arise according to natural processes which might be explained within the concepts of science? If the origin of life is seen as some kind of mutation from inanimate matter, we are still of course faced with the problem of the origin of matter – and of the laws of Nature. Here the Steady State cosmology of Fred Hoyle, Hermann Bondi and Thomas Gold is surely most appealing; though current evidence seems to point rather to a Big Bang origin, at least for what is observable to us now in the distribution of galaxies (Narlikar, 1977).

Recent ideas of how life might have originated from inanimate matter, according to the laws of physics, start with the ideas of the Russian scientist A. Oparin, in the 1920s. These ideas received support at the time from the British biologist J. B. S. Haldane. Oparin's book, *Origin of Life* (English translation, 1953), set the scene for remarkable experiments which were carried out in America, on synthesizing biochemical building-blocks of life in simulated proto-atmospheres of the earth of some three billion years ago. The first of these experiments was carried out by S. L. Miller. Miller (1959) concocted his primeval atmospheres, essentially lacking oxygen, in large glass flasks, with some water kept boiling to maintain circulation. The brew was kept on the boil for several days, while it was activated with electric sparks. Later experiments by S. W. Fox (1965) used ultraviolet light to provide the energy, in place of Miller's artificial lightning.

In an atmosphere mainly of hydrogen, methane and ammonia, and in the absence of free oxygen, it was found that hydrogen cyanide and aldehydes are formed, at the expense of ammonia. Adding sulphur, A. T. Wilson (1960) in this way produced sheets of solids as large as coins, which formed on the gas-liquid surfaces. The sulphur evidently served as a catalyst. Later simulations of conditions to be expected before life on earth, produced huge molecules with molecular weights up to 300,000. These contained 18 out of the 23 amino acids common in present living organisms. The peptide chains produced were ordered in lifelike ways.

It is worth pointing out that these remarkable experiments were initiated by pure hunch, with no theoretical basis for believing that they would work. And they were not set up to refute any existing theory or hypothesis. They have been repeated many times since, in laboratories in several countries with similar results, which are generally accepted; though of course the interpretation and implications remain open to question.

Of equal interest – though having less reliability – are the claims that not only organic molecules but even complex life-forms, are found as micro-fossils in meteorites. The claim is that these are life-forms of extra-terrestrial origin. The claim depends on showing that no contamination has taken place, either before the meteorites are examined (and many have been lying around in museums), or in the laboratories where they are sectioned and examined. Criticisms have been made of contamination by rag-weed pollen and other air-borne organisms. Only certain types of meteorite are suitable, and these are rare. In these metallic meteorites, the outer layers boil off in the earth's atmosphere, which keeps them cool exactly as for re-entering space vehicles. It is found that carbon and gypsum are preserved without the changes which occur beyond temperatures compatible with life; so it does seem possible that living structures could enter the atmosphere to seed life on earth – in which case we could, in our origin, be extra-terrestrial Minds.

The evolutions of men and women

Darwin discusses the Evolution of man from ancestral species, and the possibility of different selection, and therefore innate differences (polymorphism) between the sexes – in *The Descent of Man and Selection in Relation to Sex* (1871). (I shall refer to the revised edition of 1888.) Darwin's *Expression of the Emotions in Man and Animals* (1872) is also highly relevant, and an extremely interesting book.

The Descent of Man starts by pointing out general similarities between man and other animals – even in some characteristic similarities to insects – and especially how similar are their diseases and the effects of drugs, alcohol, and so on. This of course is the reason why veterinary studies are so important, indeed indispensable, for medicine.

Darwin is impressed throughout with sex differences in mammals. For example, speaking of vocal organs he writes:[71] 'almost all male animals use their voices much more during the rutting season than at any other time; and some, as the giraffe and the porcupine, are said to be completely mute excepting at this season'. Darwin is puzzled by this difference:

As the case stands, the loud voice of the stag during the breeding-season does not seem to be of any special service to him, either during his courtship or battles, or in any other way. But may we not believe that the frequent use of the voice, under the strong excitement of love, jealousy, and rage, continued during many generations, may at last have produced an inherited effect on the vocal organs of the stag, as well as of other male animals?

This is taken up for humans:[72]

Voice and Musical Powers – In some species of quadrumana there is a great difference between the adult sexes, in the power of their voices and in the development of the vocal organs; and man appears to have inherited this difference from his early progenitors. His

vocal cords are about one-third longer than in woman, or than in boys. . . . With respect to
the cause of this difference between the sexes, I have nothing to add to the remarks in the
last chapter on the probable effects of the long continued use of the vocal organs by the
male under the excitement of love, rage and jealousy.

This refers to an interesting discussion of singing and music, in which he points
out the importance of overtones, which are richer for deep voices.

Having discussed clear sex differences, including facial hair, and what is valued
as beautiful, Darwin discusses the sensitive question of the seeming intellectual
advantages of the human male. He was writing when this was becoming an issue
of the greatest significance, and of course he saw it from the upper middle-class
Victorian viewpoint, writing at a time when males of his social class were highly
dominant. This discussion is remarkably short.[73] Here is a key passage:

The chief distinction in the intellectual powers of the two sexes is shown by man's
attaining to a higher eminence, in whatever he takes up, than can woman – whether
requiring deep thought, reason, or imagination, or merely the use of the senses and hands.
If two lists were made of the most eminent men and women in poetry, painting, sculpture,
music (inclusive both of composition and performance), history, science, and philosophy
. . . the two lists would not bear comparison. We may also infer, from the law of the
deviation from averages, so well illustrated by Mr Galton, in his work on 'Hereditary
Genius' [1869], that if men are capable of a decided pre-eminence over women in many
subjects, the average of mental power in man must be above that of woman.

Darwin goes on to point out that 'mere bodily strength and size would do little
for victory' in the male battles of the past, 'unless associated with courage,
perseverance, and determined energy'. The male, Darwin suggests, has, through
past need, greater energy, perseverance and courage; and possesses especially
patience:

He may be said to possess genius – for genius has been declared by a great authority to be
patience; and patience, in this sense, means unflinching, undaunted perseverance. But this
view of genius is perhaps deficient; for without the higher powers of the imagination and
reason, no eminent success can be gained in many subjects. These latter faculties, as well
as the former, will have been developed in man, partly through sexual selection. . . .

This discussion ends with a Lamarckian argument: that acquired characteristics
may be transmitted to children, and that mothers should develop these characteristics of the male to the full, to help their daughters to gain these characteristics by
inheritance. Alas, this is no longer believed possible.

Commenting on this, Fraser Harrison (1977) very justly says:[74]

But, even as he [Darwin] wrote, millions of women were daily giving the lie to his
assessment of the two sexes and their relative capabilities. The working class inhabitants of
large towns undertook a struggle for existence, not dissimilar in its bleak and elemental

simplicity to that undertaken by our savage ancestors . . . thus urban conditions extracted from working-class women those precise characteristics – courage, perseverance, determination, etc. – that Darwin had nominated as peculiar to man.

We would now say that Darwin greatly underestimated the need for these qualities in the female (and surely her lesser strength might require, and so develop, them in fuller measure for survival), and he also underestimated the handicap of social constraints on women. All the same, one can be surprised at the lack of women poets, painters and musicians. Emancipation is a wonderful experiment – not yet complete – whose full results we must wait to see. The abilities of women are surely the ace, yet to be drawn, which may well win the trick. And if it turns out that there are differences, these may be of great advantage in, and to, technological societies capable of freeing women to release their potentialities.

I have gone into this here at some length, not only for its social significance but also to illustrate how scientific theory and knowledge, even when tempered by the unique genius of Darwin, can be misleading when applied to human problems of the highest importance. This provides justification for constant checks, vigilance, and careful research in the Social Sciences – even though they lack the elegance and the prestige of biology and physics. The particular point here is that though Darwin's arguments are sound, the conclusions depend on accumulation through perhaps millions of years of small and contradictory effects; and it is not clear *a priori* which effects will dominate. If the three-body problem in astronomy is unsolved, we can hardly expect social interractive phenomena to be understood and quantified to give unquestionable accounts or predictions. The question here is whether, even with modern computing facilities, science can be depended upon for problems of psychology and sociology. But even if this were so, science and philosophy may give us insights of the kinds of processes involved; this is intellectually satisfying, and useful for guiding research to issues of immediate importance. Here we may look at the last paragraph of *The Descent of Man:* [75]

Man may be excused for feeling some pride at having risen, though not through his own exertions, to the very summit of the organic scale; and the fact of his having thus risen, instead of having been aboriginally placed there, may give him hope for a still higher destiny in the distant future. But we are not here concerned with hopes or fears, only with the truth as far as our reason permits us to discover it. . . . We must, however, acknowledge, as it seems to me, that man with all his noble qualities, . . . with his God-like intellect which has penetrated into the movements and constitution of the solar system – with all these exalted powers – Man still bears in his bodily frame the indelible stamp of his lowly origin.

This was a powerful lesson – one that we have learned to our immense benefit.

There were other attempts to apply biological ideas and facts to the emancipation problem, as it was seen towards the end of the nineteenth century. A distinguished Scottish biologist, Sir Patrick Geddes (1854–1923), who studied under T.H.Huxley at University College London, wrote (with J. Arthur Thompson) *Evolution of Sex* (1889) and *Sex* (1914). Geddes argued that basic differences between men and women are due to differences of cell metabolism and that maleness is throughout the organic world characterized by expending energy, while femaleness is essentially the capacity to store energy. Thus 'the hungry, active cell becomes flagellate sperm, while the quiescent, well fed one becomes the ovum.' (This, of course, we no longer accept.) He concludes: 'What was decided among the prehistoric Protozoa cannot be annulled by Act of Parliament.' Very true; but do we know what was decided long before man, among the protozoa, that determines our fate now? It is difficult enough to see how Mind is at all related to what we know from the Physical Sciences, let alone infer essential innate differences between abilities or Minds of men and women. The similarities are far too great for differences to be explained in basic biological terms, and it is not yet even clear just what the differences are. This is so, even after such monumental works as Havelock Ellis's *Man and Woman* (1894). He concluded that 'Women are quite equal, and perhaps superior to men in devotion to truth.'

Later investigations have failed to establish more than small differences, or to separate effects of individual experience (nurture) from innate differences that might be produced by sexual selection through the long drama of development of life-forms by Evolution.

Is man beyond the evolutionary sequence?

Although we see a sequence of anatomical form and behaviour through the Evolution of species there is a remarkable gap in the fossil record, and in living species: man has a unique ability to tackle problems and pose abstract and 'useless' questions. Although other species perceive and learn, it is not clear that questions can be posed by any other species: our uniqueness lies especially in questioning the world, each other, and ourselves. Darwinian Evolution allows only small incremental steps, but here we see a huge jump in a short space of biological time. How has this occurred? And how has language developed so fast? Perhaps the most likely explanation is that language has occurred as a take-over operation of earlier, indeed pre-human perceptual processes and classifications (Gregory, 1977). Possibly Noam Chomsky's 'Deep Structure' of language is essentially ancient perceptual classifications. But what of our ability to pose questions, and tackle problems far removed from those necessary for earlier, or for our survival? Here we may suggest a kind of answer not available to Darwin, for he lived before the introduction of programmable computers.

What is so remarkable, while at the same time giving a clue, is the essential similarity of human and ancestral brain structure in spite of the huge gap in abilities. No doubt language is a most powerful tool for thinking, and we alone possess effective language, but there may well be more to it than the power of language. It could surely be that there was selective pressure to develop knowledge-based predictive brain models so that control of behaviour by reflexes became secondary – which allowed behaviour to become exploratory. Discoveries from exploration would be useless without rapidly formed and novel strategies. This is very like the development of computer programs. A change of program allows a vast change of performance without any dramatic or easily seen change of structure. Indeed, a computer intended for mundane affairs may be turned to abstract problems without essential modification. This has implications for discovering how the brain works, and especially for how we may relate structure to function. If much of the brain is 'general-purpose' and set up for a great variety of particular functions by selected programs, much as very small changes to sentences totally change their meaning, the task of relating structure to function is extremely difficult.

In his Gifford Lectures of 1975–7, Professor J.Z. Young comments that:[76]

If the essential feature of the brain is that it contains information then the task is to learn to translate the language that it uses. But of course this is not the method that is generally used in the attempt to understand the brain. Physiologists do not go around saying that they are trying to translate brain language. They would rather think that they are trying to understand it in the 'ordinary scientific terms of physics and chemistry'.

John Young goes on, before discussing brain programs in detail, to stress the significance of what has been discovered with 'ordinary' scientific procedures and aims; but he is surely right to stress the importance of brain programs for understanding man, and the extraordinary jump in powers of abstract problem setting and solving at this point in the evolutionary sequence at which we find ourselves. With programmable brains and computers, man and present technology are evolving in unprecedented ways, which are not controlled by mundane survival. This is a new leap into a future of promise free of biological fetters. Or is it a recipe for annihilation?

Part 2 Links to Mind

Oh! dreadful is the check – intense agony –
When the ear begins to hear, and the eye begins to see;
When the pulse begins to throb, the brain to think again;
The soul to feel the flesh, and the flesh to feel the chain.

Emily Brontë (1818–48) [1]

6 Links of light

If light can thus deceive, wherefore not life?

Joseph Blanco White (1775–1841)[2]

Perceiving from a distance

We are – or at least we seem to be – linked by light to objects. What is this link? How does it allow us to perceive, to know, objects beyond the reach of touch? Is perception by sight more indirect than touch, or is the light link to the eyes as immediate and as to be trusted as perception by probing fingers? Touch may seem more direct, more definite, but it has its mysteries and its limitations. It is also, like vision, subject to illusion. Our arms are lumps of matter that we can extend to other objects within reach to discover whether they are hard or soft, hot or cold, fragile or robust. We may discover distances and sizes and weights by touch, and whether objects are separate or attached to each other, by attempting to move them. We may carry out active experiments to discover many properties of objects, and subject them to taste, to touch, and to use in many situations – ultimately to destruction – to discover their parts and how they are formed and how they function. But although our arms and fingers – our entire bodies – are objects occupying the space of other objects, and sharing with them weight and warmth and other properties that we discover by touch, we do not believe that these other – inanimate – objects likewise discover us, or know each other. We may break an egg as a stone may break an egg; but we feel it though the stone does not. Whether the egg (or rather the chicken or its embryo inside) feels it, is a matter for doubt. We do not know how to find out, as none of our senses reach another creature's perceptions, and we have no theoretical model adequate for linking Minds except indeed by the still more indirect means of shared symbols, or language. But symbolic language cannot substitute for touch or sight or the other senses: they require links of light, or contact of some kind with transfer of energy, however small, to the sense receptors for symbols to register or be transmitted. We cannot read or see pictures without light, speak without compressions through the air; or touch Braille, or lovers, without contact to the surface of our bodies. Such physical links are necessary for communication by symbols, and so they are primary to all other communication or knowledge gained by experience. But even these, though primary, do not link us to any other creature's experience. We may learn that he says he is hungry, or afraid, or experiencing red or pain; but we cannot know without trusting his symbolic expression that the symbols that we receive are indeed linked to another's experience – to another's consciousness. We limit consciousness and experience to living things, and generally to living things similar to ourselves. Sticks and stones, though available to us for sensing, cannot, we believe, sense us. Some special objects, a radio for example, may utter the word 'red', or cries of pain, but still we do not accept that it experiences red or feels pain. Machines may respond to symbols conveyed by light or by touch contact – but, we believe, without sensation.

All of this is pretty obvious to us, but it was not obvious even to the most learned thinkers of the classical age. Philosophers right through from the Greeks and to René Descartes, Benedict Spinoza and Gottfried Leibniz, in the seventeenth century, have questioned whether only organisms have sensations.

Leibniz (implausible as it may seem) thought of all physical objects as 'Monads' – Minds linked by partial perceptions of each other, though powerless to interact. It was Descartes who first made a clear distinction between experiencing and non-experiencing objects: between Mind and Matter. It is still a question for debate how this distinction should be accepted. We do not need, and we do not know how, to pursue this problem here. Rather than debate Cartesian Dualism at this point, we shall look at theories of light as a primary physical link between objects and ourselves. We shall look at early ideas, because theories of the nature of light have, through recorded thought, reflected theories of perception.

Sight has generally been regarded as our main source of knowledge of the world of objects, though requiring the sense of touch as a back-up and check. This implies that vision is regarded as comparatively 'indirect': that we can touch reality, while for vision we are linked to things that we see by light. This link has always been mysterious. It may be broken by darkness; it may bend or disperse, to distort and mislead.

The Pythagoreans regarded vision as due to an invisible 'fire' coming out of the eyes to touch and so reveal shapes (and somehow colours) of objects. Sometimes the air was regarded as the medium for transmission. Democritus thought that the air received and imparted imprints.[3] He thought that the imprint was made upon the liquid on the cornea; this was probably suggested by the small images of reflected objects to be seen in the pupils of moist eyes. Our word 'pupil' (meaning student) has its origin in this thought.

Plato distinguishes three kinds of 'fire', or light: daylight from the sun; an internal 'fire' of the same kind in the eye and directed outwards to seen objects; and a different 'fire' or light streaming off seen objects, joining and interacting with the fire emitted from the eyes. It is, however, a question how far all this is meant to be taken literally: [4]

And the first organs they [the Gods] fashioned were those that give us light, which they fastened there in the following way. They arranged that all fire which had the property of burning, but gave out a gentle light, should form the body of each day's light. The pure fire within us that is akin to this they caused to flow through the eyes, making the whole eyeball, and particularly its central part, smooth and close-textured so that it would keep in anything of coarser nature, and filter through only this pure fire. So when there is daylight round the visual stream, it falls on its like and coalesces with it, forming a single uniform body in the line of sight, along which the stream from within strikes the external object. Because the stream and daylight are similar, the whole so formed is homogeneous, and the motions caused by the stream coming into contact with an object or an object coming into contact with the stream penetrate right through the body and produce in the soul the sensation which we call sight. But when the kindred fire disappears at nightfall, the visual stream is cut off; for what it encounters is unlike itself and so it is changed and quenched, finding nothing with which it can coalesce in the surrounding air which contains no fire. It ceases therefore to see and induces sleep.

Plato also says: [5]

The cause and purpose of God's invention and gift to us of sight was that we should see the revolutions of intelligence in the heavens and use their untroubled course to guide the troubled revolutions of our own understanding. . . .

Before leaving Plato we should look at his theory of colour, which includes black and white and all visual sensation, including brightness: [6]

The particles which impinge on the visual ray from other bodies are either larger or smaller than those of the visual ray itself or else the same size. If they are the same size they are imperceptible or as we say 'transparent'. If they are larger they compress the ray, if they are smaller they penetrate it; . . . we must assign these names accordingly, calling that which penetrates the visual ray 'white' and that which compresses it 'black'.

Plato's account of dazzle is dramatic: [7]

When another kind of fire with faster motion falls on the visual ray and penetrates it right up to the eyes it forces apart and dissolves the passages in the eyes, and causes the discharge of a mass of fire and water which we call a tear; this incoming fire meets a fire moving towards it, and the outgoing fire leaps out like lightning while the incoming is quenched in the moisture. The result is confusion of all kinds of colours; this we call 'dazzling' and the object which produces it 'bright' and 'gleaming'.

The various colours are then accounted for by varieties of fire 'intermediate between these which reaches the moisture of the eye and mixes with it'—like mixing burned pigments with water on a pallette.

Vasco Ronchi (1970) takes the view (and I follow him) that the internal fire was Plato's way of saying that there are psychological perceptual processes adding to what is received by light. Throughout classical discussion, two kinds of light have been distinguished: the objective physical light of the world of objects, and the subjective light of 'experience', especially the sensation of brightness. These were called respectively (though not always consistently; there are other terms) '*lux*', or '*lumen*', for external physical light, and '*fulgor*' or 'splendour' for the sensation of brightness.

Very bright light – lots of *lumen* – may hurt and damage the eyes. This was a proof of something entering the eye. A mystery for purely 'eye-emitting' theories was why very near objects should be unclear. Not until the seventeenth century AD was it realized that light is imaged on the retina, and that near vision is restricted by limited accommodation of the lens of the eye. The notion of 'light rays' was, however, developed in this first period of speculation and experiment in vision and optics. The notion of light rays may have been the origin and a prerequisite of geometry. It is also, however, possible that geometrical thinking tended to make philosophers regard the eye as a point – in relation to the vast space that it encompassed – which took attention away from its structure. Euclid reduces the eye to a mathematical point in his *Optics* (third century BC).

The structure of the eye was described by Galen, but unfortunately he regarded the lens as the light-sensitive medium and not the retina. He also adopted a physiological version of Plato's theory, supposing sensation to arise from a mixture of fluids from the brain with physical light entering the pupils of the eyes.

The notion of light reaching the eye in straight lines (rectilinear propagation) was first stated clearly by the Arab scientist Alkindi (*c*. AD 813–873), who had a wide influence in several branches of science; his full name was Abu Ysuf Yaqub Ibn Is-haq. In his *De Aspectibus* (as known in the twelfth-century Latin translation), Alkindi rejected the geometrical notion of light rays as mathematical abstractions and urged that, to have causal effects, they must have breadth. This led to searching for physical properties of light by experiment. The important mathematical notion that light (reflected from mirrors, or refracted by lenses or prisms) takes the shortest path was an ancient idea, but it was couched in static geometrical terms. Alkindi effectively challenged this.

It was not at all clear how light could convey the shapes of distant objects or structures to the eyes. This was not indeed clear before the invention of the first crude forms of camera obscura, which at last showed optical images giving geometrical projections of objects. Before this principle was discovered in the tenth century, it was a puzzle how large objects could be seen with the small hole of the pupil. How could they get inside the eye? How could equivalent extensions of light do so? Plato, Euclid and later Descartes thought of light rays as a kind of extension to touch: probing objects as a blind man can explore the world with his stick. But another and generally more prevalent view, which is still not dead in philosophy, was that objects gave out 'simulacra', or 'husks', maintaining the forms of objects as intermediaries between them and us, when we see. This idea is essentially the same as the 'sense data' of C.D.Broad (1929) and H.H.Price (1932), and is still taken seriously by philosophers in spite of Alhazen's discovery of optical images.

The Arabs were dominant in optics through our Dark Ages. They did considerably more than preserve and transmit the learning of the Greeks, after the disastrous loss of the Library of Alexandria by fire in the first century AD. The most important of the Arab scholars in this field is known to us as Alhazen, his full name being Abu Ali Mohammed Ibn Al Hasan Ibn Haytham. He was born at Basra in about AD 965. He lived in Egypt, dying at Cairo in AD 1039. Alhazen followed Alkindi in accepting light rays as 'real'. The main argument for this came from the power of the sun, as seen directly or reflected from a mirror, to hurt and damage the eyes. Alhazen described after-images following intense stimulation of the eyes by light; considering someone looking at the bright sky through a crack in a roof and then at a dark wall, or shutting his eyes: 'He will still see that shape,' from which he concludes, 'Light exerts some sort of action upon the eye.' This was crucial for seeing light as an active medium, linking the world of tangible objects to the eye, and giving perception via optically projected images in the eyes.

Alhazen realized, in effect, that information structure must be preserved by light to give visual perception of shapes of objects. He rejected the 'fluids' and 'fires' and 'husks', in favour of the view that 'vision is produced by rays emitted by the objects towards the eyes'. Thus he gave direction to Euclid's abstract geometrical rays. He studied the structure of the eye, developing an image-forming account – that only light entering the cornea at right-angles could enter – so that each ray reached a particular region or point inside. Although this account is not correct, the essentials of an image-forming eye are here, for Alhazen realized how large objects could be represented as patterns of light in the much smaller eye. This was a fundamental achievement, which, in spite of such talented men as John Peckham (c. 1230–92),[8] was not followed up until the flowering of Western science in the seventeenth century around the figure of Isaac Newton and other extraordinarily brilliant men of that time.[9]

Newton's theory of light

From antiquity, light had been compared with, or had been thought to be made of, either waves or particles. Particles fitted the observation that, except when disturbed by reflecting or refracting objects, light follows straight lines; but it was not clear how rays could cross each other without mutual interference between streams of particles. It was obvious that particles should collide, and deflect each other: why was this not so for light? On the other hand, ripples on water could, somehow, pass through each other, leaving each as they were before they met. But if light was waves, light must move in some medium analogous to water-bearing ripples; hence the ancient notion of an all-pervading ether. This was never supposed to be just like known physical fluids, such as water, for it had to pervade transparent substances, as well as space. This seemed implausible; so there were clear advantages to particle theories. Light particles reflected from mirrors might be like balls bouncing off walls. Their behaviour in dense transparent substances was, however, not so clear: why should ball-like particles be deviated by glass prisms? Why, also, should prisms produce coloured light? One theory for this was that the light particles were spun, as balls spin upon grazing a surface, and that the spinning of the particles gave light colour.

Newton did not, as is sometimes thought, discover that prisms produced spectra. This was known far earlier. What he did notice was that sunlight from a small hole in a shutter, when passed through a prism, did not produce a circular patch of light, but an elongated one, the spectral colours being spread out across the longer axis. This he first described in a communication to Oldenburg, then Secretary of the Royal Society, dated 6 February 1671/2: [10] 'In the beginning of the year 1666 (at which time I applyed myself to the grinding of Optick glasses of other figures than Spherical) I procured me a Triangular glass-prisme, to try therewith the celebrated Phenomenaon of Colours.' Newton continues:

And in order thereto having darkened my chamber, and made a small hole in my window-shuts, to let in a convenient quantity of the Sun's light, I placed my Prisme at its entrance, that it might be thereby refracted on the opposite wall. It was at first a very pleasant divertisement, to view the vivid and intense colours produced thereby; but after a while applying myself to consider them more circumspectly, I became surprised to see them in an *oblong* form; which, according to the received laws of Refraction, I expected should have been *circular*.

They were terminated at the sides with streight lines, but at the ends, the decay of light was so gradual, that it was difficult to determine justly, what was their figure; yet they seemed *semicircular*.

Comparing the length of this coloured *Spectrum* with its breadth, I found it about five times greater; a disproportion so extravagent, that it excited me to more than ordinary curiosity of examining, from whence it might proceed.

Newton then describes how he ruled out various possibilities experimentally, by varying the size of the hole, placing the prism at the other side of the hole, and so on (just as Francis Bacon would have approved), to conclude: 'But I found none of those circumstances material. The fashion of the colours was in all these cases the same.'

Newton ruled out the possibility of unevenness of the glass of the prism in a particularly interesting way:

I took another Prisme like the former, and so placed it, that the light, passing through them both, might be refracted contrary ways, and so by the latter returned into that course, from which the former had diverted it. For by this means, I thought, the *regular* effects of the first Prisme would be destroyed by the second Prisme, but the *irregular* ones more augmented, by the multiplicity of refractions. The event was, that the light, by which the first Prisme was diffused into an *oblong* form, was by the second reduced into an *orbicular* one with as much regularity, as when it did not at all pass through them. So that, whatever was the cause of that length, 'twas not any contingent irregularity.

After making measurements and calculations showing that rays from the opposite sides of the sun's disk should give but an angle of 31', though he found an angle of 2° 49', Newton considered the possibility of the rays from the prism travelling in curved paths. He was drawn to this from the notion of spinning tennis-balls:

I had often seen a Tennis-ball, struck with an oblique Racket, describe such a curve line . . . its parts on that side, where the motions conspire, must press and beat the contiguous Air more violently than on the other. . . . And for the same reason, if the Rays of light should possibly be globular bodies, and by their oblique passage out of one medium into another acquire a circular motion, they ought to feel the greater resistance from the ambient Aether, on that side, where the motions conspire, and thence be continually bowed to the other. But notwithstanding this plausible ground of suspition, when I came to examine it, I could observe no such curvity in them.

He then showed that 'Light consists of *Rays differently refrangible*'. Newton goes on in this same communication to Oldenburg to show that colours are properties of light: 'Some Rays are disposed to exhibit a red colour and no other; some a yellow and no other, some a green and no other, and so of the rest.' Also:

To the same degree of Refrangibility ever belongs the same colour, and to the same colour belongs the same degree of refrangibility. The *least Refrangible* Rays are all disposed to exhibit a *Red* colour, and contrarily those Rays, which are disposed to exhibit a *Red* colour, are all the least refrangible: so the most *refrangible* Rays are all disposed to exhibit a deep *Violet* colour . . . contrarily all those which are apt to exhibit such a Violet colour are all the most Refrangible.

This most important result he obtained from the observation that

When any sort of Rays hath been well parted from those of other kinds, it hath afterwards obstinately retained its colour, notwithstanding my utmost endevours to change it. I have refracted it with Prismes and reflected it with Bodies, which in Day-light were of other colours; I have intercepted it with the coloured film of Air interceding two compressed plates of glass; transmitted it through coloured Mediums, and through Mediums irradiated with other sorts of Rays; and diversely terminated it, and yet could never produce any new colour out of it. It would by contracting or dilating become more brisk, or faint, and by the loss of many Rays, in some cases very obscure and dark; but I could never see it changed *in specie*.

Newton states here that there are two kinds of colours, 'original or simple' and mixture colours 'compounded of these'; and that the primary colours (he listed at that time the five colours red, yellow, green, blue and violet-purple; adding orange and indigo later, to make seven) can be made by mixture of well separated colours. He then adds:

But the most surprising and wonderful composition was that of *whiteness*. There is no sort of Rays which alone can exhibit this. . . . Hence therefore it comes to pass, that *Whiteness* is the usual colour of *Light*, for, Light is a confused aggregate of Rays imbued with all sorts of Colours, as they are promiscuously darted from the various parts of luminous bodies.

To this he adds: 'Of such a confused aggregate, as I said, is generated *Whiteness*, if there be a due proportion of the Ingredients; but if any one predominate, the Light must incline to that colour; as it happens in the Blew Flame of Brimstone; the yellow flame of a Candle; and the various colours of the fixed stars.' The colours of objects were simply explained thus: '. . . the Colours of natural Bodies have no other origin than this, that they are variously qualitied to reflect one sort of light in greater plenty than another'.

Newton regarded his experiments as proving beyond all doubt that white light is made up of the colours of the spectrum. He regarded his conclusions as axioms. However, his reasoning did not go unquestioned by his contemporaries. Robert Hooke (1635–1703),[11] especially, challenged the logic of the basic experiment by which colours were identified with refrangibility. In a letter (*c*. June 1672, probably to Lord Brouncker, then President of the Royal Society), Hooke wrote:[12]

I have only this to say that he [Newton] doth not bring any argument to prove that all colours were actually in every ray of light before it has suffered a refraction, nor does his *experimentum crucis* as he calls it prove that those properties of coloured rayes, which we find they have after their first Refraction, were Not generated by the said Refraction. For I may as well conclude that all the sounds that were produced by the motions of the [? strings] of a Lute were in the motion of the musician's fingers before he struck them, as that all colours which are sensible after refractions were actually in the ray of light before Refraction.

This argument is repeated with another example:

. . . for a ray of light may receive such an impression from the Refracting medium as may distinctly characterise it in after Refractions, in the same manner as the air of the bellows does receive a distinct tone from each pipe, each of which has afterwards a power of moving an harmonious body, and not of moving bodys of Differing tones.

This was written at a time when relations between Newton and Hooke were extremely strained: their arguments became almost too personal to be acceptable to the Royal Society. It seems [13] that this letter of Hooke's did not reach Newton. There is no known reply. This argument of Hooke's is strictly correct. By changing the analogy (or the general paradigm) the same observations can look more or less cogent – so how can there be crucial experiments? How can any experiment, or set of experiments, ever *prove* anything?

Hooke considers an alternative explanation to Newton's in his *Micrographia* (1665). This was published in the year preceding Newton's prism experiment. Hooke wrote: [14]

But above all it is most observable, that here are all kinds of Colours generated in a *pellucid* body, where there is properly no such refraction as Des Cartes supposes his *Globules* to acquire a *verticity* by: For . . . the second refraction does regulate and restore the supposed *turbinated globules* unto their former uniform motion. . . . This Experiment therefore will prove such a one as our *thrice excellent Verulum* [Francis Bacon] calls *experimentum crucis,* serving as a Guide or Land-mark, by which to direct our course in the search after the true cause of Colours.

This reference to crucial (crossroad) experiments should be noted. It is often said that Newton did not like hypotheses. But he often put up hypotheses for consideration.

Writing some thirty years later in the *Opticks* (1704), Newton denies (as we are sure correctly) that the colours we see are in the light itself; but could he have been certain? 'The . . . rays which appear red, or rather make objects appear so, I call rubrific or red-making; those which make objects appear yellow, green, blue and violet, I call yellow making, green making, blue making, violet making, and so of the rest. . . .' He took great trouble to make this clear: 'In [the rays] there is nothing else than a certain power and disposition to stir up the sensation of this or that colour. For as sound in a bell or musical string . . . is nothing but a trembling motion.'

Newton wrote papers in which there are quite frankly displayed hypotheses going far and controversially beyond available evidence. For example, in a paper written at the end of 1675 he postulated 'an aetherial medium, much of the same constitution with air but far rarer, subtiler, and strongly elastic', which is 'a vibrating medium like air, only the vibrations are more swift and minute'. It 'pervades the pores' of all bodies. He attributed gravitation to this ether: 'every body endeavouring to go from the denser parts of the medium towards the rarer'. Indeed, he tried to relate his ideas of gravitational attraction between the planets and stars to his 'globules' of light (see pp. 139 *et seq.*).

It seems that Newton's inference – now universally accepted in spite of Hooke's forgotten doubts – that white light is a mixture of coloured lights each of different refrangibility, led him away from inventing the correction of chromatic aberration in lenses, by the use of elements of different kinds of glass, as was achieved later. His incorrect assumption that the same coloured rays are deviated by the same amount in all transparent substances made him give up very promising experiments on lenses of several elements of different materials, and some fluid-filled, which might have successfully countered the chromatic aberration that was a serious flaw in early telescopes and microscopes. In the letter to Oldenburg dated 11 June 1672 he describes his experiments, before giving them up for this mistaken reason:

What may be done not only by *Glasses alone* and more especially by a *complication of divers successive Mediums,* as by two or more *Glasses* or *Chrystalls* with water or some other fluid between them, all which may performe the office of one *Glasse,* especially of the *Object-Glasse* on whose construction the perfection of the instrument chiefly depends.

Giving this up, he turned his attention to reflecting telescopes, which were also invented at that time by the Scottish mathematician James Gregory (1638–75),[15] in 1661. It is perhaps fortunate that they failed to produce achromatic lenses, for the reflecting telescopes were, and remain, essentially superior for many uses, as they can be made far larger with greater light-gathering power. Correction of chromatic aberration – which transformed microscopes – was probably achieved in 1757 by the London instrument maker John Dollond (1706–61). The Swiss mathematician Leonhard Euler (1707–83) published a theoretical prediction that this should be possible in 1769, based on work of his of some fifteen years' standing. The fact, stressed by Euler, that dispersion and refraction are not simply related, is itself difficult or impossible to understand on a corpuscular theory of light, and it was not appreciated by Newton, who therefore abandoned his experiments on achromatic lenses.

I have tried here to interweave something of the history of Newton's discoveries concerning light and colour with hints as to how far these were made directly by experiment and how far they were imaginative hypotheses, more or less tested by 'crucial' experiments. Newton seems to have believed that experiments could be crucial, in the sense of deciding the truth. Newton does, however, sometimes refer to hypotheses, commenting that some are better than others (very humanly preferring his own!) in the modern manner; and saying that some hypotheses are erroneous, which may be taken to imply that he did not regard all hypotheses as misplaced.

Considering Hooke's interesting challenge to Newton's 'crucial experiment', we may be mistaken in thinking that Newton regarded experiments as much more than (as Bacon saw it) helping to make choices at 'crossroads' of conceptual possibilities. To expect experiments to do more – to be 'crucial' in the sense of deciding the truth – is to suppose that their interpretation does not depend on assumptions that may be doubted. But is this ever the case? One might well hold that doubt is only limited when imagination has run out for questioning and for putting up alternative hypotheses (see chapters 8, 12 and 20).

What then was Newton's estimate of the status of his account of light and colour? We find this in his letter written to Oldenburg (previously cited), in reply to public criticism by Robert Hooke:

I said indeed that the *Science of Colours was Mathematicall and as certain as any other part of Optiques;* but who knows not that Optiques and many other Mathematicall Sciences depend as well on Physicall Principles as on Mathematicall Demonstrations: And the absolute certainty of a Science cannot exceed the certainty of its Principles. Now the evidence by which I asserted the Propositions of colours is in the next words expressed to be from *Experiments* and so but *Physical.* Whence the Propositions themselves can be esteemed no more than *Physicall Principles* of a Science. And if those Principles be such that on them a Mathematician may determine all the Phaenomena of colours that can be caused by refractions, and that by computing or demonstrating after what manner and how much these refractions doe separate or mingle the rays in which severall colours are originally inherent; I suppose the *Science of Colours* will be granted *Mathematicall* and as certain as any part of *Optiques.*

But as we know from his letters he fumbled and tried all sorts of possibilities. And indeed the 'crossroads' of Bacon are not typically, and never strictly, crossroads involving only two choices. Newton had the genius to see so many roads to follow that it is remarkable he so seldom lost his way in the labyrinth of his uniquely creative imagination.

Light now

The modern notion of light as both particles and waves would until recently have been rejected as paradoxical. Throughout the history of philosophy and science paradox has been a principal ground for rejection of theories; but recently we have grown far more tolerant of paradoxical appearance.

The current view of light is entirely non-intuitive: nowhere does it fit how things seem to be from untutored experience. So we might say that what has changed here is not quite that paradox is no longer grounds for rejection, but rather that conflict between what things seem to be like to the senses and how they are concerned in science is no longer strong ground for rejection. The scientific account must still be consistent within its own criteria of consistency; but these may look very queer indeed.

This shift of emphasis has emasculated philosophy, for its traditional plea for common sense (especially in the Empirical tradition, which generated the science that has changed all this!) no longer applies, now that scientific accounts are hardly guided or judged by how things seem to the senses. Even the history of perspective is surprising.

The world imaged in the eyes

Geometrical perspective as we understand it was not known to the Greeks; and it was not fully developed until the Italian Renaissance. Euclid (c. 300 BC) did, however, come close to perspective, in his *Optics* and his *Catoptrics*. Being a geometer, he drew up postulates. These he based on observations from shadows that light travels in straight lines – an idea germinal for geometry as for geometrical optics. He considered that light rays are emitted by the eye, allowing those objects to be seen on which they fall. His Postulates in the *Optics* are:

1   The rays emitted by the eye travel in a straight line.

2   The figure enclosed by visual rays is a cone which has its apex at the eye and its base at the edge of the object looked at.

3   Objects on which the visual rays fall, are seen.

4   Objects which are not reached by the visual rays are not seen.

5   Objects which subtend large angles are large.

6   Objects which subtend small angles are small.

7   Objects which subtend equal angles appear equal.

8   Objects which are seen with the higher rays appear higher.

9   Objects which are seen with the lower rays appear lower.

10   Objects which are seen with rays directed to the right appear on the right.

11   Objects which are seen with rays directed to the left appear on the left.

12   Objects which are seen with several angles are seen more distinctly.

13   All rays have the same speed.

14   Objects can only subtend certain angles.

It should be noted that the essential problem for visual location of objects in space – and the point of Renaissance 'geometrical perspective' in pictures – is missing from these Postulates of Euclid. Postulates (5), (6) and (7) refer only to the sizes of objects, not to their distances; so the size–distance ambiguity of pictures and retinal images was not then realized. The Postulates in the *Catoptrics* add plain and curved mirrors but nothing further to the understanding of perspective. It seems from the preface to Euclid's *Optics,* written by Theon, that Euclid thought there would be gaps between the light rays, increasing with distance. This is indeed to be expected if they are a finite set of radiating lines. Theon explains why it may be difficult to see a needle fallen on the ground – because it may lie between the light rays shot out from the eyes and so be invisible.

The invention of the camera obscura, a dark box with a hole or lens giving an image, was crucial for recognizing and understanding perspective and the optics of the eye. It was an invention linking art, science and philosophy. Plato conceived something of the camera obscura in his celebrated 'shadows in the cave', in the *Republic:* [16]

Imagine the condition of men living in a sort of cavernous chamber underground, with an entrance open to the light and a long passage all down the cave. Here they have been from childhood, chained by the leg and also by the neck, so that they cannot move and can see only what is in front of them, because the chains will not let them turn their heads. At some distance higher up is the light of a fire burning behind them; and between the prisoners and the fire is a track with a parapet built along it, like a screen at a puppet-show, which hides the performers while they show their puppets over the top.

Now behind this parapet imagine persons carrying along various artificial objects, including figures of men and animals in wood or stone or other materials, which project above the parapet. . . .

Now, if they could talk to one another, would they not suppose that their words referred only to those passing shadows which they saw?

This has the optical representation of the camera obscura, though not its physical principle of an image produced by a pinhole or a lens. (Indeed, the fact that Plato makes no such reference is evidence that optical images were unknown at that time.)

Alhazen gave the essential principle of the image-forming camera obscura in the tenth century AD, as follows: [17]

He [Alhazen] placed several candles in front of a wall with a hole and looking at a screen placed on the other side of the wall, he saw on it as many images as there were candles. These images were found along lines which crossed at the hole. When he suppressed one of the candles he noticed that the corresponding image also disappeared, and when the candle was replaced the image came back in the same place.

It seems, though, that Alhazen did not mention that the images of the candles were upside-down. According to Ronchi, Alhazen did not understand the geometry of the situation. This may however be questioned.

26   An early nineteenth-century camera obscura in use. The lack of appreciation of optical images (in spite of shadows) before the surprisingly late invention of the camera obscura inhibited representative accounts of perception.

The pinhole camera obscura was clearly described by Giovanni Battista della Porta in his *Natural Magic* of 1558.[18] Ten years later the effect of replacing the pinhole with a lens, giving images sufficiently bright for artists to see form and colour clearly, was described by a Professor of the University of Padua, Danielo Barbaro, in *Practica della Perspectiva* (1568–9):[19]

Close all the shutters and doors until no light enters the *camera* except through the lens, and opposite hold a sheet of paper, which you move forward and backward until the scene appears in the sharpest detail. There on the paper you will see the whole view as it really is, with its distances, its colours and shadows and motion, the clouds, the water twinkling, the birds flying. By holding the paper steady you can trace the whole perspective with a pen, shade it and delicately colour it from nature.

The first cameras were rooms – '*camera obscura*' means 'dark chamber' – in which the observer sat. When the eye had been dark-adapted, the image was truly dramatic, as we can see from existing instruments at Edinburgh, Bristol and other places. Portable camera obscuras were made from sedan chairs, blacked out with a lens fitted to one end. This would be put down with a suitable view imaged on paper for the resident artist to trace. Boxes with lenses and a ground glass screen for tracing on were developed, as described by Count Francesco Algarotti, in his *Essay on Painting* (1764). Its first use is described by Leonardo da Vinci (1452–1519).[20] He probably did not, however, invent it, as is sometimes thought. It was used some sixty years before Leonardo by the Milanese architect Filippo Brunelleschi (1377–1446).[21] Leonardo recognized that the pupil of the eye is like the hole in the camera obscura, but he invoked a second imaging in the eye to provide – incorrectly – an upright image on the retina. In *The Notebooks* he writes:[22]

The images which pass to this uvea as they are outside the eye pass to it through the centre of the crystalline sphere, and having arrived at the uvea they become inverted as also are those which pass to the uvea without passing through this humour. We may surmise therefore, admitting this visual faculty to reside at the extremity of the optic nerve, that from here it may be seen in the crystalline sphere that all the objects caught by it are upright, for it takes those that were inverted in the uvea and inverts them once again, and consequently this crystalline sphere presents the images upright which were given to it inverted.

And later Leonardo says:[23] 'No image however small enters within the eye without being turned upside down, and as it penetrates the crystalline sphere, it is turned again upside down, and so the image within the eye becomes upright as was the object outside the eye.'

The photographic camera was of course produced by adding light-sensitive plates to the camera obscura. This was described by Sir Humphry Davy, writing of the half-successful researches of Thomas Wedgwood, the son of the potter, as early as 1802:[24]

When a white surface, covered with solution of nitrate of silver, is placed behind a painting on glass exposed to the solar light; the rays transmitted through the differently painted surfaces produce distinct tints of brown or black, sensibly differing in intensity according to the shades of the picture, and where the light is unaltered the colour of the nitrate becomes deepest.

But these first contact photographs faded in light; Wedgwood never found out how to fix them. This was achieved over the next thirty years by Joseph and Claude Niepce and Louis Jacques Mandé Daguerre (1789–1851). Now it was clear that the eye was not emitting light, as touch-like probing fingers, but that the eye produced perspective images by focusing, with its camera-like optics, rays of light received from objects – and that we can see distances from external flat perspective projections. Perspective began to look important. Also, with the precision of shading and colour in camera obscura images (and in high-quality black-and-white photographs) it became clear that fine texture and colour is also important, especially for seeing the forms of unusual or randomly shaped objects, for which perspective is relatively powerless. All this suggested the importance of visual 'cues' for seeing depth and form, while the classical 'shells' and so on were totally dropped – except indeed by such philosophers as H.H. Price and C.D. Broad in the 1920s and 1930s who hung on to sense data as entities lying between object reality and us. They did not discuss (in spite of the Kodak box Brownie prevalent at that time) how we see objects from images, and this is odd indeed.

Delay in the light link

Whether light arrives with no delay from light sources, or reflected from objects, was a matter for debate from the time of the Greeks to the middle of the seventeenth century. The matter was finally resolved beyond scientific debate by astronomical observations, for which the newly invented telescope was necessary. The first telescopes were just adequate to show, by indirect means, that light travels at a finite velocity. It follows that we are separated, visually, from other objects not only by space but also by time. We see what *was* happening, not what *is* happening, to distant objects. Although we learned this from astronomy, strictly it applies also to perception of earthly objects. The problems that this raises led to Einstein's question: how can ever we know that two separated events are simultaneous? This led to the counter-intuitive combining of space and time, into the space–time of Relativity. Present physics runs counter to what we seem to know, from what we take to be direct knowledge through experience.

For Aristotle, light did not travel, for he saw light as simultaneous changes in an all-pervading transparent medium, which came to be called the ether.[25] Aristotle's view of instantaneous light was largely accepted right through to the seventeenth century, by Galen, Avicenna, Descartes, to Kepler. On the other hand, the notion that light travels with a finite velocity was argued by Empedocles, who thought of light as a material substance; and also, in the eleventh and thirteenth centuries, by Alhazen and Roger Bacon (c. 1214–92), who realized the importance of magnifying glasses, though he did not invent them as sometimes claimed, and who predicted, though he did not produce, the telescope.

We see in the extensive and important writings on optics of René Descartes (1596–1650) a most curious ambivalence over the question of whether the velocity of light is infinite (giving direct and instant knowledge), or whether it is, like the velocity of known and presumably all physical objects, finite. Descartes held at least three models of light: that light is moving particles (like tennis-balls); that light is a kind of fluid passing through the pores of an all-pervading medium (like wine in a vat passing through grape pips); and that light is pressure waves, moving through an all-pervading medium. If this had no elasticity the transmission rate would be infinite. Descartes wrote *De La Lumière* between 1629 and 1633, but withdrew it from publication upon hearing of the condemnation of Galileo by the Inquisition, on 23 June 1633. Here he explains the bending of light by refraction as due to different velocities. So here he must be holding that light has finite velocity. (Actually, he argues that light travels *faster* in dense mediums – and yet he correctly derives Snell's Law.) Now, in spite of realizing that refraction is given by ratios of velocities of light, he also says as clearly as possible that its velocity is infinite. Thus in a letter written on 22 August 1634 to his friend Marin Mersenne (1588–1648)[26] he says:

> I said to you lately, when we were together, not in fact as you write that light moves in an instant, but that (which you take to be the same thing) that it reaches our eyes from the luminous object in an instant; and I even added that for me this is so certain, that if it could be proved false, I should be ready to confess that I know absolutely nothing in philosophy.

The phrase 'in an instant' surely means 'immediately', and 'without delay' – for otherwise what is he staking? He seems to change his opinion according to which physical model he is considering at the time of writing.

Descartes suggested the kind of experiment which finally showed that he was wrong in thinking of light as instantaneous. This experiment he described in the letter of 1634 cited above but not in his published works. He considered eclipses of the moon, where the earth (lying between sun and moon) interrupts light from the sun, to cast a shadow on the moon. The apparent position of the sun at the onset of the eclipse should not be on a straight line from moon to earth, if indeed there is a finite delay in the sun's light passing the earth to reach the moon, and reflected back from the moon to us. This argument is sound geometrically (as one would expect of Descartes) but the observations were too imprecise to detect passage of light over the half million miles (twice the moon's distance from earth) at its velocity of 3×10^{10} cm./sec. It would, however, have detected a much lower velocity of light.

The finite velocity of light was established by the Danish astronomer Olaus Roemer (1644–1710) from observations of the eclipses of Jupiter's four moons (which had recently been discovered) by the shadow of the planet falling upon them at regular intervals as they orbited Jupiter. This could be seen with the first telescopes, and the intervals between eclipses could be timed with sufficient accuracy to show that eclipses were sometimes 'early' and sometimes 'late'. This was not discovered by Roemer but by the first Director of the Paris observatory, the Italian Giovanni Domenico Cassini (1652–1712), who made several discoveries including a gap in Saturn's ring and white polar caps on Mars. Cassini produced tables of the movements of Jupiter's moons, from which he could predict their eclipses. His predictions were sometimes surprisingly wrong. Cassini

considered the possibility of a variable delay between the event and his observations due to the changing distance of Jupiter together with a finite velocity of
light. This was taken up by his assistant, Roemer, who derived a value within
30 per cent of the now accepted velocity of light.

The velocity of light was first measured on earth by Armand Fizeau (1819–
96) using a method essentially suggested by Galileo, who tried, by alternately
covering and uncovering a pair of distant lanterns, to measure the speed of light between two mountains. This failed, but Fizeau succeeded in 1849, with a rapidly
rotating sector disk used to interrupt a light beam and its return after reflection
from a distant mirror. The speed of light can now be measured over a few inches with pulsed lasers, and it sets practical limits for example to computing
speeds, where signals from sources of various distances even within a room, or
within an instrument or a computer, need to be compared or combined.

Delay in receiving light signals had little effect in physics at that time, except
for astronomy, where the immense distances make the delay of light signals vitally
important. It became clear that as we look at distant objects, even objects seen
with the naked eye, without a telescope, we look back into the distant past. So we
may see objects that no longer exist, at least in their observed positions.

The finite velocity of light explained a new and subtle observation – that the
apparent positions of stars shift slightly according to the earth's motion around the
solar orbit. This was discovered by James Bradley (1692–1762), the English
astronomer and third Astronomer Royal,[27] who, with his wealthy doctor friend
Samuel Molyneux (1689–1728), set up at Greenwich a very carefully constructed
vertical telescope for measuring positions of overhead stars with the greatest
possible accuracy. They were looking for parallax shifts, due to displacement of
the earth across the orbit round the sun. This was attempted earlier by Robert
Hooke, but parallax was too small to detect before photographic methods became
available. Bradley did, however, find small and entirely unexpected shifts, which
he realized required a quite different explanation. Having then ruled out various
possibilities (nutation or nodding of the earth's axis; changes in the plumb line
reference for his telescope; atmospheric refraction) he conjectured that

All the Phaenomena hitherto mentioned, proceeded from the progressive Motion of
Light and the Earth's annual Motion in its Orbit. For I perceived that, if light was propa-
gated in time, the apparent Place of a fixed object would not be the same when the Eye is at
Rest, as when it is moving in any other Direction, than that of the line passing through the
Eye and Object; and that, when the Eye is moving in different Directions, the apparent
place of the Object would be different.

There is a story that Bradley thought of this while walking in a downpour of rain.
As he walked faster, the rain – for him as a moving observer – no longer fell

vertically but came as from a source slightly in front of his line of motion. Since light has also a finite velocity, the same should happen for the positions of stars.

The second great impact of the finite velocity of light was interpreting the shifts of spectral lines. This 'Doppler' Shift is named after Christian Doppler (1803–53), an Austrian physicist who found that sounds changed pitch with change of relative velocity between sound source and observer. Doppler thought that the different colours of stars might be similarly produced. This turned out to be incorrect, however, because the spectrum extends beyond the red and the blue of the visible spectrum, so that what is normally infra-red becomes visible as red with rapid recession, and ultra-violet becomes visible as blue with rapid approach. So the observed colour remains the same, but spectral lines are shifted by Doppler's principle, as realized by Fizeau. This in turn led to Edwin Hubble's (1889–1953) discovery in 1916 – from his observation that spectral lines of galaxies have various degrees of red shift – that the galaxies are receding from us and each other, to give the current view of the expanding Universe.

So these studies scaled the Universe, and they upset the classical notion that perception is direct knowledge of objects. There is clearly uncertainty of what, where, and when as signalled by light. This uncertainty extends into the nervous system which receives signals from the world.

7   Links of Nerve

It is however easily proved that the soul feels those things that affect the body not in so far as it is in each member of the body, but only as it is in the brain, where the nerves by their movements convey to it the diverse actions of the external objects which touch the parts of the body.

René Descartes (1596–1650) [1]

It is an amazing thought that all our sensations and experience and so knowledge come from signals running to the brain down tiny cables: that the brain does not receive light, or sound, or touch or tickle, but only patterns in space and time of electrical pulses which must be read – decoded – before they can have reference to the world of objects.

It was unknown to Greek philosophers that there is anything like the complexity of links between the world and experience that are being so dramatically unravelled by physiology. Descartes had an inkling of this complexity and indirectness, but in the ancient world, from which epistemological traditions have come and remain in force, the links were supposed direct – and even to consist of the stuff of awareness, knowledge and Mind. This was the concept of the pneuma. For the Greeks, links giving sensation and links giving motion of the limbs seemed different. The latter were strings, by analogy with marionettes and the automatic theatres that were highly popular at least in Athens from the fourth century BC. We know that the afferent (input) nerves from the senses are in all essentials the same as the efferent (output) motor nerves to the muscles, controlling actions. That they are the same would surely have seemed incredible to the Greeks, for the marionettes did not have their own senses, and so there was no available mechanical model for considering links to the senses. In our terms, their view of sensory links is mystical, while their account of limb movements is too mechanical.

Pneuma in nerves

Although we know now that the brain receives electrical signals from the senses, carried by nerve fibres which conduct at about the speed of sound, none of this was dreamed of by the Greeks and was not known in any detail until this century. Accounts of perception that still carry authority were born before this basic scientific knowledge, and still live, with unfortunate results. It is necessary to backtrack into the history of nerve conduction to pin down some serious errors that can still make us stumble and lose our way. (We shall find other cases of extremely serious early errors, especially the Greek view of geometry which for over two thousand years misled logicians and gave unwarranted credibility to *a priori* knowledge: see pp. 226 *et seq.*)

The accepted concept of nerve conduction from the pre-Socratic Greek philosophers until the end of the eighteenth, or the early years of the nineteenth, century AD was that nerves are tubes conveying Vital Spirit, or 'pneuma'. This is the Aristotelian 'breath' which not only is the substance of the soul but also fills the space of the Universe. The Greek anatomist, Galen, whose authority remained unchallenged until the sixteenth century, discusses the earlier view of Erasistratus, that the Natural spirits give growth and nutrition and are distilled in the liver. They were supposed to be combined with air in the lungs (though 'air' was not as we think of it) to produce the Vital Spirits necessary for the essence of life. These were supposed to be transformed in the brain to produce the pneuma of nerves. Galen writes:[2]

> For how could the *nerve*, being simple, attract its nourishment, as do the composite veins, by virtue of the tendency of a vacuum to become refilled? For, although according to Erasistratus, it contains a cavity of sorts, this is not occupied with blood, but with *psychic pneuma* . . . thus there could never be in it a perceptible space entirely empty. And an emptied space which merely existed in theory could not compel the adjacent fluid to come and fill it.

I quote this passage, from a long discussion, to show something of the endeavour to find mechanistic explanations of nerve function along with the admittedly psychic and non-material spirit of the pneuma, and the soul.

The muscles were supposed to bulge when exerting force through being filled with pneuma from the nerves. This notion was disproved, with particularly neat experiments, by a Dutch microscopist Jan Swammerdam (1637–80) and, in 1677, by a Cambridge physician and founder member of the Royal Society of London, Francis Glisson, who went on to develop the important biological notion of 'irritability'. They found that muscles do not increase their volume when exerting force, though they look larger. This Swammerdam found by immersing muscle preparations of frog in a vessel filled with water, and noting the position of a bubble in a tube joined to the vessel containing the muscle. The bubble did not move so the muscle could not have changed in volume. Glisson found the same for living human subjects by immersing the arm in a long glass jar filled with water – which did not rise when the muscle was contracted. Since the volume remained unchanged, muscles could not be filled with Vital Spirit. This, at least, is so if the Vital Spirit or 'pneuma' is regarded as a kind of substance, occupying space. There was considerable conflict of opinion here, for the soul was often regarded as incorporeal and space-less, and yet the pneuma was occupying space in the nerve-tubes of the body. Discussions of this problem read very oddly to us, partly because it was thought (even as late as Descartes) that a vessel could not contain a vacuum without its walls caving in and touching. This is a beautiful example of technical inadequacy setting limits to theoretical concepts.

Understanding of nerve function did not change essentially before the development of the microscope, which allowed relevant structures to be described, and the discovery of effects of electricity on nerve by Galvani in the 1770s. Luigi Galvani of Bologna (1737–98) was a physician rather than physicist. Finding that when he touched a frog muscle with a pair of metal rods placed end to end, one of silver the other of brass, the muscle twitched, he concluded that the metals were conducting electric Vital Spirit – or pneuma – from the frog. This was called 'animal electricity', and gave rise to 'galvanism'. But instead of nerve action thus appearing as explicable on mechanical principles, electricity was taken to be the occult spirit of life. This was so, though these experiments led to the discovery and production of voltaic electricity, from zinc and silver plates in contact, each pair separated by leather washers to form voltaic piles.

As discussed above, many early electrical demonstrations with discharge tubes, as well as attractive forces of static electricity known as far back as Thales, must in any case have given electricity a lifelike character. It would seem to be a likely candidate as the vital force linking Matter and Mind (see pp. 142 *et seq.*).

As we all know, the Galvani experiments were misinterpreted. The acid on the skin of the frog produced the electricity, which then stimulated the nerve. But electricity was later found to be the key to understanding nerve function. This was first shown conclusively in 1843 by Emile du Bois-Reymond. He showed that a wave of electrical activity runs along nerve. The founder of modern physiology, Johannes Müller (1801–58), finally rejected the ancient 'animal spirits', while accepting that the nervous activity might be electrical: his main reason for accepting electricity was its great speed.

Electricity in nerves: action potentials
Francis Glisson put forward the notion that muscles have stored power which is released by nerve action. This notion of a small force releasing the large stored

force of the muscle is the meaning of 'irritability'. The mechanism of nerve actions was found to be similar, for the action potentials, discovered by Keith Lucas[3] and E.D. (later Lord) Adrian,[4] in the first decade of the present century, do not grow smaller however far they travel along the nerve. They are pulses of equal intensity, with varying frequency – which is the physical basis of sensation and of control of muscle. This, and the 'all-or-none' response of nerve are described by Adrian in *The Basis of Sensation* (1928) in this way:[5]

The stimulus, then, may be compared to the pressure on the trigger of a rifle: either it is strong enough to fire the bullet or it is too weak to do anything. The nerve fibre is not the only excitable tissue which reacts in this way, for the same kind of behaviour has been known for many years in the case of heart muscle. Here, too, the force of the contraction cannot be controlled by altering the strength of the stimulus, and the latter is either completely adequate or completely inadequate. These facts are best expressed by the statement that there is an 'all-or-nothing' relation between the stimulus and the activity which it produces.

The existence of the 'all-or-nothing' relation in nerve fibres means that as far as each impulse is concerned there is no possibility of gradation by changing the strength of the stimulus, but it does not mean that the total activity of the fibre cannot be graded, for it is obviously possible to control the total number of impulses which are set up and the frequency with which they recur (up to the limit imposed by the refractory period).

Adrian and Lucas made these discoveries with newly available electrical measuring instruments: the string galvanometer and early valve amplifiers. Just as telescopes revealed undreamed-of structures of the heavens, and microscopes new biological structures, electronics made it possible to record the functional activity of the nervous system. Electronics also provides concepts and models for describing the function of the nervous system. Much as the wheeled mechanisms of the Greeks provided models of the Universe of Matter, so perhaps electronics can provide models of Mind. Or can computers *be* Minds?

The recording of electrical nerve activity allowed Adrian to discover that continuous stimulation generally gives fading, and finally zero neural activity. This is adaptation to continuous stimulation, which is a mechanism favouring the signalling of change. Plato may indeed have wished to reject change from the Universe, but this is not the way of the nervous system. Adrian writes:[6]

It is easy to multiply instances of sensations fading owing to the adaptation of the receptors to a constant environment. We cease to be aware that our clothes are touching our bodies almost as soon as we put them on. This may be due partly to the diversion of our attention to more interesting topics, but even if we try to focus it on the body surface we find that there is little to feel as long as we are careful not to move. . . .

The fact that the receptors can be moved about in relation to the external world enlarges their scope enormously. To gain information about an environment there is no need to wait for it to change, for a motile animal can explore a stationary world by changing the relation of the receptors to the environment. Not only does this counteract the rapid adaptation which takes place in many of the receptors, but it enables us to extract information about the external world, not only from the exteroceptors on the surface of the body, but also from the proprioceptors – the highly efficient sensory apparatus in the muscles and joints.

The new detailed knowledge of nerve function has thus led to intimate tie-ups between physiology and sensation.

It was realized at the start of the nineteenth century that the different sensations are given according to which nerves are stimulated, and not according to which objects stimulate nerves. This distinction was realized in principle by Charles Bell in 1811, and was put forward in definitive form by Johannes Müller in 1826 and in 1836. Müller's Principle is known as 'specific energies' of nerves. The term 'energies' is, however, something of a misnomer, for as we have seen their energy is their least important attribute, as they trigger sensation from extremely low energies. (The energy consumption of the entire nervous system is probably less than 100 watts, though it contains well over 10,000,000,000 cell bodies, most of which are continuously active.) The point is not that energies are specific, but that sensations are given by nerves as they are connected to brain regions. The action potentials are virtually identical in all nerve fibres (including efferent motor nerves controlling muscle, as well as afferent sensory nerves). What matters is the brain region to which they are connected. If nerves were switched from, say, eyes to ears, we should hear sounds when the eyes were stimulated, and vice versa. The implication from neurological evidence is clear: sensations are generally given by stimulation of specific brain regions. An exception is pain. Pain is not produced by specific brain stimulation but seems rather to be given by disarray of patterns of signals.

If we consider, for example, sensations of colour, we find two further important principles. First, sensations are produced according to proportional mixtures of neural signals: all colours are given by three colour channels, as proposed by Thomas Young in 1801. Secondly, simple-seeming sensations may not be given by single signals but by complex mixtures of signals. Thus, yellow is given by a combination of red and green signals; that is, from channels tuned to red and green spectral wavelengths of light. Though yellow seems to be a simple sensation, it is physiologically complex. This implies that, in general, introspections of sensations are no sure guide to underlying mechanisms. Indeed, visual perception itself may seem to require merely opening the eyes in suitable light, but this should by no means suggest that the processes of perception are simple. From all that we now know, they are exceedingly complicated and exceedingly difficult to understand either in general principles or in physiological details. The brain (or the 'sensorium') usually gives appropriate sensation and perception in typical situations, but when receptors are stimulated in atypical ways then we are often misled. This follows directly from the specific energies characteristics of the nervous system, and yet is frequently forgotten. We should not be surprised to find atypical stimulation a source of many illusions (see pp. 404 *et seq.*).

The physics of nerve signals is now understood in detail. It is beautifully described by Bernard Katz in *Nerve, Muscle and Synapse* (1966). The detailed physics is not, however, our concern here. What I want to ask is: does this understanding place neural signals firmly within physics? Are 'messages', 'codes', 'information' and 'data' within physics? Consider a record of action potentials from nerve (fig. 27). This may be signalling colour, temperature, touch, tickle, taste – any of a very large number of possibilities. Now in order to know what it is signalling, we must know where this nerve is in the nervous system (which kind of receptor it serves) and also what stimulus situation the organism is in. We cannot read what is being signalled from the record alone – though when this same nerve, with this same activity, is signalling to our brain we do generally read it correctly. This at once implies that our brain requires 'knowledge' of which nerve this is or what receptor it serves, and also a great deal about the situation being signalled. This is typical of messages of all kinds. To understand a telegram, or a note left under the mat (let alone a book or a newspaper), background information and assumptions are required.

How is sensation related to impulse frequency? The early recordings showed that action potentials generally increase in frequency logarithmically with increase in stimulus intensity. So it may seem reasonable to assign neural impulse frequency directly to perceived intensity of sensation. But does it not follow necessarily that *sensation* is related to stimulus intensity in this way, even though

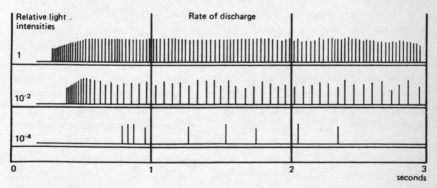

27 Nerve impulses, or action potentials. The stronger the stimulus, the greater the rate of firing. All sensory signals are of this form.

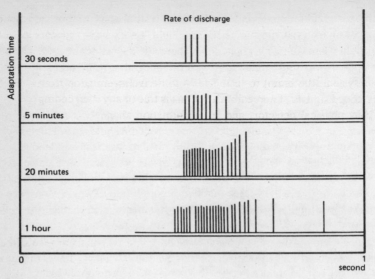

28 Dark adaptation of the eye increases its sensitivity; so the rate of firing of the optic nerve increases as for increasing stimulus (light) intensity. This is experienced as being dazzled in normal light after a period in the dark.

sensation is somehow given by the neural action potentials? It is even logically possible that increased intensity of sensation *could* be given by *reduction* in the frequency: it could be related in any way. It is an empirical question how sensation is related to frequency of nerve action potentials, even though in some way sensation is caused by this activity. It is an extraordinarily difficult relation to discover, because it is very difficult to measure intensity of sensation. It is, however, quite easy to measure sensitivities to stimuli, and to differences between stimuli. These are traditionally expressed as, respectively, absolute and differential thresholds. Although thresholds to stimulus changes are readily measured, it is still not the case that their relations can be deduced from impulse rates. Weber's Law may be due to this relation, or it may not.

There are, however, some threshold detection relations (with associated sensations of which we are aware) that can be deduced from what we know of nerve impulses, which we shall now consider.

Nerve function.
1 Time.   In order for frequencies to be estimated, a finite sample of instances is required. A nerve can fire up to 1,000 times per second. If it requires, say, ten action potentials for the nervous system to estimate the frequency, then time intervals shorter than $(1,000/10 = 100 = 0.001$ seconds) cannot be discriminated.

By comparison with corresponding electronic instruments, we are very poor at discriminating intervals of time. A flashing light, for example, appears as a continuous light when the rate is greater than about 100 flashes per second. The point is: we must expect this limitation, given the integrating time necessary for the nervous system (the brain) to 'estimate' – to derive information from – the frequency-coded signals of nerve fibres, for this is true of any such coding.

2   Any physical detector, and any information channel, is subject to some randomness, or 'noise'. We should therefore expect thresholds to fluctuate according to random variation in the frequency of firing of nerve. This is indeed the case. 'Thresholds' of detection are not sharp steps, from no detection to detection, or from no sensation to sensation. So-called thresholds are probability distributions of detection, the probability increasing with an increase in the difference between stimulus intensities, and so with an increase between frequencies of action potentials.

3   Since there must be some noise-induced action potentials at zero stimulus intensity, we should expect some sensation – for example of brightness in physical darkness. This is indeed the case. The visual field is never entirely black.

4   Since there is a finite range of frequency in nerve, and some randomness of the frequency, there must be a strict limit to the information that a nerve can convey. The limit to time discrimination mentioned in (1) above is an example of this limitation. The randomness of frequency, which 'blurs' detection of differences – see (2) above – is another limitation to information rate.

Information is defined in engineering channels (such as telephone lines) in terms of the number of changes which can be transmitted over a given time. This is obvious for the Morse code. The more dots and dashes that can be sent every second, the greater is the number of words that can be transmitted. If we now also consider the reliability of the transmission, we find more errors with increased rate, because the sample time for each dot and dash is reduced, so they are more easily confused with 'noise'. This in turn implies that unlikely messages must be transmitted more slowly than likely messages in order to be trusted. It is indeed the case that it takes longer to recognize improbable than it does to recognize probable objects, or situations. In the limit, it becomes impossible to convey highly unlikely information in a finite time – when it is more likely that the information channel is at fault than that the highly unlikely message is correct.

5   If nerve fibres back each other up as witnesses corroborate each other, then we should expect several functional fibres to be more reliable than one alone. Further, we should expect reliability to increase as the square root of the number of fibres, where all impulses are pooled equally. This theoretical square root function (which derives from statistical considerations) is indeed found for the increased sensitivity to differences of intensity of light on the retina, as the area of

the 'test' and the 'background' fields are increased. (This is known as Piper's Law, and is written for a 'test' field against darkness, or zero background, as $\Delta I \propto \sqrt{A}$. For the general case of a 'test' field, A_1, to be discriminated from a background field, A_2, this becomes – as confirmed by experiment –

$$\Delta I = 1/\sqrt{(A_1 + A_2)} = C$$

where C is the Weber constant.)

These points have been made sketchily, to indicate the problem of relating nervous activity with sensation. It is an immensely difficult problem. We can indeed only deduce limitations to sensation, and then only when we assume that the sensation is given from the known activity and not from other activity, including more central activity of the brain, which may not be closely related to what we know or can record. Considering recording activity in the brain itself: this has only recently become possible, and it is extremely important; but remember that it is only practicable to record from a very small number of cells (generally only one) at a time. It is impossible to say what else is going on – so again the data can be no more than suggestive, until we have a far more complete account of brain function and its relation to sensation.

Delay in the nerve link: the Personal Equation

Accurate mechanical clocks were not available until after Galileo's invention in about 1640 of the pendulum clock. He did not build this himself, as he was too old, but it was built by his son before he died in 1642. Astronomical observations were refined, as large quadrants and telescopes were built. The development of the mechanical clock gave a new task for the observer, who was now called upon to make new kinds of observations which taxed the human nervous system to its limit and beyond. It could have led to deep understanding of perception. As it was, it led to a drama – for drama it was – at the observatory at Greenwich. The highly romantic setting of the observatory which was founded by Charles II in 1675. At the end of the eighteenth century, when this crisis of the Science of Mind occurred, it was the centre of astronomical observation, of navigation and of time. It was here that the human observer was the key link between the positions of chosen stars and time. From the dropping of a ball high on the observatory at noon, ship's chronometers carried Greenwich time around the world for finding longitude at sea. To match the newly developed chronometer clocks, every effort was made to improve the accuracy of transit observations. It was hoped to determine time to within one-tenth of a second, using the so-called 'eye-and-ear' method invented by Newton's friend James Bradley. The first transit telescope was installed by the eighth Astronomer Royal, Neville Maskelyne (1732–1811). He was Astronomer Royal from 1765 to his death in 1811. During this period he set up a large transit telescope, which still exists,

29 The transit telescope at the Royal Observatory, Greenwich, where time was taken from the stars, and where the Personal Equation of observers was discovered, though not appreciated.

on the prime meridian (longitude 0°). Having taken every precaution to build the most accurate instrument possible, Maskelyne found that his observations did not agree with those of his assistant, Kinnebrooke. It was this disagreement which nearly led to a new paradigm for psychology – which might have been highly successful had it been followed.

The 'eye-and-ear' method required the observer to glance at the hands of the observatory pendulum clock as the star approached the critical hair-line marking the meridian, and then to count the seconds from the tick of the pendulum while he watched the transit of the star across the telescope's hair-lines. Intervals between the audible ticks, and of the visual separation of the hair-lines, could be judged and should (so it was believed) have given an accuracy of 0.1 second. To his immense consternation, Maskelyne found, upon comparing his times with Kinnebrooke's over a six-month period, that they differed by 0.8 second. This was an intolerable error. Maskelyne assumed that his observations were correct and those of his assistant were wrong. He sacked poor Kinnebrooke, saying that he had 'fallen into some irregular and confused method of his own'.

The details of the observations and comparisons were written up by Maskelyne in *Astronomical Observations at Greenwich* (1799).[7] This was noticed by the German astronomer and mathematician F. W. Bessel (1784–1846) at Königsberg. He realized its significance nearly thirty years after the event. Friedrich Bessel was highly interested in theories of measurement, and he was perhaps helped by Johann Karl Friedrich Gauss (1777–1855), who was Director of the Göttingen observatory. He was immensely important in appreciating the logical status of geometry (p. 226), and also the variation of observations. His book, *Theoria motus corporum coelestium*, appeared in 1809. Errors were now treated as a genuine scientific problem, following Laplace's earlier work – at last to receive experimental treatment by some of the best observers and physical scientists of the time, especially German astronomers.

Bessel compared his own transit times with Walbeck's, at Königsberg, in 1820. They selected ten stars, and each observed five on one night and the other five on the next night, alternating in this way for five nights. It was found that Bessel always recorded the transit earlier than Walbeck. The average difference was greater than that between Maskelyne's and Kinnebrooke's observations, being 1.041 seconds, with small variability. Perhaps fortunately, this is about the largest personal error ever recorded, though discrepancies of 0.5 second are not uncommon. Bessel continued this work by comparing himself with other skilled astronomers (Argelander and Struve), who also compared themselves with each other. In this way Bessel showed that each observer has his own individual personal error. This had some variability, but the observer differences were characteristic for each man. This led to transit observers being individually 'calibrated'

– to give their 'Personal Equations', which were applied to compensate their individual differences. So the most expert observers allowed themselves to be not only aided by instruments but also corrected by statistical averages from other observers. For the first time in science, the average took over to correct the individual testimony of expert observers.

In order to obtain more accurate Personal Equations, an experiment was set up between three observatories which were all in sight of a mountain. They could thus compare results with heliograph flashes, and also with sounds from a gun. At Greenwich, Personal Equations were applied from 1838 until automatic methods took over – to remove the human observer altogether as a link in the chain from stars to time.

The Personal Equation could correct for differences between observers, but it could not correct for absolute errors – for no one knew when the real or artificial star crossed the hair-line in fact, apart from human observation. Which, if any, of the observers was correct? This problem was not solved before the invention of the chronograph, in 1849. This was a smoked drum, rotated at constant speed, with electrically operated pens to record the times of events. But the transit of a star could not be recorded directly before the later invention of photoelectric cells, or of adequate photography. So the chronograph could not yet release the human observer as a necessary link from star to time. It did, however, allow the human response to *simulated* stars to be measured.

Physiologists had for many years been concerned with the conduction rate of nerves; but they were not able to measure this, or human 'reaction time', before the chronograph was invented. It was generally believed that nerves conducted at about the speed of light, so the reason for the enormous difference of individual Personal Equations was not at all clear. Some of the most distinguished physiologists gave 'mentalistic' explanations. This includes the father of modern physiology, Johannes Müller,[8] who wrote: 'It is well known that the sensorium does not readily perceive with equal distinctness two different impressions, and that, when several impressions are made on the nerves at the same time, the sensorium takes cognizance of but one only, or perceives them in succession.'[9] Müller considered three very different neural conduction rates, the highest being almost sixty times the velocity of light – which just shows that physics does not always help biologists! This was arrived at by assuming that the rate of flow of animal spirits in the nervous tubes and of blood in the arteries would be the same for vessels of the same size, and would vary inversely with the size of the vessels much as water moves faster in a narrower river. To accept this velocity is to us ridiculous, because we accept Einstein's upper limit of possible physical velocities as the velocity of light: 186,000 miles per second (3×10^{10} cm./sec.). Müller could not have known of this, but he did finally reject a velocity in excess of that of light as unlikely. He concluded, though, that 'We shall probably never attain the power of measuring the velocity of nervous action; for we have not the opportunity of comparing its propagation through immense space, as we have in the case of light.' He is referring here to the astronomical measures of Olaus Roemer, made a hundred and fifty years before he was writing (in 1685), which required the relatively enormous distance between Jupiter and earth for measuring light's velocity (see p. 199).

Physicists had considered how to measure the velocity of light over terrestrial rather than astronomical distances since Galileo, Dominique Arago (1786–1853), and Sir Charles Wheatstone (1802–75) in 1834 came near to it, with a revolving mirror method, but it was not accomplished until Armand Fizeau (1819–96) built an apparatus based on Wheatstone's method, but using a rotating cog-wheel (as mentioned on p. 200), which he completed in 1849. However, this method required a path length of a mile or so to get a detectable time-lag at a feasible rotational speed of the wheel, so this might not have affected Müller's judgement that the velocity of conduction in nerve could not be measured. It was, however, measured – just a year after Fizeau's measure for light – by Müller's assistant, Hermann von Helmholtz. Helmholtz (1821–94) went on to become the founder of the modern science of vision and of hearing, as well as being a major physicist.

Helmholtz measured neural condition time in 1850, first in the frog and then in a man, by stimulating him on the toe and the thigh and measuring his reaction time for these two positions, where the neural path length was longer by a known amount for one stimulus than the other. The reaction time was measured with the chronograph, which had just come into use by astronomers for measuring the Personal Equation. Helmholtz found that the velocity was not only less than for light: it was slower than sound! Müller at first rejected Helmholtz's measure, on the ground that the muscle twitch would add to the delay; but later he came to realize that the muscle twitch would be the same for either position of stimulation, and so could be discounted for the difference between the reaction times. This work was developed with more accurate measures by du Bois-Reymond, leading to detailed understanding of the form of the neural impulses, the 'all-or-none' principle, and the discovery that they are spikes of electricity, by Keith Lucas and E.D. Adrian at Cambridge, just before World War I.

Helmholtz's discovery that nervous conduction takes an appreciable time was at first resisted, but when accepted it had a profound effect on how man saw himself. Before this discovery it was thought that a person's body was all one. This is well expressed in a letter written to Hermann von Helmholtz by his father, upon first hearing of this discovery from his son: [10]

As regards your work, the results at first appeared to me surprising, since I regard the idea and its bodily expression not as successive, but as instantaneous, a single living act, that only becomes bodily and mental on reflection: and I could as little reconcile myself to your view, as I could admit a star that had disappeared in Abraham's time should still be visible.

Hermann von Helmholtz himself came to realize that the total reaction time to a stimulus is too long to be accounted for entirely by conduction time of the peripheral nerves. This led him to consider the time taken by the brain to process information (as we would say in modern terminology). So he ended up with a physical account, in terms of neural switching rates, of his teacher Johannes Müller's *mentalistic* ideas of the 'sensorium'.

The problem of the enormous variations of the Personal Equation was, however, very far from explained, for these were much too large for plausible variations of conduction time and variations of synaptic switching time in the brain. The key to this problem was not then found – and even now it is not always realized that the Personal Equation is not merely a matter of delay; for it can be *negative* in time from the moment of the star crossing the hair-line. All the earlier measures of the Personal Equation were relative between observers, for there was no independent way of knowing when the star reached the hair-line. The first absolute time measures were made by O.M.Michel, of the US Coast Survey, in 1858. Michel used a moving artificial star, with a chronograph for the time measures. His artificial star positions could be recorded at the crucial moment on the chronograph, as well as the observer's responses – so it was at last possible to measure the observer's time relation to events of the physical world. This was perhaps the first psychological simulator. Michel described his individual results as the 'absolute personality of the eye', and the 'absolute personality of the ear'; or of touch, which he also investigated.[11] It became clear with further experiments that the Personal Equation was affected by the intensity of the source (the stellar magnitude) and by many other variables, which at first were discussed in terms of astronomy but gradually became psychological as the characteristics of the observer became clear. Then attention moved to him, with the measuring tools of the Physical Sciences. Thus was experimental psychology born.

One might, however, be cynical at this point to say that in spite of this psychology was still-born. It was born with what is now to be seen as a sad deficiency: an almost total emphasis on reacting to present stimuli rather than to acting from stored knowledge. When absolute measures of the Personal Equation were first made, this distinction could and surely should have been made, and its significance realized, but this was not so.

J.Harmann found (with observations with an artificial moving star and chronograph) that the Personal Equation was not simply a variable delay. Some observers pressed the key before the star crossed the meridian hair-line. They had *negative* delay, in spite of the physiological lag of the nervous system established by Helmholtz. This showed that *anticipation* is an important part of the Personal Equation. This was a situation quite foreign to physics. It looked like an event before the cause.[12] But of course there was a cause – though this was *inside* the observer. The implication is that behaviour is largely anticipation from stored knowledge. This was, however, rejected, and has continued to be rejected by generations of psychologists.[13]

Psychologists are however traditionally interested in logical and also illogical thinking – to which we now turn.

8 Links of logic

Of science and logic he chatters,
As fine and as fast as he can;
Though I am no judge of such matters,
I'm sure he's a talented man.

W.M. Pread (1802–39) [1]

We have looked at some of the links, and their special characteristics, by which we know the physical world. Light and nerves are specially important, though we could have added the vibrations of the air giving sound, aromatic particles giving smell and taste, and the subtle transducers for touch and pain and the other proximal senses, which we might regard as the most immediate. They are, however, all distant by the neural pathways – and by the steps of inference – from stimulus to object as perceived; for how can the richness of objects, only a few of whose properties can be sensed at any moment, be given or signalled immediately? It is curious that the obvious importance of inference for perceptual knowledge has throughout the history of science been absurdly minimized, and this tendency persists strongly today.

If perceptions were simply selections of the physical world, this would be the end of it; but visual perception, and knowledge derived from vision, are far removed from the link of light and retinal images. Perceptual knowledge is almost as far removed from the physical inputs (stimuli) of touch, and the other more 'proximal' senses.

Perception (as we shall examine in detail later: chapter 12) evidently depends on elaborate and subtle inferences from neural signals. Perceptual inference is usually adequate, but it is not infallible and can lead us to make dramatic errors: illusions of many kinds. These can be used to discover the kinds of inference and assumptions on which perception depends. For now, though, we shall look at inference in philosophy and science. First of all it may be of interest to note what David Hume, writing before the study of perception became a scientific activity, accepted – that the perceptual knowledge of animals exceeds what philosophers discover by the inference of explicit logic. Hume's account represents what is still, perhaps, the generally held opinion: [2]

> . . . the animal infers some fact beyond what immediately strikes his senses, and that this inference is altogether founded on past experience. . . . It is impossible that this inference of the animal can be founded on any process of argument or reasoning by which he concludes that like events must follow like objects . . . any arguments of this nature [are] surely too abstruse. . . . Animals, therefore, are not guided in these inferences by reasoning; neither are children; neither are the generality of mankind in their ordinary actions, and conclusions; neither are philosophers themselves, who, in all the active parts of life are in the main the same with the vulgar. . . .

Hume concludes by discussing the wonder of instincts:

> But our wonder will perhaps cease or diminish when we consider that the experimental reasoning itself, which we possess in common with beasts, and on which the whole conduct of life depends, is nothing but a species of instinct or mechanical power that acts in us unknown to ourselves. . . .

We may conclude that Hume, writing before 1758, had no feel for the power of operations or procedures, which may be formalized and carried out by mechanisms. This seems hardly to have been realized before Boole and Babbage, and as a principle though not a practice by Helmholtz, at that time in the mid-nineteenth century; it was not applied to perception by computers for a further hundred years, and this is still in its infancy (see pp. 378 *et seq.*).

We are interested here in steps of inference as links to perception and knowledge of the world. It is generally agreed that there are two very different kinds of logic: Deduction and Induction. Although distinctions were drawn between them by Aristotle it has only recently become clear just how profoundly different are these kinds of inference. It was clear to Aristotle that Deductions go from general statements to less general or to particular statements, while Inductions go from individual instances to generalizations. Aristotle succeeded in presenting formal rules for some kinds of Deduction, which came to be called syllogisms. Deduction remains more 'formal' than Induction; though as we shall see, several attempts have been made to give Induction adequate rules. These, if found, might relate to how organisms learn how to predict – from generalizations of individual events as recognized by perception.

Perception is always of *particular* instances, but for predictions and understanding – indeed, for knowledge – *generalizations* are needed. Once knowledge is available as generalizations (for example, that some fruits are edible, others not; or that the sun rises at regular intervals) then behaviour can be predictive. It is possible to infer what will happen, or is likely to happen, or what the results of an action will be. This, surely, is necessary for intelligence.

Inference is generally discussed within the terms of textbooks of logic; but it is important for us to see how, if at all, the kinds of inference discussed by logicians are represented by brains, to give predictive intelligent behaviour.

Perhaps the first question to ask is whether Deduction and Induction are essentially different, or whether one can be reduced to the other. There have been many attempts to reduce Induction to Deduction – by suggesting that Inductions are incomplete Deductions, or that Deductions are exceptionally satisfactory examples of Inductive generalizations as derived from many repeated instances. John Stuart Mill tried to explain the certainty of arithmetic in this way. It is most important to realize that our current extremely sharp distinction between Deduction and Induction has by no means been accepted until recently, which makes it difficult to appreciate some of the doubts and confusions in the history of the philosophy of logic, and in the psychology of thinking and learning. This is not to say that we have no problems left – far from it – but our problems are different from those of the earlier writers such as David Hume and John Stuart Mill, who put so much of their thought into these questions and who are still very well worth reading.

We shall discuss first Deductive, and then Inductive inference, as providing links of logic from the world to Mind.

What are Deductive links?

Philosophers, we have found, generally seek certain knowledge, while scientists remain content with probable knowledge – expecting frequent revisions and occasional dramatic changes in principles and premises. Both philosophers and scientists rely on logic; but can logic be relied upon? Why should logic be trusted?

Conclusions given by Deduction are accepted as certainly true, when free from any kind of fallacy, provided the premises are true. But can Deduction give new knowledge? Can its conclusions from premises be both surprising and necessarily true? How could this be so if Deductions move from the general to the more particular – for are not particular conclusions part of, contained in, the premises? On the other hand, why should Inductions – moving as they do from particular instances to general conclusions – be trusted? It might be said that since the Greek philosophers, who started to ask such questions, the status of Deduction has fallen while the importance accorded to Induction has risen. But there are many modern philosophers from David Hume onwards who have cast grave doubt on the powers of Inductive inference. And there are some, especially Sir Karl Popper, who today deny Induction altogether as a means of gaining knowledge. Learning by animals and humans might, however, be regarded as gaining knowledge by Inductive inference, as Bertrand Russell, among philosophers, accepted, so we see here important conflicts of opinion spanning the centuries and still with us. We should look into this carefully, though we shall not need to go into much of the technical detail of modern logical studies, which are beyond my expertise.

We shall start by asking why Deduction works. To the Greeks, Deduction worked because the world was, as they supposed, constructed on logical principles. Causes linking events were supposed to be logical links in the Universe. Events were supposed to happen by logical necessity. So, since cause was a logical relation in reality, Deductive logic could reveal the structure of reality and predict future events and new properties of things with certainty. This notion was backed by the dramatic successes of geometry, as finally set out by Euclid.

For the Greeks, scientific laws were like legal laws, enforced by the will of the Gods. But how did they come to claim to know the will of the Gods? This depended on their intuition – their 'inner knowledge'. But why is this intuition to be trusted? There has indeed always been clear evidence against the trustworthiness of intuitive certainty – for individuals and schools of philosophers differ in what they claim in this way to be true. Given conflicts between some intuitions, how can any intuition be trusted? It was this deep difficulty that inspired, and has continued to breathe life into, the rival school of Empiricism. There seems to be but one alternative to some kind of Empiricism: to accept that the intuitions of some philosophers are better than conflicting intuitions of other 'lesser' philosophers. This generates arguments from authority, which are endemic to much of philosophy and all religion; and, curiously, Empiricism in some of its forms.

One can feel certainty that one understands something correctly, or that one appreciates the power of a demonstration to prove a conclusion. But when one discusses this with others, and finds that they disagree, what can one say or think, but that they are stupid? The triumph of Greek philosophy was to take the step of making arguments and assumptions explicit, so that each step and each assumption can be examined and questioned. In this, the Socratic method, the importance of clear argument and explicit analogy was realized, allowing Minds to be combined on difficult problems. It was from shared problem-solving and speculation that, from apparently irreconcilable views, individual opinions gave way to procedures of proof: to formal logic and later to scientific method.

Methods of proof could be applied automatically, by following rules. The most famous of these are the syllogisms, first described by Aristotle. They were intended to show how valid inferences can be made for deriving a conclusion from a major and a minor premise. It was recognized that if either premise was false, the conclusion could be untrue though the syllogistic form of the argument was correct; but the rules guaranteed that if both premises were true the conclusion would be true. It was claimed that, by applying the rules correctly, new knowledge could be derived. This claim is now generally abandoned.

As an example of a valid syllogism, we may take:

All philosophers are wise (*major premise*);
   Aristotle is a philosopher (*minor premise*);
Aristotle is wise (*conclusion*).

For an incorrect syllogism we may reverse the minor premise with the conclusion:

All philosophers are wise;
   Aristotle is wise;
Aristotle is a philosopher.

This is invalid because the class of what are wise is broader than the class of what are philosophers. It would hold only if one could not be wise without being a philosopher, but the major premise does not state this. It states the converse, that one cannot be a philosopher without being wise, which is not adequate for the conclusion. It is not adequate because the conclusion is more general than the premises allow. Even though the conclusion is (empirically) true, it is still a false, or invalid, syllogism.

Just as good as our valid first example is:

All men have green hair;
   Aristotle is a man;
Aristotle has green hair.

This is a correct syllogism, though the conclusion is false – because the major premise is false.

There are other correct forms of syllogism, such as:

No men have green hair;
   Aristotle has green hair;
Aristotle is not a man.

Here again the syllogism is correct though the conclusion is false – this time because the minor premise is false.

The forms of valid syllogisms are sometimes represented by a diagram known as the 'Square of Opposition' (fig. 30), where S stands for subject and P for predicate.

(1) and (2) – the two universal propositions – are in 'contrary opposition'. One at least must be false; both may be false. (1) and (4), and (2) and (3), which are propositions at the extremes of the diagonals of the square, stand in 'contradictory' opposition. One of them must be true and the other false. Each universal proposition (1) and (2) involves the truth of the particular proposition lying below it on the square: if (1) is true, (3) is true; and if (2) is true, (4) is true.

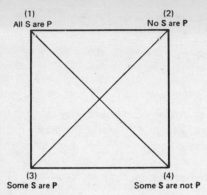

30   The Square of Opposition.

A proposition consists of linked terms, such as *'Plato* is a *man.'* Terms may be singular or plural, and they may be particular or universal. They are universal when they refer to the whole kind and particular when they refer to some instances of members of the kind. This may be stated as the distinction between classes and members of classes. It is often said that only members of classes exist, while classes are figments of Mind.

To return to our previous example of an invalid syllogism – one that breaks the rules:

> All philosophers are wise;
> Aristotle is wise;
> Aristotle is a philosopher.

The notion of wise is not necessarily 'contained in' the notion of 'philosopher', for we can imagine (and we may have read or personally met) philosophers who are not wise. To be wise is not part of the definition of 'philosopher'. So the sentence 'He is an unwise philosopher' is not self-contradictory. On the other hand, mortality would generally be regarded as a defining characteristic of being human. Then if we know that Aristotle is a man, we know that he is mortal. We may seem to be drawing an empirical conclusion from a logical step. This occurs when we fail to realize that, for example, 'mortality' is part of the definition we accept for 'man'. Having accepted it, we can argue from a single premise, 'Aristotle is a man,' to 'Aristotle is mortal.'

Now take an example where the conclusion is not part of the definition of the object in the major premise, such as:

> All philosophers are ugly;
> Aristotle is a philosopher;
> Aristotle is ugly.

Here 'ugly' is clearly not 'contained in' the meaning of the word 'philosopher' (Pierre Abélard was not ugly, at least to the eyes of Héloïse), and yet the conclusion does follow from the premises. But is this new knowledge? The best way to answer this question seems to be to say that it is new to someone who does not know both that all philosophers are ugly and that Aristotle is a philosopher. But it is not new knowledge in the sense that a new fact about the world has been discovered – for these propositions must already be accepted before the conclusion can be drawn. It is therefore useful to a particular person but it does not give new knowledge, as observation can discover new facts about the world, which can then be shared.

This remains even for such cases as:

Everybody was drowned on the *Water Sprite;*
   John Smith was on the *Water Sprite;*
John Smith was drowned.

For here again it is a matter of putting together already-known pieces of information. The conclusion is new only to those who did not know both that all were drowned, and that John Smith was on this ill-fated ship, the *Water Sprite*.

If Deductions could provide knowledge of the world beyond the knowledge stated in premises, then we would be faced with having to explain how Deductions can be both logically certain and empirically new. That Deductions can be both has generally been believed by philosophers. There are some who still hold to this position; but the evidence for this belief was based on the most profound mistake in human history: that geometry gives unquestionably true *a priori* knowledge of the world. We shall go on to discuss this, arguing that the case for Empiricism has been distorted throughout the history of philosophy by this cardinal error. It is, however, still difficult to be clear just what we should mean by 'empirical' knowledge (see p. 226).

To give true conclusions, Deductions must start from true premises. Where are premises to come from that can be accepted as true? For Aristotle, they come from wisdom. Wisdom is for him knowledge of the essence and first causes of things. This notion of 'cause' is mentalistic (or theistic), as Aristotle describes it in *Metaphysics:* [3]

The more a science is concerned with causes, the more instructive it will be: for an instructor is one who explains the causes of a thing . . . the most knowable things are first principles and causes, for it is through and from these that other things are known, and not they through the particulars falling under them. The most authoritative science, reigning supreme over the subsidiary, is that which knows for what purposes every act takes place, i.e. the final cause, the good in each particular instance, and in general the *summum bonum* in nature as a whole.

We see from this that Aristotle preferred to argue from the general to particulars deductively rather than inductively – as for him it is through first principles and causes that 'other things are known, and not through the particulars falling under them'. Aristotle's preference for Deduction dominated philosophy, until Francis Bacon began to formulate Inductive principles in the *Novum Organum* (1620); and even today there are distinguished philosophers who follow David Hume's doubts of Induction, especially Sir Karl Popper, who tries altogether to deny that knowledge can be gained by Induction. All this stems and follows (so it seems to me) from the philosopher's perennial quest for certainty.

It should at once be pointed out that we seldom argue or draw conclusions in this way in normal life, and we seldom demand certainty. In normal life arguments are, rather, of the form:

Most men are stronger than women;
    Albert is a man;
Albert is probably stronger than his sister.

This would not have been allowed as a valid syllogism by Aristotle, but it is very much how we derive conclusions that are useful and likely to be true if the assumptions on which they are based are true.

It is useful to recognize the forms of chains of statements deriving or suggesting conclusions. But no one now believes that we can deduce such things as the number of planets. Such facts at least are now regarded as contingent, only to be known by observation, and then perhaps but indirectly.

Deduction requires definitions. Further, by changing the definitions of the terms, an argument may change from Inductive to Deductive, or from Deductive to Inductive. This can be a trap for the unwary – and even for the wary. A famous example is the case of white and black swans. If we say 'All swans are white,' this may be a generalization from observations of birds, or it may be, or imply, a definition of the word 'swan'. If a non-white, but in other ways swan-like bird were denied the term 'swan', then the statement that 'All swans are white' would at once be seen to be a tautology. We know that it is logically necessary that all swans are white, because the term 'swan' is not accepted for any non-white bird. White is thus a defining characteristic of 'swan'. If white is not a defining characteristic, then a swan-like bird can be called a 'swan' even if it is black, and even if no black examples have been found. The term 'black swan' is self-contradictory in the first case, but acceptable (though perhaps highly unlikely to be used) in the second case. When black swan-like birds were discovered in Australia, they were called 'black swans'. At that moment we learned – or decided – that white is not a defining characteristic of 'swan'. From that moment 'Swans are white' was established as being an Inductive generalization, and not merely a definition.

Though it had held for two thousand years since Aristotle, it was abandoned as a Deduction, and as a true Induction – for black swans were allowed to be exceptions.

To take another example, Freud's term 'hysteria': when he applied it to men he was laughed at, as it refers explicitly to the womb; but here, as in many other cases, definitions change as use changes. Since Freud, we cannot deduce that a sufferer from hysteria is non-male, and so female. This illustrates clearly how conventional use of words may set what can be deduced.

Why should tautological Deductions be useful? There seem to be two reasons. In the first place, because our intelligence is limited a chain of explicit Deductions can lead to conclusions that surprise us. In the second place, Deductive mathematics and language structures may reflect structures of the world of objects. Mathematical examples of this are the relation between the diameter and the circumference of circles (pi); another is the Pythagoras relation ($h^2 = a^2 + b^2$) for right-angled plane triangles. Unlike the 'black swan' case, we believe that it is logically impossible to be confronted with any circle not obeying pi, or any right-angled plane triangle not obeying Pythagoras' Theorem. These may be useful as they seem to reflect basic features – basic facts – of the world in which we live. Although we can only measure actual circles with an accuracy of a few decimal places, we believe that the series as generated by automatic mathematical procedures (3.141592654 . . .) is a basic property of circles in our Universe. But pi may, in the last resort, be a constant in the logical structure of the way in which we describe the Universe, rather than of the Universe itself. This move, from assumed reality to our description of what we believe to be the case, allows us to regard mathematics as useful and certain and yet as dependent on empirical – ultimately observational – premises, which are never certain. There is some uncertainty in the application of all mathematics on this account.

One may say that pure mathematics can give certain answers, within its own terms, and that applied mathematics is formally certain, though it may give inappropriate answers when its constants and empirical data are incorrect, or not sufficiently precise. So, ultimately, we must allow the possibility that neither Pythagoras' Theorem nor pi will hold perfectly in a space that is not defined in such a way that they must hold true as part of the accepted definition of space.

It is most important to recognize as essentially different what is observed as a fact and what is defined as necessarily true. Induction applies to the former and Deduction to the latter. How, then, can geometry tell us about the world of objects as we also find them to be (though less accurately) by measurement? This question has the greatest historical importance, as for almost the entire history of philosophy the accepted answer was incorrect, and prevented acceptance of Deductions as tautologies – for it appeared that geometry, though Deductive, gave surprising knowledge of the world. This apparent example of empirical-looking knowledge from Deduction justified Rationalist metaphysics, which would not have seemed at all plausible if only non Euclidean geometries had been discovered, or invented, earlier, and preferably by the Greeks. But this did not happen until just after the lifetime of Immanuel Kant (1724–1804); so Kant and all previous philosophers based much of what they said on a most serious epistemological misconception.

How is geometry linked to the world?

For Euclid, geometry was linked to the world by axioms, which were intuitively certain and could not be questioned. Perhaps the most important example is the axiom that parallel lines never meet. Euclid derived useful properties of the forms of things from a few axioms and Deductive proofs. The proofs could be seen 'by inspection' to be valid by any man trained to think in logical form. Someone claiming to see the axioms as uncertain or the proofs as unsound would be labelled as mad, or an unreasonable fool. So, for axioms of such seeming certainty there was no effective criticism or contradiction. This state of affairs held for geometry from Euclid (c. 300 BC) until about AD 1830. Through all this time it was impossible to deny *a priori* knowledge. This was destroyed at a stroke when mathematicians, including the great Johann Karl Friedrich Gauss (1777–1855), started to write alternative geometries having non-Euclidean axioms. Gauss, it seems, had the basic concepts for twenty years, but kept them to himself, fearing, as he wrote in a letter, 'the hue and cry of blockheads' upon the announcement of the discovery. Some twenty years later, Nikolai Lobachevski (1793–1856) and Janos Bolyai (1802–60) developed self-consistent geometries as alternatives to Euclid's, and based on different axioms. This led George Riemann (in 1854, 1868) to speak of 'hypotheses' where his predecessors had spoken of 'axioms'.[4] This was the essential change and revolution of non-Euclidean geometry. Descriptions of space were now seen as hypotheses about the basic structure of the Universe, and not as certain bases of observation and measurement. Both Empiricism and Rationalism lost their foundations.

With the realization that descriptions of space are a matter of choice – though some may well be more convenient, more simple, or in various ways more useful than others – the Kantian theory of organization of perception, observation and measurement, by 'intuitive' classifications and innate knowledge of space, died. More generally, there remained no clear case to demonstrate any *a priori* knowledge. It does not so much matter to us whether or not Euclidean geometry should turn out to be 'true' or 'false', or even whether 'true' and 'false' can have their normal meanings in this context. Non-Euclidean geometries cast light on the logical status of axioms and on the relation between Deduction and Induction. This is central to understanding logic and thinking.

'Axioms' are the foundations of Deductive systems such as geometry. The problem here is: are they given *a priori* (as Kant believed for Euclid's axioms), or are they somehow empirical, given by observation? To consider one of the most interesting of Euclid's axioms, the postulate of the unique parallel, this states that through any one point not on a given line, one and only one line can be drawn parallel to a given line; or, in other words, parallel lines never meet. This is especially interesting because it can never be verified by experiment. It may be suggested by observation, and it may be regarded as some kind of extrapolation, or Induction (though not 'mathematical Induction'), but whatever one says, it goes beyond what can be known from observation; yet it is supposed to describe both the real world and the 'ideal' world of Euclidean geometry. Nowadays the word 'axiom' (or 'axiomatic') is often used in its original meaning of 'necessarily true', or *a priori* true. The discovery of non-Euclidean geometries has changed the technical meaning, if not yet the popular meaning. It is difficult now to accept that there is *a priori* certain knowledge of the world (as Kant among modern philosophers believed there to be), or to accept that there is certain empirical knowledge gained by observation. We would, however, add to Locke's 'knowledge through the senses', knowledge inherited from ancestral experience. There is no claim for this to be certain, in the sense that it cannot be doubted, as it was until a century ago thought that Euclid's axioms could not be doubted. We shall come to suggest that biologically inherited knowledge may have more significance than is commonly realized in current philosophy; but this is very different from the kind of intuitive certain knowledge that was accepted in philosophy before the insight of non-Euclidean geometries, which reduced the status of axioms to hypotheses that can be questioned.

The axioms of geometry were examples of our supposed intuitive knowledge of the essence of things – things or objects of the world being (in logical terms) subjects with characteristics that are predicates. This is a central feature of the Aristotelian subject–predicate logic, which dominated thinking on these questions and had profound effects on all thinking, in philosophy and science, until its demise in this present century.[5] (The traditional term 'subject' is confusing – it might be better in many cases to say 'object'. What has happened is that 'object' has changed its meaning with modern theories of perception. The ancient conception is of an object getting in the way (hence 'objection'), while we think of objects as things perceived or potentially perceived. I have kept, as is usual, to the classical terminology here and hope that it is not confusing.)

How is Aristotle's logic linked to the world?
Aristotelian and indeed all Greek logic was limited to what we call subject–predicate logic. This itself is linked closely to the notion that objects have

essences and properties. For example, it was held that being a man is to have (or to be made of) the essential essence of manhood. Men differ from each other, but only in 'accidental' ways: they share the common 'essence' (or 'essential') which constitutes them as being men. So it would be meaningful, though false, to say 'Aristotle is a green-haired man' because (we may assume) hair colour is not of the essence of man. To say, however, 'Aristotle is an unreasoning man' may be nonsense. It is nonsense if ability to reason is of the essence of man; then lack of reason in any x precludes x being a man.

We have, in Aristotelian logic, subject man and predicate reason (or green hair or whatever). Syllogisms are concerned to show which properties can be predicated to which subjects. The subjects may be individuals or classes. For Aristotle, existence requires a substance. ('Substance' is close to 'essence'.) Difficulties arose over imaginary entities and over 'empirically' false propositions or statements. For how could properties be predicated, or attributed, to non-existent subjects? How could a non-existent Aristotle be said to have green hair, or indeed brown hair? This was in part the reason, and the justification, for Plato's other world of Ideal Forms, which it was thought could provide the subjects needed for imaginary and false statements to have meaning.

The meaning of words was generally regarded as part of the essence of the object denoted by the word. Thus the word 'sun' would be as much part of the sun as the heat and brilliance of the sun. (This was fairly easy to maintain while there was, in Greece, only one common language. This is the basis of Nominalism.)

Aristotelian logic was limited to the subject–predicate form, but this is no longer so. We now include relations and class inclusion, which were not covered by Aristotle. Thus to say 'This dog is larger than the cat' is to state a relation ('larger than') which is not a property of the dog or the cat *per se*. Similarly, to say that 'Most men have jobs' is not to predicate a property to men, or to jobs, but rather to assign membership to a class: the class of those that have jobs. Criteria for class inclusion may change without change of description of the members. Here, it will depend on what is allowed as a 'job'; so the range and flexibility of logic have, in this century, been greatly increased from Aristotle's conception which virtually limited Deduction to syllogistic inferences. Much of the new power is given by developments in operating on symbols – hence 'symbolic logic'.

How is symbolic logic linked to the world?

Symbolic logic is characterized by extensions beyond the subject–predicate logic of Aristotle, to relations and classes, and to formulations that avoid ambiguities and other inadequacies of Natural language. Symbolic logic formalizes, and makes explicit, step-by-step procedures of thinking which are outside the grammar of language. It was invented by George Boole (1815–64), who invented class inclusion logic. This was developed by John Venn (1834–1923) for handling probability frequencies, as envisaged by Boole. Boole saw his system as exemplifying laws of thought. His emphasis on binary operations is now built into computer technology.

Principal logicians following Boole are Giuseppe Peano (1858–1932), who attempted to show that arithmetic was a direct extension of logic; George Cantor (1845–1918), who worked out how continuous functions and infinities can be expressed; and Gottlob Frege (1848–1925),[6] who distinguished between 'sense' and 'reference' of a proposition and defended the objectivity of numbers and relations, above all showing how symbolic logic can be made to reveal the sequential operations and hidden structure (as well as inadequacies) of Natural language. Bertrand Russell pulled this together, especially with his Theory of Descriptions which showed how proper names work and how language and logic can specify individuals. Wittgenstein provided the most subtle linguistic analyses, and 'language games', and devised Truth Tables for revealing general logical restraints.

All of this immensely important work has fundamentally affected philosophy. It may well be the basis of future high-level Artificial Intelligence, starting from George Boole's *The Mathematical Analysis of Logic* (1847). Boole regarded the study of logic as a study of Mind. What he did was to emphasize the importance of operations, which might apply to any number of different problems. Thus:[7]

> That which renders Logic possible, is the existence in our minds of general notions – our ability to conceive of a class, and to designate its individual members by a common name. The theory of Logic is thus intimately connected with that of language. A successful attempt to express logical propositions by symbols, the laws of whose combinations should be founded upon the laws of mental processes which they represent, would, so far, be a step toward a philosophical language.

For Boole, appropriate laws of logic are rules by which the Mind works. It is thus possible to discover how the Mind works by looking at the most effective rules of logic.

I am inclined to believe that such arguments must be treated cautiously, for it might be that explicitly formulated rules of logic are powerful just because the Mind does not normally work in this way – so it gets a special boost. Just this is so for hand tools: they are so effective because the hands are not like them. If our hands had screwdriver-like fingers, we would not need screwdrivers. So possibly Boole's and later logic are useful just because the Mind does not normally work this way. (The distinction may be described as 'multiplying' compared with 'adding': see p. 308). However this may be, symbolic logic is useful, and it suggests that simple mechanizing of at least some kinds of thought processes is possible.

The logician who is carrying out operations set out in formal terms is essentially acting as a defined machine: what he does may be done by a defined mechanism. This is the start of logical and mathematical machines. This notion is, indeed, to be found in Wittgenstein's *Tractatus* (1922), although he leads to it gradually. Wittgenstein suggests initially that we see the truth of logical relations intuitively (*Tractatus*, 5.13): 'That the truth of one proposition follows from the truth of other propositions, we perceive from the structure of the propositions;' and (5.131) '. . . these relations are internal, and exist as soon as, and by the very fact that, the propositions exist'. But this is related to meaning, and this is where the difficulties come in. As Wittgenstein puts it (5.14): 'If a proposition follows from another, then the latter says more than the former, the former less than the latter.' Then (5.21): 'We can bring out these internal relations in our manner of expression, by presenting a proposition as the result of an operation which produces it from other propositions (the bases of the operation).'

Wittgenstein distinguishes between logical (analytic) and empirical (synthetic) statements, by whether we can see 'from the symbol alone' whether the proposition is true:

It is the characteristic mark of logical propositions that one can perceive in the symbol alone that they are true; and this fact contains in itself the whole philosophy of logic. And so also it is one of the most important facts that the truth or falsehood of non-logical propositions can *not* be recognised from propositions alone [6.113].

But Wittgenstein holds that whether a proposition belongs to logic (i.e., is a tautology) can be calculated, by

calculating the logical properties of the *symbol*.

And this we do when we prove a logical proposition. For without troubling ourselves about a sense and a meaning, we form the logical propositions out of others by mere *symbolic rules* [6.126].

This is summed up (6.1262): 'Proof in logic is only a mechanical expedient to facilitate the recognition of tautology, where it is complicated.'

For Wittgenstein, then, we see logical necessities (which are tautologies, and cannot be surprising) intuitively, but mechanical processes may be needed for complicated relations where many steps are involved. It seems that the mechanical steps make explicit what we see for simple examples intuitively; so, it seems that (1) logical intuition can be mechanized, and (2) we are not as good as simple machines at carrying out several steps of logic.

Deductive logics are 'closed systems', for they do not for their truth require or at all depend on facts of the world. How then are they linked to the world? Wittgenstein tried to answer this question in the *Tractatus* by holding his 'picture theory' of language: that the structure of logic and language reflect the structure of the world. He later gave this up. It is expressed in the *Tractatus* (6.12): 'The fact that the propositions of logic are tautologies *shows* the formal – logical – properties of language, of the world.' This notion of the world as being logical in its structure is typical of Greek thought and central to Rationalism. On an empirical account, names become the links to the world, rather than the structure of logic or language.

Names have, however, always proved difficult for logicians to handle. It is here that Russell made a highly significant contribution with his Theory of Descriptions. This links logic and language to the world and also shows how it is possible to speak meaningfully about fictitious objects without having to hold (as Plato did) that they in some way exist. We can point to most physical objects. Their names can 'point to' them, almost as our fingers can point to them, when – in Russell's terminology – they have been given ostensive definitions from being pointed at and named. Particular objects can thus be given 'definite descriptions', having been ostensively defined from sensory experience, from being pointed out. But classes and abstractions cannot be experienced by the senses – and so cannot be defined ostensively from perceptual acquaintance. How, then, can they be described or named?

Russell's Theory of Descriptions was developed when he came to realize that on Plato's account the world is intolerably overcrowded with imaginary objects, and objects underlying all false (but meaningful) statements. 'Logic', he said (1919), 'must no more admit a unicorn than zoology can; for logic is concerned with the real world just as truly as zoology, though with its more abstract and general features.' [8] How can something that does not exist have a name? Until this is cleared up, there is a strong tendency to believe that because some 'thing' has a name, it must exist in the same way that objects that can be touched exist. Russell tackles this problem by using quantifiers. His favourite example is: 'Scott was the author of *Waverley*.' This becomes: 'There is an x such that x wrote *Waverley;* such that, for all y, if y wrote *Waverley,* y is identical with x, and such that x is identical with Scott.' For Russell, proper names (such as 'Scott') are substitutes for descriptions that the speaker has in Mind.

On this account it is meaningful to say that anything exists which satisfies the value of a variable. Indeed, as W. v. Quine has said: 'To be is to be the value of a variable.' [9] This leads away from the Aristotelian dependence on substance for existence, and away from the distinction between subject and predicate, towards purely predicative language; that is, a language of properties, and an account of objects as bundles of properties. This in turn makes us consider anew what we mean by saying that objects exist, or what we mean by 'physical objects'. As, for example, numbers have properties, numbers may be said to be objects – though not 'physical' objects, as tables are physical objects, because numbers cannot be seen or touched. It may be useful to speak of 'concrete' and 'abstract' objects. It is, however, controversial how best to distinguish between them. There is a case for saying that numbers, classes, and so on are objects because we can refer to them and because they have definable properties.

There is also a strong case for saying that we do not know physical objects such as tables anything like as directly by perception as is often assumed both in philosophy and common sense. Further, some 'physical objects', such as electrons, are often referred to by physicists as 'logical fictions', or as 'hypothetical entities'. Such considerations as these are highly important for comparing what we may call *conceptual* hypotheses with *perceptual* hypotheses, and for considering what we mean by 'objects'.

Aristotle and Euclid put their trust in intuitive knowledge of the essence of objects, and a few particular axioms concerning the structure of things. We no longer accept either as certain. The premises of Deductions now all look like – or are – hypotheses which might be incorrect. It may, however, be impossible to question some of our premises; for, as Wittgenstein puts it (*Tractatus*, 6.51), 'For doubt can only exist when there is a question; a question only where there is an answer, and this only when something *can* be said.'

So the only certainties are where there are not meaningful questions! But perhaps these certainties are *ipso facto* meaningless too; hence Wittgenstein's notion that things may *show* which cannot be stated (*Tractatus*, 7.00): 'Whereof one cannot speak, thereof one must be silent.'

Should Deduction be trusted?

We can all make mistakes. We have seen that seventeenth-century physics ousted the observer from descriptions of the physical world partly because observations are unreliable; and sought knowledge beyond what can be known by 'direct' sensory experience, from the use of instruments. I think that there is a similar trend in the history of inference. By demanding formal proofs, with the use of written symbols and, later, computer programs, inference as well as perception is removed as much as possible from human judgement. Ideally, inference should on this view be entirely automatic – independent of brains. Whether this is done by transforming written symbols according to 'rules of inference' or by mechanical or electronic means does not much matter.

There is something of a paradox in the philosophers' attitudes to inference, for they tend to think of it as the highest expression of human intelligence while, at the same time, making enormous efforts to mechanize it with symbols and rules – and so to remove it from human judgement. They are in effect mechanizing thought, preferring mechanically derived proofs to conclusions of human thinking. As we have seen, this started with Aristotle's syllogistic forms of argument and it has progressed (following lengthy stagnation of logical ideas) to the mathematical logic of Gottlob Frege, and Russell and Whitehead's *Principia Mathematica* (1910–13), to modern computing. A dramatic example is the recent proof, largely by machine, of the celebrated 'Four-Colour Theorem' which has defied the human ability of the best mathematicians for a hundred and fifty years. After hundreds of hours of computing, the machine arrived at a proof that four colours are indeed sufficient, for any two-dimensional map, for separating any adjacent countries or areas of whatever shapes.

This raises several questions, including: why should we trust the machine? Suppose it made a slip: would we recognize the mechanical error? If Deduction is mechanical, could there not always be mechanical slips producing errors? If so, how can Deductions (whether carried out by brains or machines) ever be certain? When intuition was accepted, it might have been thought that a proof was somehow internally certain, though each time it was carried out and expressed there might be mistakes in even the simplest operations. But intuition perhaps relies on mechanically followed rules, and these are clearly required for steps of inference.

It seems that we need a *caveat:* that a Deduction is only dependable on the assumption that the rules have been followed without error: so in this sense, Deduction needs Inductive faith. This does not, however, destroy the important essential distinctions between Deduction and Induction: that Deductions go from the general to the particular (Inductions vice versa) and that only Deductions depend on definitions.

It has, however, recently become apparent that there may be an empirical element in Deductions: for how can we know, or believe that we agree with others, the criteria for tautology? For example, the word 'bachelor' means the same as 'unmarried man'; but could one know this without experience of the world – experience of men, and marriage and so on? One might look the words up in a dictionary, or a thesaurus, and discover from such a source that these words are exact synonyms; but this is merely taking the editors of the books on trust, for basing the certainty of Deductions. Why should they know? If they define the meanings as the same, they are not necessarily reflecting general usage; and dictionaries can only recommend usages. So it seems that when Deductive (analytic) logic ceases to be 'closed' (as in formal examples with symbols having no reference) then – when it has links to the world – an empirical element enters in: to give the symbols meaning and set the criteria for the tautologies that give Deduction certainty.[10]

Finally, when the Austrian logician Kurt Gödel (1906–78) showed, in 1931, that no Deductive system can contain its own proof of consistency, the spectre of an endless set of wider and wider systems necessary for providing proof of consistency for the contained nested systems presented itself, and it is with us still.[11]

What are Inductive links?

Alexander Dumas (1824–95) once said: 'All generalisations are dangerous, even this one.'

Inductions move from particular instances to general conclusions. Aristotle appreciated the importance of deriving generalizations from particular instances. He stressed the importance of knowledge from experience of many instances with a remarkably modern example. Considering seeing an eclipse, he says: [12]

> . . . Even if we had been on the moon and seen the earth cutting off the sun's light from it, we should have perceived the present fact of the eclipse, but not at all why it happens, because the act of perception is not of the universal. Still, as a result of seeing this happen often we should have hunted for the universal and acquired demonstration; for the universal becomes clear from a plurity of particulars.

He concludes from this: 'The universal is valuable because it shows the cause, and therefore universal knowledge is more valuable than perception or intuitive knowledge. . . .'

Aristotle seems here to say that we appreciate causes in generalizations. He thinks of things and their causes as having essences, which can be appreciated by us from seeing instances. [13] For Aristotle, wisdom is the seeing of causes. [14] He gives due weight to experience, as experience produces science, inexperience chance. You have art where from many notions of experience there proceeds one universal judgement applying in all similar cases. Thus the judgement that a certain remedy was good for Callias, Socrates, and various other persons suffering from a particular disease, is a matter of experience; but judgement that such and such a remedy is good for all men of similar constitution (for example phlegmatic or bilious), suffering from such and such (a burning fever, for instance), belongs to science.

As a shorthand, philosophers are apt to speak of 'sequences of events' for forming Inductive generalizations. This can be misleading, though, for events are strictly unique and so cannot be repeated.

It is clear that although Aristotle is considered most for his work on Deduction, and especially the syllogism, he by no means ignored Induction, though formulations of Inductive procedures had to wait for the seventeenth and eighteenth centuries – to which time we now jump.

Induction according to Francis Bacon

Inductive procedures for science were first effectively discussed by the founder of the philosophy of science: Francis Bacon (1561–1626). Francis Bacon, Lord Verulam, Viscount of St Albans, was the son of Sir Nicholas Bacon, whom Queen Elizabeth 1 on her accession to the throne made Keeper of the Great Seal. Francis in his turn took over the great Office of Lord Chancellor, in which he was highly distinguished, though his career ended in tragedy as he was imprisoned for bribery. It is generally considered that he merely acted as was the custom, except that the bribes he accepted did not affect his judgements: he would accept the bribe but still convict!

Bacon's books on scientific method as he saw it – and indeed scientific method is very largely his creation – are *The Advancement of Learning* (1605) and, most fully worked out, the *Novum Organum* (1620), meaning the 'new instrument' for generating knowledge. Bacon saw science as an instrument giving power to men. Immensely ambitious – not only politically but also in his aim to demolish the current prestige of classical learning and erect a new kind of understanding through controlled and classified observations – he rejected Greek philosophy as 'the talk of idle old men to raw young fellows'. His principal objection was that, as he saw it, the Greeks made few useful discoveries. Bacon stressed the potential usefulness of science, for medicine, navigation and all kinds of industry. He might even be said to be a martyr to his practical science, for he died of a chill, caught on Hampstead Heath, after experimentally refrigerating a chicken with snow. He did not see science as important only for applications to industry but saw methodical experiment as the principal and perhaps the only way to discover the nature of things.

Bacon continually urged the importance of repeating observations, and observing phenomena under various conditions. This could include controlled conditions for a crucial experiment, though these were rarely performed in his day. By stressing the importance of repeated observation and experiments, he departed dramatically from the practice of the philosophers of his time (and, it must be admitted, of ours also). Repetition of observation, and controlled experiment in defined conditions, are the hallmarks of science, not of philosophy.

Bacon's procedures for gaining knowledge by science may be summarized thus:

1   To accept as data only the results of repeated observations.

2   To classify these data, by what he called 'tables of invention'.

3   To use the 'tables of invention' for comparing data, for establishing associations and predictions.

4   To derive 'minor generalizations' for inferring scientific laws.

5   To test these laws, by seeing whether they predict new phenomena.

This sounds like a statement of scientific method as we now understand it. Bacon is, however, often criticized at present, perhaps unfairly. He gave too little weight to the importance of imaginative leaps for generating hypotheses, but his insistence on repeating observations and experiments in many situations, and his dissatisfaction with Induction by Simple Enumeration (which is not always recognized by commentators on Bacon, though perfectly clear in many passages of the *Novum Organum*), and, above all, his emphasis on testing generalizations and supposed laws by setting up critical situations, are the very essence of modern science.

Since Bacon is important, interesting, and at present unduly neglected and sometimes misrepresented, I shall now give space to his own words. Of the senses he does not have much to say, except that all knowledge comes ultimately through the senses. He is concerned with their inadequacy as vehicles of truth, and is more concerned with illusions than the pure philosophers were, or indeed are. He even suggests (in *New Atlantis*) that there should be a House of Illusions for experiencing and understanding errors. He expresses his doubts of the senses in the *Novum Organum*: [15]

But by far the greatest hindrance and aberration of the human understanding proceeds from the dullness, incompetency, and deceptions of the senses; in that things which strike the sense outweigh things which do not immediately strike it, though they may be more important. Hence it is that speculation commonly ceases where sight ceases; insomuch that of things invisible there is little or no observation.

He says a little later that instruments such as the microscope are not a great help:

For the sense by itself is a thing infirm and erring; neither can instruments for enlarging or sharpening the senses do much; but all the truer kind of interpretation of nature is effected by instances and experiments fit and apposite; wherein the sense decides touching the experiment only, and the experiment touching the point in nature and the thing itself.

To try to counter errors of the senses Bacon proposed the bold step of requiring many confirming instances, and even more important, critical experiments: [16]

In general, however, there is taken for the material of philosophy either a great deal out of a few things, or a very little out of many things; so that on both sides philosophy is based on too narrow a foundation of experiment and natural history, and decides on the authority of too few cases, diligently examined and weighed, and leaves all the rest to meditation and agitation of wit.

He charges Aristotle with this fault: [17] 'The most conspicuous example . . . was Aristotle, who corrupted natural philosophy by his logic: fashioning the world out of categories.'

One might expect from this that Bacon would espouse the Empiricism of his day: far from it. He is more than frank as to its inadequacies, though he might approve of how it has subsequently developed – indeed in large part through the influence of the *Novum Organum*: [18]

But the Empirical school of philosophy gives birth to dogmas more deformed and monstrous than the Sophistical or Rational school. For it has its foundations not in the light of common notions (which though it be faint and superficial light, is yet in a manner universal and has reference to many things), but in the narrowness and darkness of a few experiments.

After admitting William Gilbert (1540–1603)[19] in his work on magnetism (*De Magnete*, 1600) as a possible exception, he continues:

I foresee that if ever men are roused by my admonitions to be-take themselves seriously to experiment and bid farewell to sophistical doctrines, then indeed through the premature hurry of the understanding to leap or fly to universals and principles of things, great danger may be apprehended from philosophies of this kind, against which evil we ought even now to prepare.

He complains further of the lack of reliable data in his time:[20]

Now for grounds of experience – since to experience we must come – we have as yet had either none or very weak ones; no search has been made to collect a store of particular observations sufficient either in number, or in kind, or in certainty, to inform the understanding, or in any way adequate. On the contrary, men of learning, but easy withal and idle, have taken for the construction or for the confirmation of their philosophy certain rumours and vague fames [rumours] or airs of experience, and allowed these the weight of lawful evidence.

Later he says:[21]

Now my directions for the interpretation of nature embrace two generic divisions: the one how to educe and form axioms from experience; the other how to deduce and derive new experiments from axioms. The former again is divided into three ministrations; a ministration to the sense, a ministration to the memory, and a ministration to the mind or reason. ['Educing' is deriving Inductive conclusions.]

On Induction he adds here: 'We must use induction, true and legitimate induction, which is the very key of interpretation.'

Bacon emphasizes his belief in the power of selective Induction – though not, be it noted, merely Induction by Simple Enumeration, which he explicitly and in several places disavows. He describes Induction as following and requiring the recognition of forms in Nature. This is very important, an essential part of his development of Induction for science from Simple Enumeration which he finds inadequate; though various kinds of Induction are essential for knowledge:[22]

The investigation of forms proceeds thus; a nature being given, we must first of all have a muster or presentation before the understanding of all known instances which agree in the same nature, though in substances the most unlike. And such collection must be made in the manner of a history, without premature speculation, or any great amount of subtelty.

He then gives an example, from many instances, of heat: [23] 'Secondly we must make a presentation to the understanding of instances in which the given nature is wanting; because the form . . . ought no less to be absent when the given nature is absent, than present when it is present.' He then raises an interesting difficulty:

But to note all these would be endless. . . . . The negatives should therefore be subjoined to the affirmatives, and the absence of given nature inquired of in those subjects only that are most akin to the others in which it is present and forthcoming. This I call the *Table of Deviation*, or of *Absence in Proximity*.

After examples of this, he considers his third principle, which we would now call correlation: [24]

Thirdly, we must make a presentation to the understanding of instances in which the nature under inquiry is found in different degrees, more or less; which must be done by making a comparison either of its increase or decrease. . . . For since the form of a thing is the very thing itself, and the thing differs from the form no otherwise than as the apparent differs from the real, or the external from the internal, or the thing in reference to man from the thing in reference to the universe, it necessarily follows that no nature can be taken of the true form, unless it always decrease when the nature in question decreases, and in like manner always increase when the nature in question increases. This Table therefore I call the *Table of Degrees* or the *Table of Comparison*.

After examples of this third principle, Bacon has more to say of his notion of forms. He stresses that these are not the structure of human classification (before science), or of how words categorize, or how language is structured and used. These can, however, confuse and make difficult the discovery of 'forms in nature': [25]

For when I speak of forms, I mean nothing more than those laws and determinations of absolute actuality which govern and constitute any simple nature, as heat, light, weight, in every kind of matter and subject that is susceptible of them. Thus the form of heat or the form of light is the same thing as the law of heat or the law of light.

So for Bacon laws, like objects, are facts of the world. To discover them, negative instances may be much more important than confirming instances, given by Simple Enumeration, which he many times dismisses as trivial except perhaps for confirming observations. We find again the importance of negative instances stressed later: [26] 'In the process of exclusion are laid the foundations of true induction, which however is not completed till it arrives at an affirmative.' Bacon regards his Tables as only the first stage of gaining knowledge by Induction. He points to various kinds of phenomena and observations that should be favoured. This introduces the notion of analogy, and follows the idea of forms favoured.

We may end this account of Bacon's philosophy of Induction, with what he calls 'fingerposts'. These are what we would call signposts: 'set up where roads part, to indicate the several directions.' Bacon's point here is fundamentally important: the notion of critical experiments, or observations, to decide between theoretical alternatives. He points out that 'Sometimes these instances of the fingerpost meet us accidentally among those already noticed, but for the most part they are new, and are expressly and designedly sought for and applied, and discovered only by earnest and active diligence.'[27] By this Bacon means the setting up of critical, or *crucial* experiments. Bacon hopes that his Tables and Methods of Induction may 'in very truth dissect nature, and discover the virtues and actions of bodies, with their laws as determined in matter; so that this science flows not merely from the nature of the mind, but also from the nature of things.'

Perhaps Bacon, rather than Locke, should be called the founder of modern Empiricism. He is surely the first to state a modern view of scientific method, and the first to appreciate the importance of Induction and analogy, and of critical experiment. One would, however, criticize him for undervaluing the importance (surely clearly demonstrated in his time) of instruments – such as aids to the senses given by the microscope and the telescope. This is surprising, because Bacon was well aware of the power of illusions to deceive. This is clear from his notion of the House of Illusion, designed to teach how we can be deceived. He suspected the telescope. After saying how it had shown Galileo new stars, the star structure of the Milky Way, and the moons of Jupiter, Bacon says of telescopes:[28] 'But, indeed, I rather incline to suspect them, because Experience seems wholly to rest on these few Particulars, without discovering, by the same means numerous others, equally worthy of search and enquiry.' Bacon's objection to the microscope is that only small objects can be seen with it. His emphasis on the importance of many instances for Induction and testing analogies led him astray here to reject the importance of the very technology he saw as the fruit of science.

Induction according to J.S. Mill

John Stuart Mill (1806–73),[29] the son of the distinguished Scottish philosopher James Mill (1773–1836), wrote extensively on Induction, and effectively codified Bacon's principal Methods. Unfortunately, however, Mill did not distinguish between Induction and Deduction as essentially different. We see this as a basic and most serious mistake. I shall now set out Mill's Methods of Induction, as they are a useful formulation.

1 Method of Agreement. If we can either find or produce the agent A in such varieties of circumstances that the different cases have no circumstances in common except A, then whatever effect we find in all our trials is indicated as the effect of A. This Mill defines in the *First Canon:*

If two or more instances of the phenomenon under investigation have only one circumstance in common, the circumstances in which alone all the instances agree is the cause (or effect) of the given phenomenon.

2 Method of Difference. Mill gives as an example: 'When a man is shot through the heart, it is by this method we know it was the gunshot wound which killed him, for he was in the fullness of life immediately before, all circumstances being the same except the wound.'
Second Canon:

If an instance in which the phenomenon under investigation occurs and an instance in which it does not occur have every circumstance in common save one, that one occurring only in the former, the circumstance in which alone the two instances differ is the effect, or cause, or an indispensable part of the cause, of the phenomenon.

Mill describes both of these as 'methods of elimination'.

3 Joint Method of Agreement and Difference. This applies 'when the agency by which we can produce the phenomenon is not that of one single antecedent, but a combination of antecedents which have no power of separating from each other and exhibiting apart'.
Third Canon:

If two or more instances in which the phenomenon occurs have only one circumstance in common, while two or more instances in which it does not occur have nothing in common save the absence of that circumstance, the circumstance in which alone the two sets of instances differ is the effect, or the cause, or an indispensable part of the cause, of the phenomenon.

4 Method of Residues. This applies where there are several terms left after applying the previous methods. Mill says of it: 'The method of residues is one of the most important among our instruments of discovery. Of all the methods of investigating laws of nature, this is the most fertile in unexpected results, often informing us of sequences in which neither the cause nor the effect were sufficiently conspicuous to attract themselves the attention of observers.'
Fourth Canon:

Subtuct [subtract] from any phenomenon such part as is known by previous inductions to be the effect of certain antecedents, and the residue of the phenomenon is the effect of the remaining antecedents.

5 Method of Concomitant Variations. This applies where conditions cannot be removed, but can only be varied. It is essentially correlation.
Fifth Canon:

Whatever phenomenon varies in any manner whenever another phenomenon varies in some particular manner is either a cause or an effect of that phenomenon, or is connected with it through some fact of causation.

Mill adds a warning which has only recently been fully accepted, and its importance realized:

The last clause is subjoined because it by no means follows, when two phenomena accompany each other in their variations, that the one is the cause and the other the effect. The same thing may and indeed must happen supposing them to be two different effects of a common cause; and by this method alone it would never be possible to ascertain which of the two suppositions is the true one. The only way to solve the doubt would be . . . by (ascertaining) whether we can produce the one set of variations by means of the other.

Mill takes an example to establish such asymmetries in Nature: 'By increasing the temperature of a body we increase its bulk, but by increasing its bulk we do not increase its temperature.'

Mill was by no means satisfied with his or any available account of Induction. He expressed his doubt vividly in a passage in *A System of Logic:*[30]

Why is a single instance, in some cases, sufficient for a complete Induction, while, in others, myriads of concurring instances, without a single exception known or presumed, go such a very little way toward establishing a universal proposition? Whoever can answer this question knows more of the philosophy of logic than the wisest of the ancients and has solved the problem of Induction.

Why indeed are some analogies powerful, others wildly misleading? This question strikes at the heart of the problem of Induction. We are still confronted with analogies that seem powerful and yet cannot be tested, for example consciousness in our friends, and in animals. Laplace put this well in *A Philosophical Essay on Probabilities* (1820):[31]

Analogy is based upon the probability, that similar things have causes of the same kind and produce the same effects. This probability increases as the similitude becomes more perfect. Thus we judge without doubt that beings provided with the same organs, doing the same things, experience the same sensations, and are moved by the same desires. The probability that the animals which resemble us have sensations analogous to ours, although a little inferior to that which is relative to individuals of our species, is still exceedingly great; and it has required all the influence of religious prejudices to make us think with some philosophers that animals are mere automatons.

Laplace then sets a limit to the analogy (but without being able to specify in general terms what this limit should be): 'Although there exists a great analogy between the organization of plants and that of animals, it does not seem to me sufficient to extend to vegetables the sense of feeling; but nothing authorizes us in denying it to them.'

Mill tackles the problem by suggesting that Induction and analogy work only when they reflect deep structures of the nature of things – as suggested also by Bacon. Mill says: [32]

We must first observe that there is a principle implied in the very statement of what Induction is; an assumption with regard to the course of nature and the order of the universe, namely, that there are such things in nature as parallel cases; that what happens once will, under a sufficient degree of similarity of circumstances, happen again, and not only again, but as often as the same circumstances recur. This, I say, is an assumption involved in every case of Induction. And, if we consult the actual course of nature, we find that the assumption is warranted. The universe, so far as known to us, is so constituted that whatever is true in any one case is true in all cases of a certain description; the only difficulty is to find what description.

Mill regards this general 'uniformity of Nature' as itself inductively derived – not as an initial Induction but rather as a final Induction, based on the success of innumerable small Inductions from the past.

Mill says rather more about assumptions behind Induction. He compares two similar-looking examples. The first is that 'All swans are white' (which was 'disproved' by the discovery of black swans in Australia, two thousand years after Aristotle gave 'All swans are white' as an example of a safe Inductive generalization, from which 'the next swan I shall see will be white' can be deduced). His counter-example is, that 'All men's heads grow above their shoulders and never below;' in spite, he adds, 'of the conflicting testimony of the naturalist Pliny'. Mill puts this:

As there were black swans, though civilized people had existed for three thousand years on the earth without meeting with them, may there not also be 'men whose heads do grow beneath their shoulders' notwithstanding a rather less perfect unanimity of negative testimony from observers? Most persons would answer, no; it was more credible that a bird should vary in its colour than that men should vary in the relative position of their principal organs. And there is no doubt that in so saying they would be right; but to say why they are right would be impossible without entering more deeply than is usually done into the true theory of Induction.

He then applies the same consideration to the general Uniformity of Nature notion:

Again, there are cases in which we reckon with the most unfailing confidence upon uniformity, and other cases in which we do not count upon it at all. In some we feel complete assurance that the future will resemble the past, the unknown be precisely similar to the known. In others, however invariable may be the result obtained from the instances which have been observed, we draw from them no more than a very feeble presumption that the like result will hold in all other cases.

An example that Mill gives of a safe Induction from but a single instance is when a trusted chemist announces the properties of a newly discovered substance: 'We feel assured that the conclusions he has arrived at will hold universally, though the Induction be founded but on a single instance. He goes on: 'Here, then, is a general law of nature inferred without hesitation from a single instance, a universal proposition from a singular one.' One might add that what he says here of single instances holds for perceptions, or reports of other's perceptions. These we accept as 'true' – and predictive (for example a warning of an approaching car, or of an attractive girl). But we do not trust all perceptions. Perhaps we only trust Inductions and perceptions when they agree with familiar, seemingly analogous, cases. There are, however, many analogies that have been accepted for centuries but which have turned out to be grossly misleading; for example, Aristotle's notion of the pituitary body in the brain (which we see – following Baconian-type experiments – as a gland producing important controlling substances) was seen as the source of the nasal phlegm which, with the rest of the brain, served to cool the blood. This analogy was so powerful that until after Vesalius (1514–64), the notion still survived that the pituitary body secretes 'pituita' to the nose.

Induction in practice

Does Induction, as Bacon and Mill saw it, work in practice? Some recent philosophers, as we shall see, have come to deny that Induction without hypotheses or 'models' has any power in science. Before considering this – which if true would have vital consequences for how we should look at cognitive processes as well as science – it may be useful to consider in some detail an example from the history of science where the procedures of Bacon and Mill had a striking success before there was any working hypothesis. Consider the discovery of the Periodic Table of the Elements by Dmitri Ivanov Mendeleev (1834–1907), who had no initial hypothesis for guiding the generalizations or for refuting, by fact or argument. The available data were very far from complete: his Inductions filled in vitally important gaps in the limited data.

Mendeleev started with John Dalton's suggestions, made in 1808, that every element has its own characteristic atomic weight. The problem then was: how do the properties that make them alike and different flow from that single constant? We now have theories of atomic structure; but Mendeleev tackled the problem purely by Inductive means, with no theory to guide him. He started by writing the names of the then-known elements on cards, and playing a classifying game with them – which his friends called his 'patience'. This is well described by Jacob Bronowski (1973) as follows: [33]

Mendeleev wrote on his cards the atoms with their atomic weights, and dealt them out in vertical columns in the order of their atomic weights. The lightest, hydrogen, he did not really know what to do with and he sensibly left it outside his scheme. The next in atomic weight is helium, but luckily Mendeleev did not know that because it had not yet been found on earth – it would have been an awkward maverick until its sister elements were found much later.

Mendeleev therefore began his first column with the element lithium, one of the alkali metals. So it is lithium (the lightest that he knew after hydrogen), then beryllium, then boron, then the familiar elements, carbon, nitrogen, oxygen, and then as the seventh in his column, fluorine. The next element in order of atomic weights is sodium, and since that has a family likeness to lithium, Mendeleev decided this was the place to start again and form a second column parallel to the first. The second column goes on with a sequence of familiar elements: magnesium, aluminium, silicon, phosphorus, sulphur, and chlorine . . . and sure enough, they make a complete column of seven, so that the last element, chlorine, stands in the same horizontal row as fluorine.

Evidently there is something in the sequence of atomic weights that is not accidental but systematic.

This worked out as columns of seven, until he came to gaps, due to elements not discovered at that time. Mendeleev stuck to his 'octave' scheme, leaving gaps that he expected to be filled later. They *were* filled – so his predictions on the basis of his Inductive scheme were justified. He predicted with accuracy the properties of entirely unknown substances which were later found. What more can one ask?

Here one may note a bizarre counter-example – using a hypothesis! Ten years before Mendeleev put forward his Periodic Table of Elements, a paper appeared in *Chemical News* of August 1865, entitled 'The Periodic Law' by a J.A.R.Newlands. Newlands arranged the elements in columns of eight, but not for Inductive reasons. Newlands had a theory, a theory so bizarre that he was laughed at and his table forgotten. He suggested that the elements form columns of eight by analogy with musical notation. This seems ridiculous to us, though Pythagoras in his time argued persuasively that the harmony of the heavens was related to the harmonies of vibrations of the strings of musical instruments. We can only conclude that Induction can sometimes work without any effective hypotheses; and that hypotheses and analogies can be highly misleading.

The humiliation of Induction

David Hume cast doubt on all knowledge by pointing out that Induction is needed for justifying Induction. Scientists are generally content to gain from Induction uncertain knowledge, which may be modified later. Philosophers, on the other hand, have tended to seek and demand no less than certainty. This has its origins in equating Induction with Deduction (usually by regarding Inductions as incomplete Deductions) and led to what must surely be regarded as sterile scepticism. This was especially the case with David Hume. In his *An Enquiry Concerning Human Understanding* (1758), Hume argues that we cannot justify Inductive inference, for, as he puts it:

> You must confess that the [Inductive] inference is not intuitive; neither is it demonstrative; of what nature is it then? To say it is experimental, is begging the question. For all inferences from experience suppose, as their foundation, that the future will resemble the past, and that similar powers will be conjoined with similar sensible qualities. If there be any suspicion that the course of nature may change, and that the past may be no rule for the future, all experience becomes useless, and can give rise to no inference or conclusion. It is impossible, therefore, that any argument from experience can prove the resemblance of the past to the future; since all these arguments are founded on the supposition of that resemblance.

The issue seems to turn on what we mean by 'justify' and 'demonstrative'. If Hume (like Mill) is trying to make Induction the same as Deduction, then he is right to be pessimistic. But why should Induction be like Deduction? Why should Inductive procedures be supposed to produce certain rather than probable knowledge? Hume gave up the 'problem of Induction' in despair, coming to regard knowledge as an 'irrational faith'. But for him the 'rational' had to be certain.

He distinguished between logical and psychological aspects of Induction. Although he rejected the logical (because it is not like Deduction), he allowed the psychological: that we (and other animals) do gain useful knowledge by repeated experience. Bertrand Russell sums up Hume's position with the terse comment: [34]

> Hume's philosophy . . . represents the bankruptcy of eighteenth-century reasonableness. . . . It is therefore important to discover whether there is any answer to Hume within a philosophy that is wholly or mainly empirical. If not, there is no intellectual difference between sanity and insanity. The lunatic who believes that he is a poached egg is to be condemned solely on the ground that he is in a minority. . . .

Although Russell himself departs from Hume and from Mill in their attempts to reduce Induction to Deduction, Russell develops a view of Induction related to Mill's notion that 'there are such things in nature as parallel cases. . . . The Universe . . . is so constituted that whatever is true in any case is true in all cases of a certain description; the only difficulty is to find what description.' Russell also follows Hume in distinguishing between logical and psychological aspects of Induction. In *The Problems of Philosophy* (1912), Russell writes: 'A horse which has been often driven along a certain road resists the attempt to drive him in a different direction. Domestic animals expect food when they see the person who usually feeds them.' [35] Russell then points out that such Inductions may be disconfirmed, and often are: 'The man who has fed the chicken every day through-out its life at last wrings its neck instead, showing that more refined views as to the uniformity of nature would have been useful to the chicken.'

This is surely a most important point. We do not expect any Induction (except indeed for the most general laws, those of physics) to continue or be a guide to prediction indefinitely. One's cigarette-lighter works for many consecutive times, but then it runs out of fuel. We both expect it to work (so that we use it) and we also expect that sooner or later it will not work — when it runs out of fuel or flint. Similarly, we expect to see the sun rising tomorrow (and that there will be a tomorrow) but also that, at some distant time, it will die and fail to rise. This point is sufficient to show that our expectations for Induction are very different from those for Deductions. It is merely silly to expect Induction to give certain knowl-edge – or certain predictions – for any generalization when there are other (usually wider) generalizations that set limits by running counter to them. The broader generalizations will also have their limits – until we reach the most general laws of physics, which seem to have no exceptions but to hold for ever.

Russell asks whether there are indeed general laws that have no exceptions. This to him is, or would be, the 'Uniformity of Nature'. He writes: [36]

> Science habitually assumes, at least as a working hypothesis, that general rules which have exceptions can be replaced by general rules which have no exceptions. 'Unsupported bodies in the air fall' is a rule to which balloons and aeroplanes are exceptions. But the laws of motion and the law of gravitation, which account for the fact that most bodies fall, also account for the fact that balloons and aeroplanes can rise; thus the laws of motion and the law of gravitation are not subject to these exceptions.

Russell adds that science has been 'remarkably successful' in finding 'uniformit-ies, such as the laws of motion and the law of gravitation, to which, so far as our experience goes, there are no exceptions'. He concludes, however, that although the Inductive principle cannot be disproved by experience, neither can it be proved. So Russell accepts the 'psychological' aspect, but rejects the attempt to make Induction logically certain, except by assuming the 'Principle of Induction'. This Russell develops in his last major philosophical work, *Human Knowledge: Its Scope and Limits* (1948). Here, Russell distinguishes between two kinds of knowledge: knowledge of facts, and knowledge of general connections between facts.[37] He suggests that, closely connected to this, is the distinction between knowledge that may be said to be 'mirroring' (as for Leibniz's Monads) and knowledge that is the capacity to handle or change (whose extreme form is Prag-matism). He continues to maintain Hume's distinction between the 'logical' and 'psychological' aspects of Induction and knowledge; thus: '. . . the psychology of general propositions is something very different from their logic. The psychology is what does take place when we believe them; the logic is perhaps what ought to take place if we were logical saints.'[38] It is fascinating that Bertrand Russell – the joint author of *Principia Mathematica* – concludes: 'what really constitutes belief in a general proposition is a mental habit'. He also allows that it may be acceptable to assign equal initial probabilities to alternative possibilities when we have no evidence for their probabilities. Thus, each of two possibilities of which we know nothing would be assigned a prior probability of 0.5. This idea was used by John Maynard Keynes. It implies – surely oddly – that *lack* of knowledge can be a basis or premise for inference. (As a student at Cambridge in 1948, I asked Bertrand Russell about this, and he said he was more doubtful than he allowed in *Human Knowledge,* which had just appeared.)[39]

Russell also considers a very different kind of principle for giving initial probabilities: the Postulate of Limited Variety. This is given somewhat implicitly by J.S.Mill, and explicitly by Keynes. It is a statement that the Universe is so constituted that Induction is possible. It attempts this by giving finite initial probabilities to observations by supposing that the Universe has only a limited number of states.

At the end of the last century, there were supposed to be ninety-two essentially different kinds of atom, imparting essentially different properties to all objects. Similarly, species were regarded as examples of Limited Variety, to justify classification and Inductive inference in biology. But with further knowledge of the structure and function of genes in biology, and of the structure of atomic particles in physics, this version of the Limited Variety Postulate no longer appears as a basic principle or basic structure of the Universe. Russell ends by rejecting the Postulate of Limited Variety as a postulate of scientific inference. What he does come to accept as a fundamental postulate is the notion of causal lines. He describes causal lines in the following way:[40]

A 'causal line', as I wish to define the term, is a temporal series of events so related that, given some of them, something can be inferred about the others whatever may be happening elsewhere. A causal line may always be regarded as the persistence of something – a person, a table, a photon, or what not. Throughout a given causal line, there may be constancy of quality, constancy of structure, or gradual change in either, but not sudden change of any considerable magnitude. . . .

That there are such more or less self-determined causal processes is in no degree logically necessary, but is, I think, one of the fundamental postulates of science. . . .

Russell continues:

It is in virtue of the truth of this postulate – if it is true – that we are able to acquire partial knowledge in spite of our enormous ignorance. That the universe is a system of interconnected parts may be true, but can only be discovered if some parts can, in some degree, be known independently of other parts. It is this that our postulate makes possible.

This notion of underlying structures of the Universe – allowing some Inductions to be viable, others not – may (I believe) be found in Francis Bacon's *Novum Organum* and may lie behind the notion of critical experiments which he introduced. Thus:[41]

> In establishing axioms, another form of induction must be devised than has hitherto been employed, and it must be used for proving and discovering not first principles (as they are called) only, but also the lesser axioms, and the middle, and indeed all. For the induction which proceeds by simple enumeration is childish: its conclusions are precarious and exposed to peril from a contradictory instance; and it generally decides on too small a number of facts, and on those only which are at hand. But the induction which is to be available for the discovery and demonstration of sciences and arts, must analyse nature by proper rejections and exclusions; and then, after a sufficient number of negatives, come to a conclusion on the affirmative instances.

And a little later: 'And this induction must be used not only to discover axioms, but also in the formation of notions. And it is in this induction that our chief hope lies.'

There could hardly be a clearer statement. Similar accounts are developed by recent writers on Induction, as we shall now see.

H. Reichenbach's formulation is that we set up what he calls 'posits', some of which are found to work. Russell at first rejected this notion, but later came to accept it – but only when applied to 'non-manufactured' situations or predicates. This is an important qualification. Reichenbach states his notion:[42]

> . . . a way out of Hume's scepticism can be shown when knowledge is conceived, not as a system of propositions having determinable truth value or probability value, but as a system of posits used as tools for predicting the future. The question of whether the inductive inference represents a good tool can then be answered in the affirmative by means of considerations which do not use inductive inferences and therefore are not circular.

It is generally agreed that if this is a workable way of looking at induction, it can only apply (as Russell argued in *Human Knowledge*) to some situations – for one can invent any number of 'artificial' situations in which it clearly will not work. It works when the 'posit' matches relevant features of the deep structure of the world. Kantian philosophers hold that we know something of this hidden structure innately.

Intuitive theories

Rudolf Carnap (1964) holds that we have an 'intuitive sense of inductive validity'. This is of course open to the objections of all Intuitionist positions: why is one man's intuition different from another's if any can be accepted? The fact that so many theories have been put forward, and that there are so many contradictory

Inductions, suggests that we cannot trust our intuitions on the nature of Induction. This is so for all intuitions. We might, however, look at this as a problem to be decided by experiment. The experiments may concern learning in animals or men; or in machines designed to derive Inductive conclusions.

Meanwhile we may consider the philosophical view that there is no such thing as Induction. I shall consider Sir Karl Popper's views in some detail, including his notions of hypotheses and verification of hypotheses, for his views are highly influential and bear directly on a principal concern of this book: the status of hypotheses, both perceptual and conceptual.

Should Induction be trusted?[43]

Deduction is inferring from the general to the particular: Induction is inferring from the particular to the general. The celebrated 'problem of Induction' is to justify this procedure, to show that Inductions can be necessarily true. In a weaker form, it is to show how probable, not certain, general conclusions can be derived from instances.

To risk an Induction at this state: philosophers have a predilection for certain knowledge, while scientists are content with ever-changing hypotheses which they hope will gradually asymptote to the truth. The question for them is not 'How can Inductive conclusions be certain?', but rather 'How can Inductions, from many instances, give any more knowledge than do single instances?' There are, as we shall see, distinguished philosophers of science, especially Sir Karl Popper, who deny that Induction can do this – or that it is any use. This is a strong recent challenge to the founder of the philosophy of science, Francis Bacon.

Gaining knowledge by refuting hypotheses: Popper's view

We have seen something of Bacon's view of scientific method and we have looked at his and later methods of Inductive inference. It is time now to look in more detail at how conclusions (or hypotheses) are derived in science. We may then be in a position to ask, with some hope of an answer, 'Does the brain work like science works?' or: 'Is Mind like Science?' Central to this question is the status of Induction.

Bacon held that scientific knowledge is Inductive generalizations given both by Simple Enumeration of instances and by 'crucial' decisions at choice-points for deciding between alternative possibilities. He did not distinguish between Deduction and Induction as we now do (which is why I have given considerable space on this issue) but in other ways his views are indeed the foundation of modern philosophy of science. His views are, however, challenged in some quarters – even to denying all Induction – and there has been a great deal of vigorous discussion on what it is to verify or to refute a hypothesis. All this is in the Positivist tradition, following the French philosopher Auguste Comte (1798–1857).[44] Since Comte, there has been a growing appreciation of the importance of hypotheses – to the extent of saying that perceptions and all knowledge are hypotheses. I shall discuss Popper's view in this light.

Sir Karl Popper has had a major impact on the philosophy of science. His position on Induction is radical in denying not only – as also did Hume, as we have seen (pp. 244 *et seq.*) – that Induction can be justified as a procedure for deriving certain knowledge, but more extreme, that Induction is powerless to provide any scientific knowledge. He extends this to denying that Induction provides knowledge for men or animals in learning. He thus denies both Hume's 'logical' and Hume's 'psychological' Induction. This denial extends to saying that when we believe that we have derived knowledge by Induction, this belief is a kind of 'illusion'.

Popper does not, either, believe that knowledge can be derived by Deduction. What Popper does hold is that scientific knowledge can be gained only by putting up 'risky' hypotheses – and refuting them. Hypotheses remaining, Popper holds, may be true: though scientific knowledge is, for Popper, always tentative. Hypotheses are regarded as limited in power and scope. He does not allow that hypotheses have predictive power, or that they have antecedent probabilities. They are regarded as statements that may or may not correspond with fact. If they do not correspond they are in danger; and they are a danger to, and may disconfirm or refute, conflicting statements or hypotheses. This is part of what Popper calls an 'objective' theory of knowledge, where beliefs of observers are replaced by operational statements. The observer is essentially left out. Although Popper is a kind of Empiricist, he quotes with approval Parmenides' dictum: 'Most mortals have nothing in their erring intellect unless it got there through their erring senses.' Popper does not however have much to say about perception, as he rejects observers. (This is a problem for us, as we come to see perceptions as hypotheses.)

Popper's philosophy of science has been well received by many highly distinguished scientists. It has been especially welcomed for its emphasis on the importance of framing hypotheses so that they might be refuted. Freud (for example) is often criticized for advancing ideas that cannot be tested by any conceivable contradictory data. Even the attempt to criticize may be 'explained away', as a 'symptom' of aggression, or some such. Consider birth trauma as an explanation of adult problems: could one use Caesarian births as controls for this Freudian theory? Apparently not, for 'birth trauma' gets redefined: Caesarians can suffer, though they are not born in the usual way, from 'the shock of fist experience', or something similar. Thus the possiblity of finding controls – and so refutation – slips away. With this may slip away the power of the original hyothesis.

Popper generally takes his examples from physics. His favourite examples of 'good' hypotheses are Newton's and Einstein's accounts of astronomical space. To Popper, the vulnerability of Newton's theory to such tests as the bending of light round massive bodies (measured from shifts of star images near the sun during a total eclipse) makes this a 'good' hypothesis – provided it is not immediately changed to encompass the new data. Thus, hypotheses are not for Popper 'open-ended'. Like boys playing soldiers they must know when they are dead, and not answer back when 'killed'. This emphasis on the importance of framing hypotheses so that critical aspects are made explicit, and designing experiments to attack at weak places, has undoubtedly had good effect on recent biological science. So, even if Popper's extreme form of Refutationism does not survive, his influence is highly significant.

Considering Induction, Popper starts with Hume's doubts. He accepts as fundamentally important Hume's distinction between 'logical' and 'psychological' aspects of Induction. However, he then departs radically from this – and almost every other writer – on these questions, with his Principle of Transference: 'What is true in logic is true in psychology.'

The Principle of Transference is designed to 'translate all the subjective or psychological terms, especially belief etc., into objective terms'. Popper continues: [45]

> Thus instead of speaking of a 'belief', I speak, say, of a 'statement', or of an 'explanatory theory'; and instead of an 'impression', I speak of an 'observation statement' or of a 'test statement'; and instead of the 'justification of a belief', I speak of 'justification of the claim that a theory is true', etc.

Popper confesses that his Principle of Transference is 'admittedly a somewhat daring conjecture in the psychology of cognition or of thought processes'. But Popper is nothing if not daring. Here he is acting on his own precepts – by putting up a refutable hypothesis. In this spirit, it is up to us now to try to refute it, or at least to test its strength. He could hardly object, as this is his own advice.

What, exactly, is the Principle of Transference, on which Popper puts so much weight, namely that 'What is true in logic is true in psychology'? It attempts to transfer 'subjective' to 'objective' knowledge. By this step, Popper claims to have transferred the logical problem of Induction as Hume saw it to the psychological problem, and then claims that 'it will be clear that my principle of transference guarantees the elimination of Hume's irrationalism', continuing (using H_{ps} for Hume's psychological problem): 'If I can answer his [Hume's] main problem of induction, including H_{ps}, without violating the principle of transference, then there can be no clash between logic and psychology, and therefore no conclusion of our understanding is irrational.'

Popper's most radical notions (the Principle of Transference and his rejection of Induction) are stated in *Objective Knowledge* (1972):

. . . since Hume is right that there is no such thing as induction by repetition in *logic*, by the principle of transference there cannot be any such thing in *psychology* (or in scientific method, or in the history of science). . . . The idea of induction by repetition must be due to an error – a kind of optical illusion. In brief *there is no such thing as induction by repetition*.

Does this not look like a magician's amazing disappearing rabbit? If the principle of transference were accepted, does the 'problem of Induction' disappear? Is there, indeed, any such problem to worry us – if we do not expect *certainty* from Inductions – or try to use Induction to justify itself? But Popper is being far more radical: he is denying all Inductions.

What Popper is doing here is trying to make knowledge 'objective', by saying that 'What is true in logic is true in psychology' with the related statement, 'There can be no clash between logic and psychology, and therefore no conclusion of our understanding can be irrational.' But consider the kinds of statements on Induction that Popper objects to. They are products of people (and so in this sense 'psychological') but they are not all regarded by Popper as rational. Can we not use this very example as a *reductio ad absurdum* refutation of Popper's position? This brings out one of the principal problems of a radical Refutationist philosophy: namely, that it is not always by any means clear what does or could refute the most explicit statements. This is so especially if antecedent probabilities and predictions are denied to hypotheses, as Popper denies them. I must confess to feeling like turning his own argument round, and using his conclusions as evidence against his premises – is this playing the game? The outcome is important for it is possible that Popper's position has dangers greater than the 'open-ended' kind of science that he attacks. If neither are refutable he cannot claim the victory – and open-ended science has the advantage that it can grow without cataclysmic revolutions at each hitch, jolt or surprise (and it may well be preferred as an *account* of, rather than a *rejection* of, Mind).[46]

Russell takes a diametrically opposite view of the 'psychological' aspect of Induction, holding that, strictly, logicians would not even know what one meant by a hen, or other common object, given a category name – without its being introduced by pointing to examples. Is this a logical issue, or is it a question for psychology? Popper rejects Induction for all learning.

It is surely a scientific question whether we find sufficient evidence for or against Popper's notion that learning is limited to hypothesis rejection. Popper says that the idea of Induction by Enumeration 'must be due to an error – a kind of optical illusion'. This is saying that what appear to be facts are not facts. But how can there be illusion or error in Popper's terms? Are these not sometimes clear examples of 'irrational' beliefs – which Popper denies are possible by his Transference Principle?

Most of us, and perhaps all scientists concerned with learning, take the Popper-banned view that much of learning is Inductive generalization. Are we irrational, or just plain factually wrong? Does a philosopher (as a non-expert on the psychology or physiology of learning) have the right to legislate here? We may at least doubt that philosophy has such power. It is surely an empirical question to be decided by experimental observations of behaviour to discover the strategies that animals and men adopt for gaining knowledge; in this sense the problem of Induction is scientific rather than philosophical.

Popper's most celebrated idea is perhaps that knowledge can only be gained by refuting hypotheses. As we have seen, for Popper, hypotheses cannot be confirmed but only refuted by evidence. It is important to realize at once that his notion of the uses and powers of hypotheses is not what is usually accepted by 'hypotheses'.

For Popper, hypotheses (1) do not have prior probabilities (so none are more likely than any others); (2) cannot predict to new situations; and (3) cannot confer probabilities to other hypotheses – that is, for Popper they have no predictive power. One may wonder if such emasculation has not killed the golden goose (if the implied change of gender may be excused!), for what power is left to hypotheses? But, however this may be, Popper's emphasis on framing hypotheses so that they are vulnerable to attack by evidence has been extremely well received by the most highly distinguished scientists – and clearly it has served to make us critical of many dubious claims. The biologist Sir Peter Medawar[47] and the physicist Sir Hermann Bondi were quick to see its importance. Perhaps the only criticism that they make (and this is muted) is that theories should be to a degree 'open-ended', to accommodate new data and ideas. As Bondi says:[48]

If, then, we have a theory that is in some sense an open theory, then we can accommodate at least some new discoveries. The theory will retain its utility even after the discovery of new things . . . take Newton's second law of dynamics – that the rate of change of momentum of a body equals the force applied. It is a perfectly precise statement, but it leaves it entirely open to you to put in under the heading of 'force' any force so far discovered. And if you find some new kind of force there is no reason why you should not put that in. The theory has a ready-made place in it for putting in something new and unexpected. That does not mean that the theory cannot be disproved; indeed, we know that Newtonian dynamics has been superseded by relativistic dynamics. It does not mean that the theory does not say anything; everything that you can disprove says something. But nevertheless it is not a closed theory. More than that, I regard it as an essential of any scientific theory to have room for putting in what one does not know yet.

Also, surely, if a scientist gives up at the first – or indeed even after many – difficulties or apparent objections, he will not develop his ideas or discover and correct artefacts, errors or illicit assumptions. Science does not have such precise rules as games, where one knows when one has 'lost'. There are many cases in science of 'losers' later judged to be 'winners'. Theories have been 'refuted', only to be reinstated later. For example, that the earth is a sphere rather than flat was a hypothesis generally held in classical Greece, then dropped through the Middle Ages, to be resurrected around the time of Columbus. At some times there has seemed to be irrefutable evidence for rejecting the round earth hypothesis – that we do not fall off, that people on the other side would have terrible headaches, that the oceans would drain away – and yet these objections, which may have looked crucial, melt and disappear with a change of paradigm. This admittedly rather trite example may serve to show that open-ended theories should be encouraged, and that we cannot always (and indeed perhaps ever) judge with certainty when a hypothesis is 'refuted'.

There is more to be said of Popper's claim that the only way to gain knowledge is to put up hypotheses, which may be refuted by even a single observation, but never confirmed by any number of observations. In the first place, since observations themselves are not certain – because there can be illusions, errors of interpretation and so on – individual observations cannot be held capable of refuting with certainty. If we now allow several observations, the added observations cannot help the refutation unless they confirm each other. But this, again, is just what Popper has rejected. In rejecting Induction by instances, and rejecting the notion that multiple observations tend to be more reliable than single observations, Popper tries to throw out much of what Bacon emphasized as vital to successful science and he rejects the whole of statistical methods. Can Popper justify this radical departure from what is universally regarded as essential for justifying scientific claims?

Further, if we cannot assume inductively that our apparatus remains in calibration from day to day, and that our methods remain appropriate, we have no power to refute by experiment or observation. Refutation by observation depends on just the Inductive beliefs that Popper rejects.

Considering now confirmation by multiple observations: this is standard practice in all sciences. In general (when errors are random) precision increases as the square root of the number of readings taken, and this gain is vital for detecting small effects, let alone making numerical estimates where averages are usually preferred to single readings. Summing signals is the basic way of gaining information of weak signals masked by 'noise'. It is essential for detecting weak sources in radio astronomy, as it is for establishing the electrical patterns of brain activity by the technique of Evoked Potentials, which have to be repeated many times and summed by computer. The point is that the noise disturbance (which underlies all detection, whether by instruments or sense organs) is varying positively and negatively around a mean value, but a signal is asymmetrical; so by integration in time or with several instances the signal increases faster than identically integrated noise. Therefore, with integration or summed repetitions the wanted signal emerges through the unwanted masking noise. This is extremely familiar to any experimental scientist and essential in astronomy and electrophysiology.

Summing of readings or signals is a simple kind of Induction by Simple Enumeration, and it is very powerful. For more subtle use of multiple observations, we may cite a particular experiment which is one of Popper's favourite examples (though for different reasons): the 1919 eclipse expedition which provided data contrary to Newton's theory but compatible with Einstein's, by detecting apparent shifts of stars as predicted by Einstein's notion that light is bent to a certain amount by gravitational fields. This experiment relied on sophisticated measures of photographs taken with telescopes near the limits of their resolution, and under difficult conditions of daylight turbulence and sudden temperature drop during the exposure. Not just one star but as many stars as possible near the sun were photographed, and their positions measured and averaged for the final result. Any single measure could show a shift due to atmospheric turbulence or some such, so averaging was essential.

Errors and mistakes of very many kinds occur even in the best laboratories. Sometimes, indeed, they lead to important new findings (such as the contamination of a culture slide that led Alexander Fleming to discover penicillin).[49] There may be a large chance element in the generation of hypotheses, as for all inventions. The role of chance and the importance of averaging many readings – which at once make the concepts of 'observation' and of 'experimental test' complicated – are surely underplayed by too many philosophers of science who are not familiar with what is needed. To practising scientists the importance of repeating observations is as certain as anything they ever discover about the world. To deny the gaining of knowledge by multiple instances runs counter to the first principles of the experimental techniques that have generated the Natural Sciences. What philosophical arguments could prevail over this evidence?

Popper does on occasion admit that refutation cannot be certain. In the first place, there is the logical point: how can an observation (or set of observations) refute a proposition? He may answer that he has converted his observations into statements that are propositions; but is this adequate? This point is, however, of more formal than practical importance. More serious is the doubt as to whether observations are 'neutral'. If N.R. Hanson (1958), Imre Lakatos (1970) and T.S. Kuhn (1962) are right – as we have every reason to believe that they are – observations and experimental tests are highly theory-laden. A great deal has to be accepted before observations can be made at all, or their relevance and significance appreciated. To take an example: there is a vast network of theory before spectral lines can be 'seen' as representing energies of quantum jumps, or electron spins in atoms. Even the most precise measures of spectral lines are meaningless and have no power without a vast network of assumptions. Popper does accept at least some of this. In *Logic of Scientific Discovery* (1959), he writes: [50]

In point of fact, no conclusive disproof of a theory can ever be produced; for it is always possible to say that the experimental results are not reliable or that the discrepancies which are asserted to exist between the experimental results and the theory are only apparent and that they will disappear with the advance of our understanding.

This has been discussed – to lead to distinctions between 'Falsificationism' and 'Naive Falsificationism'. Thus Thomas Kuhn comments on this aspect of Popper's position: [51]

Having barred conclusive disproof, he [Popper] has provided no substitute for it, and the relation he does employ remains that of logical falsification. Though he is not a naïve falsificationist, Sir Karl may, I suggest, legitimately be treated as one.

These may seem hard words, but the issue is too important to sweep under the carpet. To reject almost the whole of Baconian Induction – substituting positing and refuting but never confirming hypotheses – is undoubtedly a daring and exciting move. But it is essential to decide whether this is indeed the *only* way that science and organisms can gain knowledge.

Historically, as we have seen, it is clear that Bacon did not accept Simple Enumeration as adequate for Induction. He emphasized the importance of 'crucial' experiments. He was clear that Induction required prior selection of what was likely to be relevant and important. He held that many positive instances could be useful; but he clearly preferred negative instances for ruling out possibilities, or areas of possibility. When the dust has settled I think that Bacon will appear essentially correct: not as he is often represented, but as he expresses himself in the *Novum Organum,* where he both emphasizes tests and allows Induction by instances and analogies an important place in learning and science.

Can knowledge be objective?

Popper rejects subjective or psychological probability, and attempts to develop an epistemology that is free of observers holding beliefs. This is the essential thrust of his Principle of Transference: that 'What is true in logic is true in psychology.'[52]

Popper rejects prior probabilities, conditional probabilities, including inverse probabilities, and the predictive power of hypotheses. This can hardly go unchallenged. One might, indeed, ask: if all this is denied to hypotheses, what power is left for them? What interest can they have? Consider, for example, that if hypotheses have no prior probabilities, then no one hypothesis is more likely than any other conceivable alternative. Suppose that someone puts up the hypothesis:

'The moon is made of Danish Blue cheese.' Would support be found for gaining knowledge by refuting this hypothesis? It would surely be given too low a prior probability to justify the cost of refutation; but according to Popper it deserves refutation as much as any other possibility that may be put up. There are many interesting borderline cases at the present time: intelligent messages from outer space; telepathy; psychokinesis; gravity waves: Although exciting, these command little support simply because they or their evidence is judged unlikely.

The probabilities that scientists assign to hypotheses change through the history of science. Popper considers this, but discusses it in terms of 'mistakes'. This has been commented on by Kuhn in an interesting passage where he compares his position with Popper's: [53]

> In our view then, no mistake was made in arriving at the Ptolemaic system, and it is therefore difficult for me to understand what Sir Karl has in mind when he calls that system, or any other out-of-date theory, a mistake. At most one might wish to say that a theory which was not previously a mistake has become one or that a scientist has made the mistake of clinging to a theory for too long.

Surely this tangle is generated by Popper's rejection of subjective probability. More importantly, if one does not adopt a probability approach to what is *likely* to be worth looking at, one is most unlikely to see anything significant – or, in a deeper sense, to see anything at all.

Rejection of subjective (observer's) probabilities is linked to Popper's rejection of prior probabilities and the predictive power of hypotheses. He does not accept that we can say that one hypothesis is more likely than another, or that hypotheses suggest predictions. Having refuted a set of hypotheses, the probabilities of remaining hypotheses are not, on this view, increased. The most that Popper allows is that *'The assumption of the truth of test statements sometimes allows us to justify the claim that an explanatory universal theory is false.'* [54] He does not, however, allow that remaining candidates have a higher probability after the demise of their rivals. He sometimes speaks of theories, or hypotheses, as having a 'fitness to survive' – something like individuals or species in the Darwinian paradigm of Evolution by the survival of the fittest. This occurs apart from the observer (as we judge from fossils left from times long before there were at least human observers), and so, in this sense, Darwinian Selection is 'objective'. Popper wishes to make knowledge objective along Darwinian lines. He takes a particularly apt analogy, comparing scientific theories with biological species. This analogy is implicit in the following quotation: [55]

> Instead of discussing the 'probability' of a hypothesis we should try to assess what tests, what trials, it has withstood; that is, we should try to assess how far it has been able to prove its fitness to survive by standing up to tests. In brief, we should try to assess how far it has been 'corroborated'.

The point is then made explicitly on the Darwinian metaphor with a comment on the above passage in *Objective Knowledge*:[56]

Nobody expects that a species which has survived in the past will therefore survive in the future: all the species which ever failed to survive some period of time t have survived up to that time t. It would be absurd to suggest that Darwinian survival involves, somehow, an expectation that every species that has so far survived will continue to survive. (Who would say that the expectation of our own survival is very high?)

Thus Popper holds that good theories are like fit species and yet says that it is 'absurd' to expect fit species to survive. This is consistent with the account of death as caused by species still more fit to survive, and of refutation by rival theories having more power. What does seem unclear is what this power of theories is – if hypotheses are powerless to predict.

The notion of hypotheses being born perhaps by accidents, and living until killed by rivals, is attractive, and there is now a school of Epistemology based on this analogy with Natural Selection. The best account, with a most impressive bibliography of early versions of this concept, is the American philosopher and social scientist Donald Campbell's paper *Evolutionary Epistemology* (1974), and his unpublished William James Lectures. Related ideas have been worked out in detail by Stephen Toulmin, in *Human Understanding, Vol. I: The Evolution of Collective Understanding* (1972), and in other papers cited by Campbell.

This analogy may well be appropriate and useful without the claim that it characterizes the only way of gaining knowledge; though this is Popper's position. Whether or not this is so, the account seems to me to have the following problems, or questions requiring answers:

1   What 'kills off' hypotheses? If they are killed by lack of practical usefulness we have a form of Pragmatism, which Popper would reject.

2.   What initiates new 'mutations'? Are these purely by chance; or are they given by suggestions from existing knowledge, or by human intentions or aims? Popper seems to deny both of these, so I am not sure what his position is here.

3   Is the gaining of knowledge as 'blind' as Evolution: is no guiding wisdom allowed to us? Popper does speak of some hypotheses put up for testing as more worthwhile than others; but I am not sure how this fits with his rejection of probability and prediction.

4   What sets the life-span (or generation time) of hypotheses? (This becomes, surely: how tough are hypotheses against 'refutations' or whatever it is that kills them?)

5   Do we know enough about the processes of Natural Selection to develop a sufficiently rich analogy for the gaining of knowledge, by analogy from how the design problems of organisms have been solved in the Evolution of species? (Here one may be worried by R. C. Lewontin's doubts expressed in 1974 that evolutionary processes, and their relation to genetics, are known in detail.[57]

The possible application of mechanisms of Evolution should however give epistemology an empirical flavour, which could be healthy. These comments of mine are no more than notes. They hint at issues which I shall go on to discuss.

As Donald Campbell says: 'An evolutionary epistemology would be at minimum an epistemology taking cognizance of and compatible with man's status as a product of biological and social evolution.'[58] The question, surely, is whether we have moved so far from our biological origins, in our knowledge and our ways of gaining knowledge, that the evolutionary model blinds us to what we now are, and to what we are capable of discovering, inventing and doing. Granted that the development of life has required problem-solving beyond us to understand – it by no means follows that our problem-solving, and the development of our hypotheses formulated in symbols of language and mathematics, are like Natural Selection. This analogy is itself a hypothesis, which may be true or false – and if true may not describe the *only* way we can gain knowledge. Flowers are very different from the earth in which they grow!

Conclusions on Popper's philosophy of Science and Mind
We shall now summarize what we have learned, by considering especially Popper's views in the context of earlier writers. This will lead us to try to derive an account of scientific method to serve as the basis for a paradigm for thinking about intelligence: for how the brain achieves perception, learning and thinking.

1   That some scientific knowledge may be true; but that it is always tentative
This is the least challenging statement from Popper to be considered. It does, however, have implications for his views on Hume's dissatisfaction with Induction. Hume considered that to be 'rational' we must have certain knowledge – such as is found in some Deductions, especially in mathematics. Popper claims to have 'solved the problem of Induction', but he evidently does not claim to make, or to see, scientific knowledge as certain. So it is not quite Hume's problem that Popper claims to have solved, by his Refutationism, in which he rejects Induction as regarded by Bacon *et al*. But the point here is that Hume's worries about lack of certainty in empirical matters, we might say, softly and suddenly vanish away, with the invention of geometries alternative to Euclid's – when geometry and all applied mathematics became open to doubt.

2   That scientists should put up risky hypotheses, which should 'live dangerously'
It is this that has been very generally accepted as excellent advice. It will be accepted here. But do the scientifically untutored brains of animals and men

function in this way? Does, or can, science succeed *only* by the deaths of risky hypotheses? Is this the whole of learning?

3   That there is not, and cannot be, Induction by Enumeration of instances

To deny that knowledge can be gained by Enumerative Induction – by organisms or by science – is to make a general statement that seems to be an empirical statement. If true, it would be of the first importance for theories of learning; for the philosophy and conduct of science; and for Artificial Intelligence. If it is an empirical statement it must, in Popper's terms, be a hypothesis; and if a useful hypothesis it must be open to challenge by facts.

It might be called a negative hypothesis, for it suggests what cannot be. Are we allowed to try to refute a 'negative' hypothesis with 'positive' facts? On Popper's premises, I assume that even one good example of learning by Induction would refute his negative hypothesis. Similarly, the negative hypothesis 'Bricks cannot be made without straw' would be refuted by one batch of bricks made without straw – provided these were allowed as 'bricks'. Or 'Intelligence requires a brain' would be refuted by one example of an admittedly intelligent computer-machine (unless a computer is accepted as a 'brain'). We may look, then, at animal and human learning for evidence against Popper's hypothesis, or claim, or stricture, that 'Induction is mere illusion.'

It has been suggested that we should distinguish between Induction by Simple Enumeration, and Induction by selective enumeration. Popper's position would, however, reject both. We shall proceed to look at learning in animals, to try to decide whether untutored brains adopt only trial-and-error procedures which might be regarded as putting up and refuting hypotheses, as Popper allows, or whether they follow Inductive procedures, which Popper rejects for epistemology and the psychology of learning. Experiments on animal learning should have philosophical interest if they provide evidence for alternative theories of how knowledge can be gained.

4   That scientific knowledge is 'objective'

This is bound up with the notion that 'belief', 'impressions', and so on should be translated into 'test statements' that might settle issues without observers. Observers drop out with the notion of 'objective knowledge'. It is therefore a daring extension of physics (but is also a version of solipsism).

This is closely related to several other cardinal positions in Popper's philosophy, especially his denial that hypotheses have prior probabilities or predictive power, his denial of applications of Bayes's theorem of inverse probabilities, or of any subjective probabilities. The essential feature is that experience is rejected, or discounted. It thus bears close relations to the Behaviourism especially associated with J. B. Watson and B. F. Skinner (see pp. 279 *et seq.*)

5   That 'What is true in logic is true in psychology'

This is Popper's Principle of Transference. He admits it to be 'a somewhat daring conjecture in the psychology of cognition and thought processes'. Why, given this, is it so difficult to be 'rational' if we are 'incapable of irrationality'? Life might be simpler if we were, but on any acceptable criteria for reason we seem all too often to depart from it. Since Popper denies Induction, he must mean, by 'logic', Deduction; but what is 'irrational' Deduction?

The point of this notion is to make descriptions of psychology 'objective', as descriptions of physics are 'objective'. It seems, however, to be riddled by paradoxes. How can one philosopher object that another is irrational? Do we know how to describe human behaviour, or for that matter problem-solving, in terms of logic? Certainly Hume and Russell thought not. This principle is surely difficult to defend. I do not see how it can be accepted or used. At best, we are only generally 'logical' when (as now) this is our explicit aim. Many psychological accounts (especially Freud) emphasize illogical thinking and what seems to be irrational behaviour as more or less typical of human beings.

6   That refuting scientific hypotheses does not increase probabilities of alternative hypotheses

Given this, what is the point of rejecting or refuting hypotheses for gaining knowledge? If remaining alternatives do not gain prior probabilities for acceptance, where is the progress of science? This, again, is rejection of Bayesian strategies: it rejects the notion that our probabilities change as possibilities are ruled out or introduced, or themselves change in probability of acceptance. Would he apply this strategy to 'Twenty Questions'? (Imagine not learning, from previous answers, to 'home in' on now more likely answers.) Is it supposed to apply only to scientific hypotheses? If so, why is strategy appropriate for them so essentially different from strategies efficient for 'Twenty Questions' and other thinking and discovery?

One might say that 'Twenty Questions' strategies work only in finite situations, where there is Limited Variety, and where it can be ordered (see p. 247). Perhaps the Universe is infinite, with unlimited variety. But this, surely, is a guess about the nature of the Universe: a guess that may be wrong. If, indeed, we found that our predictions improve as we refute more and more hypotheses, we might come to *explain* our increasing success with a Limited Variety Postulate.

7   That scientific hypotheses are not predictive

Although Popper rejects prediction as a power of hypotheses, he recommends that they should be made to 'live dangerously'. This may seem contradictory; but Popper means by 'living dangerously' that they should be framed in such ways that they *can* be incompatible with accepted facts. Failure to predict is not, for Popper, grounds for refuting or rejecting hypotheses; but perhaps it is not any too clear why we should bother to posit and develop them if they are not predictive.

Strictly, for a hypothesis framed at time t_1 to be compatible with t_2 facts, the implication is that the hypothesis at least might have been adequate for prediction at t_1. There must have been lawful continuity between t_1 and t_2, so why should not the hypothesis (if a 'good' one) have been predictive? One can point to any number of cases where hypotheses have every appearance of being predictive. One may also find cases of hypotheses being abandoned for failure to predict. For example, Harvey's theory of the circulation of the blood predicted what were discovered considerably later – named 'capillaries'. These could not be seen at the time – and their apparent absence was potentially 'refuting' for Harvey's hypothesis, but special efforts were made to observe them with microscopes, as they were clearly predicted, until they were found. The discovery of oil from seismographic data by following generalizations and hypotheses from previous cases may serve as a second example of prediction. One could add discovering planets from perturbation data, brain tumours from clinical symptoms and properties of new compounds or molecules from chemical theories. Surely applied science depends almost entirely on predictions from theories – science would be no use without prediction.

There may be philosophical doubts as to the basis and the trust to be placed on prediction. This is a heritage especially from David Hume. But if we are to reject the predictive power of hypotheses, what can we accept? To communicate, or accept another's existence, is to accept predictive hypotheses. To reject prediction is to court the philosophical death of solipsism.

Biologists universally hold that learning extracts predictive regularities from the environment. This is accepted for primitive animals (even some without nervous systems) and for man. According to Karl Popper this is mistaken: knowledge can be gained only by positing and refuting hypotheses. Is this how organisms extract predictive generalizations? Is this indeed the only way that animals and man can learn from Nature? We are confronted with a philosophy making an empirical statement – that learning by Induction is not possible and so does not take place in animals or men. Is this a modern case of philosophy pronouncing beyond its self-acknowledged limits? This forces us to question what is philosophy and what is science. In this instance, it leads us to look at the scientific evidence of what we know about learning, to judge whether this is a question for the philosophy of Mind or the Science of Mind. Does currently accepted scientific knowledge of animal and human learning run counter to Popper's view that learning is never Inductive, and is always given by hypothesis rejection? We shall look at this by considering memory learning, and other links in Mind.

Part 3 Links in Mind

'What is Matter?'
  'Never mind.'
'What is Mind?'
  'It doesn't matter.'

Old joke

9 Links of memory

A little learning is a dang' rous thing;
Drink deep, or taste not the Pierian spring:
There shallow draughts intoxicate the brain,
And drinking largely sobers us again.

Alexander Pope (1688–1744)[1]

Ancient notions

The present is a razor-edge of time poised between the remembered or assumed past and the expected future. Or so it is for us; what it is for other animals may be very different.

Aristotle thought of memories – and indeed of all thinking – as depending upon images. These are 'icons': images, pictures or models that are like in form to what is remembered. An account of this kind has the immediate difficulty of how to deal with abstract notions: how can they be pictures or models? Aristotle considered two kinds of icon: the first is a simple copy or model; in the second the icon stands for the abstraction. Thinking of a triangle, Aristotle holds that we have an image of a particular triangle, but that we can ignore its particular size to see the general or abstract qualities of triangularity.

Aristotle is particularly ingenious when considering how we may know the age of memories. He takes an analogy to seeing objects at different distances, when the further will be smaller at the eye, to saying that earlier memory images are smaller than later memory images; so time is given by (as we might put it) a kind of internal temporal–distance perspective of mental images.[2]

The Greeks and especially the Romans invented elaborate techniques to enhance memory. Before printing (and note-taking seems to be a recent custom), detailed memory was vitally important for business, politics and scholarship. Greek orators would construct their speeches in sections, each associated with a certain place, or architectural feature: so the structures of their cities became, indeed, imbued in their Minds. Greek (and later) ideas for improving memory are brilliantly described by Frances Yates in *The Art of Memory* (1966). The idea starts with Simonides of Ceos (556–468 BC) with an episode recounted by Cicero that suggested to Simonides the importance of place for memory.[3] Simonides chanted a poem in honour of his host, Scopus, at a banquet. Scopus offered to pay him only half his fee, saying that the Gods Castor and Pollux, who were also praised in the poem, would pay the rest. Simonides was told that two young men were waiting outside; thinking that these were in some form Castor and Pollux, Simonides left the banqueting hall to see them – and then the roof caved in, killing everyone within. The bodies were too mutilated for identification, but Simonides found that he could identify each from his memory of where they were seated when the catastrophe occurred. Cicero concludes: 'The images of things will denote the things themselves, and we shall employ the places and images respectively as a wax writing-tablet and the letters written on it.' This is the basis of all mnemonic systems and is central to much that has been written on memory in twentieth-century experimental psychology, especially by F.C. Bartlett in his important *Remembering* (1932).[4] Bartlett develops this idea at great length, but generalizes it from places to any imposed structures. In his view, memory is as creative (and susceptible to many of the same kinds of errors) as perception.

We should not be surprised to find that Greek orators aimed at ideal structures for remembering their long speeches and that these were associated with what is good and prudent. Cicero writes in *De Inventione:* 'Prudence is the knowledge of what is good, what is bad and what is neither good nor bad. Its parts are memory, intelligence, foresight. . . .'[5] The Greek method for artificial memory is described a century after Cicero by Marcus Fabius Quintilian (b. *c.* AD 30–5) in *Institutio Oratoria:*[6]

Places are chosen, and marked with the utmost possible variety, as a spacious house divided into a number of rooms. Everything of note therein is diligently imprinted on the mind, in order that thought may be able to run through all the parts without let or hindrance. The first task is to secure that there shall be no difficulty in running through these, for that memory must be most firmly fixed which helps another memory. Then what has been written down, or thought of, is noted by a sign to remind of it. This sign may be drawn from a whole 'thing', as navigation or warfare, or from some 'word'; for what is slipping from memory is recovered by the admonition of a single word. However, let us suppose that the sign is drawn from navigation, as for instance an anchor; or from warfare, as, for example, a weapon. These signs are then arranged as follows. The first notion is placed, as it were, in the forecourt; the second, let us say in the atrium [central court or hall]; the remainder are placed in order all round the impluvium [square basin in the centre of the atrium, usually filled by rain]; and committed not only to bedrooms and parlours, but even to statues and the like. This done, when it is required to revive the memory, one begins from the first place to run through all, demanding what has been entrusted to them, of which one will be reminded by the image. Thus, however numerous are the particulars which it is required to remember, all are linked one to another as in a chorus nor can what follows wander from what has gone before to which it is joined, only the preliminary labour of learning being required.

What I have spoken of as being done in a house can also be done in public buildings, or on a long journey, or in going through a city, or with pictures. Or we can imagine such places for ourselves.

We require therefore places, either real or imaginary, and images or simulacra which must be invented. Images are as words by which we note the things we have to learn, so that as Cicero says, 'we use places as wax and images as letters'. It will be as well to quote his actual words: 'One must employ a large number of places which must be well-lighted, clearly set out in order, at moderate intervals apart, and images which are active, which are sharply defined, unusual, and which have the power of speedily encountering and penetrating the mind.'

Quintilian continues this most suggestive and informative passage with a (typically academic!) personal attack: [7] 'Which makes me wonder all the more Metrodorus can have found three hundred and sixty places in the twelve signs through which the sun moves. It was doubtless the vanity and boastfulness of a man glorying in a memory stronger by art than by nature.' For Quintilian, Metrodorus of Scepsis was a charlatan, but Cicero described Metrodorus as having 'divine' memory, probably meaning that he shared his memory with the Gods, or at least the places and patterns set up by the Gods in the stars.

The emphasis upon mental pictures, or mental images as important for memory may have reinforced the notion that mental images (especially visual images) are the stuff of Mind, and causally important for producing behaviour. For many later writers on perception (such as Berkeley), sense data are the stuff of perception, woven to structure our experience, from the fleeting sense impressions given by the eyes and the other senses. The problem now for neurology is to decide whether 'mental images' and 'sense impressions' are indeed data: whether behaviour is controlled purely by physical brain activity, with 'mental images' merely tacked on, or somehow part of sensation, but not causally important.

If these orators, who after all had great experience and need of accurate memory, found strong visual images to be so useful (as most though not all apparently did), then should we now accept this as evidence for mental images cohering to form strong linked memories? Perhaps not; for the links could be of, or between, physical brain events representing or in some way producing the mental events. We might then say that the mental events known to us serve to let us know that physical memory links are likely to be formed. If so, seeking strong visual connections may be useful for forming strong memory sequences, though it is not the visual sensations that are the memory or the links between memories. Could this, indeed, be a use for consciousness, awareness, though the processes of memory and perception are physical brain states and activities?

Modern notions
Human learning

Experimental psychology started with a home of its own in 1879, when Wilhelm Max Wundt (1832–1920) founded the first laboratory of experimental psychology at Leipzig.[8]

These first experimental studies were greatly influenced by prevailing philosophical schools and by the aims and techniques of Natural Science. This is hardly surprising, as the first experimental psychologists were, by training, philosophers or physiologists. This produced an initial duality of approaches, the first being holistic accounts from the prevailing influence of Georg Wilhelm Friedrich Hegel (1770–1831), who developed aspects of Kant's idealism, stressing the unity and wholeness of reality. (This no doubt had profound political implications and effects, as he convinced many that the whole was more important than the parts composing it, including human societies. It is the basis of Gestalt psychology.) The tradition of physics and physiology stressed, on the other hand, the need to analyse into small 'atomic' units. In the study of memory, this was undertaken at this time by Hermann Ebbinghaus (1850–1909), publishing his important *Über das Gedächtniss* in 1885. Ebbinghaus set out to study the 'higher mental processes' with methods of the exact Physical Sciences, following the lead of Gustav Theodor Fechner (1801–87), and to this end he studied learning and memory with nonsense syllables. It is extremely interesting that he found it necessary to use meaningless material to study memory, when from classical times it was clear that memory is highly bound up with meaning. But, of course, meaning is not part of the Natural scene.

Ebbinghaus produced quite lawful-looking learning curves with his nonsense syllables (2,300, selected mechanically). These were single syllables having a vowel with initial and final consonants. He wrote out all the possible combinations of vowels, and 19 consonants for beginning and 11 for ending typically German-sounding words, on slips of paper, and shuffled them. Later lists rejected familiar words, and alliterations or other preformed 'associations'. It soon became clear that memory was complete for short spans of syllables, up to about 7, but many repetitions were required to exceed this span. The increase tended to be lawful and fairly smooth – at least when many trials or subjects were combined. Many relations were found between time to learn and (1) the length of the list; (2) intertrial spacing; (3) the rate of presentation; and so on. These results (which I shall not describe in any detail) were highly appealing to Natural Scientists: it was the kind of experiment and the kind of result that suited a mechanism of Mind.

We may now look at Bartlett's comment on this, for it was Sir Frederic Bartlett who had the courage to accept the difficulties and to design experiments to investigate memory of meaningful material (*Remembering*, 1932): [9]

It is worth considering in some detail what Ebbinghaus actually achieved. He realized that if we use continuous passages of prose or of verse as our material to be remembered, we cannot be certain that any two subjects still begin on a level. Such material sets up endless streams of cross-association which may differ significantly from person to person. It is an experiment with handicaps in which the weighting is unknown. Provided the burden of explanation has to be borne by the stimulus, this is obviously a real difficulty; for the stimuli have every appearance of varying from one person to another in ways incalculable and uncontrollable. There appears an easy way of overcoming this obstacle. Arrange material so that its significance is the same for everybody, and all that follows can be explained within the limits of the experiment itself. . . . Now, thought Ebbinghaus, with great ingenuity, if all the material initially signifies nothing, all the material must signify the same for everybody.

Bartlett summarizes three criticisms, and makes further points of great importance for psychology, and for considering its relation to the Natural Sciences. He writes: [10]

So far as the stimulus side of his method goes, Ebbinghaus's work is open to the following criticisms:

(*a*) It is impossible to rid stimuli of meaning so long as they remain capable of arousing any human response.

(*b*) The effort to do this creates an atmosphere of artificiality for all memory experiments, making them rather a study of the establishment and maintenance of repetition habits.

(*c*) To make the explanation of the variety of recall responses depend mainly upon variations of stimuli and of their order, frequency and mode of presentation, is to ignore dangerously those equally important conditions of response which belong to the subjective attitude and to predetermined reaction tendencies.

Bartlett goes on to say that the difficulties for the response side are even greater: [11]

There is, however, only one way of securing isolation of response, and that is by the extirpation or paralysis of accompanying functions. This is one of the perfectly legitimate methods of the physiologist. It can be argued that the psychologist, who is always claiming to deal with the intact or integrated organism, is either precluded from using this method, or at least must employ it with the very greatest caution . . . it is certain that such isolation is not to be secured by simplifying situations or stimuli and leaving as complex an organism as ever to make the response. What we do then is simply to force the organism to mobilise all its resources and make up, or discover, a new complex reaction on the spot.

Here in a nutshell is a principal difficulty in psychological experimentation, but the difficulty springs from this characteristic of organisms to adapt and change with each challenge and each restriction. Add predictive power to this, and the large contribution of internally stored data (so that applied stimuli may be relatively unimportant), and one begins to see how difficult a subject is experimental psychology. Ebbinghaus made a brave effort to simplify the situation and make it compatible with normal scientific method but, as Bartlett says, the organism remains complicated and inconstant, and refuses to accept input stimuli as separate.

Bartlett (in direct opposition to Ebbinghaus) presented his subjects with pictures, sentences and stories that were as normal as possible – to discover systematic effects of meaning. He summarized his discoveries in the statement that the organisms have an *effort after meaning*. For Bartlett, organisms seek meaning, and they create meaning, from whatever is presented. For Bartlett, both perception and memory are thus highly active with major contributions from the organism – which structures its sensory inputs, and by interpolation and extrapolation fills in gaps in space and in time to give, while awake, continuous experience and behaviour. He pointed also to distortions of memory produced by this effort after meaning. Distortions tended towards the familiar: the unfamiliar is difficult both to perceive and to remember correctly, if only because perceptions and memories are built very largely from what is stored from the past, and organized into current belief of what is important and true. This Bartlett found to hold for memory of pictures and stories.

We might say, from the classical accounts of improvement of memory (artificial memory) and from Bartlett's criticisms of Ebbinghaus's nonsense-syllable experiments, and the success of his own very different experiments – in which meaning and structure are all-important – that the 'atomistic' view of memories as separate from each other and stored like 'bits' in a computer is totally false for the brain. There are, however, a few cases of exceptional individuals that make us pause at this point. These cases of exceptional memory worried Bartlett and they should worry us. I shall mention one example, which is particularly dramatic, described by the highly distinguished Russian neurophysiologist, the late Alexander Romanovich Luria, in his remarkable book *The Mind of a Mnemonist* (1968). Luria describes a man whom he studied for thirty years who had, effectively, complete memory. This man (known as S) was able to recall long lists of numbers and even nonsense words after sixteen or more years, following a single presentation. Luria describes how he did it (or believed himself to learn and recall) with superb clinical acumen. Luria discusses his first experience with S:[12]

I gave S a series of words, then numbers, then letters, reading them to him slowly or presenting them in written form. He read or listened attentively and then repeated the material exactly as it had been presented. I increased the number of elements in each series, giving him as many as thirty, fifty, or even seventy words or numbers, but this, too, presented no problem for him. . . .

The experiment indicated that he could reproduce a series in reverse order . . . just as simply as from start to finish; that he could readily tell me which word followed another in a series, or reproduce the word which happened to precede one I'd name. He would pause for a minute, as though searching for the word, but immediately after would be able to answer my questions and generally made no mistakes.

It was of no consequence to him whether the series I gave him contained meaningful words or nonsense syllables, numbers or sounds; whether they were presented orally or in writing. All he required was that there be a three-to-four second pause between each element in the series, and he had no difficulty reproducing whatever I gave him.

At least at first sight, this seems quite contrary to Bartlett's emphasis upon structuring and meaning for memory. It turns out, though, that S *did* structure the material, and in ways remarkably similar to Simonides two and a half millennia before. Luria reports S as having remarkably vivid and complex synaesthesia. Thus in a record of November 1951: [13]

To this day I can't escape from seeing colours when I hear sounds. . . . If, say, a person says something, I see the word; but should another person's voice break in, blurs appear. These creep into the syllables of the words and I can't make out what is being said.

Then, in a record of June 1953: [14]

For me 2, 4, 6, 5 are not just numbers. They have forms. 1 is a pointed number – which has nothing to do with the way it's written. . . . 2 is flatter, rectangular, whitish in colour, sometimes almost a grey. 3 is a pointed segment which rotates. 4 is also square and dull; it looks like 2 but has more substance to it, it's thicker. 5 is absolutely complete and takes the form of a cone or a tower – something substantial.

Luria comments that S had no distinct separation of hearing, vision, or touch or taste: 'they left their mark on his habits of perception, understanding and thought, and were a vital feature of his memory'. The images for familiar names were complicated and detailed, though S later simplified them, when he deliberately developed his remarkable gift as a professional mnemonist. Perhaps most remarkable of all, he needed to place his images in 'good light', very much as Quintilian described it. Here is a vivid description by S of placing images and the need for a good, internal, light. Luria describes S's occasional defects of memory as defects of perception, illustrated with the following account: [15]

I put the image of a *pencil* near a fence . . . the one down the street, you know. But what happened was that the image fused with that of the fence and I walked right on past without noticing it. The same thing happened with the word *egg*. I had put it up against a white wall and it blended in with the background. How could I possibly spot a white egg up against a white wall?

As S developed his memory, he overcame this difficulty: [16]

I know that I have to be on guard if I'm not to overlook something. What I do now is to make my images larger. Take the word *egg* I told you about before. It was so easy to lose sight of it; now I make it a larger image, and when I lean it up against the wall of a building, I see to it that the place is lit up by having a street lamp nearby . . . I don't put things in dark passageways anymore. . . .

Mr S was forced to learn how to forget. Forgetting is almost as important as learning.

For S, memory was dominated by particular visual scenes, which seemed to handicap his abstract understanding, and visually associated images led his thinking into inconsistencies, to make simple problems difficult. He would also confuse reality with his imagery – for example, not realizing when a child that the hands of a clock had changed position, hours after seeing it and storing it in his memory. He was dominated by puns, by associations between sounds of words so that intended meanings and implications could be lost in irrelevant associations.

There is no doubt an art to forgetting. How far we can decide to forget, or plan to forget some things and remember others, is hardly at all known. There is however a large body of clinical data on abnormalities of memory – which are generally forced and unintentional forgetting or inability to lay down memories: see Whitty and Zangwill (1966).

Forgetting

Possibly forgetting is a phenomenon as remarkable and as little understood as remembering. Why do we remember some things and forget others? Freud held that everything is remembered unless rejected for 'psychological' reasons. For Luria's subject S, not forgetting became a serious problem, and he developed special methods to forget, and to isolate and put away his memories: [17]

I'm afraid I may begin to confuse the individual performances. So in my mind I erase the blackboard and cover it, as it were, with a film that's completely opaque and impenetrable. I take this off the board and listen to it crunch as I gather it into a ball. That is, after each performance is over, I erase the board, walk away from it, and mentally gather up the film I had used to cover the board. As I go on talking to the audience, I feel myself crumpling this film into a ball in my hands. Even so, when the next performance starts and I walk over to that blackboard, the numbers I had erased are liable to turn up again.

What are memories?

This case of Luria's and other such reported cases do not after all seem to run counter to Bartlett's notion of effort after meaning and the importance of rich associations for memory. They do, however, pose questions about whether images are causally important for memory, or whether the associations and storage are to be thought of purely at the neuronal level. This takes us to the most difficult question of all: what *use* is consciousness? We shall leave this until later (chapter 15). Meanwhile, is it possible that though memory is physically based in brain mechanisms, these vivid images somehow help in discovering memories? Do they somehow allow highly efficient strategies for memory and recall, though the physical basis is no different in these people than for the rest of us? Here we reach the end of our understanding, and even of our ability to pose questions, for we do not know anything of the physical basis of memory. This is the most striking gap in our present physiological and psychological knowledge. It might indeed seem quite easy to discover what physical changes occur with learning to store memory; but so far we do not even know whether it is neuronal, molecular, or global as in a kind of hologram in which perhaps the patterning of almost the entire brain may serve to store all memories.

What are memories? Is Aristotle right that memories are mental images? This implies that memories are similar to perceptions, as we experience the present with the senses. I argue that perceptions are hypotheses of the present and immediate future. Like perception, memory depends upon gap-filling, and guessing from inadequate (stored) data. It seems appropriate to suggest that memories are hypotheses of the past. They are thus closely related and linked to perceptions. On a Representative view of perception, this offers no special problem; but Direct Realist theorists of perception would have to say (as indeed Ayer does at one point suggest while discussing Russell) [18] that memories *are* bits of the past. On this account, memories are not links with reality: they are samples of past reality, which thus in some sense still exists. On Bartlett's account memories are constructions from bits and pieces of brain-stored data from the individual's past perceptions. Whatever account we accept, the status of memory can hardly be considered apart from theories of perception, which we shall discuss in more detail later (chapter 12, where we shall develop the notion that perceptions are hypotheses).

Most recent experimental work on memory has been conducted with animals as subjects. This has the advantage that the individual's past can be known in detail and controlled, but has the grave disadvantage that verbal reports are not available – and so memories of individual events are impossible or at least extremely difficult to obtain. So experiments on animals reveal the learning of skills, but not the learning and recall of individual events, though it is these that constitute our awareness in memory of the past. [19]

It seems important to point out that we are aware – conscious – of individual events, and sequences of events in memory; but not the processes of skills that we have learned. One may indeed remember falling off a bicycle while learning to ride, but one has no 'internal knowledge' of how one rides a bicycle in consciousness, though the nervous system has learned what to do. So if the learning of skills and generalizations is Inductive (if Popper is incorrect on this issue) then we may say that Induction is unconscious. If Popper is correct, all learning of skills is by positing and rejecting hypotheses (see pp. 249–62).

When a philosopher says that such-and-such a way of gaining knowledge is 'impossible', he is saying that this such-and-such is not, and never will be, found in any actual or conceivable organism, or present or future computer. This is a bold claim. It could be supported only by a firmly held theory of knowledge. But do we have such a secure theory of knowledge that such a claim can be made with confidence, or adequately supported? It is worth pointing out again that philosophers have, through the ages, set out to show what is 'inevitable' or 'certainly' true; and also to set limits to what is 'possible'. But on both counts – of what is supposed certain and what is impossible – philosophers have a poor track record. This might be countered by saying that the same is equally true for science, but there is a difference. Science is self-correcting, and self-generating, to discover new problems and questions through the power of experiment. Philosophers may indeed discover holes and paradoxes in arguments, but perhaps their premises are not sufficiently secure for these drastic 'certain' and 'impossible' claims concerning the nature of the world and of Mind.

If Popper is correct, all learning is positing and rejecting hypotheses. Could Pavlovian Conditioning be like this? Or is Pavlovian Conditioning Induction by Enumeration? Here science – studies of learning – should help philosophy.

Are Conditioned Reflexes Inductions?

While I was a student, it seemed to me clear that Conditioning is building up Inductions by Enumeration of instances. They are so like what one would expect, if something like Mill's Methods of Induction were being carried out to discover predictive features, and the onus is surely on those who wish to reject this to give reasons for denying that much of learning is deriving generalizations by noting instances and associations, very much as Bacon and Mill described Induction. There could also be hypothesis-testing; but could simple organisms set up hypotheses and refute them, as Popper supposes?

Animal learning

Pavlovian (Classical) Conditioning

The Russian physiologist, Ivan Petrovitch Pavlov (1849–1936)[20] came across (it is perhaps not precisely true to say that he discovered) Conditioning when investigating salivation in dogs. It was earlier reported by the American psychologist E. B. Twitmeyer (Hilgard, 1948). The term 'Conditioned Reflex' was first used (it might be better translated as 'Conditional Reflexes') by Tolochinov, in 1903. The work of Vladimir M. Bekhterev (1857–1927)[21] a few years later is also of first importance; his book *Objective Psychology* (which appeared in instalments from 1907 to 1912) made the conceptual leaps from what had started as a frankly physiological technique for investigating autonomic functions, to a theory of psychology. Bekhterev's theory became known as reflexology, and was based on association reflexes as the units of behaviour. Having studied with Wundt, Bekhterev was aware of more general psychological problems, and was able to put across the importance of Pavlov's and his own work to European and American psychologists. It was grasped with special enthusiasm by the founder of Behaviourism, John Broadus Watson (1878–1958).

The essential feature of the Conditioned Reflex is that an innately linked stimulus (such as salivation upon presentation of food) becomes generalized to other stimuli, when those are presented at about the same moment as the original Unconditioned Stimulus. Thus, after several repetitions, the famous bell or buzzer becomes a Conditioned Stimulus for salivation. Second-order reflexes may be established, by using the new Conditioned Stimulus, the bell, as the Unconditioned Stimulus for another response.

The usual need for several instances does seem to suggest that Conditioning is essentially like Induction by Enumeration. There are many subtle similarities, especially that Classical Conditioning is upset by 'distractions' – which could be like the upset of Induction-building when further possibilities are introduced. Also, some relations are easier than others to establish, which may correspond to Induction's dependence on hidden structures for plausibility. Some stimuli are specially effective, though others are present for forming reflex links. This is known as 'response-specific stimulus preferences'. The evidence for this is recent; for example, honey-bees can be trained to come to feeding places by olfactory or colour stimuli, or by particular patterns. Learning, however, occurs more readily with olfactory than with colour stimuli, and least effectively with pattern stimuli. These differences are not simply related to stimuli that are most directly physiologically effective. It seems that what is effective for learning obeys its own laws. This is complicated, for which stimuli are selected for learning associations may depend on the response or on the task. With further experiments, these differences might reveal a kind of 'deep structure' of stimulus selection, and preferred associations. There is growing evidence that some learning is far easier than others. (Is 'Deep Structure' in Chomsky's sense (p.417) perhaps related to Russell's 'causal lines' for Induction?)

Conditioning requires several trials in almost all cases. Exceptions are avoidance of pain or of danger. These may be learned by a single trial. The 'lesson' may last for life. This strongly suggests that the usually large number of trials is an Inductive strategy for developing or discovering reliable relations. Since one-trial learning is possible, we know that the nervous system has this capability. So it is hard to see why Conditioning normally requires many trials, if it is not Inductive strategy, somewhat like Mill's Methods. For dangerous or damaging situations, the cost of building up reliable links by many instances must finally become too high – not worth the risk of death or mutilation by more than one or two trials to learn about dangerous situations.

A further example of very rapid learning is 'imprinting', which occurs in the young of many species, as studied especially in nest-reared (nidifugous) birds. R. A. Hinde and J. Stevenson-Hinde describe it in their useful book *Constraints on Learning* (1973): [22]

> The young of many nidifugous birds readily come to direct their filial response to a foster parent. In particular, they will learn to follow a novel model with little resemblance to their natural parent. The age during which this is possible is, however, limited to a single sensitive period extending from shortly after hatching to a few days of age – the precise limits varying with the species, conditions of rearing and testing, and other factors [e.g. Bateson (1966)]. In this case the end of the sensitive period is related to the development of the fear responses to strange objects, 'strangeness' demanding an acquired background of the familiar before it becomes effective. . . . In some species the learning of the characteristics of the sex partner depends on learning that is similarly limited to a sensitive period, though rather later in development.

It is then stated that recent research has shown that the limitations of the critical period during which such rapid learning is possible are affected by preceding experience. Objects that may be 'imprinted' may be very different from the parents which have normally to be recognized. Suitable artificial objects are generally fairly small and fast-moving, or coloured flashing lights or revolving coloured sectors. Human beings may be accepted as 'parents', for example by grey-legged geese; there are, however, limits to what is acceptable for this rapid learning to occur. It is likely that some insects learn very rapidly which leaves on a tree to accept as food, but the range is limited to the point of rejecting other leaves, even though they would save an insect from starvation, in the absence of its normal or accustomed food. Nature can prefer certain death to dangerous discovery!

The normal requirement of many trials to establish the link is exactly what Francis Bacon advocated for useful scientific induction. The fact that animals can learn associations by a single trial strongly suggests, I believe, that the usual requirement of many trials for associative learning is due not to physiological limitations – inability to form associations with only a single trial – but rather to the need for several trials to establish relevant predictive power. This is necessary because most stimuli that occur are not associated with particular events, and so have no predictive power. It seems important to determine how animals select what is predictively useful from what is not, and whether they have *a priori* (innate) selections or strategies for selecting predictive stimuli. Unfortunately, not much is known of this. It is also not clear how far animals learn that learning is worthwhile, from previous success in gaining reward through learned associations.[23]

Modern theories of learning based on animal experiments start from the work of the American, Edward L. Thorndike (1874–1949).[24] This is described in detail in his *Animal Intelligence* (1898). This is the original 'stimulus–response' (or 'bond' or 'connectionism') theory. It is essentially John Locke's Atomistic account of linked sensations, in which sense impressions are supposed to become associated with impulses to action; but the links are no longer sensations, for we know nothing of these in animals. The basis is random trials, which when successful become connected with actions neurally, to give integrated appropriate behaviour. Thorndike saw learning as problem-solving but hardly as hypothesis rejection. Was he correct?

Thorndike's puzzle boxes

Thorndike carried out experiments in which cats were placed in boxes locked with simple catch mechanisms. The imprisoned hungry cat was allowed to see food placed outside, through vertical slots which also allowed the cat to be observed by the experimenter. The interest of this experiment is the procedures by which the problem of how to escape to the food was solved by the cat in this clearly defined situation, in which the animal was unconstrained except by the problem. (This sounds like the ideal academic life!)

The door could be opened by, for example, pulling a string, or pressing a latch or lever. At first the cat would try to reach the food directly through the vertical slots of the box but the food was too far away to be reached. It would then move about the box, scratch its sides, and behave with more or less random activity. Sooner or later the cat would pull the string, or press the catch, or whatever was the necessary action to escape. The animal was then returned to the box by the experimenter. In later trials, it would direct its activity to the region of the string, or catch, eventually moving straight to it and releasing itself in one movement. It is interesting that the releasing movement was not stereotyped: the catch might be moved with a paw, or a hind leg, in a variety of ways. This variety of responses could develop after the initial movement that served to operate the mechanism. The cat thus learned to identify what was necessary for escape: to move the lever or pull the string, in any way.

Thorndike showed that entirely arbitrary movements or gestures could be called up and used as solutions, though they had no direct mechanical linking with the catch; for example, cats were taught to escape by licking themselves, or by shaking hands with the experimenter. So we find here no evidence of mechanical causal chain assumptions. But the experiments might, from this point of view, be criticized for being too 'difficult' for cats, which could not be expected to understand, or have any knowledge of such complex mechanisms as levers and catches or strings and pulleys. The cats were incapable of forming adequate hypotheses of the situation: they responded without understanding.

This type of Conditioning – Instrumental Conditioning – has since been developed by B.F. Skinner (1972). Skinner improved Thorndike's technique by immediately rewarding the animal in the box, without its having to escape and be returned. This allowed the processes of learning to be studied continuously and without interference by the experimenter. Progress of learning tends to be measured by rate of responding (pressing a lever) which is easily recorded automatically, though possibly at the cost of not recording the various initial failures which may give the key to trial-and-error learning procedures.

Thorndike was concerned with procedures of learning, rather than with any physiological basis. He supposed that there were links of some kinds forming associations, but his account was purely psychological until that most seductive basis for psychology – physiological mechanism – was suggested by Pavlov's Conditioned Reflexes. Could these be the underlying links of learning? What could be more attractive than physiological atoms and molecules of behaviour?

Physiological links

The Conditioned Reflex was essentially introduced into American psychology by J.B.Watson,[25] in his Presidential Address to the American Psychological Association, in 1915: 'The place of the conditioned reflex in Psychology'. It was Watson who introduced Bekhterev's reflexology ideas as well as Conditioning into psychology. What particularly appealed to Watson was the possibility of building long chains of Conditioned Reflexes by secondary conditioning. In fact, it was exceedingly difficult to go beyond three conditioning 'links'; but Watson supposed all learned behaviour to be built of sometimes very long chains of Conditioned Reflexes, all of which are anchored from their original innate Unconditioned Reflexes. So the whole of civilized behaviour is attached to primal innate responses! Watson built his account on Classical Pavlovian Conditioning, but in time this gave way in importance to the Instrumental Conditioning earlier envisaged by Thorndike.

It is often pointed out that the Conditioned Response is not in all ways the same as the original Unconditioned Response (even the chemistry of the saliva may be different), and most important, Conditioned Reflexes tend to be anticipatory. With stimuli regularly spaced in time, the 'response' eventually appears *before* the stimulus, and appears even at full force when the stimulus does not occur. This at once makes the physiological basis of Conditioning seem too complicated to be a simple unit of behaviour and it shows that the physical stimulus from the world is less important than the prediction that is built up by the organism. It shows indeed that the term 'stimulus–response' is essentially misleading. This is at least so for Classical or Pavlovian Conditioning. For this and other reasons its importance for accounts of learning gradually succumbed to another form of Conditioning: Instrumental Conditioning. This became the basis of Skinner's Behaviourism.

Instrumental Conditioning depends on each Conditioned Response; so the behaviour can become subtly adapted to the situation, and can generate solutions to at least simple problems. There are four kinds of Instrumental Conditioning: (1) reward training, where the Conditioned Response is followed by a reward, such as food; (2) escape training, where the Conditioned Response is followed by the termination of a noxious stimulus, such as electric shock; (3) avoidance training, where the Conditioned Response prevents a noxious stimulus; and (4) secondary reward training, where the Conditioned Response is followed by a stimulus that has become a token, or symbol, through acquiring reward-value from previous associative learning.

Instrumental reward training goes back to C. Lloyd Morgan (1852–1936).[26] It was this that was described in 1894, and (in 1896) by Thorndike, the inventor of the puzzle boxes for cats as 'trial, error, and accidental success'. Lloyd Morgan combined the trial-and-error concept with a view of active creative Mind. He was consistent, for he held a Principle of Parsimony (Morgan's Canon) such that 'In no case may we interpret an action as the outcome of the exercise of a higher mental faculty, if it can be interpreted as the exercise of one which stands lower in the psychological scale.'[27] He would, surely rightly, have seen whether trial and error suffices before demanding hypothesis generation and testing for gaining knowledge.

Trial-and-error discovery of relations and generalizations became the basis of the extremely influential Behaviourism of B.F. Skinner.

Links in Behaviourism
B.F. Skinner realized around 1930 that accounts of behaviour limited to responses to stimuli must be inadequate, for very often in animal and human behaviour there are no identifiable stimuli, and yet behaviour continues. After various accounts of stimuli as present but not observable, this led him to develop a new Behaviourism. Skinner's experiments and theories were stated in his important book, *Behaviour of Organisms* (1938), where he also described his techniques for investigating conditioning in rats using bar-pressing, in what came to be called the 'Skinner box'. Later, pigeons were used in highly ingenious experiments with pecking

responses. Skinner related his experiments and radical theoretical position to human behaviour in *Science and Human Behaviour* (1953).

Some stimuli are clearly identifiable, for example in the knee jerk or the pupillary response to changing light intensity at the eye. But if all behaviour were of this kind, there would be no behaviour without prior eliciting stimuli, which Skinner came to realize to be absurd. In this he departed from the earlier Behaviourism of J. B. Watson. Skinner's solution was to propose two kinds of response: elicited and emitted responses. Responses such as the knee jerk, or pupil reflex, which are responses to known stimuli, he called 'respondents'. Responses that are not and need not be related to any stimulus he called 'emitted responses'. Emitted responses are generally called 'operants'. There are two views of the status of operants: either there are eliciting stimuli but these have not been recognized, or there are no eliciting stimuli – which is Skinner's position. He holds that most human behaviour is operant: it becomes 'shaped', to be appropriate to the situation, rather than a rigid set of Pavlovian reflexes elicited by stimuli. Skinner considers two kinds of reflexes, to cover these two cases. The first is Respondent Conditioning (Type S), in which the response is correlated with and elicited by stimuli; the second is Operant Conditioning (Type R), in which the significant feature is learning by reinforcement. Reinforcement, such as food reward, is conditional upon the response, such as eating. This is supposed to become in general more and more selective, so that reinforcement may become limited to, for example, a particular food. This is a very different account from Pavlov's, as taken up by Watson, and leads to a highly sophisticated behaviour theory which has had immense influence in education and in the treatment – by using Operant Conditioning techniques – of various psychological disorders, including addictions (see H. J. Eysenck, 1964).

Skinner regarded his Type R Conditioning as more important than Pavlovian Type S, and it is indeed possible to question whether Type S exists. Skinner considered that Type S occurs, but only for the autonomic nervous system – skeletal behaviour being governed entirely by Type R Conditioning.

The techniques that Skinner devised allowed rich measurements of learning under controlled conditions that were less artificial than those of Pavlov or of Ebbinghaus. Once an operant response is set up, a great variety of reinforcement schedules can be introduced and their effects on behaviour studied. The frequency of reinforcement can be varied experimentally and the effects of irrelevant stimuli can be investigated. The course of extinction (the gradual loss of the reflex) can be measured, as also can discriminative stimuli, such as different colours to which an experimental pigeon learns to respond differentially. All this gave experimental psychologists the opportunity to carry out respectable-looking (because quantified) experiments, which yielded more repeatable results than almost any others in psychology.

Skinner worked towards an account of behaviour sufficiently rich to include humans, partly by taking the concept of drives away from physiology, as well as from psychology. This he did by stating what, in his view, drives are not:

A drive is not a stimulus.

A drive is not a physiological state.

A drive is not a psychic state.

A drive is not simply a state of strength.

Drives such as hunger, thirst or sex are investigated by withdrawal of their satisfaction for controlled periods of time. Drives are not explained as stimuli from stomach contractions or whatever, or from mental or emotional states. He does not allow *any* physiological intermediary, or any kinds of 'logical construct' processes: he simply makes the measurements and describes drive in terms of what he observes and measures. This brings out a central feature of Skinner's position, which philosophically speaking is a systematic Operationalism, following the Positivism of Comte and Carnap, with their requirement of verification to give meaning to statements, as put forward with such effect in A.J.Ayer's *Language, Truth and Logic* (1936). But Skinner's rejection of the need for models of underlying processes is curiously like Babylonian astronomy right at the start of science – which provided systematic observations but did not give any account of what might lie behind or produce the observed phenomena (see chapter 1). To the Sumerians, the wandering stars presented systematically changing patterns, and this was enough. To the early Greeks, they were Gods endowed with intelligence and purpose. Since Galileo, the moon and planets have been seen as roughly earth-like worlds moved by forces found on earth. The Operationalism of Skinner is intellectually therapeutic, in disposing of flabby mentalistic speculation; but it might be too extreme, for it does not allow explanation as explanation is accepted in other sciences. It is as though Galileo did not infer from his observations of the moons of Jupiter that they are bodies, rotating round the parent planet. He could have merely described the changing patterns as seen through his telescope; but from these observations he gave us a new conceptual insight, which finally displaced the earth from the centre of the Universe (see pp. 107–12).

Behaviourism has been described, in so many words, as a science of pointer readings. Owing to its emphasis on measurements, it may on the surface have the respectability of physics, though it misses the point by refusing to use its measurements and observations to suggest and test hypotheses of why phenomena occur.

It is strange, but brains get left out of Skinner's account of behaviour. Although learning is most important in this account, memory for individual events is rejected – for memory of events cannot be established for animals by these experimental techniques on which the account is based, even though one can ask someone what they had for breakfast, or whether he remembers an argument in a book.

The physicist using a measuring instrument generally has a fairly clear notion of what it is connected to, or what it is measuring. He does not write his papers giving just the pointer readings: he uses these (following the calibration of his instruments, and assumptions or knowledge of what they indicate) to give information on states of affairs otherwise hidden from us. The problem for psychology is, I think, to learn how to do the same for behaviour and experience. The problem is how to read from behaviour and recorded physiological events, as well as speech, what they mean in terms of underlying processes. This would be something like developing and using conceptual models to appreciate underlying structures and processes of the world of physics.

Having said this, it is possible that psychology cannot be like the Natural Sciences – perhaps even to the extent that conceptual models for explanation cannot be found. Could it be because there is nothing to be found? This is almost the implication of extreme forms of Behaviourism.

It does, however, seem clear that something essential to behaviour is to be found in every intelligent organism: a knowledge base. Skinner saw very clearly that accounts of behaviour in terms of real-time inputs (stimuli) are inadequate, if only because controlling stimuli are not continuously available though behaviour remains generally appropriate to the prevailing situation; but he did not take the step of suggesting that the animal is guided by its knowledge. To say this, we do not imply that this knowledge is like school knowledge, expressed in propositional forms. For example, knowledge can be represented as instructions, or as maps, or relations, in many different ways. Let us consider how an animal learns the spatial structure of its environment.

Links in Conditioning – or cognitive maps?

Laboratory rats are good at learning their way around mazes. The way that they learn mazes provides important evidence against the notion – believed by John Locke and most Behaviourist psychologists – that all animal learning is a building up of associative links. The evidence is that animals such as rats do not learn mazes by chains of Conditioned Responses and sub-goals, but rather that they build up internal 'maps' of the maze, with the goal given as a position in their cognitive map. This is shown, among other ways, by a greater number of errors made towards, rather than away from, the maze's goal. Mazes are learned by exploration which need not be rewarded. A goal is such because the animal finds reward there; but a maze can be learned by 'latent learning' before any reward (generally food) is placed in the maze. This learning without specific reward or reinforcement is shown by leaving the animal for some time in an unfamiliar maze with no food or other reward to provide a 'goal' or specific reinforcement. When food is then added at some place, to define a goal, the animals that have this free exploration reach it more rapidly than control animals denied previous exploration. So learning can occur without specific stimulus–reward relations.

For many animals it is vitally important to find the way back to base, or to feeding grounds. D.S. Olton (1977) describes interesting laboratory experiments with rats in mazes that have radiating arms from a common centre. They use spatial position as their primary strategy, but this fails when lesions are produced in the hippocampal region of the rat's brain. It would seem that the hippocampus, which is also associated with the sense of smell, is specially involved in the rat's cognitive map of its surroundings.

The theorist who developed the cognitive map notion most effectively was the American psychologist E.C. Tolman. Some of the implications of this approach are well put by Ernest Hilgard, in his excellent *Theories of Learning* (1948). He writes: [28]

> Stimulus–response theories, while stated with different degrees of sophistication, imply that the organism is goaded along a path by internal and external stimuli, learning the correct movement sequences so that they are released under appropriate conditions of drive and environmental stimulation. The alternative possibility is that the learner is following signs to a goal, is learning his way about, is following a sort of map – in other words, is learning not movements but meanings.

Although many situations give the same predictions on either theory of behaviour, there are at least four features showing cognitive map following: (1) expectancy of reward from objects that have been hidden, but whose positions have been remembered; (2) going to the same place, though the behaviour pattern has been upset in various ways; (3) latent learning, which occurs while the animal is 'exploring' the maze without special reward (it seems to learn its way around, and can use this knowledge when particular places are given significance with food rewards placed to determine goals); and (4) the tendency to make mistakes in the direction of the goal (here, as elsewhere in psychology – if not in physics – we can learn by accepting some kinds of mistakes as significant data).

Psychologists generally base 'schools' of behaviour theory on such distinctions as these between stimulus–response and cognitive accounts. I see no reason for it here: I see no objection to accepting that Conditioning (which might be described as 'stimulus response') provides the first layers of generalizations from events – essentially by Induction – and that these data are used with more sophisticated organization (and meaning) to give cognitive behaviour and understanding. This again is very much what has happened in the history of science.

Hypothesis learning: Krechevsky

Conditioning certainly looks like Induction by Enumeration of instances. It seems almost inconceivable that the links are formed by testing all manner of hypotheses, if only because the search for the appropriate response would be like looking for a needle in an almost infinite haystack of possibilities. The evidence for cognitive maps suggests that an internal representation of space is built up gradually, and not necessarily by explicit rewards as required for Conditioning. There is, however, good evidence that animals *can* learn by hypothesis-testing.

A common and just criticism of Classical Conditioning as an experimental technique, is that the animal is extremely restricted – both in what stimuli it receives and in what responses it can make. Restricted situations are often essential to isolate and study phenomena, but they do have their dangers, if only because normally important features may be rejected by the restricted situation of the experiment – which may effectively act as a filter to give atypical and perhaps highly misleading results.

To avoid difficulties of this kind, learning situations in which the animal has a wide choice of what to learn, and a wide choice of behavioural activities, are needed. How choices are made of what to look for, or what to learn about, should provide crucial data for discovering whether animals build up and test hypotheses. This was the kind of question asked by the American psychologist I. Krechevsky. (He later called himself Krech.) Krechevsky (1932, 1933) noted that rats do not select alternatives randomly, or by chance, but rather tended to adopt systematic selection strategies – as though trying out or testing hypotheses. This may be contrasted with the animal's running off a previously learned behaviour sequence. This situation is described by Ernest Hilgard (1948) in these words: [29]

The alternative is that the original behaviour is *not* the running off of earlier habits in the new situation, but is a genuine attempt at discovering the route to the goal. Past experience is used, but in a manner appropriate to the present. Such an interpretation makes the original adjustment a *provisional try*, to be confirmed or denied by its success or failure. What is here being called a provisional try corresponds to what Tolman and Krech have called 'hypothesis' behaviour. The theory supposes that a provisional behaviour route is kept in suspension until its consequences change its provisional status; if it is confirmed it is an appropriate path of action to be followed under like circumstances.

It is perhaps surprising how little has been done to follow up this lead of forty years ago. It would be very interesting to know just how animals form, and test, behaviour hypotheses in a wide variety of situations. Exploratory behaviour (and latent learning) strongly suggest that animals have curiosity. This essentially means that they pose questions, though the questions may be pre-linguistic. Possibly the first questions in Nature are seen in behaviour as these non-random, systematic selections of what to learn or associate, which seem to have received too little attention by psychologists or students of animal behaviour since the work of Krechevsky.

Looking again at Thorndike's puzzle-box experiments in this light suggests that his cats were trying out hypothesis-suggesting and testing strategies – but the problems were too difficult for the cats. This kind of thing is true also of much of human behaviour, for how many people have adequate hypotheses of what makes their car go faster when they press the accelerator, or stop when they press the brake pedal? How many appreciate how the needle of a sewing-machine makes stitches? Such considerations suggest that we usually get by with minimal hypotheses – minimal understanding of the world around us. From the *behaviour* of a motorist or a seamstress, it might appear that they have full understanding, though they have only crude correlations, adequate for their skill. This suggests that it is easy to read too much into the performance of animals. It suggests that it is a very difficult experimental problem to discover just what are the hypotheses that are used for controlling behaviour.

Machines that learn

The concept and realization of machines that can be taught skills, or learn for themselves, is new and highly important both for practical reasons and for illuminating the nature of learning and intelligence. If we are correct in thinking that ancient technology deeply affected the development of concepts of philosophy and physics, we may expect much the same for psychology and epistemology with the advent of machines displaying Artificial Intelligence. As a first step, it seems entirely clear that simple learning machines do provide crucial exceptions to Popper's strictures against gaining of knowledge by Induction, for machines capable of just this are easily made and clearly work. It is remarkably easy to make simple circuits that build up generalizations from instances – which produce Inductions by Enumeration. It is surely impossible to argue that these produce generalizations by putting up and refuting hypotheses, for (unlike the case of brain circuits) we know exactly what these circuits are and how they work. The major early development of Inductive learning machines was done by W. Grey Walter (1953)[30] and A. M. Uttley in the early 1950s. These were analogue devices: learning was achieved by charging capacitors with pulses from signals indicating events of various kinds. The relative frequencies of the events produced related charges on the capacitors which were read, or controlled output devices showing the conditional probabilities and predictions assessed by the machine. The same can

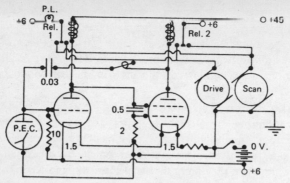

31 The circuit of Grey Walter's mechanical animal, *M.Speculatrix, c.* 1950. Looking like a tortoise, it moved around on wheels and had a scanning photo-cell eye. The adapting circuit, provided by the combination of the d.c. connection of the two thermionic valves (or 'tubes'), together with the 0.5 mF a.c. coupling with long time-constant to the grid of the second valve, gave a marked response to changes of illumination (transients) and reduced steady-state responses to continuous signals. *M.Speculatrix* demonstrated how lifelike behaviour could be attained even with very simple circuits. (From W. Grey Walter, *The Living Brain,* Gerald Duckworth and Co. Ltd, 1953; and with the permission of W. W. Norton and Company, Inc. Copyright © 1963, 1953 by W. W. Norton and Company, Inc.)

of course be achieved with computer programs, but this work was first undertaken in the infancy of electronic computers. These analogue solutions are delightfully simple, and how they work is easy to see without technical knowledge. Even if the nervous system works very differently – if its circuits are very different from these or any other learning machines – the analogue systems are still important for showing what can be done by defined physical systems.

Uttley suggested (1959) that there are two mathematical principles that may underlie the organization of the nervous system: classification and conditional probability. He writes: 'The suggestion is based on the similarity of behaviour of these formal systems and of animals.'[31] Uttley sets out four conditions for machine classification:

1 Each input channel must be always in one of two states, active and inactive.

2 The inputs must be combined in as many ways as possible – ideally in all possible ways.

3 There must be a unit for every combination of inputs, which indicates if every input of the combination is active. A combination or set of inputs is said to define a pattern of activity. The connections between inputs and units are called counting connections.

4 If temporal patterns are to be distinguished, each input must pass through a series of delays; the output of each delay must provide a separate input to the system of indicating units.

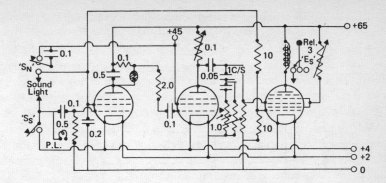

32 The circuit of Grey Walter's *M.Docilis*, giving Conditional Response learning with simple analogue circuits. (From W.Grey Walter, *The Living Brain*, Gerald Duckworth and Co. Ltd, 1953; and with the permission of W.W.Norton and Company, Inc. Copyright © 1963, 1953 by W.W.Norton and Company, Inc.)

The main point is that simple devices can be made that classify and compute conditional probabilities to build up Inductive generalizations without hypotheses, or hypothesis refutation.

It is easy to make circuits have the main characteristics of Pavlovian Conditioning.[32] No one, of course, suggests that the components are the same in brain and machine, or that the circuits are in detail similar; the point is that these functions can be reproduced by machines whose functions we understand, which shows that such understanding is possible for the brain without invoking uniquely special properties. It may always turn out that animal learning is very odd indeed – but the existence of these machines places the onus on those who put animal learning in a special class, or who argue that learning is never Inductive. The 'in principle' argument that learning is *never* Inductive collapses because these machines clearly do learn by Inductive Enumeration of instances.

These simple devices stand at the very beginning of Artificial Intelligence. By exhibiting learning, they allow aspects of learning to be studied in isolation from brains. With recent computer programs which start to show intelligence and perception, more and more 'mental' phenomena can be studied in isolation from complexities. Perhaps now a great new sweep of understanding, similar to that which followed alchemy's invention of apparatus for isolating, refining and acting upon matter, which led to chemistry and biochemistry, may lead via intelligent machines to a successful science of Mind.

The physical basis of learning
Among psychologists and physiologists there seems to be a greater urgency to find a physical basis for learning and memory than for other 'mental events'. Why

should this be so? One reason is perhaps the availability of familiar mechanisms such as tape recorders and computers which have excellent memories, though their memories are rather different from ours. Another reason has its roots in philosophy: events can be recalled in memory long after they were experienced, without intermediate mental events of which we have awareness. This suggests that there is some kind of enduring substance in which memories are maintained. We generally think of enduring substances as physical (indeed the notion of substance is bound up with continuity in spite of disturbing forces and so on), so we are almost forced to think of memory as having an underlying physical substance. There is strong reason to suppose that the continuity of mental events is given by stability of patterns of physical brain substance.

We might suggest that there is something odd about this or any other theory based on analogy from physical to mental characteristics. If physical and mental are supposed essentially different, how can analogies linking them be drawn? If they are not supposed essentially different, why do we not say that, at bottom, physical and mental are the same? This we shall discuss later (chapter 16). Meanwhile, we may look at the problem of finding the physical basis of memory.

It is remarkable that we do not even know *where* memories are stored in the human brain. And there is no account that is at all accepted of how they are stored. This is particularly odd by comparison with related questions, such as the physical basis of perception or emotions, because we have highly efficient technologies for storing information. Technical means for retrieving information according to need are, however, relatively crude. I shall now discuss why it is so difficult to establish the physical basis of memory.

Plasticity of nerve

Although it is extremely difficult to establish changes of neural connections related to particular experiences or memories, there is evidence of such changes. Perhaps the most striking evidence is from regeneration from sectioning, or from misrouting, of nerve fibres in animals such as salamanders, frogs and toads, which show far more regeneration than do mammals and especially primates. It is particularly important to establish whether fibres 'find their way' to form appropriate connections on the basis of signals, or information. If 're-wiring' can be set up by transmitted information (rather than, for example, by chemical gradients) then there could be a direct clue to a physical basis of memory. The classical experiments are due to Roger Sperry (1945, 1959), who transplanted limbs or eyes to various parts of the bodies of amphibians and found that, for example, a part of an eye implanted into the back of amphibians would produce changes in surrounding cells so that an almost complete eye would form. It would not, however, function:

there would not be the normal optic nerve–brain connections. So a more relevant kind of change for our present concern is, what happens when the normal eye is rotated?

Generally speaking, in amphibia the behaviour remains inappropriate: the tongue moves down instead of upwards when food is offered, and so on, without adaptation or learning. In man, as we know from the famous prism rotation experiments first made by G.M.Stratton (1897), actions are disturbed by this image rotation, and everything seen is reported as appearing upside-down, as might be expected; however, after about five days, behaviour starts to become normal and appearances become more nearly, if never quite, normal. Detailed studies of neural connections following eye rotations of the frog by R.M.Gaze (1970) give evidence that connections can be made as a result of stimulation, or transmitted information. It seems, however, that this depends greatly on which species is considered.

Colin Blakemore and his colleagues (1973, 1975) reared cats in the dark except for periods when one group of animals was allowed to see only horizontal stripes, and another group was allowed to see only vertical stripes. Later, the first group had impaired vision for vertical, and the other for horizontal, contours. There is evidence (and some other counter-evidence) that specific cells responding to particular orientations develop in the striate visual cortex, according to early experience of contours of given orientation. These may be rather small changes, for the final 'homing in' and 'tuning' of connections and response characteristics to frequently encountered stimuli, and perhaps information.

These changes are too sweeping to be typical of memory. Memory as usually considered refers to particular events, or particular skills, rather than, say, perception of contours of *any* object depending only on orientation. A curious kind of visual associative memory (if this is the right description) was, however, found in human vision, by Celeste McCollough (1965). She found that when coloured stripes of one orientation were alternated several times with stripes of a different orientation and a different colour, then black-and-white stripes of the two orientations appeared to be coloured, appearing with the complementary colours to their 'adapting' stimulus colours. Although the neural basis for this striking effect is not known, it appears to be limited to the visual system, for related effects do not occur between sound, or touch and visual stimuli: though other pairs of visual stimuli (such as colour and motion) can become associated in this way. It is worth noting that these associations are negative (like negative after-images) and so are not like normal associative learning. They seem to be generalized compensations to maintain perceptual calibration. Interesting as such effects are, they tell us little about the physical basis of learning, except to make it clear that generalized physiological adaptations do occur and can be studied. Much as for 'memory molecules', we do not yet know how to relate particular memories to particular neural structures.

A related class of theories of memories are neural trace theories, particularly associated with the highly influential view of the Canadian psychologist Donald Hebb expressed in *Organisation of Behaviour* (1949). This is the most systematic and brave attempt to give a theory of learning and memory in terms of physiological processes (see pp. 373–5).[33]

Where are memories stored?

It is well known that following a severe accident, and especially concussion, memory is usually not available for a period of some minutes preceding the accident. This is usually attributed to the time taken for experiences to be registered in 'long-term' memory. It is usually supposed that there is a separate active short-term memory store, which holds memories for about twenty minutes, and transfers them into the long-term store. Since long-term memories are not lost, though all electrically recorded brain activity has virtually ceased in severe accident or anaesthesia, it is assumed that the long-term store must be passive. Certainly, by analogy with computers, this is a convincing argument. Experiments in which the brain is entirely or locally frozen to remove activity in a controlled way confirms this distinction. There is also evidence of specific short-term (presumably dynamic) memory buffer stores for hearing and vision, which may be necessary for encoding serially received information before it is laid down in recoverable memory. These are functional characteristics of memory; but where in the brain are the mechanisms that give rise to these functions?

This question assumes of course that there is a physical basis of memory. By this one means that memories are stored by some kind of physical code, perhaps rather as printed letters store information in books. This at once suggests that information in the nervous system is (1) received, by the senses; (2) edited, for relevance; (3) written, in some kind of code; (4) read, by a retrieval system which can select what is needed, often for very different situations; and (5) incorporated and used for ongoing perception and behaviour.

We have distinguished between two kinds of memories which are, conceivably, encoded, stored and read in different ways and in different places. The two kinds of memory are of individual items and of skills. Considering again learning to ride a bicycle, one may be able years later to remember the event of buying the bicycle; and perhaps also various occasions when one fell off it; but although one has learned the skill of riding one cannot recall, as individual events, the movements and so on that led to the skill, yet the skill lasts for the rest of one's life. And it is indeed astonishing how soon such learned skills return after years of disuse.

Animal experiments are virtually limited to memory of the second kind. It is extremely difficult and usually impossible to discover whether an animal can recall a particular event, such as what it had for its last meal. A highly traumatic event may give one-trial learning; but this is rather different, for we do not know that the event is remembered – only that later performance has been modified by one trial rather than the many usually required to 'shape' behaviour and develop skills. The difficulty is lack of language. Language is almost essential for knowing individual memories in other organisms (in practice, only in fellow humans having a shared language), and it may be almost as important for recalling one's own individual memories.

Perhaps the only way that we can establish that an animal has individual memories is by noting whether it returns to food it has buried, or whether it can go straight to food (or some other reward) that the animal has seen being hidden. The duration of such event memory can be measured. The classical experiments on this are O. L. Tinklepaugh's (1928) on monkeys, who show individual memory for where objects have been hidden; the same may be observed in dogs. As memory for individual events is far more difficult to investigate in animals than the development of behaviour changes following repeated events, it is the physical basis of this *skill* kind of memory that is primarily considered. The theories may be divided into (1) active processes, such as circulating electrical activity, and (2) passive structural changes, such as chemical changes or connections formed between cells.

One kind of chemical theory supposes that there are 'memory molecules', and that if these could be transferred from one brain to another, the owner of the second brain would receive memories from the experience of the first brain's owner. It is very well worth considering just why this is such a difficult hypothesis to test. Here we shall follow the excellent discussion by Richard Mark in his most useful book, *Memory and Nerve Cell Connections* (1974). Mark (1974) compares the problem of identifying memory molecules with identifying microbes responsible for specific diseases. [34]

Microbial organisms were identified by following a procedure worked out by the German bacteriologists, Freidrich Gustaf Jakob Henle (1809–85) and Robert Koch (1843–1910). They set down the following criteria for establishing that an observed microbial infecting organism (or molecule) is uniquely related to a given identified disease: [35]

1   The infecting organism must be found in all cases of the disease.

2   Its distribution should accord with the lesions observed.

3   The infecting organism should be cultivated outside the body in pure culture for several generations.

4   The isolated infecting organism should reproduce the disease in other susceptible organisms.

We can ignore condition (2) for 'memory molecules'; and presumably they could not be cultured, so we can ignore condition (3). Why, then, are conditions (1) and (4) so difficult to achieve for memory molecule experiments? This may indeed seem odd, given the dramatic success of the bacteriologists using this experimental paradigm. The principal difficulty for the memory case is that there is no clear control which does *not* have memories. There are (fortunately!) plenty of cases of animals and humans to be found free of microbial disease which can serve as control groups; but how do you stop an animal or a human from learning? You can restrict its experience, and you can certainly give it particular kinds of experiences. These may be either quite general (such as rearing animals in darkness) or particular (such as specific Conditioning) but how do you identify specific memory molecules with the presence, or the absence, of experiences? This is another needle-in-the-haystack situation – which extends also to looking for neural connections for memories. The trouble, basically, is that there are so many memories, and learning cannot readily be switched off.

An approach of some promise is to make use of the fact that the brain has two similar hemispheres, and sensory signals can be routed to one or the other. In birds, this is relatively easy: it is possible to feed information from one eye to one hemisphere (and if necessary from the other eye to the other hemisphere). It is therefore possible to give one hemisphere much more to learn, or more special things to learn, than the other hemisphere. This means that each individual bird can be its own experimental control. The hemispheres can be examined for structural changes or chemically assayed separately, to look for differences following asymmetrical experience and learning. The best study along these lines, using imprinting stimuli to one eye, is by P.P.G. Bateson, G. Horn and S.P.R. Rose (1972) and is considered in later papers. They find evidence of a change in ribonucleic acid (RNA) with the imprinting learning. This is a start, but of course its relation to learning may be quite indirect; and presumably might be due either to 'memory molecules' or to structural changes of brain connections. There is some evidence that brain connections become richer with learning, but again we are far from associating learning, or memory, with increase in connections.

Is it possible to ablate individual memories? The evidence is, generally, that small localized lesions of brain tissue (ablations) do not remove specific memories, though they may modify or disrupt skills. The classical experiments are due to Karl Lashley (1929, 1950). Part of the problem here is that memories for individual events cannot usually be established by animal behaviour experiments.

In human brain damage, it is typically classes of memories that are lost, or made difficult to recall, and recall often improves in time after the injury. Can *stimulating* small regions of the human brain produce specific memories? The Canadian brain surgeon Wilder Penfield has carried out remarkably interesting experiments to this effect (Penfield and Roberts, 1959), but again the results are difficult to interpret. Vivid streams of memory may indeed be elicited, but this is equally true for less bizarre stimuli, such as smell. Perhaps more awkward, it is found that quite different memories may be elicited by identical re-stimulation of the same region (see pp. 474–6).

A great deal of work has been done on memory in the octopus by J.Z.Young (1965, 1966, 1976), who has identified two small precisely defined memory regions for the octopus's touch and visual systems. But he is the first to admit the immense difficulty of the next step: to identify what changes with memory. He is convinced, on general biological grounds, that memory is given by some kind of reconnecting, rather than by memory molecules.[36]

Is memory a 'Mass Action' of the brain?

When Karl Lashley (1950) ablated small regions of the rat brain, he found not specific behavioural losses but, rather, a general reduction in learning ability. This showed up with difficult learning tasks: the operated-on rat might be normal with simple mazes but impaired for mazes with more choice points. This led him to the celebrated notion of 'Mass Action', or 'Equi-potentiality', for the brain. The notion is that functions are not localized, but that every brain region partakes in almost all functions. (The sensory projection areas and the motor control areas were found to be exceptions, where small lesions did have specific effects.) There are, however, other ways of looking at the data. Suppose there *are* specific regions for specific memories, and other functions; but for skills several regions are involved. Then removal of one, or a few, might have little or no effect – perhaps suggesting Mass Action, though in fact each function is specifically localized. There is evidence for this from changing particular features of the maze, or removing some of the sensory input to the animal, or forcing it to move differently as by flooding the maze, so that it has to swim although it has learned the maze by walking. These have effects similar to the brain lesions. In short, Lashley's results might be due to the redundancy of the memories, through being represented in several regions, rather than to the brain working as a whole by Mass Action or Equi-potentiality.

In species higher on the evolutionary ladder than the rat, there is more evidence of specific losses of memory. Possibly redundancy becomes less. Unfortunately, for man, recovery of function after brain lesions is at its lowest. It is also conceivable that the appearance of Mass Action may be produced by unaffected brain regions taking over the function of damaged regions, though at any time function is localized. This is least plausible for particular memories; and there is evidence of this in lesions of the motor cortex, especially in animals below man on the evolutionary scale.

Another holistic theory is the notion that memories are stored like photographic holograms. These are like Lashley's Mass Action, or Equi-potentiality, in that if part is removed the entire picture is preserved, though with some overall loss of resolution. This impressed the American brain surgeon and neurophysiologist, Karl Pribram, to commitment to this account of memory (1966). Again, though, it is extremely difficult to test by any available kind of experiment. What evidence it has is similar to Lashley's idea; but as we have seen, his Mass Action notion is open to the alternative interpretation of multiple specific localization of memories and skills. The physical basis of memory remains a mystery.

10 The nature and nurture of intelligence

Intelligence is as much an advantage to an animal as physical strength or any other natural gift, and therefore . . . the most intelligent is sure to prevail in the battle of life. Similarly, among intelligent animals, the most social race is sure to prevail, other qualities being equal.

Francis Galton (1822–1911)[1]

What is intelligence? Can intelligence be measured? Is its power, or kind, inherited? Are animals intelligent? Are plants intelligent? Can intelligent machines be made? These are all questions about intelligence, and there are many more that are often asked, including whether the sexes, or races, have genetically different intelligence. Answers are not easy to find and there is a notable lack of expert agreement on how to define intelligence, or whether the available data are meaningful.[2]

High intelligence is regarded, at least by those claiming it for themselves, as a most precious endowment which makes almost everything, including understanding, possible. Intelligence seems to come in many forms – to bless musician, writer, inventor, scientist and statesman differently – and yet it is measured along a single continuum, the IQ scale, which it is claimed sets limits to our aspirations, whatever they may be.

This is a rather recent notion, having its origins in the nineteenth century in the writings of Herbert Spencer (1820–1903)[3] and Sir Francis Galton. Their notion of a general ability, necessary for specific abilities, received strong support, or so it appeared, from the statistical methods of Karl Pearson (1857–1936),[4] who suggested the method of factor analysis used by Charles Spearman (1863–1945) to derive General Intelligence (g) from cross-correlations of measured Specific Abilities (s). This use of factor analysis is almost certainly statistically fallacious, but I shall not discuss the point here or (with one exception) call upon statistical concepts. The IQ (Intelligence Quotient) scale was devised by Alfred Binet

(1857–1911)[5] and Théodore Simon, at the Sorbonne, while under pressure from the French Government (in 1904) to find a method for distinguishing between lazy and stupid backward children. This has had profound effects on psychology, sociology, and, indeed, politics. If this is fallacious or illusory, it has a place in history comparable to the powers of ancient magic, in affecting how we judge people and make decisions of great importance for frighteningly absurd reasons.

The intelligence—knowledge paradox

To introduce the kind of intellectual uneasiness that this notion produces, we may look at what is surely a paradox of intelligence as it is conceived in this kind of way:

1 Intelligence is not supposed to be increased by education.

2 Abilities are supposed to be increased by education.

3 Intelligence is measured by abilities.

How can these three propositions be reconciled? Are only certain abilities, which are not increased by education, used for measuring intelligence? We shall discuss this paradox of intelligence later (p. 313).

Definitions of intelligence

Generally, accepted definitions reflect current theories. I shall consider first psychologists' definitions, and then (since these are surprisingly unsatisfactory) suggest a rather different kind of definition for intelligence.

Traditionally, the word 'intelligence' has two subtly related meanings: possessing knowledge, and solving problems or creating knowledge.

We see the first meaning, which is the older, in *Macbeth:*

Say from whence
You owe this strange intelligence? or why
Upon this blasted heath you stop our way
With such prophetic greetings?

This is the meaning of Military Intelligence: it is possessing and handling knowledge, rather than creating it. The 'intelligence' may be secret and protected, or it may be collated with other 'intelligences' and passed on to whom it concerns. The second, and newer, meaning is largely the product of psychologists, and refers to the capacities of humans, or animals, to perform tasks that are more or less difficult. In general terms this is the ability to solve problems. The psychologists' definitions are not in terms of what has to be solved – why some tasks or problems are more difficult than others – but rather on individual characteristics of Mind supposed to be associated with the ability to solve problems.

The psychologists' definitions do not take into account Artificial Intelligence (AI). They assume qualities of Mind in people, and to a lesser degree animals, and – because they compare people – they have very tricky social implications, which I shall try to bring out as we go along. We are throughout treading on treacherous ground here: on human sensitivity to comparison and criticism.

The following quite representative definitions are given in the *Penguin Dictionary of Psychology*:

Intelligence: 1 The relating activity of mind; *insight* as understood by the Gestalt psychologists; in its lowest terms, intelligence is present where the individual, animal, or human being is aware, however dimly, of the relevance of his behaviour to an objective.

2 The capacity to meet novel situations, or to learn to do so, by new adaptive responses.

3 The ability to perform tests or tasks, involving the grasping of relationships, the degree of intelligence being proportional to the complexity, or the abstractness, or both, of the relationships.

The first definition is in terms of the awareness (presumably the conscious state) of the organism. This cannot be compared with AI definitions of intelligence, for we know nothing of the consciousness of machines, beyond indeed the assumption that they are not conscious. On this definition, it would not be possible to say that machines can be intelligent, which seems unfortunate.

The second and the third definitions are very different: they are both in terms of performance. Definition (2) is the least theoretical (and it is applicable to simple adaptive devices). Definition (3) includes a word difficult to incorporate into a machine description: 'grasping' relationships, if this is taken to mean being consciously aware of relationships. One might accept 'grasping' in machine terms but I suspect that this would not be acceptable to the writer of the definition. The phrase 'the ability to perform tests or tasks' is purely behavioural; but the further suggestion that the intelligence of the performance depends on the grasping of relationships makes this theoretically loaded, and difficult to assess, as 'complexity' and 'abstractness' are difficult to pin down. However, let us try.

Are 'complex' and 'abstract' subjective (characteristics that are relative to particular Minds) or objective (characteristics of the world, or of tasks to be performed or problems to be solved, without reference to the thinker or the observer)? We should also ask whether *difficulty* is related to our knowledge and skills; surely it is. We have now jumped in the deep end, to ask questions bordering upon philosophy, but I think that this is necessary.

Consider a pair of relations: 'A is longer than B' and 'A is larger than B'. The first seems simpler than the second. Why should this be so? To work out the size of something, we must compute or estimate two or three dimensions; but for length only one dimension is involved. So the task of estimating size requires more steps than for length. It is therefore a more complicated mental, or computer, task. Now, one might imagine cases where the complexity depends on how the task is performed. Complexity would then have a subjective component. For an example, we may cite a famous mathematical joke, which is probably true, concerning the distinguished mathematician John von Neumann and the fly problem.

There are two cyclists, a mile apart, cycling towards each other and each going at 10 miles per hour. A fly flies from the nose of one cyclist to the nose of the other, backwards and forwards between them, until the cyclists meet. The fly flies at 15 miles per hour. How far has the fly flown when it gets squashed by the cyclists' noses meeting?

When he was asked this by a friend, the mathematician Neumann thought for several moments before coming up with an answer, adding that it was an interesting mental exercise. It turned out that he had performed a remarkable feat, by not seeing the easy way. He computed the distance that the fly flew as the limit of a series (which is beyond almost anyone to do mentally) rather than noting that the time to nose-touching is 3 minutes. So in 3 minutes at 15 miles an hour it will fly $\frac{3}{4}$ mile. With one method this calculation is extremely complicated, with another it is simple. So we must not ignore the subjective component of 'complexity' – from which it follows that one person's simple problem may for another be complicated. This at once complicates how we should think of intelligence: was Neumann being incredibly intelligent in being able to use his complicated method which is beyond us, or was he being stupid in not appreciating that there is a simple method?

The general point is that relations and problems of all kinds may be complicated on one knowledge base, or available strategies and simple with other knowledge or strategies. This raises the important question (which we shall discuss later): is intelligence increased by better strategies and more knowledge – or do these make problems easier but without increasing intelligence? This is a difficult question to answer; we shall creep up on it gradually.

Turning now to 'abstract': can we consider abstractness as a characteristic of the world (or of intelligence test tasks) or is this also a mental term? Let me suggest a general reason why this matters. Suppose we want to measure the strength of a man's arm: we get him to lift a weight. The heavier the weight he can lift, the stronger, correspondingly, he is found to be. Now the weight he has to lift is objectively defined, in physical terms, so we have a clear objective standard by which to measure his strength. Similarly an eye chart allows objective measurement of the eye's acuity, in defined conditions. But suppose we want to estimate someone's aesthetic sensibility or 'good taste': we have no such objective criteria, because aesthetics lie (so far as known) very much in the Mind, and depend greatly on how we look at things from our past experience and so on. Now if intelligence were measured in terms of the intrinsic difficulty of problems (by analogy with weights to be lifted), the measures should be objective, as measures of visual acuity and muscle power are objective, though referring to individuals. We have found that 'complexity' also has a subjective component, in that how complicated a problem is depends on the skills and knowledge that we bring to bear. Now let us look at 'abstract' in this light.

We may start with a definition from the *Oxford English Dictionary*: '*Abstract*: Separated from matter, practice, or particulars; ideal; abstruse. Opposite to *concrete.*' We may take it that the degree of abstractness is given by how far generalizations are removed from the instances, or the data, from which they are made. And we may take it that generalizations are not in the world: they are created by us, and perhaps by computers.

We conclude that complexity is subjective because it depends on accepted assumptions and methods. Abstractness, also, is subjective, for nothing abstract exists, as concrete objects exist, but only by virtue of human intellectual activity. These definitions of intelligence therefore fall short of what we may judge as intelligent definitions: they are circular, they do not point to criteria by which 'intelligence' can be measured apart from individual differences, and they refer to vague notions of Mind not clearly related to tasks.

This leaves intelligence tests as comparisons between individuals' performance, on tasks that are not clearly related to mental processes or characteristics of problems supposed to reflect basic abilities. This seems a flimsy basis for comparing and judging people, and claiming that there are (or are not) differences of intelligence between sexes and races, from test tasks that often appear trivial.

Can we define intelligence objectively?

For this we must refer to characteristics of tasks or problems, without reference to how the tasks are performed or the problems solved. We should seek a definition that can allow machines to be intelligent, so reference to consciousness must be avoided: we should be allowed to talk about intelligent machines even if we doubt the possibility that machines can be intelligent.

Consider two features that stand out:

1 An intelligent solution must have some *novelty*, at least for the person (or animal or computer device) that produces it. Merely reproducing what already exists does not display intelligence, at least as I wish to use the term here.

2 An intelligent solution must be in some degree *successful*.

I suggest, then, that we accept problem-solving or inventing as intelligent when it produces *successful novelty*. This makes no comment on *how* the solutions are obtained. This is a matter for cognitive psychology and computer programmers; it may well reveal mechanisms of intelligence.

In practice, it may be difficult to estimate success and novelty, but at least we know what to try to measure with this definition. We may also wish to say that *understanding* requires and demonstrates intelligence. We may therefore tentatively define intelligence as more or less: *creation and understanding of successful novelty*.

I started by pointing out that the word 'intelligence' has two traditionally accepted meanings: roughly, possessing knowledge and creating knowledge. I shall retain this kind of distinction, suggesting that it is important to include in our notion of intelligence available knowledge, which can be called upon to solve current problems and give understanding; as well as the generation of current novelty, to solve problems for which adequate solutions are not available in memory store, or derivable purely by following available heuristic rules. I shall call the first 'Potential Intelligence', and the second 'Kinetic Intelligence'.

IQ tests attempt to measure what I am calling Kinetic Intelligence without reference to stored knowledge. We should consider whether this is possible; and whether the attempts have created confusions, and even paradoxes in how we think of the nature and nurture of intelligence.[6]

Traps of IQ

The Intelligence Quotient (IQ) scale was devised by Binet and Simon to 'show that it is possible to determine in a precise and truly scientific way the mental level of an intelligence, to compare the level with a normal level, and consequently to determine by how many years a child is retarded'. For some people, IQ tests are the best justification of psychology as a useful science; to others they are a scientific and social disaster. The Intelligence Quotient is defined as (mental age) × 100/(chronological age). Mental age is multiplied by 100 to get rid of the decimal point. IQs are measured with a battery of tests of 'mental ability', the proportion of tests passed being used to assess individual IQ. Different tests are used for different age groups, to avoid, for example, young children's being penalized for not being able to read. The average of the population across the age range is given conventionally as 100 IQ. The scores have a normal distribution. A markedly non-Gaussian distribution would promote a change of the tests, as would a marked deviation from a mean of 100 for the population.

Children who are five years old and adults who are twenty-five years old having the same score will, of course, have very different abilities, for the test and other tasks. It follows that IQ is not a straightforward measure of abilities. IQ scores are, however, supposed to be measures of intelligence: so intelligence is not being defined simply in terms of abilities. Chronological age is introduced as a weighting constant, so that individuals of different age may be compared although performance improves with age. The constancy of IQ with age implies a roughly constant slope of development for all, or at least most, individuals. There is, however, no independent measure for rate of development so it is meaningless to claim that intelligence increases, decreases, or remains unchanged with age, because its measure, the IQ, is defined as unchanging across the age range of the principal development of skills. In other words, because average IQ is defined to remain constant over this age range it is logically impossible to say that it increases, or in any way changes, although abilities change most rapidly over this age range. There is no great harm or danger in this provided one is aware of the logic of the situation: that it is beyond empirical power to claim changes of average IQ in spite of the changes of abilities, having defined IQ as remaining constant on average.

The tasks set in IQ tests appear trivial, but are supposed to measure the limits of our capabilities to understand and create. This is a bold claim; one that needs justification, perhaps on grounds of working predictively in practice, or of having a theoretical basis in our understanding of the nature of intelligence. Can either be supported? We shall look now at what seem to be difficulties, which may be fallacies and paradoxes, in the basis of IQ tests, and how they are applied to socially significant and sensitive issues including comparative IQs of the sexes and races. We shall not be concerned with particular claims that have been made, but rather with what seems to be the logical basis for claims that differences between mean scores imply corresponding differences of average intelligence: between the sexes, between races, or whatever. I suggest that there are surprisingly difficult problems to be resolved before such claims can be made, and that if precautions (which may be difficult or even impossible to implement) are not carried out, the procedures are strictly logically fallacious. In other words, if these precautions are not carried out, the means do not mean what they seem to mean.

As an initial shot across the bows, to remind ourselves how we can be misled by averages or means, consider the average family of, say, 2.5 children. This does not at all imply that there will be any family with half a child. It is so obvious to us that half a child isn't a child that we are not misled; but this statement might mislead the unwary in situations where the answer is not obviously ridiculous. Now we are not concerned with individuals, but only with groups at this point of our discussion (the sexes, races and so on), so this fallacy of reasoning from means will not concern us; but the example is a warning that arguing from means has traps, and when results are not obviously ridiculous we may not take the trouble to see that conclusions that are asserted do not follow from the data presented.

1   That the selection of test tasks favours groups arbitrarily
Let us take the example of mean IQs of girls and boys. It is commonly admitted by test designers that there are sex differences of ability for many of the test tasks that are used. Girls are supposed to be better at language and (though these are

difficult to score) social skills, while boys are generally better at spatial and mechanical problems. Now clearly a test that is weighted with a preponderance of spatial and mechanical problems will give boys a higher average score, while a test with more language or social problems will favour girls – so that they will then come out with a higher mean score for the second test than the first, which favours the boys. There is no escaping this conclusion. So what a test designer must do, as a minimum requirement to justify 'objectivity' of his results, is to justify the weighting of the test tasks known to be sex-linked to ability. The point is that it is possible to 'steer' the means to some extent, simply by changing the proportion of the boy-favouring and the girl-favouring tasks given in the IQ test. How much the means can be steered in this way may not be known, but this could be discovered by inspection of published scores on the individual tasks.

As things stand, it seems entirely possible that claims of sex or race differences have no validity: the results may be due to the proportion of the favouring tasks given in the tests. This may be quite arbitrary, chosen by 'feel' – or by deliberate intention to produce a wanted result – such that boys have a higher IQ than girls, or whatever. What is needed, in each case, is an analysis to show that the tasks have been chosen neutrally; this could be very difficult, for there is an essential circularity in this situation, as there are no objective criteria for balancing the various 'favouring' tasks. We do not know *a priori* whether boys have more, or less, or the same intelligence as girls.

It is not to be doubted that there are marked differences between individuals: in height, colour and abilities of all kinds. This is obvious from common experience, and variation is a feature of biology. Whatever intelligence may be, it would be amazing if there were not marked individual differences in this also. The question that concerns us is, however, not this, but rather whether IQ tests give valid evidence for or against group differences of IQ associated with race, sex, education, class or whatever. This cannot be shown if there is no objective way of weighting the test tasks so that they are on balance 'neutral' for the groups being compared.

2   That favouring scores can be weighted to avoid bias

We have seen that weighting constants for chronological age are applied in the very definition of IQ; for since older children can perform many tasks better than younger children of the same IQ, the constancy of average IQ across the age range has to be obtained by applying a weighting constant derived from the average improvement in performance at the selected tasks with increasing age, up to about sixteen years. Clearly if this weighting constant were incorrect, we would find that children increased, or decreased, in measured IQ as they got older – but this is not allowed. The weighting constant is adjusted to keep average IQ the same although abilities improve with age. The criterion for setting the weighting constant is simply that average IQ remains unchanged with age although abilities improve.

How can appropriate weighting constants be applied for differences of education, diet, motivation, social expectations, anxiety – and all the other variables that can clearly affect performance? In situations where disturbing factors may cancel each other out, this problem may be solved by simply taking many cases to average them out; this will not do here, for effects of education, diet, motivation and the rest may be systematically associated with the groups of people. For example, women may be more anxiety-prone than men in test situations. How could the appropriate weighting constant be found to correct this – to get a valid average IQ for women, by comparison with men who have less anxiety? It is extremely difficult to know how to weight results or compensate for such impairing effects, which affect the score but are not accepted as part of intelligence.

In practice, weighting constants are not applied much, if at all, except for age. Not only is it difficult (if not impossible) to assess what the correction should be; but if the weighting adopted became known, the cat would be out of the bag on all sorts of socially sensitive issues – for the weighting factors must reflect the different social or sexual handicaps as assessed by the test designers. So, if the weighting factors employed became known (and it is likely that they would), the assumptions or prejudices of the test designers would become known on sensitive social issues. So we conclude that appropriate compensations are not applied.

3 That test tasks can be assumed neutral

The alternative to *correcting* task-favouring biases (by balancing the proportion of the various favouring tasks given, or by applying weighting constants to each task, or to adjust the final mean) is to select *neutral* tasks, which do not on average favour any of the groups being compared. The problem is how to select and know that these tasks are neutral for the groups. The problem is complicated by there being, as recognized by test designers, many aspects of test tasks that affect performance but are not regarded as directly related to intelligence. This is indeed an important reason for not equating performance with intelligence; for there are many psychological variables that affect performance without, it is said, being part of intelligence: variables such as motivation and attention, and kind and level of education. Now how can intelligence tests be devised that isolate those aspects of performance supposed to be directly related to intelligence, without contamination by these other features?

The efforts of test designers have been mainly in trying to find tasks that do not require special knowledge, and so are not unduly affected by education. It is indeed hoped that the tests will measure intelligence quite independently of the advantages and handicaps of educational differences. Intelligence is therefore thought of as something separate from abilities given by education. Tasks are chosen that avoid particular factual knowledge (and so they are not quiz questions), and where knowledge is needed this is supposed to be so general that it will be shared equally by the groups or else supplied in the preambles of the tests. It is fairly easy to avoid questions of specific fact (such as 'Which is the second highest mountain?' or 'What does the differential gearing in a car's back axle do?'), but it is not so easy to avoid the use of strategies, which may be acquired by special learning and which may be highly useful. Indeed, we shall include strategies and rules and so on as a most important kind of knowledge. This extends not only to short-cuts and rules of inference, but to having learned to sit still and concentrate for an hour or so, which is not required in all societies and is a mark of age development in ours. Without this ability the candidate is hopelessly handicapped from the start.

What is it to think without knowledge? If we ignore, for now, specific factual knowledge: are not rules for problem-solving knowledge – knowledge developed by education and experience? Again, strategies for even the simplest tasks may surely be gained from the environment and by education, such as the skill of concentrating on the test, being confident and yet self-critical – and guessing what kinds of answers are needed.

If education did not have such effects, it would be useless – and this is a conclusion that few educational psychologists would accept. So, the claim that IQ tests can be freed of education and other biases by suitable choice of tasks seems ill-founded. It is indeed deeply misconceived, for intelligence requires and surely is in large part effective deployment of knowledge.

These considerations bring out how very difficult it is to believe that the tasks used for testing IQs can be neutral between groups. But if they are not neutral, and we do not know the biases that their non-neutrality introduces, then how can we rely on their use for measuring intelligence or for showing that groups, such as the sexes or races, differ in intelligence?

So, we seem to have three strong objections to the use of averaged IQ scores to show that groups of people have different intelligence. The objections are:

1   That the proportions of the tasks given that admittedly favour various abilities of girls and boys and perhaps races (at least with their current social and educational differences) are not balanced without circularity; and perhaps cannot be balanced, for we have no independent criteria except the performance of the groups for determining when they are correctly balanced. So the balancing act is likely to be circular.
2   That corrections can be made to the scores for various test tasks, to correct favouring effects. But, again, how can we know what weighting to apply, without circularity?
3   That test tasks would have to be neutral, in not favouring the groups apart from what is

specifically their intelligence. But this requires tasks that require only intelligence, or tasks from which variables such as motivation and, above all, education have no significant effect. This seems impossible on both counts.

We may turn now to some particularly tricky but important issues, not over the application of results from averaging IQ scores to try to establish group differences, as we have been discussing, but on very different problems: applying the IQ notion to individuals and especially individuals with handicaps, or with special gifts, or aids to intelligence or performance. I shall start by considering handicaps, and then move on to aids, with the hope that these discussions will illuminate how intelligence is conceived and show whether there are confusions in these conceptions. If there are, we may expect social confusion to follow.

Handicaps
Physical handicaps

No one would wish to say that handicaps such as blindness reduce a person's IQ, and yet blind people have great difficulties over many tasks that the rest of us do without thought, such as crossing the road. If blind people were required to do visual shape-matching in an IQ test, this would be regarded as unfair, and the results would be discounted as due to blindness, not lack of intelligence. The IQ tests are designed to preserve intelligence though performance is impaired, not by introducing explicit weighting functions to compensate for socially sensitive handicaps (which could be embarrassing), but rather to avoid tasks affected by the handicap: the problem is therefore hidden.

The method of preserving intelligence through handicap is different, then, from its preservation to give a constant IQ through age change. It is acceptable to apply a weighting function based on normal human development, but far less acceptable to define explicit compensations for physical (or some mental) handicaps. So we see very creditable social reasons for some of the confusions of intelligence measurements and what they mean. To summarize, IQ may be preserved against handicaps in three ways:

1   Applying a weighting constant. (This is used only when the weighting constant can be applied explicitly without social embarrassment.) Since we all develop as children, it is acceptable for keeping IQ age-constant. It is not used for handicaps.
2   Balancing group-linked tasks. (This is used where an explicit weighting function would admit that there are group differences.) This makes claims for equal average scores between groups meaningless, because the procedure is tautological unless it is shown that the tests are 'objectively' balanced or that the 'steering' effect of changing the balance of the individual test tasks is small. This has not been done.
3   Selecting group-neutral tasks. (This is used to preserve measured intelligence (IQ) in spite of handicaps.) This limits the range of acceptable tasks for intelligence tests. If

'handicap' is extended to socially or educationally deprived individuals (which is generally so), then the range of acceptable tasks is too narrow to include effects of special learning or specially developed abilities. It is in any case difficult to represent these in tests, for it is particularly difficult to weight unusual skills.

The assumption is that these 'neutral' tasks reveal some kind of intelligence common to all individuals, whatever their training or special skills may be. This has deep and not altogether fortunate implications, as we shall now see.[7]

'Protecting' intelligence

Protecting human intelligence from simple direct performance measures seems to have two principal dangers – one social, the other philosophical. By separating intelligence as measured from much of the behaviour and skilled performance of normal life, intelligence becomes what we might call a 'Golden Egg' in the basket of each individual's appreciation of himself. Most children and all intellectuals treasure their Golden Egg of intelligence – which is socially preserved from almost everything except, paradoxically, individual 'intelligence', which sets limits to what we can understand and do. Possibly this notion of a central intelligence that can be preserved against impairment of performance is an illusion – generated by the very procedures that are designed to protect handicapped people. Perhaps also – and this is the philosophical danger – the selection of socially neutral tasks for measuring IQ has generated the mythical notion that intelligence is a kind of entity underlying performance. There may be no Golden Egg of intelligence – but only individual abilities.

If IQ tests are indeed designed to minimize individual differences associated with education, race, sex and many kinds of handicaps, then how should we think about performance differences that are found to be related to socially sensitive variables? The trouble is that differences become suspect – and yet they include the extremes of creative genius, and much that makes individuals special, memorable and exciting. By selecting only 'neutral' tasks for measuring intelligence, psychology has unwittingly blown up, like a balloon, our most ordinary characteristics – and so praised them that they have become our Golden Egg of precious intelligence, while at the same time intelligence is attacked by IQ tests which appear trivial. The tests have been developed to stream children in schools, and they aim to serve the need in industrial society to measure individual power to gain and use skills and understanding. Special skills are, however, left out of the psychologist's notion of 'intelligence', if only because they are too difficult to quantify.

We have considered physical handicaps; what of mental handicaps? Are these cases where intelligence itself is impaired – and cannot be 'saved' by plausible IQ test designs?

Mental handicaps

We have argued that intelligence is 'saved' by psychologists when performance is impaired by socially sensitive physical handicaps. Obvious examples are blindness and deafness. We have no wish to regard blind or deaf people as having impaired intelligence, though they find some tasks difficult or impossible which are easy for the rest of us. We have also suggested that people with unusual endowments, such as great strength or agility, can perform feats that we find difficult or impossible – but we do not wish to think of them as more intelligent than us. This applies to many, but not to all, kinds of abnormalities and subnormalities. We do think of geniuses as especially intelligent, and we accept that some people have subnormal intelligence. How is this distinction from physical handicaps made?

This is not a simple issue of physical versus mental differences. For example, dyslexia is generally regarded as a 'mental' problem, and yet children are excused for poor writing and spelling if they are dyslexic, without these poor performances being counted against them intelligence-wise. Their 'intelligence' is preserved, though these are mental handicaps. Autism is different. This may be and is generally thought of as impaired intelligence. Just why these distinctions are made is far from clear. They seem to be based on ill-defined models of intelligence, and the classifications change quite rapidly as to which abnormalities are impaired intelligence and which are handicaps impairing performance without loss of intelligence. There is a great deal to be thought about here and I do not pretend to have the answers. Ultimately, can we distinguish adequately between 'physical' and 'mental'? Clearly a distinction is implied by speaking of physical and mental handicaps, but the distinction is not spelled out in psychology and remains highly controversial among philosophers.

I shall try to explore this issue, not by recourse to philosophy, but rather by considering how intelligence can be increased – with aids of various kinds. Some aids are thought of as clearly physical aids, such as spectacles, which, we say, increase performance though not intelligence; and others, such as language, mathematics, formal logic – as well as external aids such as writing, calculators, and computers – which are doubtfully physical. They may be regarded as mental aids, in which case we are further along the road to understanding 'mental', and distinguishing between mental and physical. Do such aids increase intelligence, or do they make existing intelligence more effective? Are we more intelligent because we have language, writing, books and computers? Or do these serve as aids for increasing abilities though without increasing intelligence? It is the cardinal assumption of intelligence testers that intelligence is not increased even by education. This is held though intelligence is not regarded purely as some kind of limit to possible individual performance. On this view, all aids, including education, increase performance without amplifying intelligence. This suggests that we should examine kinds of aids, and try to classify them. A consistent theory of intelligence should come to terms with issues such as these, and would no doubt be clarified in the process.

Aids to intelligence

There are many kinds of aids to human problem-solving. The question now is: do aids make the task easier, or do they increase our intelligence? This turns out to be a tricky distinction; but considering aids or tools in this way may throw light on the handicap argument that I have raised – for useful tools are counter-handicaps – and it may clarify our thinking about the nature and nurture of intelligence.

Although handicaps are always impairments to performance, similar considerations apply to improvement of performance by tools, and unusual physical endowments. For example, a man who is unusually strong may need less intelligence to perform a task, such as changing the wheel of a car, than a man or woman with only average strength. We would say that such problems are easier for him, because of his unusual physical strength, rather than saying his intelligence is greater. The weak man equipped with tools may have an easier task than the unaided strong man, and the tools or the strength reduce the need for intelligent solutions.

What should we think of mental aids? Consider a man working with a number-crunching calculating-machine. With his calculator, he can perform tasks and solve problems beyond his reach without the machine. With it he may, indeed, be superhuman. Then, if we did not know that he had the calculator to command, we might attribute to him superhuman intelligence. Once we realize, however, that he is aided by the calculator he becomes human again. But do we say that the calculator has made the problem easier, or do we say that the aid has increased his intelligence?

It seems important to analyse the notion of aiding. Accepting as aids anything that improves performance, what kinds of things can aids be? Performance can clearly be improved if the task is simplified, or made easier; I shall not consider this now, but rather ways of improving ability for a given task. What are these ways? We are primarily concerned with mental skills, but it may help, first, to consider physical examples, as these are easier to think about. Consider the concept of amplifying force, or ability, or whatever.

Multipliers

Suppose we need to lift a heavy weight, a weight too heavy for our unaided muscles. A lever, or block and tackle, would serve as aids that multiply the available muscle force. Are there also amplifiers that multiply mental abilities?

Additionals

Suppose we choose not a lever, nor a block and tackle, but instead a motor-driven crane to lift our weight. This does not increase the force of our muscles: it replaces them, with a different power source. So this is an additional and not a multiplier. The crane may work entirely without human intervention, an autonomous machine which replaces us. An intelligent machine is the 'mental' equivalent of the autonomous crane.

Matchers

The concept of matching is subtle and perhaps not entirely clear. In engineering, matching transformers may be electrical or mechanical, and may be of many forms. Matching may improve transmission of power, or of information. The concept might be extended to language translation: a French-speaking person requires a translation into French from English, to match his language skill. Recoding is a rather special kind of matching, but perhaps it is particularly important here. Matching is extremely important for efficiency.

Can we distinguish in practice between 'multipliers', 'additionals' and 'matching' aids? I think that we might, but these are not surface distinctions: they require appreciation of inner processes. We may draw an analogy here from something very familiar: the performance of a car, as given by the power of its engine and the frictional drag the engine has to overcome. We may increase performance by reducing the load (making the task easier); by increasing the power of the engine (amplifying it); by adding another source of power; or by improving the matching by selecting the appropriate gear. In each case the car goes better. Purely on performance we are unable to make these distinctions; but the distinctions can be made when we know enough about the internal workings of the car, and its relation to its environment.

Consider selecting the most appropriate gear. What this does, of course, is to match the performance characteristics of the engine with the task that it has to do. If its power peaks at, say, 5,000 revs/min, then the best gear for maximum performance is the gear ratio that allows the engine to run at around 5,000 revs/min. A low gear is selected for steep hills for the engine to perform at its best in the hilly environment. (Something of this applies to the hardware–software distinction in computers, for the software and the hardware must also be matched for effective performance.) In human skills, knowledge and external aids should match the problem. It follows that intelligence measures taken in non-optimum matching situations (or by ill-chosen questions) must underestimate intelligence.

'Matching', as used in engineering, is a rather subtle notion: it differs from 'amplifying' in various ways, including the fact that matching may be optimized (and so has a limit), while amplifiers and additionals may be increased without any essential limit. It is an interesting question whether it is most appropriate to regard language as a new and unique addition, or as improving the matching of Mind to Matter. It is hard to discover just how language, writing, mathematics or whatever, increase human performance. It is harder still to discover in which way they increase or add to intelligence.

However this may be, it seems clear that these and previous considerations make intelligence tests appear inadequate. This is borne out by considering programming computers to perform human intelligence tests. Here the first thing that we find is that it is far more difficult to program machines to do what are for us very simple tasks, such as recognizing objects, than to solve difficult human intelligence tests, which are rather easy for machines. And of course the machine does not get bored, or disturbed, as we do by being judged by tests that may appear to lack practical significance or theoretical justification. It seems that we need to look deeper into what we may take to be intelligence before we can accept that the kinds of problems used in intelligence tests give valid measures of intelligence by which we may be judged. What is this 'intelligence', which may be multiplied, and needs to be matched to problems? In what ways is it different from 'mental' aids, which help as additions to our abilities or intelligence?

It seems important to realize at this point that the notion of intelligence is generally applied only to some kinds of activity. It is not seen as appropriate for the basic life processes: metabolic activity is not regarded as intelligent. Perhaps curiously, the take-up of oxygen by our red corpuscles and its distribution throughout the body is not an 'intelligent' activity, though mending a clock or writing an essay is supposed to require intelligence. Plant activity is mainly restricted to metabolic processes, and it might be only just acceptable to ascribe intelligence to plant behaviour, such as climbing or fly-catching. Why should this be so? The use of intelligence seems to be restricted to handling information in ways that we see as problem-solving, or coming to understand problems. This is perhaps part of the distinction between 'activity' and 'behaviour': either may require information, but perhaps only behaviour is using information for problem-solving. In plants, the problems of metabolism, photosynthesis and so on are so great that they largely defeat human chemists, but they have already been solved in earlier generations of plants – though not exactly *by* plants. So they do not present problems that are solved by plants now. It is, I think, for this reason that we do not call plants (or our red corpuscles and so on) intelligent, though what they do is exceedingly clever – or at least it is exceedingly hard for us to understand how they do it. Plants have, it would seem, too much inbuilt skill to be judged intelligent! To begin to resolve this paradox we must broaden the usual (the psychologist's) concept of intelligence.

We shall now discuss our distinction between the processes of problem-solving and *solutions* to problems which have already been solved, and we will include the latter as a kind of intelligence. I shall call the first 'Kinetic Intelligence' and the second 'Potential Intelligence'. I shall suggest that we cannot assess Kinetic Intelligence (which is roughly what psychologists accept as the intelligence they try to measure) without knowing the contributions made by stored intelligent solutions – Potential Intelligence. I shall now develop this notion, considering first Potential Intelligence. This will lead to a resolution of the paradox from which we started this discussion.

Two kinds of intelligence: 'Potential' and 'Kinetic'

The distinction here is between solutions (or part-solutions) already available, and solutions that have to be discovered. The first might be called the 'design' of the organism (and may include the design of tools, etc.) and the second might be called the 'processing' required to solve the problem. So one might call them 'Design' and 'Processing' Intelligence; but I shall use the terms 'Potential' and 'Kinetic' Intelligence, by rough analogy with potential and kinetic energy. The analogy is, I think, sufficiently suggestive to justify these terms. (Potential energy may be created by kinetic energy, and stored in such ways that work may be done at a later time with kinetic energy, needed merely to release the stored potential energy. The analogy breaks down, though, in that potential energy is lost by being used, but Potential Intelligence is not lost by use.)

Potential Intelligence

Consider any well designed object (which may be active or passive) such as the roof of a house, or scissors. We may say at once that they are results of intelligence.

The roof of a house is a solution to a problem. It keeps out rain. It functions usefully and efficiently although it is a passive structure. Dynamic structures, such as scissors, are useless except when moved in appropriate ways. Roofs and scissors were both designed by intelligences, and they are intelligent solutions to problems. They may solve problems (set by rain or snow or our having to cut paper or cloth) without requiring further thought, or intelligence, on our part. We can forget the roof, and we can ignore the lack of cutting ability of our hands, when we are provided with these intelligent solutions. We live off intelligence stored in artefacts designed by our ancestors. These solutions are of enormous potential use for our problem-solving. We no longer have to invent roofs or scissors, when we know about these things. So education increases our 'internal' Potential Intelligence, through giving knowledge of what problem-solvers or aids are available.[8]

Kinetic Intelligence

Kinetic Intelligence is on this account mainly gap-filling, necessary where solutions are not adequate as stored Potential Intelligence.

Most problem-solving is done by following rules, which may generate successful novelty. But existing rules are on this account Potential, not Kinetic Intelligence. We do not equate the generation of successful novelty with Kinetic Intelligence alone. The point is that Potential Intelligence is useful as tools are useful, for producing successful novelty. This in its turn may well become stored Potential Intelligence, which may be generally used for routine performance and may occasionally be used to produce more successful novelty.

For many situations, available knowledge and rules and the rest, of Potential Intelligence may be entirely adequate; then Kinetic Intelligence would not be required. But selection of appropriate knowledge and heuristic rules is generally necessary, and this may require Kinetic Intelligence. It is commonly said that the greatest difficulty in thinking is to ask the right question, or to see what method to use or what sort of answer or solution is needed. The point is that these may require Kinetic Intelligence; though once appropriate analogies have been made and adequate rules selected, thinking and behaviour may cruise along as though they are as thoughtless as vegetable metabolic processes. So we see Potential Intelligence as part-solutions of new problems, which often need only a small component of Kinetic Intelligence for their solution. As civilization advances, Potential Intelligence becomes more and more important, for it provides ever larger chunks of part-solutions for problems which generally require ever smaller components of Kinetic Intelligence for their solution.

As a result of this we may judge ourselves more intelligent and more creative than our ancestors, but this is only because we have the benefit of a vast store of Potential Intelligence which was created by their Kinetic Intelligence. When this transfers positively to our problems we feel intelligent and take the credit: the fact is, of course, we are only superior because we are standing on our ancestors' shoulders. Potential Intelligence is rich, and gets even richer, as it is the ever-available harvest of thousands of years of individual Kinetic Intelligence stored in forms such as books – so education is (or should be!) ever more rewarding. The fact is, we should be more and more creative with less and less need of Kinetic Intelligence – except when situations radically change: then Potential Intelligence becomes inappropriate and we have new gaps to bridge, requiring solutions that can only be produced by Kinetic Intelligence. The moral is that change inspires new ideas – creativity – but if change is too fast the gaps between available Potential Intelligence and needed solutions become too great for our Kinetic Intelligence to bridge; then we fail.

The novelty of intelligence may be old novelty or new novelty – depending on whether we are considering Potential or Kinetic Intelligence. Intelligence in any case need be novel only to the individual. Potential Intelligence is created mainly by Kinetic Intelligence, to be used later and shared as social knowledge and technology. On this account, we cannot hope to assess Kinetic Intelligence without knowing the contribution of Potential Intelligence. The psychologist's attempt to measure 'pure' intelligence, as in IQ tests where knowledge is discounted, is for this reason clearly doomed to failure.

The intelligence–knowledge paradox resolved

We may now resolve the paradox that we started with: that high intelligence is credited to those with little knowledge, though added knowledge improves performance – while performance scores are used to measure intelligence.

In our discussion we have made use of both the old sense of the word 'intelligence' as 'available information' (as in 'Military Intelligence') and also the newer use, which is more like the ability to solve and understand problems. The first (stored knowledge and functional processes and structures) we call Potential Intelligence. This may be called upon to solve current problems by Kinetic Intelligence, which is active information-processing capable of leaping gaps.

When there is adequate Potential Intelligence, a psychologist would say that no intelligence is required. This would be said of metabolic processes (which may require information from monitoring conditions, and may involve servo-loops), and for reflexes and stored knowledge when run off without modification for the particular circumstances. However, we are accepting that all this is a kind – and a very important kind – of intelligence. It provides most of the answers to the problems that we can solve, and so it is stored potentially useful intelligence.

To measure IQ we need to know the contribution of the available Potential Intelligence to the solving of the IQ test problems. The greater this contribution, the less Kinetic Intelligence required, or exhibited. Kinetic Intelligence fills the gap between what is available in Potential Intelligence and what is needed to solve the problem or to cope adequately with the situation. Problem-solving or skilled performance may be improved by either greater Potential or greater Kinetic Intelligence. Our paradox of intelligence is resolved once we allow that stored knowledge, and anatomical structures, innate and learned behaviour patterns, and so on, are part of intelligence. Potential Intelligence is the useful results of past problem-solving; and Kinetic Intelligence jumps the gaps (usually small) from what we know, to what we need to know, to solve the problem or perform the task.

To measure intelligence usefully we must know the contribution made by Potential Intelligence. This is extremely difficult to estimate. Nevertheless, the paradox is resolved by realizing that knowledge is intelligence, of an important kind, which cannot be entirely separated from the information-processing of Kinetic Intelligence, for this is what it uses.

We may now ask how intelligence has been created, and how it is that intelligence is creative. This leads to asking how our intelligence may be applied to increasing our intelligence, and, finally, to creating intelligence in machines. I shall throughout make use of the Potential–Kinetic Intelligence distinction, as well as the definition of intelligence as creating successful novelty. We may regard intelligence also as *understanding* novelty, but I shall not discuss this here.

How is intelligence created?

Let us start with Potential Intelligence. Potential Intelligence we define as available solutions (or part-solutions) to current problems. When the built-in Potential Intelligence is adequate, the problem is already solved. Then it may not even be seen as a problem; although of course it was a problem, and it would be a problem now without the inbuilt stored solution, which is in memory or structure as part of the organism's design. Potential Intelligence is thus Kinetic Intelligence frozen from the past, which is applicable now. Potential Intelligence may be 'internal', or it may be 'external' – in artefacts. To take an 'external' example: scissors are available solutions, which solve the problem of cutting paper or cloth. Hands are in our sense 'internal' designs (part of the organism), though not in the brain.

How is Potential (Design) Intelligence created? Where does it come from? It is a product of past experience and problem-solving. We know (apart from theological accounts) of two generators of Potential Intelligence. So we know of two very different embodiments of biological intelligence: namely (1) Darwinian Natural Selection, from random mutations; and (2) problem-solving by organisms, especially brains. Darwinian Natural Selection is 'blind', while we are supposed to have look-ahead goals and aims towards which we direct our intelligence. So although we ourselves are the result of blind processes of evolutionary development as a species, individually we create deliberately planned designs. This is so odd that Gods with planning intentions have often been postulated to give symmetry to the situation. Another tack is to deny intention in us by supposing that we are puppets, entirely reflex-controlled by stimuli without power to initiate or decide. What sets the directions of human intelligence?

It is tempting to imagine that not only has the physical brain, with the rest of the organism, developed by randomly occurring quantal jumps of genes, but that novel *ideas* may also be generated by random generators initiating novelty in thinking and perception. There might then be *symbolic* processes of selection analogous to the selection by trial-and-error of novel variations by Natural Selection against Nature. Conceptual selection is not only by symbolized facts, but also by criteria of consistency, or aesthetic elegance, and other more or less academic criteria for acceptance or rejection. They are tested against knowledge, as stored and organized in the individual nervous system, rather than against events affecting the organism as in Natural Selection. But, surely, most problem-solving by organisms is through following rules and adopting analogies from past situations. This is different from the intelligence of Natural Selection: it is odd to say that Evolution *occurs* by rule-following other than trial and error though Natural Selection has *created* rules for intelligent behaviour and solving problems.

If adequate heuristic rules can be carried out by man-made mechanisms – though they are very different from the physiology of brain mechanisms – Artificial Intelligence is possible, and for all we know will come to overtake us. This would be the extreme example of 'additionals' to our intelligence. The essential concepts are applying rules and transferring knowledge appropriately. We shall now look briefly at rules and analogies as essential, together with chance mutations of concepts, for brain intelligence.

Rules and analogies: heuristics

Almost always, problems are solved by following rules, which may be more or less explicitly formulated, and by reading the present from the past by analogies drawn from experience. This is so both for individuals and for science. Assumptions of similarity to and from familiar situations are generally made so effortlessly that we are unaware of this use of knowledge in behaviour. Rules, also, are generally followed without effort or awareness. This is so for all manner of situations: social encounters, walking, driving a car, and perhaps the majority of professional decisions. There are, however, exceptional cases (which have prominence because it is these that we stop and think about) which are not solved by accepting analogies from familiar situations for 'automatic' rule-following.

When Potential Intelligence is transferred to a somewhat different situation, it may contribute to successful novelty. It is indeed often said that inventions are generally only new combinations of old elements. This is familiar in the psychological terminology of 'transfer' of knowledge or skill. Transfer is 'positive' when it is useful (and so intelligent or 'smart') and is 'negative' when a handicap (and so stupid or 'dumb').

It is important to note that we can appreciate the effects of transfer of knowledge, or transfer of the know-how of skills, though we may not understand the mechanisms of learning and performance. In this sense Potential Intelligence is not a physiological concept: it is *what* is transferred that matters, rather than *how*.

Here we approach another paradox in conventional ways of thinking about intelligence. It seems paradoxical that when we have adequate rules (such as in mathematics), all we have to do is follow them blindly and, if we avoid mistakes, the answer follows as a matter of course. But at the same time, mathematicians are in general regarded as rather specially intelligent people. They retain this prestige though computers now do much of their hack work – the rule-following. In fact, computers are far better than men at following rules without error. What they are relatively poor at is deciding which rules to follow. It is these decisions, as well as inventing new rules, for which we compliment mathematicians on their intelligence. This is their Kinetic Intelligence which we respect. Available rules are examples of Potential Intelligence which allow current problems to be solved from the stored past without requiring Kinetic Intelligence. So again a paradox disappears by accepting the Potential Intelligence notion. On the conventional view, intelligence seems to vanish just when the greatest intellectual progress is being made – when there are adequate analogies and rules to follow – which is very odd. We may say that previous Kinetic Intelligence has produced adequate Potential Intelligence, to solve the current problem by providing analogies and rules to follow.

We come now to a curious question: are there analogies and rules for selecting analogies and rules? If so, then Kinetic Intelligence would disappear. It would disappear if there were sufficient rules – for selecting rules – for selecting rules – for selecting rules. . . . But this would require an impossible number of nested rules: so Kinetic Intelligence is necessary.

I think that we can see other and less formal reasons why there cannot be adequate rules and analogies for all situations. For example, new situations and new problems do arise. These may be new combinations of familiar elements, which produce effects beyond available analogies or rules to predict, appearing as 'Emergent properties' (see pp. 86–92).

It seems clear from studies of perceptual illusions (Gregory, 1970) that rules for deriving perceptual hypotheses (perceptions) from the available sensory data are in some situations inappropriate; they then generate curious discrepancies or illusions. In illusions we experience mis-matches, between perception and conceptual knowledge, which are highly revealing. Above all, illusions show the power of inappropriate rules to generate answers (perceptions) which, though clearly deviant from what we accept intellectually as true, yet we cannot correct.[9] This is a kind of failure of intelligence which may be quite general. The problem for measuring the intelligence of individuals, and for comparing individuals or species, is that we do not know how to estimate the Potential: Kinetic Intelligence ratio for individuals or species except in the crudest terms. It may be interesting to look at extreme ratios of Potential: Kinetic Intelligence, and the extreme intelligence of genius.

Genius
We have discussed handicapped people and the problems of distinguishing between mental and physical handicap, and the implications of these to how we may think of intelligence. We may now consider the converse: the extremely high intelligence (at least on our definition of 'successful novelty') of genius.[10]

The word 'genius' is interesting, meaning originally to beget, or to come into being. This is related to classical pagan belief in a tutelary god or attendant spirit allotted to every person at birth. 'Genial' comes from the state of Mind engendered by feasting the family genius at ritual feasts. Our present meaning dates from the eighteenth century. The powers of genius can be so astonishing that it is hardly surprising that they are often given supernatural origins.

How should we regard genius in terms of our Potential–Kinetic Intelligence distinction? Acknowledged geniuses generally make remarkable contributions to their society; so they add significantly to the shared external Potential Intelligence, which Popper calls World 2. This idea can actually be traced back nearly two centuries, to a remark by the Swiss-born British painter Henry Fuseli (1741–1825)[11] who wrote, in his *Lectures on Painting*: 'By genius I mean that power which enlarges the circle of human knowledge; which discovers new materials of Nature, or combines the known with novelty.'

There is a tradition to think of genius, rather than mere talent, as creating originality without special knowledge. Thus the English novelist Henry Fielding, in *Tom Jones* (1749), writes: 'By the wonderful force of genius only, without the least assistance of learning'. This suggests that genius is attributed to high Kinetic Intelligence – and that it creates special Potential Intelligence, which can be used or appreciated later. It is worth noting that prodigies are found mainly in music and mathematics and other spheres where wide knowledge is not essential (as it is for history or biology), and this does perhaps confirm the popular notion of genius being, if not a 'wonderful force', at least sometimes almost 'without the least assistance of learning'. This would be exceptional Kinetic Intelligence.

I shall now consider the cases of extreme ratios of Potential/Kinetic Intelligence. The examples may seem somewhat bizarre, but may help to clarify these issues by their being extreme cases, and so clear and dramatic. I shall consider first plants and then intelligent machines: Robots.[12]

Extreme ratios of Potential: Kinetic Intelligence
High Potential: low Kinetic (plant intelligence)
On our account we should say that plants do have intelligence – if only Potential Intelligence. This is not in the form of external artefacts (such as our tools) but in their structures. Similarly, of course, *we* have a great deal of Potential Intelligence, which can immediately be applied to solving our problems, in the design of our hands, our reflexes, and so on. In addition we gain Potential Intelligence by learning, which probably is not so for plants. Plants are cases where Potential Intelligence is obviously far greater than Kinetic Intelligence, which may indeed be zero.

Plants do have remarkable Potential Intelligence. The humblest plant exhibits photochemical processes that are not yet fully understood by the cleverest human chemists. Their structures for giving strength, accepting light, moisture and so on, are brilliantly clever designs adapted to the problems that the plants have to face. They do not, however, show much ability (compared with animals) in dealing with novel situations. But some plants do have genuine behaviour.

The stalks of some climbing plants only twist in one direction (some right-handed, others left-handed), while some rare-genus geniuses of the plant world can twine either way, and change direction as the need arises. The stalks of runner beans (*Phaseolus multiflorus*) coil only counter-clockwise. Their stalks, unlike tendrils, are not touch-sensitive and are somewhat mysterious in their functioning; but they have a sense of gravity, for they climb only vertical or near-vertical supports, entirely ignoring rods, poles, or perches that are horizontal.

Climbing plants show some animal-like behavioural capabilities. The Venus fly-trap (*Dionaea muscipula*) has logic circuits, activated by six hairs inside its death-trap, which distinguish the walking of a fly from raindrops. The hairs must be touched in fly-walking sequence for the trap to close. The passion flower (*Passiflora*) and the white bryony (*Bryonia dioica*) reach out effectively to seek support. It is worth describing this in more detail. Tendrils of the young passion flower first grow straight out, then they curve in a fantastically perfect exponential helix, evidently seeking a support to grasp and cling to. Support for the plant being found, the later tendrils are different: they also grow straight outwards initially, but then form tight cylindrical springy coils. After about five turns the spiral reverses in direction, forming a neat loop in the process, and this sequence of ten turns and reversal may continue several times, until the tendril finds and clings to a support. It recognizes a possible support with its sensitive tip, which transmits information, by some unknown mechanism, to a region behind the tip, which bends to coil round the support. The tendrils respond to rubbing against objects, but they ignore other tendrils and the inconsequential kiss of raindrops. So there is some sensory capacity to cope with particular situations.

Are these reversals of the spiral initiated by the plant, or are they a geometrical necessity? If the reversals of the growing spiral occur after the tip is anchored, then there must be an equal number of right and left turns; but if the reversals occur before the tip has found its anchor point, then the reversals must be due to commands from the plant. Darwin (in 1880) considered that the reversals occur only when both ends of the tendril are fixed, one end to the plant, the other anchored to the support it has found.[13] But with great respect, it seems on my observations that they often grow with reversals before finding anything to cling to. Many never do find anything on which to anchor, and they become long spirals with up to five or more reversals, which must be initiated by the plant. Why should the climbing passion flower have this strategy in its behaviour? As Darwin pointed out, the tendrils are springy, so they do not break when the plant is blown about by the wind. The reversals may serve to prevent the tendrils twisting up.

Now the same trick could be used with wires, by telephone engineers. In modern telephones the wires to the hand-set are amazingly similar spiral coils, often with reversals every so many turns, just like the reversing coils of the passion flower. Did Post Office engineers learn this trick from the plant, or is this a separate invention? Was it first invented millions of years ago in the evolution of climbing plants, or a few years ago by someone in the Post Office? If the former, man took over the Potential Intelligence of the plant (and needed but little Kinetic Intelligence to do so); if the latter, he created this invention by his own Kinetic Intelligence. Note that, although this is far beyond the ability of individual plants, it is clearly not beyond the Kinetic Intelligence of Evolution, which invented many structures and processes that we still do not understand and cannot reproduce.

Actually, I am sure that we have landed here upon a joke – for the reversals of the telephone wires seem not to be put there by the telephone engineers at all, but to result from users twisting the phone more in one direction than the other – and so unwinding sections of the original coils! This may result fortuitously in a wire that will not tangle so readily, and so is an improvement. This kind of chance event is part of the conditions necessary for invention by Natural Selection. So we might imagine Natural Selection sometimes working blindly in technology!

It does seem clear that plants produce reversals of their tendril coils, and we assume that (unlike for the phone-owner) this has developed as a useful behavioural strategy. It seems also that there is some real-time adaptation of plant behaviour according to circumstances as sensed by them. Their repertoire of solutions to select from, and their individual novelty, are, however, extremely limited, so we should say that their Kinetic Intelligence is low by comparison with animals, and far lower than ours.

If we now go on to consider how we might measure a plant's intelligence – on a human intelligence scale – we might learn just why it is so difficult to measure human intelligence in IQ terms. What would be the IQ of a plant? As we have seen, plants have remarkably 'intelligent' processes of photosynthesis and so on, though these are not 'behaviour' because they are not controlled in at all the way that animals deal with the world by their behaviour. Yet some plants do sense environmental situations and respond appropriately, so we may be tempted to call this Kinetic Intelligence in plants. This is, however, vanishingly small by comparison with the enormous contribution of their inherited Potential Intelligence.

As I have said, it would require superhuman intelligence to design a plant. The problem beats any human and the whole of science. So if 'Design a plant' was set as an intelligence test problem no human (or the whole of science) would pass the

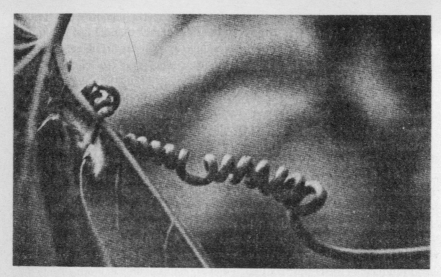

33   The tendril of a passion flower. As it grows it coils to form a spring; so when pulled by the wind it does not lose its grip. The coils reverse about every five turns.

test. Yet plants are answers to this problem. Plants, though, would not begin to solve even the simplest problems that psychologists set us for measuring our IQ! So here we see the Potential/Kinetic Intelligence ratio problem dramatized in Nature. It is extremely difficult to estimate Kinetic Intelligence when so much is given in the Potential Intelligence, of plants or of people.

The situation is similar to the point made earlier about handicapped people. Although plants have remarkable inbuilt Potential Intelligence, this is much less than ours: they are in these terms extremely handicapped in their Potential Intelligence. So to perform simple human tasks with their limited attributes would require superhuman (let alone super-plant) Kinetic Intelligence. The plant would have to invent aids to add to its absurdly inadequate physical endowments and inbuilt strategies, to do what we do without difficulty.

We can now see a conceptual justification for the psychologist's tendency to protect IQ scores against loss of performance through socially sensitive handicaps. There is, indeed, a case for giving *higher* IQ scores when human Potential Intelligence is impaired by handicap – for greater than normal Kinetic Intelligence is then required, when available aids are not adequate to overcome the handicap.

Low Potential: high Kinetic (machine intelligence)[14]

We may say at once that AI computers – and their ancillary equipment such as TV eyes, hands and so on – embody Potential Intelligence as contributed by the human

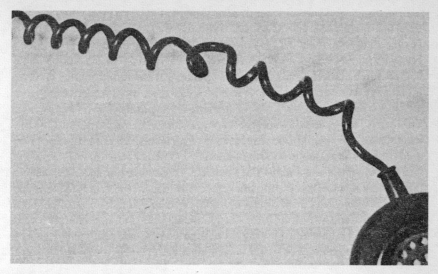

34   The wire of a telephone hand-set. This is coiled like the passion flower tendril. If this is a (human) intelligent design, does not the passion flower embody intelligence?

designers of the machine. The first point to make, then, is that for 'intelligent machines' (if such exist) the inbuilt Potential Intelligence is transferred from our Potential Intelligence by our Kinetic Intelligence, as we design and create the machine, while *our* inbuilt Potential Intelligence was created 'blindly' by Natural Selection.

The major claim for A I is, in our terms, that it has in addition to its Potential Intelligence (which indeed is imparted to all machines made by us) some Kinetic Intelligence, which it deploys by itself. It is this that is special about the A I project, and it has implications for considering human intelligence.

While discussing intelligence in man we found complicating moral considerations, not all of which apply to Artificial Intelligence. This is not to say, of course, that there are no moral considerations concerning A I; but these only apply insofar as the machines affect us. The point is that we are free to judge and to criticize machines as we please; but criticizing human beings and their intelligence is dangerous. We might take advantage of the 'moral neutrality' of machines, including imaginary or real A I machines, to clarify questions of human intelligence.

We have suggested that human intelligence, including I Q, is preserved against loss of performance due to many kinds of impairment, and especially socially sensitive impairments such as deafness and blindness. Now there is no need to protect a machine's intelligence for any such reasons. However, in order to measure its Kinetic Intelligence, it is necessary to establish its receptor and motor performance, and establish whether there is impairment of its originally designed Potential Intelligence. This is because Kinetic Intelligence is for machines, as for us, gap-filling from what is given by Potential Intelligence to what is needed to perform the task or solve the problem. The less the Potential Intelligence built into the receptor and motor mechanisms, the greater is the required Kinetic (Processing) Intelligence of the machine. If these are damaged, more Kinetic Intelligence will in general be required, and so the damage impairment must be assessed. Conversely, for example, an object-recognizing machine that is allowed to have radar ranging or a pair of camera eyes to give stereo depth has a much easier task – with less information processing required – to obtain three-dimensional form from its two-dimensional camera-eye images. The gap to be bridged is smaller, so it needs less Kinetic (Processing) Intelligence. This at once tells us that it is a cheat to give the machine all the information strictly required. If this is done, there is no problem left, and so no need for Kinetic Intelligence, or A I. The problem is already solved by its built-in Potential Intelligence provided by the designer, so there is no gap to be bridged by A I. This implies that A I machines cannot be judged simply by their performance.

It is the key assumption of A I that machines can be humanly intelligent. From what we have said we should expect to find difficulties in estimating machine intelligence, and especially in comparing it with ours. The deep difficulty is that the sensory and motor control systems of existing and planned Robots are in many ways different from ours, so how can they have human-like intelligence? Is it possible to have our human Kinetic Intelligence in a body incorporating such different Potential Intelligence?

This is the converse of plants. Plants are extremely 'handicapped': so handicapped that all conceivable aids could not make them comparable with us. Robots might have vast computing ability but are unlikely to have bodies like ours, so some problems will be inherently more difficult for them, though others will be easier. The point is that their different bodies *require* them to have different Kinetic Intelligence from ours; so that they will not be quite human in their thinking, however their computer brains may be designed. In some situations these differences may not be apparent: possibly they will enter into our moral world of rewards and taboos in situations where they are comparable with us. Then, no doubt, special allowances would be made for their inbuilt inadequacies, much as we reduce certain expectations for handicapped people. We already have, of course, experience of many superhuman abilities in existing machines.

It is interesting to consider scientific techniques and devices which solve problems beyond human competence; for example, the structure of D N A was discovered by a few brilliant people from highly ambiguous and inadequate data.[15] This required human Kinetic Intelligence of a very high order. The crucial evidence from X-ray crystallography (a technique originally suggested by Sir Laurence Bragg) [16] was essentially ambiguous, because phase information was lost, so the atomic positions were not indicated directly but only by patterns which looked quite different from the molecular structure and had to be read though they had many possible meanings. Very recently it has proved possible to derive and display phase information, so now unambiguous pictures of molecules can be given by machines which display the structure that humans had to build up by years of intolerably laborious work, for which they rightly gained Nobel Prizes. But do we consider that the new machines should receive prizes for their achievements, which are comparable with and even superior to those of the world's best scientists? Further, will scientists go on receiving prizes for discovering molecular structures, now that machines are available to make the task comparatively easy?

This is a case where aids (amplifiers, additionals and matchers) have dramatically increased the Potential Intelligence that can be applied to such problems. So far, at least, the machines are excluded from our human reward schemes, though of course their inventors are rewarded occasionally with the highest honours. What is rewarded is clear evidence of the highest human Kinetic Intelligence, when this produces a marked increase in the available Potential Intelligence of science, so that new problems can be solved that were previously too difficult or not even appreciated as problems or possibilities of invention or discovery.

What will happen when *machines* design new techniques and instruments that make important discoveries? It is already true that computers are vitally important for designing new machines – including more advanced computers. So we see the Potential and Kinetic Intelligence of machines applied for increasing the Potential Intelligence available to us. And yet these machines are not rewarded: machines are thus outside our moral system.

It is still possible to think of even these machines as merely aids (of the three kinds that we have considered) and not as initiators of new ideas or inventions. But this may not continue. It is entirely possible that machines will come to ask new questions and find ways of answering them. It is indeed possible that only machines will be able to understand answers that they discover or theories that they create.

Scientific theories are, for human preference, simple and aesthetically elegant, but neither description need apply to the kinds of answers that machines will provide. Machine criteria of 'simple' and 'aesthetic' may be very different from ours (and indeed neither are significant for existing computers), so we can expect future science to be even more non-human than it is now – possibly to the extent of being incomprehensible to anyone. Machine-derived theories may, however, match Nature better than our thinking can ever do. This might be so even though we use as aids the most high-powered multipliers of our intelligence. It may be impossible, even with every aid, to match human understanding to the deep structure of the world from which we have evolved by the blind steps of Evolution, though conceivably man-made machines will succeed.

Intelligence in computers?

The recent attempts to program computers to be intelligent – to find surprising answers, and occasionally in ways surprising to their designers – has brought questions of the nature of intelligence into a new focus, though there is still no consensus of opinion or agreed theory of Artificial Intelligence. Among the most illuminating comments are those of Marvin Minsky. In his paper, 'Steps towards artificial intelligence' (1961), Minsky questions whether intelligence is anything very special, writing: 'To me "intelligence" seems to denote little more than the complex of performances which we happen to respect, but do not understand . . . . Once the proof of a theorem is really understood its content seems trivial.'

This comment, that human or computer intelligence seems to disappear as we come to understand the underlying processes, seems to apply with special force to the processes of Darwinian Evolution by Natural Selection. What was, and indeed still is, so shocking about this is its rejection of creative intelligence as anything more than randomness and selection, in the design even of ourselves. It seems at least to many people to make our Minds trivial. If it turns out that the processes of our own creative intelligence are similar, will this make us seem trivial?

Minsky adds (in brackets): 'Still, there may remain a sense of wonder about how the proof was discovered.' This, in our terms surely reflects the present limited understanding of Kinetic Intelligence. Minsky goes on to describe how intelligent programs are seen by their designers:

Programmers, too, know that there is never any 'heart' in a program. There are high-level routines in each program, but all they do is dictate that 'if such and such, then transfer to such and such a subroutine'. And when we look at the low-level subroutines, which 'actually do the work', we find senseless loops and sequences of trivial operations, merely carrying out the dictates of their superiors. The intelligence in such a system seems as intangible as becomes the meaning of a single common word when it is thoughtfully pronounced over and over again.

But we should not let our inability to discern a locus of intelligence lead us to conclude that programmed computers therefore cannot think. For it may be so with *man*, as with *machine*. . . .

The designers of AI programs are not generally concerned to make their programs reflect, mimic, or simulate human thinking processes or strategies. One reason for this is a lack of understanding of the processes of human intelligence; another reason is, perhaps, the differences of functional structure ('architecture') of computers and brains. For the present, computers work almost completely serially while brains almost certainly work very much by parallel processing. This difference in their design (in their inbuilt Potential Intelligence) must distort machine–brain parallels. There is, however, one highly distinguished body of work which is explicitly aimed at comparing human with machine problem-solving. This is the GPS (General Problem-Solving) program of Allen Newell, H. A. Simon and J.C. Shaw. They present humans with clearly defined problems in symbolic logic. They provide their subjects with a set of rules, and ask them to describe the steps by which they arrive at their answers. This is compared with the performance of a computer. This does not describe its inner workings, and is serial, not parallel, in its operations; but these differences are carefully considered by the investigators.

These man–machine thinking experiments are described in several papers, including: 'GPS: A Program which Simulates Human Thought' by Allen Newell and H.A. Simon (1961). In this paper (in which the joint work of J.C. Shaw is acknowledged), Newell and Simon write:

> We may then conceive of an intelligent program that manipulates symbols in the same way that our subject does – by taking as inputs the symbolic logic expressions, and producing as outputs a sequence of rule applications that coincides with the subject's. If we observed this program in operation, it would be considering various rules and evaluating various expressions, the same sorts of things we see expressed in the protocol of the subject. If the fit of such a program were close enough to the overt behaviour of our human subject – i.e., to the protocol – then it would constitute a good theory of the subject's problem-solving.

A major difference – or at least an apparent difference – between the creation of solutions by the Natural Selection of Evolution and by human problem-solving is the directedness of human thinking. We set up goals; while Darwinian Evolution lacks look-ahead, to be entirely blind. The GPS program does have goal-seeking: indeed this is an essential feature of the program. Newell and Simon say: 'Basically, the GPS program is a way of achieving a goal by setting up subgoals whose attainment leads to the attainment of the initial goal.' This is done by matching symbolic patterns and seeking to remove discrepancies, according to various schemes or 'methods' available to it, for various types of subgoals:

> These methods form a recursive system that generates a tree of subgoals in attempting to attain a given goal. For every new difficulty that is encountered a new subgoal is created to overcome this difficulty. GPS has a number of tests it applies to keep the expansion of this goal tree from proceeding in unprofitable directions. The most important of these is a test which is applied to new subgoals for reducing differences. GPS contains an ordering of the differences, so that some differences are considered easier than others. . . . GPS will not try

a subgoal if it is harder than one of its supergoals. It will also not try a goal if it follows an easier goal. That is, GPS insists on working on the hard differences first and expects to find easier ones as it goes along. The other tests that GPS applies involve external limits (e.g. a limit on the total depth of a goal tree it will tolerate) and whether new objects or goals are identical to ones already generated.

Here we see hints on how direction of thinking can be given by rules, though the problem being considered is unique to the computer, or the brain, and before there is a hint of the final answer or the direction, in conceptual space, in which it will be found.

If machines come to succeed us as the intelligence over and above Natural Selection, will they accept our goals and commands? Or will the machines have to 'go it alone', as we fail to understand what they invent and discover? We should establish not only the limits of possible future machine intelligence, if this is possible, but also our intelligence limits; for when machines pass our understanding we are enslaved – if not by the children, then by the machine-built grandchildren of our intelligence.[17]

Taking stock

1   We have concluded that intelligence test scores are not direct measures of ability, if only because IQ is kept constant though abilities improve with age at least up to adolescence. This constancy of IQ with improving abilities is achieved by applying various kinds of weighting constant, and by selection of test tasks that do not show too much improvement with age. They are also chosen to be as independent as possible of special knowledge. Further, IQ scores are a single continuum though there is an enormous variety of abilities: IQ therefore cannot be equated with abilities, but is supposed to measure something unitary underlying all mental skills.

2   We have questioned the validity of applying measured mean differences between groups (such as the sexes, races, or people with different education) to show that they have correspondingly different average intelligence. This we question for the following reasons:

(i)   That since the specific test tasks used are admitted to favour groups (perhaps especially the sexes) selectively, the mean scores depend on the balance of the various 'favouring' tasks. Since there is no independent criterion for adjusting the balance of the favouring tasks, attempts at balancing these effects all suffer from circularity, and the means can be shifted, arbitrarily or to attain a desired result, by selection of the tasks to favour the groups differentially.

(ii)   That sex-neutral or race-neutral tasks may be impossible to find, or to validate as such without circularity.

(iii)   That although weighting constants are applied to maintain the average IQ at 100 (and other measured IQs correspondingly constant) though abilities improve with age at least up to adolescence, weighting constants are not applied on socially sensitive issues to compensate differences of environment, race or sex – for how they are applied would reveal what could be socially embarrassing assumptions of the testers, and in any case would be difficult to justify without circularity.

3   Since the IQ scale is a single continuum but abilities are highly varied, it cannot be a direct measure of abilities: intelligence is usually conceived as something underlying abilities, though estimated by measuring abilities at chosen tasks. The selection of the tasks used seems to have only dubious justification.

4   Tasks are chosen to be as free as possible of special knowledge; but skills that intelligence is normally credited with depend very much on effective deployment of knowledge and making effective use of education, though educational effects are (for clear reasons) minimized in the test designs.

5   Intelligence as conceived by psychologists has produced the socially unfortunate notion that what they claim to measure is very special – a kind of 'Golden Egg' – which we have come to treasure greatly, although it is demonstrated, measured and challenged by trivial-looking tasks, which do not (and are not intended to) stimulate the imagination. Worst of all, it is supposed that this intelligence, underlying and necessary for skill and understanding, cannot be developed by any efforts that we may make, though it has become a Golden Egg that we regard as most precious. It seems more than possible that the social hassles, the individual hurts to confidence, and vital decisions on education and job opportunities are based on an illusory Golden Egg of intelligence which has been created by the mystique of IQ tests.

6   Considering aids to intelligence as a useful ploy for clarifying and getting to grips with these issues, it is suggested that there are three logically different kinds of aid:

(i)   Intelligence amplifiers, which increase intelligence. Most people would accept education, access to facts, and to rules of inference as examples of intelligence amplifiers; but these are discounted by test designers, who regard IQ as essentially fixed throughout life, whatever our education or the skills that we learn.

(ii)   Intelligence additionals, which contribute to abilities without increasing the individual's intelligence. This seems to be how test designers regard education, etc. (as above). They would deny the first category but accept this one. We would include AI machines in this category, if they are themselves intelligent and used to augment our intelligence.

(iii)   Intelligence matchers, which make more effective use of existing intelligence by matching available skills, knowledge, or whatever, to requirements of the task in hand. Examples could include language, for language matches brain processes to tasks, by representing relevant features and coding and simplifying situations, so that they can be handled effectively.

There is much more to be learned. Our discussion on these issues is only a start, but I hope a helpful start, to understanding intelligence and applying whatever it is more effectively, with the help of aids of very different kinds. By discovering just what aids do, we should be able to argue backwards, to discover what are the processes of intelligence that are being amplified, or augmented, or matched. They can be tools for understanding intelligence.

7   It is far from clear how we should distinguish between 'physical' and 'mental' handicap, or how to justify the different approaches adopted to what are accepted as 'mental' handicaps, such as dyslexia and autism. Some are 'saved' or protected from loss of IQ scores, while others also categorized as 'mental' are not; though reasons for this socially and individually important distinction seem to be based on confused notions of the nature of intelligence.

8   It is suggested that the early double meaning of 'intelligence', both as stored knowledge that can be applied to present situations and as ability to derive new solutions, should be retained and included as essential to an account of intelligence. These are named, respectively, 'Potential Intelligence' and 'Kinetic Intelligence'.

9   It is suggested that what is usually taken by psychologists as 'intelligence' (which is roughly our Kinetic Intelligence) cannot be measured without assessing the contribution – which is always present and usually dominating – of Potential Intelligence. This is extremely difficult, if only because the contribution of Potential Intelligence is so large that the Kinetic Intelligence may be hardly significant – almost at the 'noise level' for measurement, so that its contribution is difficult to discover or estimate.

10   It is suggested that intelligence is essentially success at creating and understanding appropriate novelty. The greater the novelty and the greater the success, the more we should credit intelligence. Intelligent solutions are produced by :

(i)   The higher organisms, especially by brain activity. This may create and be guided by knowledge and understanding based on analogies.

(ii)   Natural Selection, as in Darwinian Evolution. This has created highly intelligent solutions to problems that we can only partly understand. It has succeeded without the guidance of understanding; in this sense the Kinetic Intelligence of Evolution is 'blind', though we are not entirely blind.

11   It is suggested that Kinetic Intelligence generates Potential Intelligence, which is stored in organic forms internally in memory and externally in artefacts.

12   It is suggested that the main role of our Kinetic Intelligence is to deploy knowledge, including rules, largely by analogies. These are structures of knowl-

edge containing facts including generalizations and heuristic rules. While rules are being followed, no Kinetic Intelligence is required (and rule following is easy to mechanize). Kinetic Intelligence is needed to build and select analogies and rules – which can never be quite adequate in novel situations. So it must also occasionally invent solutions beyond what can be derived by rule-following. This is presumably a random change-and-selection process like Natural Selection, except that clashing symbols instead of the life–death battles of organisms are the characters on the stage of their drama, which is the drama of creative Mind.

Is intelligence inherited?

In our terms this will be: 'Is Kinetic Intelligence inherited?' Considering Potential Intelligence, however: there is obviously an enormous contribution from heredity – though it is another question again whether there are individual differences – for all of the design features of organic structure and unlearned reflexes mediating behaviour, which solve ancient problems, are examples of Potential Intelligence. There may be a lot of genetically inbuilt knowledge, but only of situations that have been significant for an extensive period of biological time, which may extend through many species. It is entirely possible that we have inbuilt knowledge that was genetically coded in subhuman species and perhaps even in species that are long extinct, and even for situations that no longer exist on earth. It is now almost universally believed that individually learned knowledge does not modify the genetic code; so that learned knowledge dies with each individual, unless indeed it is transmitted by external means such as by speech, writing or pictures. The first known paintings are Palaeolithic, c. 25,000 BC (found at La Ferrassie, France, depicting animal figures and symbols on stone blocks, in a layer of the Aurignacian II period); so we may assume that external transmission of knowledge is not much older than, say, 30,000 years, which is short in biological time. Since the rejection of Lamarckianism (which has had a lingering death, with occasional revivals even today), all genetically coded long-term biological gain of knowledge is held to be through Darwinian Evolution by Natural Selection, or by some modification of Darwin's theory. Individual learning is also extremely important in higher organisms, though its results are not transmitted genetically and so are lost. Evolution by Natural Selection, and technological changes, have inbuilt 'ratchets' which ensure generally continuous improvement, though this depends on the criteria accepted for 'improvement'. Individual learning is not like this, for it is simply lost with the death of the individual; though in what he contributes to the shared Potential Intelligence by the mutations that he transmits – and the innovations that he may contribute to his society's knowledge and technology – he can attain a kind of individual immortality.

Individual learning, in our terms, is an increase in internal Potential Intelligence by individual experience. This, as we have seen, is strictly analogous to increase in the externally stored knowledge of the Potential Intelligence of libraries – except of course that it cannot be read directly by other individuals as books can be read, and so is private. Although it is not known what are the physical changes of brain structure that occur during learning, it is generally accepted that there is some physical basis to learning and memory. So we may think of these changes of structure as representing new knowledge, analogous to tools and books except that it is private. Like increases in any other Potential Intelligence, it has the important property of reducing the gap between a problem and its solution: the gap to be bridged by Kinetic Intelligence. Clearly, gain of knowledge by learning can very greatly reduce the need for Potential Intelligence in situations where stored knowledge can be immediately applied.

Since we do not accept that IQ tests measure anything directly related to the overall spectrum of abilities, we cannot accept that the statement that 'IQ is inherited' has much significance. It is, however, entirely possible that there are sex and race differences in various abilities. To say that these are inherited intelligence differences begs questions of the nature of intelligence which we have tried to raise, and especially whether it can be measured divorced from knowledge. The technical problem is to separate Kinetic from Potential Intelligence for measurement; however, this remains to be done.

The future of intelligence

Although I have rather heavily criticized intelligence tests, on what I take to be logical grounds, it could, however, be that the tests used are predictive and so useful. This could be so even if the rationale behind them is entirely mistaken. There are many examples of this in the history of science, including medicine. It is difficult here, though, to guard against their being self-fulfilling prophecies, since the results of intelligence tests have for many years been used to select people for educational opportunities and jobs which may have rather automatic advancement once held. In other words, high scores favour individuals and low scores handicap them – so the scores are not only predictions but also tend to make the predictions come true. If this is not realized, their power to predict passively will be overestimated and so they will be given undue credit.

It remains entirely possible, however, that these tests are genuinely predictive and so useful for some purposes, even though they have the unfortunate social and individually harmful effects that we have considered. It is, for example, entirely possible that individual nervous systems work at different rates, and that some are better organized at quite fundamental levels than other people's nervous systems. Possibly such differences do show up in IQ tests. But if so, it does not follow that these are absolute limits, which are reached in practice. We see this from our discussion of intelligence aids. Consider matching aids: if knowledge, language coding or some such is not optimally matched to the situation, the nervous system cannot be expected to perform near its absolute limits. The same consideration applies to computers with non-optimum programs. A better program can make a limited computer far more efficient; indeed the problems of AI are essentially programming problems. Given the power of language and of skills for thinking and doing, it seems likely that absolute limits are seldom reached, though it is of course necessary to have the ability to accept (and understand, whatever this may mean) analogies and rules which confer power to thinking. This is surely the point of education, apart from the sheer fun of understanding and thinking. It is entirely possible that some people have innate or acquired handicaps for learning – but if so this is very different from the IQ story, which asserts that learning and knowledge are not important for intelligence. It is this that we have particularly questioned. We have questioned whether intelligence can be considered separately from deployment of knowledge. This led us to describe intelligence as 'Potential' stored knowledge (including procedures, or rules for thinking and so on) which is called upon by the Kinetic Intelligence which may produce novelty beyond available knowledge. If this could be measured independently of the individual's knowledge base, then something like IQ tests might be acceptable; but this seems highly doubtful, if only because knowledge (including heuristic rules, which may be implicit) is so important. This we have surely learned from current attempts to program computers to be intelligent.

The unitary notion of intelligence implied by IQ tests appears to be a simple mistake and to have unfortunate consequences. Neither Kinetic nor Potential Intelligence should be thought of as unitary. The situation might be compared with abilities in athletics. Some physiques are appropriate for some sports (that is, they match the requirements) and not for others. Thus a physique suitable for long-distance running may be hopeless for the shot-put, and this might be judged early on. Further, since there are acknowledged sex and national differences of physique, it should come as no surprise that there are sex- and race-related abilities in athletics. But, most significantly, in spite of the enormous prestige and competitiveness of athletics and sports, these differences do not evoke at all the same traumas as the furore over sex and race differences of IQ-measured intelligence. If we were not to think of intelligence as a unitary something underlying all mental abilities, but rather regarded mental abilities more as we now regard physical skills, then we should be free of the trauma and the social dilemmas and confusions surrounding IQ tests and current views of intelligence.

There might, however, be certain kinds of task (perhaps simple tasks) that give performance scores highly predictive of the development of mental skills. Returning to our computer analogy: a basically slow computer will be handicapped for any task requiring many operations in limited time. So any test of its speed should predict limits for a wide range of abilities. Indeed, these characteristics of computers are cited by the manufacturers with just such thoughts in mind. They do set absolute limits to performance, except for optimally matched programs.

This account of limits is not at all the same as how intelligence is currently conceived. And just as computers may be upgraded in their hardware, and performance dramatically improved by better programs, so improvements in environment and education might produce dramatic improvements for us. Although we do no doubt have absolute limits, these may never in practice be reached, for our matching may never be optimal. Similarly, performance in athletics goes on improving with better training procedures and few, if any, absolute limits have been established.

If we allow our other two kinds of intelligence aids – multipliers and additionals – to augment intelligence, then the limits are pushed much further. We may think of bicycles as matching human muscles so that we can in suitable environments ride faster and further than we can walk. Space rockets are incredibly powerful additionals to our athletic powers, so that high-jumpers appear puny by comparison with astronauts. Pure muscle-power multipliers are not so common; but servo-controlled multipliers have been constructed by which a man becomes a giant, his every movement producing movements of artificial limbs with superhuman strength. If these analogies hold for athletic abilities we can see no limits to intelligence, at least for creating; though there may be very different and definite limits for understanding.

We have considered intelligence as more or less successful creation and understanding of novelty. It remains a possibility that our various intelligence aids will become more powerful for creating than for understanding novelty. This is a frightening prospect, which can only be countered by education. The education might, however, extend to AI machines: possibly they will come to understand things opaque to us. Then we shall be back in the situation that our ancestors fancied themselves in, with powerful and omniscient Gods. Perhaps we shall have to placate our AI machines as the Greeks poured libations to the Gods to tap their mysterious intelligence beyond our understanding.

Here we have allowed ourselves to float almost free of the literature; to consider intelligence in a somewhat new way, perhaps. We shall return now to the writings of some of the greatest thinkers – to consider the basis of knowledge.

11 Links of Mind

All the mighty world
Of eye, and ear – both what they half create,
And what perceive.

William Wordsworth (1770–1850)[1]

Chains of knowledge
Chain makers of rationalism
The Rationalist holds that we can obtain knowledge of the world by thinking, apart from observation or sensory experience. Intuition and mathematics – especially geometry – have the highest standing and are supposed to reveal knowledge directly and indubitably. Perception tends to be relegated to a second-class status. The Rationalist School is associated with Europe rather than Britain, which is the home of Empiricism. The Rationalist philosophers number several men of outstanding intellect. Though dubbed 'Rationalists,' they are capable of daring speculations on how Mind can be in a mathematically described Universe.

Benedict Spinoza
Benedict Spinoza (1632–77) was born in Amsterdam, of Jewish parents. He was a humble, totally sincere man. He accepted the physics of his time, and combined philosophy, following Cartesian ideas, with lens-grinding. His main book is the *Ethics* (1677), which is a Euclidean-style exposition of a metaphysical position, based on a single basic substance, 'God or Nature' (*deus sive natura*). Thought and extension are for Spinoza just two aspects of God. Minds and bodies are one – avoiding the Dualism of Descartes, and reverting to a Monism that we may associate with the much earlier notions of Parmenides, and indeed the most ancient mythological conceptions. He also returns to Greek notions of law and logic by holding that it is logically impossible for events to occur otherwise than they do – except perhaps for sinful acts, which are for him logical errors. Explanation is for Spinoza similar to Aristotle's causes as logical necessities; but Spinoza questioned

how essentially different substances could interact – hence his single Mind–Matter substance. For Spinoza the Mind is not free – but everything works out for the best.[2]

So, as for Kant, psychology and physics are ultimately one. By studying Mind, and especially one's own Mind, one discovers the nature of the world. This is the antithesis of Empiricism, which holds that we are observers separate from the world, except for sensed chinks of light which we may accept as signals – to be read or misread, according to our understanding or lack of it.

Gottfried Leibniz

Gottfried Leibniz (1646–1716) was born in Leipzig. He is generally regarded as one of the world's greatest intellects, excelling in philosophy and physics as well as mathematics, being joint discoverer with Newton (though neither would admit this!) of the differential calculus. His main works are *Theodicée* (1710) and *Monodologie* (1714). Most interesting are the *Discours de la Métaphysique* (1846; transl. 1952) which he wrote secretly; and the correspondence with Arnauld and Clarke.

Leibniz's basic though seemingly bizarre notion is that substance is an infinite number of non-material isolated Monads, each of which reflects the world from its point of view, and forms a hierarchy. They do not affect, and they are not affected by, material phenomena; but they are synchronized – as two clocks beat time together though unconnected causally. This is Leibniz's famous clocks analogy, used to suggest that Mind and Matter are separate but synchronized (see chapter 16). Substance, for Leibniz, 'is a being capable of action'. Simple substance has no parts: compound substance is a combination of simple substances, or Monads.

The simpler Monads have *perception,* which is 'the inner state of the Monad representing external things'; and some of them sometimes have *apperception,* 'consciousness, or the reflective knowledge of the inner state'. Each Monad makes a living substance; so there is life everywhere, but when a Monad has special organs, such as a lens of the eye, the perceptions are heightened and this may give *sensations*. The organs of living things come from *preformed seeds* and are transformations from pre-existing living things. There are small animals in the seeds of large ones, which they appropriate, and so 'pass onto a wider stage'. There is, strictly, no death: instead, 'casting off their masks or their rags, they merely return to a more sensible scene, on which, however, they can be as sensible and as well-ordered as on the greater one'.

Nothing happens without its being possible to give a sufficient reason 'to determine why it is thus and not otherwise'. For the Universe itself, he realizes that sufficient reason cannot be found within the Universe; so ultimately the reason for existence is God's reason – and God is the Top Monad.

Leibniz designed and had made a wheeled calculating machine, for which he was elected a Fellow of the Royal Society in 1673. He not only invented the notation still used for the differential calculus, which he invented independently of Newton, but he also conceived a perfect logical language, and may indeed be regarded as the founder of symbolic logic (see note 1 for chapter 16, p. 599).

Immanuel Kant

Immanuel Kant (1724–1804)[3] is acknowledged as the most important Rationalist philosopher of modern times. He is exceedingly difficult to understand, at least for readers outside his cultural tradition.

Kant started as an astronomer, or rather as a cosmologist, and as a student he was inspired by Newton. All his life he revered him. A hundred years ahead of time, Kant produced a theory of the origin of the solar system, based on the concept of accretion of masses by gravitational attraction, together with distribution of matter through vortices, which he based explicitly on Newton's principles. He wrote *Universal Natural History and Theory of the Heavens* (1755) when he was thirty-one, hoping to command the attention of Frederick the Great, King of Prussia; but Kant's publisher went bankrupt, and almost all the copies were lost. It never reached his King. This accident very likely turned him away from science and to philosophy. He seems to have been concerned with how mechanical accounts of the origin and dynamics of the Universe could be reconciled with God; and possibly this was the stirrup that lifted him into the saddle of philosophy, though later he was spurred on by considerations of the place of man and the nature of knowledge. He did not believe man to be unique in the Universe, but thought it likely that the other planets were inhabited.

Kant's account of space and time is central to questions of the nature of perception and knowledge; and his account differs in important aspects from the Empiricist position that we perceive objects directly, and infer our notions of space and time from how objects appear. Kant's main argument is that we can only perceive objects by already knowing – *a priori* – the nature of space: that it is three-dimensional and Euclidean. This, he thinks, is somehow derived from our inability to imagine any other kind of space. He was of course writing before non-Euclidean geometries were considered. As Stephan Körner (1955) puts Kant's position:[4]

We cannot, for example, imagine spatial and non-spatial elephants in the way in which we can imagine elephants which are grey and elephants which are not. If we perceive something in space we cannot imagine it not in space. Therefore the notion of being in space, and space itself, are not abstracted but *a priori*.

Kant has similar arguments for time, given by *a priori* categories of human perception – clearly expressed in his *Prolegomena to any Future Metaphysics* (1783).

The position seems to depend on the difficulty, or rather impossibility, of imagining anything other than three-dimensional Euclidean space. It is from this inability of ours that Kant derives his fundamental categories of space and time. They are derived from our inadequacy, rather than from our ability to understand. This is a very curious argument: why should our limitations be evidence, or reason, for beliefs about the world? This is surely cogent only if we suppose some deep relation between us and the world – between Mind and the nature of things. I think that just this is the basis of Kant's philosophy. I think that Kant is arguing from our psychological make-up to the deep nature of the world. For him, our Minds are crucial evidence of how the Universe is constructed. This is, surely, the basis of Rationalist philosophies.

In the *Critique of Pure Reason* Kant says: 'The immediate consciousness of my existence is at the same time an immediate consciousness of the existence of things outside me.' This he tries to justify with five main arguments:

1   That experience necessarily exhibits succession through time.
2   That consciousness has a unity, in spite of its flow and changes through time, which implies a roughly corresponding unity of the external world.
3   The 'Thesis of Objectivity' – that we can make judgements distinguishable from each particular experience, implying that there is some kind of external continuity to be judged. It must therefore, he holds, exist apart from our experiences.
4   The 'Spatiality Theses' – these objects of judgement (and perception) are ordered in a single space.
5   The 'Theses of the Analogies' – there must be permanence and causality in the external world of things in space and time.

There are marked similarities between Kant's and Berkeley's notions of the status of the world of objects. Both hold that there is an intelligent Creator, who imposes logical and mathematical structure, so that we can understand only by appreciating this structure and following its rules. These features are beyond, or are 'transcendental' to, experience, but must be in our 'intuition' for perception to be possible. Like Berkeley, also, Kant finally comes to the view that 'things in themselves' do not exist apart from being perceived. But does 'God's' perception allow a permanent Universe, for Kant as for Berkeley?

It is quite extraordinarily difficult to be sure of Kant's view of perception and its relation to his notion of 'things in themselves'. These are to him essentially beyond experience, or 'appearance'; and yet he accepts that there are 'things in themselves', and that these are required for our experience – especially for perception. The distinction here between appearance and external reality takes its starting-point from Locke's distinction between Primary and Secondary Characteristics: that, for example, objects are not themselves coloured though they appear with our nervous system to be coloured. But Kant considers also that space and time are part of 'appearance'; though objects are supposed to 'affect' us, to give perception of sequences of events, in time and space. It is extremely difficult (and the different accounts of his many commentators are evidence of this) to be clear as to Kant's distinctions between objective and subjective, or reality and appearance. It is correspondingly difficult to decide whether he held to a Realist or a Representational view of perception – if either. His subjective status for space derives from his – then generally held – account of geometry as depending on axioms which are known *a priori* and which are true beyond immediate experience. But this was exploded with the realization by Gauss that the Euclidean axioms are not special and may be replaced with more or less appropriate alternatives (see chapter 8). It is hard to see why Kant's views in these matters still attract so much attention when the demolition of the foundation of the problem as it appeared to him and to his predecessors is now clear. There is no reason to believe that Kant would have had these concerns had he appreciated the hypothetical nature and empirical content of geometrical axioms. So I shall regard these arguments, and this position, as having only historical interest.

It is, however, possible to see much of Kant's arguments as psychological statements about the ways and limits of human perception and understanding. To Kant, the idea of space is not derived from the experience of objects, but must be given *a priori*, or innately. Now this could be true whatever the ultimate status of the axioms of geometry by which we describe space mathematically. Kant may well have hit upon critically important limitations and categories of human thought and understanding; and these may be, or may be related to, limitations and strategies of perception. Something of this is emerging in an unexpected way in current problems and approaches in Artificial Intelligence, where it seems necessary to set limits and assign strategies for machines to make effective use of 'sensory' data for 'perceiving' or 'recognizing' objects. The emphasis here is upon the importance of rule-governed strategies for finding likely solutions to problems whose solutions cannot be strictly deduced from the available data. Strategies and rules have to be available before there can be perceptions derived from sensory data.

We can now provide machines with such heuristic rules. Is it inconceivable that organisms have developed perceptual processes incorporating inherited rules? There seems to be no logical or biological objection to this. If the world is very much as it has been over the biological time-span, ancient rule-following procedures could still be appropriate. There is no question here of inheritance of particular experiences, or of particular ancestral knowledge or skill. What is supposed to be inherited is, like functionally appropriate structures such as hands, developed by the trials, errors, and rejections of Natural Selection to become incorporated into the genetic code.

This is a grossly simplified account of the complexities and subtleties of Immanuel Kant's *Critique of Pure Reason*. By stressing Gauss's view of geometry, and recent hints from Artificial Intelligence for using Kant-like rules for perception, I have taken Kant as something of a springboard. I can only hope that Kantian scholars do not too much object.

Chain makers of empiricism

Empiricism is essentially the notion that all knowledge comes from the senses; so accounts of perception are central for theories of knowledge – for epistemology. Visual perception is seen as the link from light to Mind. Just what perceptual links are, remains controversial, not only in philosophy but also in physiology and psychology.

The founder of modern Empiricism is John Locke.

John Locke

John Locke (1632–1704) was born in the charming village of Pensford, near Bristol, in the West of England. He was a Fellow of Christchurch College Oxford until 1684 when he was expelled by order of Charles II. Locke was deeply attracted to the Natural Science of his time, which was at its most exciting stage with the work of Newton, and the founding of the Royal Society in 1660. Locke was made a Fellow of the Royal Society in 1668. He carried out medical and meteorological experiments. His philosophical work started from an evening with friends who decided they could get no further with their work until they could see 'what objects our understandings were and were not fitted to deal with'.[5] Locke spent twenty years in the endeavour to discover what we are and are not capable of understanding.

His life in Oxford was rudely shattered when Lord Shaftesbury (to whom he was secretary for a time) fell from royal favour as a suspected Papist. Charles II ordered that Locke be removed from his Fellowship, and he went into hiding to Holland. Locke became Commissioner of Trade, resigning in 1700 to live the rest of his life quietly in the country. He did not publish the *Essay concerning Human Understanding,* or the two *Treatises,* until 1690, when he was fifty-eight, though the *Essay* was completed twenty years before.

Locke set out to discover how we attain knowledge and what limits there may be to scientific and other knowledge. He regarded the Mind as passive, like a seal accepting impressions; and he thought that the Mind of the child was a blank sheet, to be written upon by experience, and that all knowledge was attained by the senses. He thought that the senses gave simple ideas of sensation, and complete ideas by reflection. He claimed that in perception we experience sequences of 'simple ideas' such as sensations of colour. (But this is most doubtful: we experience objects and realize that they are coloured, and so on.) He was not concerned with perception except to say that it reflected the world as mirrors reflected things. For him, the characteristics of perception were in the world, not in us.

Locke held that we have ideas of three sorts of substances: of God, of finite intelligence and of consciousness. This last, for Locke, gave self-identity, a view that David Hume was to challenge. Locke is described by Bertrand Russell (1946) as a most fortunate philosopher: 'Not only are Locke's valid opinions but even his errors were useful in practice.'[6] Russell was himself a neo-Lockian.

For Locke, then, the Mind started as a blank sheet of paper to receive impressions throughout life ('impressions' being by analogy with a seal on wax) by experience. In his own words:[7]

Let us suppose the mind to be, as we say, white paper, void of all characters, without any *ideas*; how comes it to be furnished? Whence comes it by that vast store, which the busy and boundless fancy of man has painted on it with an almost endless variety? Whence has it all the materials of reason and knowledge? To this I answer in one word, from experience: in that all our knowledge is founded, and from that it ultimately derives itself.

Locke distinguished between sensations and ideas. We think, he supposed, by means of 'ideas', but all our knowledge comes as 'sensation'. We are aware of sensation, and also of our Minds (by what Russell calls an 'internal sense').

For Locke, knowledge could not precede experience. In his own words, perception was 'the first step towards knowledge, and the inlet of all the materials of it'. Thus Locke took the crucial step of denying all *a priori* knowledge. This is the hallmark and the central doctrine of Empiricism.

The immediate problem which confronted Locke, and which still challenges us today, is why we should believe that some sensations are of external objects. This is the start of the problems confronting Empiricism. Locke himself assumed that we know with certainty that some of the mental occurrences that he called sensations have causes outside themselves – outside the Mind – these causes being physical objects or events of the physical world. He also held that some sensations were part of, or at least directly related to, objects. Some sensations, he supposed, were only indirectly related, while other sensations of sense were characteristics of objects themselves. In modern terminology, he held both a Representationist and a Direct Realist theory of perception. To Locke, both operated at once, though for different characteristics of objects. These he called 'Primary' and 'Secondary' Characteristics.

Primary Characteristics, though properties of objects, may be known directly. They are such characteristics as solidity and extension. They, and their interactions, are the principal concern of engineering and the Physical Sciences. Primary Characteristics might be described as 'objective', though they are part of experience. By holding this, Locke was a Direct Realist. Secondary Characteristics, according to Locke, were sensations such as brightness, colour, musical tones, and so on. He sometimes called these sensible qualities. They were not supposed to be in objects, but to depend upon and be in the observer. They were thus 'subjective'.

In addition to his Primary and Secondary Qualities, Locke also spoke of 'powers'. These, he supposed, caused changes. 'Powers' might produce changes between Primary Qualities (as when a bridge falls down) and they might also produce changes in us. Primary Qualities could thus be known through Secondary Qualities. Locke distinguished here between Secondary Qualities being immediately and mediately perceived. It seems best to look at his own words: I shall quote here extensively from the *Essay Concerning Human Understanding:* [8]

22   . . . I hope, I shall be pardoned this little excursion into natural philosophy, it being necessary in our present Enquiry, to distinguish the *primary* and *real qualities* of Bodies, which are always in them (viz. Solidity, Extension, Figure, Number, and Motion or rest, and are sometimes perceived by us, viz. when the Bodies they are in, are big enough singly to be discerned) from those *secondary* and *imputed qualities*, which are but the powers of several combinations of those primary ones, when they operate, without being distinctly discerned: whereby we also may come to know what *ideas* are, and what are not resemblances of something really existing in the Bodies, we denominate from them.

So Locke tried to maintain that perception gave some knowledge directly – his Primary Qualities – while conceding that this was only a small part of what could be known by science. He then distinguished three kinds of qualities of things:

23   *Three sorts of qualities in bodies* – The qualities then that are in bodies, rightly considered, are of three sorts:

First. The bulk, figure, number, situation, and motion or rest of their solid parts: those are in them, whether we perceive them or no: and when they are of that size that we can discover them, we have by these an idea of the thing as it is in itself, and is plain in artificial things. These I call *primary* qualities.

Secondly. The power that is in any body by reason of its insensible primary qualities, to operate after a peculiar manner on any of our senses, and thereby produce in us the different ideas of several colours, smells, tastes, etc. These are usually called *sensible* qualities.

Thirdly. The power that is in any body, by reason of the particular constitution of its primary qualities, to make such a change in the bulk, figure, texture, and motion of another body, as to make it operate on our senses differently from what it did before. Thus the sun has a power to make wax white, and fire, to make lead fluid. These are usually called 'powers'.

He made his distinction between Primary and Secondary Qualities:

24   . . . the idea of heat or light which we receive by our eyes or touch from the sun, are commonly thought real qualities existing in the sun, and something more than mere powers in it. But when we consider the sun in reference to wax, which it melts or blanches, we look

upon the whiteness and softness produced in the wax, not as qualities in the sun but effects produced by powers in it; whereas, if rightly considered, these qualities of light and warmth, which are perceptions in me when I am warmed or enlightened by the sun, are no otherwise in the sun than the changes made in the wax, when it is blanched or melted, are in the sun. They are all of them equally powers in the sun, depending on its primary qualities. . . .

25   . . . For though, receiving the idea of heat or light from the sun, we are apt to think it is a perception and resemblance of such a quality in the sun, yet when we see wax or a fair face receive change of colour from the sun, we cannot imagine that to be the perception or resemblance of colours in the sun itself. . . . But our senses not being able to discover any unlikeness between the idea produced in us and the quality of the object producing it, we are apt to imagine that our ideas are resemblances of something in the objects. . . .

As mentioned above, he considered that some Secondary Qualities are directly (or 'immediately') perceivable, while others are but indirectly (or 'mediately') available to the senses:

26   Secondary qualities twofold: first, immediately perceivable; secondly, mediately perceivable – To conclude: Beside those before-mentioned primary qualities in bodies, viz. bulk, figure, extension, number, and motion of their solid parts, all the rest whereby we take notice of bodies, and distinguish them one from another, are nothing else but several powers in them depending on those primary qualities, whereby they are fitted, either by immediately operating on our bodies, to produce several different ideas in us; or else by operating on other bodies; so to change their primary qualities as to render them capable of producing ideas in us different from what before they did. The former of these, I think, may be called secondary qualities immediately perceivable; the latter, secondary qualities mediately perceivable.

Locke goes on to discuss perception. He starts with what we call 'attention'. Selective attention is to Locke the only activity in perception. Perception to Locke, as to Empiricists in general, is regarded as passively accepting selected knowledge from the external world. The Mind, however, is supposed to be actively comparing and combining ideas (Locke describes knowledge as 'the perception of the agreement or disagreement of two ideas'). So thought may create new possibilities, while perception is but passive acceptance of what is.

Although perception is regarded as passive, Locke stresses the importance of previous experience – of learning. Here again it is best to read his own words: [9]

8   (*Ideas of sensation often changed by the judgment*) – . . . When we set before our eyes a round globe, of any uniform colour, e.g. gold, alabaster, or jet, it is certain that the *idea* thereby imprinted in our mind is of a flat circle variously shadowed, with several degrees of light and brightness coming to our eyes. But having by use been accustomed to

perceive, what kind of appearance convex bodies are wont to make in us . . . the judgment presently, by an habitual custom, alters the appearances into their causes: so from that, which truly is variety of shadow or colour, collecting the figure, it makes it pass for a mark of figure, and frames to itself the perception of a convex figure, and an uniform colour; when the *idea* we receive from thence, is only a plane variously coloured, as is evident in painting.

Locke then asks, following Molyneux:

Suppose a Man *born blind, and now adult, and taught by his touch to distinguish between a Cube and a Sphere of the same metal, and nighly of the same bigness, so as to tell, when he felt one and t' other, which is the Cube, which the Sphere. Suppose . . . the Blind Man to be made to see. Quaere, Whether by his sight, before he touchd them, he could now distinguish, and tell, which is the Globe, which the Cube.* To which the acute and judicious Proposer answers: *Not. For though he has obtaind the experience, of how a Globe, how a Cube, affects his touch; yet he has not yet attained the Experience, that what affects his touch so or so, must affect his sight so or so; Or that a protuberant angle in the Cube, that pressed his hand* unequally, shall appear to his eye, as it does in the Cube. [10]

This is a clear prediction of what 'should' or 'should not' happen given this philosophical theory. A scientific type of prediction is put up for testing a philosophical theory. Can scientific (or clinical) observation test Empiricism?

Locke is clearly worried about the certainty of knowledge as given by experience – and this is ultimately for him all knowledge, since he allows knowledge of our own existence by intuition; of the existence of God by demonstration; and of other things by sensation. He gives several arguments for accepting objects by perception. [11] Considering complex ideas, given by combinations of and differences between sensations, Locke names them (1) modes; (2) substances; (3) relations. [12] Modes do not exist of themselves, but, like *triangle* or *murder*, depend on existing objects. The ideas of substances represent particular objects. Ideas of relations consist in comparing one idea with another. So although knowledge is accepted passively by the senses (with very little innate) the Mind is active in comparing ideas, and generating complex from the simple ideas given by sense. Curiously, though objects in external space are provided by God, movements are nothing to do with God, because He is infinite.

Locke's theory of knowledge is based on 'agreement and disagreement of ideas'; but it is more subtle, for he considers words, signs and propositions. So he comes to include a kind of representative theory of knowledge: [13]

(*A right joining or separating of signs; i.e. ideas or words*) – *Truth* then seems to me, in the proper import of the word, to signify nothing but *the joining or separating of Signs, as the Things signified by them, do agree or disagree one with another.* The *joining* or *separating* of signs here meant is what by another name, we call Proposition. So, Truth properly belongs only to Propositions: whereof there are two sorts, *viz. Ideas* and *Words*.

We must, I say, observe two sorts of Propositions, that we are capable of making.

First, Mental, wherein the *Ideas* in our Understandings *are,* without the use of Words, *put together, or separated* by the Mind, perceiving, or judging of their Agreement, or Disagreement.

Secondly, Verbal Propositions, which *are Words,* the signs of our *Ideas, put together or separated in affirmative or negative Sentences.* By which way of affirming or denying, these Signs, made by Sounds, are as it were put together or separated from one another. So that Proposition consists in joining, or separating Signs, and Truth consists in the putting together, or separating these Signs, according as the Things, which they stand for, agree or disagree.

Locke regards reasoning as not entirely dependent on knowledge of formal logic, though it may depend largely on words, or other man-made symbols: 'God has not been so sparing of men to make them barely two legged creatures, and left it to Aristotle to make them rational.'

David Hume

The second great Empiricist of the British tradition, David Hume (1711–76), somewhat like Descartes, doubted everything until he came to his senses.

Hume was born in Edinburgh, and started his career in commerce, living in Bristol for a while; but this was not a success. He went to France in 1735, to live a frugal scholarly life. He was ambitious for literary fame, which he finally found; but when *Treatise on Human Nature,* which he wrote in his early twenties, appeared he lamented that 'Never was literary attempt more unfortunate; . . . it fell deadborn from the press without reaching such distinction as even to excite murmur among the zealots.'

He was advised to apply for an Edinburgh Chair with the glorious name of Professor of Ethics and Pneumatic Philosophy; but he was rejected on the grounds of atheism. ('Pneumatics' was sometimes used as a synonym for 'psychology' at that time, reflecting the ancient notion of Mind as air, or ether, and genius as inspiration.) Hume became Librarian to the Faculty of Advocates, at Edinburgh. He published various historical books and essays on politics and religion which were well received. His most important book is *An Enquiry Concerning Human Understanding* (1758). There is also a charming autobiography.

Hume's aim in the *Treatise* was to apply the methods of Natural Science to human nature. The resulting science of man should, he thought, be the most basic of all sciences, and all knowledge should be based on it. In this Hume followed Locke and Berkeley: they too had faith in science to solve all problems and answer all questions. They all took as their field of study and source of facts the contents of consciousness. When Hume found that he was unable to justify the Physical Sciences and their methods from his psychology, he became a sceptic. He became doubtful of the certainty of perception, though he regarded perception as the sole source of knowledge, and he came to doubt the certainty of any scientific knowledge. In *An Enquiry* he wrote: [14]

(*Of the Impressions of the Senses and Memory*) – As to those *impressions*, which arise from the senses, their ultimate cause is, in my opinion, perfectly inexplicable by human reason, and it will always be impossible to decide with certainty, whether they arise immediately from the object, or are produced by the creative power of the mind, or are derived by the Author of our being.

Hume is even more worried about Locke's distinction between Primary and Secondary Characteristics, rejecting especially the notion that Primary Characteristics are, or are of, the substance of things. On this Hume writes: [15]

(*Of Modes and Substances*) – I would fain ask those philosophers, who found so much of their reasonings on the distinction of substance and accident, and imagine we have clear ideas of each, whether the idea of *substance* be derived from the impressions of sensation or reflection? If it be conveyed to us by our senses, I ask, which of them: and after what manner? If it be perceived by the eyes, it must be a colour: if by the ears, a sound: if by the palate, a taste: and so of the other senses. But I believe none will assert, that substance is either a colour, or a sound, or a taste. The idea of substance must therefore be derived from an impression of reflection, if it really exist. But the impressions of reflection resolve themselves into our passions and emotions: none of which can possibly represent a substance. We have therefore no idea of substance, distinct from that of a collection of particular qualities, nor have we any other meaning when we either talk or reason concerning it.

The idea of a substance . . . is nothing but a collection of simple ideas, that are united by the imagination, and have a proper name assigned them. . . .

So again we have an Empiricist basing all on sensory experience but worried about its status and its reliability as the basis of all knowledge. There are also, as we have seen, related worries about the status of Induction. These, also, persist to this day. What is so appealing about these founders of Empiricism is their candour. They are genuinely concerned about the status and nature of human knowledge, and they are not over-happy with their conclusions. This does not, however, quite apply to Berkeley: he sounds a little too pleased with himself.

George Berkeley

The modern philosopher who has made the greatest impact specifically on the problems of perception is the Irish philosopher George Berkeley (1687–1753). He did his creative work as a young man in his twenties, and became Bishop of Cloyne in 1734. His main works are *A New Theory of Vision* (1707), *A Treatise Concerning the Principles of Human Knowledge* (1708) and *Three Dialogues between Hylas and Philonous* (1713).

Although Berkeley wrote as a philosopher, he frequently cited specific scientific facts (though not always reliably) to support his position. He was especially concerned with perception of space: with how we come to perceive distances and sizes though they cannot be given directly to the sense of sight. This problem (and implications of suggested solutions) is still very much alive in present-day experimental psychology and physiology.

In *A New Theory of Vision*, Berkeley starts by distinguishing between sight and touch, and by considering whether there are common ideas though the sensations are very different for the various senses.

I My design is to show the manner wherein we perceive by sight, the distance, magnitude and station of *objects*. Also to consider the difference there is betwixt the ideas of sight and touch, and whether there be any idea common to both senses. . . .

Then (as does Locke) Berkeley introduces the importance of previous experience for perception:

II It is, I think, agreed by all, that *distance* of itself, and immediately cannot be seen. For *distance* being a line directed end-wise to the eye, it projects only one point in the fund of the eye. Which point remains invariably the same, whether the distance be longer or shorter.

III I find it also acknowledged, that the estimate we make of the distance of *objects* considerably remote, is rather an act of judgment grounded on *experience* than of sense. For example, when I perceive a great number of intermediate *objects*, such as houses, fields, rivers, and the like, which I have experienced to take up a considerable space; I thence form a judgment or conclusion, that the *object* I see beyond them is at a great distance. Again when an *object* appears faint and small . . . I instantly conclude it to be far off. And this, it is evident, is the result of *experience*. . . .

Berkeley distinguishes this 'inference' from intellectual judgement:

XII But those *lines* and *angles*, by means whereof *mathematicians* pretend to explain the perception of distance, are themselves not at all perceived . . . I appeal to any one's experience . . . whether he computes [an object's] distance by the bigness of the *angle* made by the meeting of the two *optic axes?* . . . Everyone is himself the best judge of what he perceives, and what not. In vain shall all the *mathematicians* in the world tell me, that I perceive certain *lines* and *angles* which introduce into my mind the various *ideas of distance;* so long as I myself am conscious of no such thing.

Then: 'XIV . . . *Lines* and *angles* have no real existence in nature, being only an *hypothesis* framed by *mathematics*, and then introduced by *optics*, that we may treat of that science in a *geometrical* way.' He goes on to say that the connection is not necessary but is psychological:

XVII Not that there is any natural or necessary connexion between the sensations we perceive by the turn of the eyes, and greater or lesser distance. But because the mind has by constant *experience* found the different sensations corresponding to the different disposition of the eyes, to be attended each with a different degree of distance in the *object:* there has grown an habitual or customary connexion, between those two sorts of *ideas.*

Berkeley also discusses the importance of learning by considering a blind man. Locke considers a blind man 'made to see': Berkeley here describes the case of a man denied sight, asking whether the concept of space can be derived without direct experience of distant objects:

CXLVIII Suppose one who had always continued blind, be told by his guide, that after he had advanced so many steps, he shall come to the brink of a precipice, or be stopped by a wall; must not this seem to him very admirable and surprising? He cannot conceive how it is possible for mortals to frame such predictions as these, which to him would seem as strange and unaccountable as prophecy doth to others. Even those who are blessed with the visive faculty [are sighted] may . . . [have] analogous prenotion of things, which are placed beyond the certain discovery and comprehension of our present state.

Whether generalizations can be perceived, whether they are experienced, and how they develop are questions frequently discussed. Locke held that generalizations can be 'held in the mind'; but Berkeley says robustly of Locke's opinion on the matter (CXXV) that it is 'made up of manifest, staring contradictions'.

As is very well known, Berkeley set out to show that objects do not exist unless they are being sensed. But he never showed how they come into being by the act of perceiving. Also, he allowed that they could be brought into being by God's perceiving, which for us takes away the interest of this account. He may, however, have been anticipating notions of verification – developed only recently. Is it possible that he was anticipating modern ideas of verification as necessary for meaning? Demand for verification from experience or experiment was the crucial step that Galileo took, which led to the method of science as proposed and set out by Francis Bacon. (This is discussed in full in chapters 13 and 20.)

For Berkeley, perceptions were sensations. But he accepted the importance of learning for seeing, and his theory was not that perception was purely passive but rather that it required activity of Mind. He did not think of this as at all like problem-solving, if only because problem-solving is conscious, and whatever it is that leads to perception is not known to us in consciousness. Philosophers have systematically been misled by taking consciousness as virtually synonymous with Mind, and also for minimizing the complexities of processes carried out by physiological mechanisms. The immense complexity and subtlety of the logical processing required (as is now clear from attempts to make machines perceive) is only now becoming evident to physiologists and experimental psychologists.

The structure, or organization, of perceptions cannot be given directly from the world, because what are accepted as separate objects depend on familiarity, and so on learning, and on use. This was clear to David Hartley, who was a key figure in developing Associationist accounts of Mind, and who was quite prepared to consider analogies from mechanisms.

David Hartley

David Hartley (1705–57) was a practising physician, who finally settled at the fashionable resort of Bath. His Associationalism followed the thinking of John Locke (see pp. 338 *et seq.*). This became the mental equivalent of atomic accounts of matter – giving rise to what might be called 'mental chemistry'. Hartley was much influenced by Newton's notion of vibrations, described in *Opticks* and *Principia*. Hartley's starting-point is this passage of Newton's, at the end of book 3 of *Principia*, which follows a discussion of the fundamental nature of gravity:

And now we might add something concerning a certain most subtle spirit which pervades and lies hid in all gross bodies; by the force and action of which spirit the particles of bodies attract one another at near distances, and cohere, if contiguous; and electric bodies operate to greater distances, as well repelling as attracting the neighbouring corpuscles; and light is emitted, reflected, refracted, inflected and heats bodies; and all sensation is excited, and the members of animal bodies move at the command of the will, namely, by the vibrations of the spirit, mutually propagated along the solid filaments of the nerves, from the outward organs of sense to the brain, and from the brain to the muscles. But these things cannot be explained in a few words, nor are we furnished with that sufficiency of experiments which is required to an accurate determination and demonstration of the laws by which this electric and elastic spirit operates.

David Hartley takes this up in his *Observations on Man* (1749), and concludes that thought is given by just such physical brain activity. Thus Hartley's Proposition II:

The white medullary Substance of the Brain is also the immediate Instrument, by which Ideas are presented to the Mind: or, in other words, whatever Changes are made in this Substance, corresponding Changes are made in our Ideas; and vice versa.

The evidence for this proposition . . . is from the writings of physicians and anatomists; but especially from parts of these writings which treat of the faculties of Memory, Attention, imagination, etc. and of Mental disorders. It is sufficiently manifest from hence, that the perception of our Mental Faculties depends upon the perfection of this Substance; that all injuries done to it affect the trains of Ideas proportionably; and that these cannot be restored to their natural course till such injuries be repaired. . . .

For Hartley, the Mind is associations of ideas of sensation. It is not entirely clear whether he thought the links to be mental or physical. E. G. Boring (1950)[16] describes Hartley as a Dualist, but I think that this is questionable. He replaced

'Vital Spirits' in nerve with Newton's vibratory motion of 'small, and as one may say, infinitesimal, medullary particles'. These he supposed vibrated longitudinally, to give sensation, and perhaps activating still smaller particles in the brain, giving thoughts and pleasures and pains. He might have thought of the vibrations affecting Mind by a kind of resonant coupling (which would be a Dualist account), but some passages suggest that he held what we would now call an Identity view of Mind and Matter. Consider this passage from *Observations on Man:*[17]

. . . I suppose, or postulate, in my first proposition ['The white medullary Substance of the Brain, spinal Marrow, and the nerves proceeding from them, is the immediate Instrument of Sensation and Motion'] that sensations arise in the soul from motions excited in the medullary substance of the brain. I do indeed bring some arguments from physiology and pathology, to show this to be a reasonable *postulatum,* when understood in a general sense; for it is all one to the purpose of the foregoing theory, whether the motions in the medullary substance be the physical cause of the sensations, according to the system of the schools; or the occasional cause, according to *Malebranche;* or only an adjunct, according to Leibniz. However, this is not supposing matter to be endowed with sensation, or any way explaining what the soul is but only taking its existence, and connection with bodily organs in the most simple case, for granted, in order to make further enquiries. Agreeably to which I immediately proceed to determine the species of the motion, and by determining it, to cast light on some important and obscure points relating to the connection between the body and the soul in complex cases.

It does indeed follow from this theory, that matter, if it could be endued with the most simple kinds of sensation, might also arrive at that intelligence of which the human mind is possessed: whence this theory must be allowed to overturn all the arguments which are usually brought for the immateriality of the soul from the subtilty of the internal senses, and of the rational faculty. But I do no ways presume to determine whether matter can be endued with sensation or no. . . .

Hartley thought of the vibrations continuing after stimulation of nerves as suggested by after-images, described by Newton:[18]

And when a Coal of Fire moved nimbly in the circumference of a circle, makes the whole circumference appear like a circle of fire; is it not because the Motions excited in the bottom of the eye by the Rays of light are of a lasting nature, and continue til the Coal Fire in going round returns to its former place? And considering the lastingness of the Motions excited in the bottom of the Eye by Light, are they not of a vibrating nature?

Newton also considers that perhaps:[19]

harmony or discord of colours arise from the proportions of the Vibrations propagated through the Fibres of the Optic Nerves into the brain, as the harmony and discord of Sounds arise from the proportions of the Vibrations of the Air? For some Colours, if they be viewed together, are agreeable to one another, as those of Gold and Indigo, and others disagree.

Newton explains the transmission of what we would call neural signals through nerve by vibrations analogous to the great distances that ripples travel from a stone dropped into a pond – the ripples also continuing for some time. Here we have physical analogies for aspects of Mind, with uncertainty as to how far the analogies can be continued – until perhaps Mind, as an entity, entirely disappears.

Hartley, influenced as he was by Newton, set up 'ripples' that continue now in discussions of physiological psychology and philosophy of Mind.

Thomas Reid

The Scottish philosopher Thomas Reid (1710–92) wrote with extreme common sense nearly forty years after his fellow countryman David Hume, whose scepticism he tried to refute. He wrote *Inquiry into the Human Mind* (1764) and *Philosophy of the Intellectual Powers* (1785).[20]

Reid comments on Aristotle's view of perception in the *Intellectual Powers:*[21]

An object, in being perceived, does not act at all. I perceive the walls of the room where I sit: but they are perfectly inactive, and therefore act not upon the mind. . . . Nor could men ever have gone into this notion, that perception is owing to some action of the object upon the mind, were it not that we are so prone to form our notions of the mind from some similitude we conceive between it and the body. Thought in the mind is conceived to have some analogy to motion in a body: and, as a body is put in motion, by being acted upon by some other body; so we are apt to think the mind is made to perceive, by some impulse it receives from the object. But reasonings, drawn from such analogies, ought never to be trusted. They are, indeed, the cause of most of our errors with regard to the mind. And we might as well conclude, that minds may be measured by feet and inches, or weighed by ounces and drachms, because bodies have those properties.

A little later, Reid continues:

I know that Aristotle and the schoolmen taught that images or species flow from objects, and are let in by the senses, and strike upon the mind: but this has been so effectually refuted by Descartes, by Malebranche, and many others, that nobody now pretends to defend it. . . . To what cause is it owing that modern philosophers are so prone to fall back into this hypothesis, as if they really believed it? . . . Mr. Hume surely did not seriously believe that an image of sound is let in by the ear, an image of smell by the nose, an image of hardness and softness, of solidity and resistance, by the touch. For, besides the absurdity of the thing, which has often been shown, Mr. Hume, and all modern philosophers, maintain that the images which are the immediate objects of perception have no existence when they are not perceived; whereas, if they were let in by the senses, they must be, before they are perceived, and have a separate existence.

Reid then gives an argument, from real and apparent magnitude, to show that because there can be apparent visual magnitude (as to be expected on geometrical principles), this is evidence for, and not against, the existence of objects, for 'it is evident that the real magnitude of a body must continue unchanged, while the body is unchanged though with changes of viewing position'. Reid points out pitfalls of language in this connection:[22]

Language is made to serve the purposes of ordinary conversation; and we have no reason to expect that it should make distinctions that are not of common use. Hence it happens, that a quality perceived, and the sensation corresponding to that perception, often go under the same name.

This makes the names of most of our sensations ambiguous, and this ambiguity hath very much perplexed philosophers.

. . . Thus, if it is asked, whether the smell be in the rose, or in the mind that feels it, the answer is obvious: that there are two different things signified by the smell of a rose; one that there is in the mind, and can be nothing but in a sentient being; the other is truly a property of the rose . . . there can be no sensation, nor anything resembling sensation in it. But this sensation in my mind is occasioned by a certain quality in the rose, which is called by the same name with the sensation, not on account of any similitude, but because of their constant concomitancy.

All the names we have for smells, tastes, sounds, and for the various degrees of heat and cold, have a like ambiguity. . . . They signify both a sensation, and a quality perceived by that sensation. The first is the sign, the last the thing signified.

Reid continues a little later:

Pressing my hand with force against the table, I feel pain, and I feel the table to be hard. The pain is a sensation in the mind, and there is nothing that resembles it in the table. The hardness is in the table, nor is there anything resembling it in the mind. Feeling is applied to both; but in a different sense; being a word common to the act of sensation, and to that of perceiving by the sense of touch.

I touch the table gently with my hand, and I feel it smooth, hard, and cold. These are qualities of the table perceived by touch; but I perceive them by means of a sensation which indicates them. This sensation not being painful, I commonly give no attention to it. It carries my thought immediately to the thing signified by it, and is itself forgot, as if it had never been. But, by repeating it, and turning my attention to it, . . . I find it to be merely a sensation, and that it has no similitude to the hardness, smoothness, or coldness of the table, which are signified by it. . . .

But, had Mr Locke attended with sufficient accuracy to the sensations which he every day and every hour received from primary qualities, he would have seen that they can as little resemble any quality of an inanimated being as pain can resemble a cube or a circle.

Reid challenges here the power of unchecked perception to give deep knowledge of the nature of things – of Primary Characteristics. He uses argument to check perception much as arguments and measurements are used to check, and sometimes reject, perceptions in science.

We may indeed see the development of scientific method as a move away from the strictly Empiricist belief that we gain understanding simply through perception: understanding of objects or of our own minds. Perception has been demoted in science, though it still reigns supreme in the arts. This may be part of the current conflict between science and art. How deep is this conflict? Is it due to misunderstandings that we can resolve?

I think it is fair, if unkind, to say that philosophers of this present century have been remarkably unsuccessful in their attempts to characterize perception. Most have held to what can only be described as 'Naïve Realism' (and this they frequently apply to themselves – or at least to each other). This should, in our terminology be, rather, 'Naïve Direct Realism', for it is the relation between the world and the Mind which is the issue. The notion that perception is directly related is held by many Empiricists. If it were direct there would be no problem to discuss, and sensory knowledge would be certain; but the situation is neither so simple nor so dull. There are two notable exceptions to my unkind stricture that recent philosophers do not say interesting things about perception: Bertrand Russell and Ludwig Wittgenstein.

Bertrand Russell

Bertrand Russell (1872–1970)[23] is generally considered to have been the greatest philosopher active in this century, especially in logic and the philosophy of Mind. His Theory of Descriptions, and development of symbolic logic following Frege, were true breakthroughs; and indeed they serve as excellent examples (if rare!) of a philosopher discovering or inventing useful results, or tools for thinking and further discovery. His writings range from logic to psychology, and to education and social issues. He combined profundity, extreme honesty, and wit.

Russell may be said to be essentially Lockian, though he did change his position on several important issues two or three times during his long philosophical career. I shall be concerned mainly with his later writings. In *Our Knowledge of the External World*, Russell writes: 'Philosophy may be by many roads, but one of the oldest and most travelled is the road which leads through doubt as to the reality of the world of sense.' He adds, after a few pages:[24]

The first thing that appears when we begin to analyse our common knowledge is that some of it is derivative, while some is primitive; that is to say, there is some that we only believe because of something else from which it has been inferred in some sense, though not necessarily in a strict logical sense, while other parts are believed on their own account, without the support of any outside evidence. It is obvious that the senses give knowledge of the latter kind: the immediate facts perceived by sight or touch or hearing do not need to be proved by argument, but are completely self-evident. Psychologists, however, have made us aware that what is actually given in sense is much less than most people would naturally suppose, and that much of what at first sight seems to be given is really inferred. This applies especially to our space perceptions. . . . Thus the first step in the analysis of data, namely, the discovery of what is really given in sense, is full of difficulty. We will, however, not linger on this point: so long as its existence is realised, the exact outcome does not make any very great difference in our main problem.

We, however, will be very much concerned with what Russell dismisses here as not his 'main problem'. He has a great deal to say, in various books and papers, on Locke's view of perceptual knowledge, and also on the other great problem for Empiricists: knowledge by Induction. This Russell discusses especially in his philosophical book, *Human Knowledge: Its Scope and Limits.* His denial of the importance of psychological details of perception is very generally shared by philosophers today.

Russell develops Locke's distinction between Primary and Secondary Characteristics. He does this by supposing two kinds of perceptual knowledge: knowledge by acquaintance and knowledge by description. A few short quotations will bring out his essential ideas (which, however, change from time to time).

In a famous paper, 'Knowledge by Acquaintance and Knowledge by Description' (1911), Russell wrote:

I say that I am *acquainted* with an object when I have a direct cognitive relation to that object, i.e. when I am directly aware of that object itself. When I speak of a cognitive relation here, I do not mean the sort of relation which constitutes judgment, but the sort which constitutes presentation.

A year later Russell wrote in *The Problems of Philosophy* that acquaintance involves not knowledge of truths, but knowledge of things. This he expands as follows: [25]

We have acquaintance in sensation with the data of the outer senses, in introspection with the data of what may be termed the inner sense – thoughts, feelings, desires, etc.; we have acquaintance in memory with things which have been data either of the outer senses or the inner sense. Further, it is probable, though not certain, that we have acquaintance with Self, as that which is aware of things or has desires towards things.

(Russell later gives up this idea of the Self as a thing we perceive with an 'inner sense'.) The above passage continues:

In addition to our acquaintance with particular existing things, we also have acquaintance with what we call *universals,* that is to say, general ideas, such as *whiteness, diversity, brotherhood,* and so on. . . . Awareness of universals is called *conceiving,* and a universal of which we are aware is called a *concept.*

Russell allows that the world has three kinds of entities: (1) Minds, (2) material things, and (3) sense data. Later he questions whether Minds are entities and comes to consider that 'material things' are best described by scientific accounts. The status of sense data gives him great concern.

Russell's main worry about sense data as existing entities is their complexity. He was fond of his example of sense data of a dog which he called a 'canoid patch'. The 'canoid patch' sense datum had to include: brown, hairiness, sounds of whimpering – and perhaps expectations and also assumptions of the dog's past – for all these are inextricably bound up with the perception of a dog. Russell's point is that perceiving a dog includes, for example, awareness that it may bite: but how can such non-sensed characteristics be sense data? Just what is known by 'acquaintance', and what by 'description'? Can we be 'acquainted' with the past, or the future? If not, though, are we significantly acquainted with anything? Here we meet the difficulties of Direct Realist accounts of perception.

Russell distinguished between propositional and non-propositional knowledge, as indeed did Locke. For Russell, direct awareness ('acquaintance') is *de re* (non-propositional) rather than *de dicto* (propositional). He considers various kinds of relations between sense data and objects, but never seems to have satisfied himself, or to have provided a satisfying account, of what this relation could be.

There was a period when Russell proposed an elaborate theory of perceptual knowledge invoking what he called 'perspective spaces', having six dimensions for each observer, filled with 'sensibilia', somewhat like Leibniz's Monads. These sensibilia were supposed to lie in six-dimensional perspective space. But, as A.J.Ayer has pointed out (1972),[26] there seems to be a serious circularity in deriving objects as logical constructs from sensibilia, and regarding sensibilia (in the Lockian Direct Realist tradition) as characteristics of objects. Russell very soon abandoned this theory, which, frankly, now seems interesting mainly as evidence that he had serious doubts of Locke's Primary Characteristics, and remained uncertain how to reconcile classical Empiricism with a Direct Realist account of perception, though he never gave up this tenet of classical Empiricism.

Russell's comments were so often pointed and witty. In *An Enquiry into Meaning and Truth* (1940) he writes: 'Naive realism leads to physics, and physics, if true, shows that naive realism is false. Therefore, naive realism, if true, is false; therefore it is false.'[27]

Sigmund Freud

Sigmund Freud (1856–1939) was born in Freiburg, Moravia and died in Hampstead, London. During his lifetime psychology discovered its Galileo, as Freud discovered immensely exciting ways of thinking about Mind. Freud's work is so well known, and so fully discussed and creatively criticized, that he appears in this book only here and there with no attempt to describe his ideas fully or assess his contribution adequately. In any case this does not seem timely, as there is now no sure or accepted conceptual basis for psychoanalysis, but rather many very different sets of techniques, each only partly based on discernible theoretical ground. Freud was the Galileo rather than the Newton of psychology, for although he left a unique contribution he hardly produced a system to be built upon without revision of its fundamental tenets. While Newton's system lasted essentially unchanged for two centuries, Freud's was never generally accepted, and is seen now as a monumental attempt, by a man of undoubted genius, rather than a working system or philosophy of Mind.

Inspired by Goethe's *Essay on Nature,* Freud studied medicine at Vienna, where he undertook research on spinal cells in a primitive fish, working at Trieste. He showed that cells of primitive organisms are essentially the same as in higher organisms. He failed though, as did Aristotle before him, to find the testes in the eel!

As a student he was greatly influenced by Ernst Brücke, (1819–92), who directed the Institute, and whose *Lectures on Physiology* (1874) compared organisms with machines, emphasizing the interplay of forces, as known in physics, for organisms. Ernest Jones (1961) points out that 'The spirit and content of these lectures correspond closely with the words Freud used in 1926 to characterise psychoanalysis in its dynamic aspect: "The forces assist or inhibit one another, combine with one another, enter into compromises with one another, etc." ' Brücke remained for Freud a kind of Superego – a standard of scientific integrity – and an image of his steel-blue eyes would appear in Freud's mind, years later, when he was tempted to cut corners. His respect for Brücke was entirely justified: with Emile du Bois-Reymond (1818–1896), Carl Ludwig (1816–1895), and Hermann von Helmholtz (1821–1894) he succeeded in replacing Vitalism with the physics-based physiology of today. This they set out to do as deliberate policy, following the Introduction of Du Bois-Reymond, *Animal Electricity* (1848), which rejected the 'vital force'. Du Bois-Reymond wrote:

If one observes the development of our science he cannot fail to note how the vital force daily shrinks to a more confined realm of phenomena, how new areas are increasingly brought under the dominion of physical and chemical forces. And physiology will one day be absorbed into the great unity of the physical sciences.

Whether the same is true of psychology remains a pertinent question. The answer must depend in part on what is accepted as 'Physical Science'.

Freud clearly accepted wish fulfilment as an explanatory concept, and his writings are full of teleological ideas. But it can be argued that this is compatible with physical determinism (Boden, 1972a). Indeed as we have said (pp. 79–82) servo-control systems even as simple as a windmill aiming into the wind, or a

Watt's governor, or a thermostatically controlled hot-water system, maintain and seek goals – and they may be disturbed almost as we are when the goals cannot be attained – though they are fully determined according to mechanical principles.

As we have seen, although servo-control has a history back to the Greeks it lacked theoretical appreciation, until too late in Freud's life for him to use concepts that might surely have been of great use to him. As mentioned earlier the theory of servo-control was not formulated until World War II (Weiner, 1948), though it is of course now a central concept in physiology and behaviour. Although servo-control theory originated from engineering, especially guidance systems and automatic pilots, it is successfully applied to ecology and to societies (indeed, presumably Marx was as handicapped as Freud in this respect). However this may be, Freud developed his account of Mind without what is to us now an essential concept for describing dynamic systems, and also without the possibility (or at least the likelihood) of appreciating the power of computing, and the importance of programs for controlling information-handling and decision-taking. Perhaps for these reasons, his theoretical accounts are more Mentalistic than he would have wished them to be.

The various dramatic and extremely interesting episodes, such as the realization with Joseph Breuer that hysteria may be cured by recalling painful memories with hypnosis, and his close collaboration and later dissociation from Carl Jung, are so well known that I will not detail them here. His most accessible account of psychoanalysis is surely *The Introductory Lectures* (1916–17).

Freud's influence is of course immense. He removed much blame and guilt, with his unconscious desires, and innate complexes such as the Oedipus complex. These are, however, themselves characters – part- or whole-personalities as are the characters in myth – and in this his accounts of Mind are perhaps unduly Mentalistic and anthropomorphic. Freud's mental entities might be criticized as like Plato's atoms – having just the properties they are supposed to explain; as Plato's atoms to explain sharpness or hardness were themselves sharp, or hard, or whatever. By contrast Democritus had the amazing insight (so long ignored by everyone apart from Lucretius, who conveys it with such fervour to us) of combinations of individually neutral atoms providing the variety of the world. By analogy, then, one might feel that Freud's account of Mind is essentially as unsatisfactory as Plato's account of Matter, for they both invoke what they try to explain. This is the trouble with anthropomorphic accounts of man and Mind.

By discussing sexual problems openly Freud succeeded, if indirectly, in reducing the very problems he saw as basic and innate, for example the son's Oedipus jealousy of the father. This social change is surely an immense gain; but it does suggest that Freud's principles are not as fundamental, or absolute, as he supposed. This is good news, for it gives promise of not-too-drastic therapies working

– which they hardly could if the ills of Mind, and especially sexual problems, are inherited and are as deep-seated as Freud considered. The current tendency is to tackle sexual problems rather as one is helped to improve one's game of tennis, by coaching and practice, which has been made possible by the open discussion largely initiated by Freud. But the very success of these new therapies (and psychoanalysis is almost too expensive to be practicable) is evidence against the unchanging absoluteness he introduced into much of his philosophy of Mind.

Psychoanalysis is often criticized for carrying on regardless of strong evidence that it works, while being impervious to criticisms or tests (Eysenck, 1952). It is all too easy for an analyst to dismiss criticism as a symptom, having a clinical basis. Freud was challenged in the early 1930s by an American psychologist, S. Rosenzweig, who, having carried out critical experiments, received the following letter from Freud: [28]

My Dear Sir
I have examined your experimental studies for the verification of the psychoanalytic assertions with interest. I cannot put much value on these confirmations because the wealth of reliable observations on which these assertions rest makes them independent of experimental verification. Still, it can do no harm.
Sincerely yours
Freud

One does see the point of Popper's advice, and the strength of his current influence, for this letter could hardly be written now. The dangers of the clinical situation for deriving data are especially great because the patient is open to suggestion. Freud did, however, discuss this trap, taking the view that psychoanalysis is in the same situation as astronomy: it relies more on observations than interference with the situation. Clark Glymour (1974) discusses this by comparing how Kepler's theory of planetary orbits can be tested with how Freud's theory might be challenged. Glymour concludes that there are general statements or generalizations which can be confirmed or refuted, though the theoretical structure is far looser than Kepler's.

The basis of Freud's scheme of the human Mind is that it has three levels. The first is the conscious level, of emotions and available memories; the second is the preconscious level, in which memories are placed but not accessible except under special conditions, such as pressure from maintained questioning; and the third is the subconscious, or unconscious, in which memories are locked that can only be released by hypnosis or (as Freud came to prefer) by the method of free association, in which patients talk more or less at random and with minimum suggestion. Dreams are also highly important; but they have to be 'read' or 'interpreted'.

The evidence of dreams and free association, it seems, led Freud to divide the

Mind into the Ego (awareness of Self), the Superego (self-criticism), the Id (unconscious instincts), and the libido (emotional energy, associated with sex).

Freud started with the hope of a purely neurological account of psychology, in terms of the activity of brain cells (as is known from an abandoned draft paper, 'The Project'); but giving this up, he then, it seems, realized that a psychology in terms only of conscious experience would not do, if only because the data are so fragmentary, as consciousness is intermittent: as William James puts it, like a bird flying, perching, and flying off again. Possibly this intermittency of consciousness led him to lay such weight on unconscious processes, which are not physical and not available to consciousness. One might say that this is something like astronomers sampling planetary positions; or indeed the usual relation between perception and the object world, which is sampled only occasionally. But the relation between the 'perceptions' of the unconscious events or states is by no means clear – even less clear than for observations in normal life or of physical objects in physics. A problem is that the psychological observations are even more dependent on consciousness, and this is deeply mysterious. The content of the unconscious is only available in symbolic form, and the symbols – dreams or whatever – must be read. This is not in principle so different from interpreting any data in science (such as spectral lines to the constitution and temperature stars), but the signals from physical instruments are read with the highly articulated theories of physics and the other Natural Sciences. The problem for gaining knowledge of unconscious processes of Mind, rather than Matter could be simply the relative lack of context theory or understanding in psychology.

It is hardly an objection to say that Freud tries to observe things that are not material, for physics makes all sorts of observations, of fields of force, inertia, numbers and so on, that are not material.

Freud had the great advantage over physicists that he could talk to his patients, while talking to the Universe went out with the Gods of the stars. Freud emphasized the significance of blocks and mistakes of language. Here there is some similarity to the use of visual and other illusions for investigating mechanisms of perception (see pp. 404 et seq.)

Speech problems might indicate breaks in associations, or blocks due to protective 'complexes'. There is a difficulty here which is endemic to neurology as a whole; for lack of a response may be due to absence of a signal or to active inhibition. This applies to the physics of the nervous system as it does to Freud's account of Mind. The discovery of anatomically located 'speech centres' by Carl Wernicke (1848–1905) and Paul Broca (1824–80) followed the general pyramidal scheme of control of brain function set out by the great John Hughlings Jackson (1835–1911).[29] Freud sometimes looks for the anatomical location of mental processes, and this seems to receive support from relations between local-

ized brain damage and aphasias; though the relations between physiological function and speech disorders, or any behavioural disorders, may be highly indirect
and difficult to read. (If this is so for the localization of functions and the effects of
mechanical or electronic malfunction in machines, we might expect the situation
to be even more difficult to unravel for the brain.) The problem does not seem to
be that Freud was attempting something outside science (and surely his emphasis
on unconscious processes should appeal to physicists, as they have virtually nothing to say about consciousness); rather, it seems to lie in the lack of clearly
associated data. We seem almost to be in the situation of the Babylonians *vis-à-vis*
astronomy – with their disconnected and not very reliable observations, and a lack
of any adequate conceptual scheme for reading the data. Freud has the immense
distinction of making the most serious and sustained effort so far to look into the
murky depths of Mind; though perhaps he saw himself, reflected.

12   Links of brain

It seems to me that one who knows anything is perceiving *the thing he knows, and, so far as I can see at present, knowledge is nothing but perception.*

Plato (*c*. 428–348/347 BC)

It is impossible to have scientific knowledge by perception.

Aristotle (384–322 BC)[1]

Brain copies: Empedocles and Theophrastus

The first known statement of an argument for perceptions being representations is given by Theophrastus (*c*. 372–286 BC), in which he criticizes Empedocles for holding that perceptions are particles entering the senses. (On this account sight and hearing would be like smell, as we understand smell.)

Theophrastus inherited the Aristotelian library, including Aristotle's manuscripts. He criticizes Empedocles for saying that perceptions are merely copies in this way:

> . . . with regard to hearing, it is strange of him [Empedocles] to imagine that he has really explained how creatures hear, when he has ascribed the process to internal sounds and assumed that the ear produces a sound within, like a bell. By means of this internal sound we might hear sounds without, but how should we hear this internal sound itself? The old problem would still confront us.

Theophrastus is pointing out that an internal copy would be but the first step to an infinite regress, without end or point. (A photograph, though in perfect perspective and in full colour, is not a perception. A photograph may of course be perceived – and may sometimes serve as an adequate substitute for objects, of which it is a kind of copy – but it is not itself a perception.) This is taken up by René Descartes, two thousand years later in the seventeenth century.

Brain images: René Descartes

For Descartes, the nerves were tubes through which Vital Spirits conveyed sensations from all the senses to the pineal gland. United there, they were supposed to give the single 'common sense' (*sensus communis*) of experience. Descartes saw the eye as an image-forming optical device (familiar to us as a camera) and he saw the brain as having to make sense of the eyes' images.

Descartes was confused over the fact, which he knew, that the retinal images are inverted. This confusion is surprising, for it does not in the least matter that they are upside-down with respect to the objects that they image – for there are, so to say, no eyes looking at the retinal images: they are signal sources, related geometrically to the objects imaged, and related (via the transducer characteristics of the retina) to the neural signals, which go through all manner of changes and convolutions which do not appear in perceptions.[2] Descartes thought of the nerves as tubes carrying 'pneuma', but he had no reason to suppose that perceptions should be inverted because the image is inverted – unless he equated the image itself with the final perception. He may have started thinking in this way, but he came to a far more sophisticated conception, as we shall see.

The central region (the fovea) was regarded by Descartes as far more important than the rest of the retina; this seems to be because he was worried that the inverted image should give inverted vision. At this point he explained right-way-up vision as given by the movements of the eyes building up perceptions from many fixations of the fovea. Descartes sometimes, however, thought of the image actually being the perception, as we see from his theory of birthmarks, which he saw as like photographs:

> I could go still further, to show how sometimes the picture can pass from there through the arteries of a pregnant woman, right to some specific member of the infant which she carries in her womb, and there forms these birthmarks which cause learned men to marvel so.

However, Descartes finally held a sophisticated view of how the retinal image could give perception as a representation, not as a picture or a copy. He says in the *Sixth Discourse:*

> Now although this picture, in being so transmitted, always retains some resemblance to the objects from which it proceeds, nevertheless, as I have already shown, we must not hold that it is by means of this resemblance that the picture causes us to perceive the objects, as if there were yet other eyes in our brain with which to apprehend it . . . it is necessary to think that the nature of our minds is such that the force of the movements in the areas of the brain where the small fibres of the optic nerve originate cause it to perceive light; and the character of these movements cause it to have the perception of colour . . . there need be no resemblance between the ideas the mind conceives and the movements which cause these ideas.

Descartes seems to be referring back here to this interesting passage in the *Fourth Discourse:*

> We must at least observe that there are no images that must resemble in every respect the objects they represent – for otherwise there would be no distinction between the object and its image – but that it is sufficient for them to resemble the objects in but a few ways, and even that their perfection frequently depends on their not resembling them as much as they might.

He continues by noting how effective imperfect pictures are:

> You see that engravings, being made of nothing but a little ink placed here and there on the paper, represent to us forests, towns, men and even battles and storms, even though, among an infinity of diverse qualities which they make us conceive in these objects, only in shape is there actually any resemblance. And even this resemblance is a very imperfect one, seeing that, on a completely flat surface, they represent to us bodies which are of different heights and distances, and that following the rules of perspective, circles are often better represented by ovals rather than by other circles; and squares by diamonds rather than by other squares; and so for all other shapes. So that often, in order to be more perfect as images and to represent an object better, they must not resemble it.

We have to wait a long time to find anything as sophisticated as this. Descartes continues immediately to apply these ideas to the brain:

> Now we must think in the same way about the images that are formed in our brain, and we must note that it is only a question of knowing how they can enable the mind to perceive all the diverse qualities of objects to which they refer; not of knowing how the images themselves resemble their objects; just as when the blind man . . . touches some object with his cane, it is certain that these objects do not transmit to him except that, by making his cane move in different ways according to their inherent qualities, they likewise and in the same way move the nerves of his hand, and the places in his brain where these nerves originate. Thus his mind is caused to perceive as many different qualities in these bodies, as there are varieties in the movements that they cause in his brain.

Descartes was quite clear that the perceived size of objects is not given simply from the sizes of their retinal images, in the *Sixth Discourse:*

> Their size is estimated according to the knowledge, or the opinion, that we have of their distance, compared with the size of the images that they imprint on the back of the eye; and not absolutely by the size of these images, as is obvious enough from this: while the images may be, for example, one hundred times larger when the objects are quite close to us than when they are ten times further away, they do not make us see the objects as one hundred times larger (in area) than this, but as almost equal in size, at least if their distance does not deceive us.

He generalizes this to shape:

And it is also obvious that shape is judged by the knowledge, or opinion, that we have of the position of various parts of the objects, and not by the resemblance of the pictures in the eye; for the pictures usually contain only ovals and diamond shapes, yet they cause us to see circles and squares.

There are plenty of discussions in current psychology textbooks far more confused and confusing than this account, written over three hundred years ago!

Descartes saw perception as knowledge-based, and yet providing new knowledge. The notion of perception as being given by inferences, from sensory data and stored knowledge of the world, was not developed until the nineteenth century, initially by Helmholtz.

Brain inferences: Hermann von Helmholtz

Hermann von Helmholtz (1821–94) combined highly creative work of lasting importance in physics, physiology and psychology. He was both an experimental and a theoretical scientist. Perhaps his greatest achievement was in visual perception. The *Treatise on Physiological Optics* (1856–67) is the Bible of the subject. The *Sensations of Tone* is almost the equivalent for hearing and sound.

Helmholtz was born at Potsdam. He was Professor of Physiology at Königsberg (1849), Bonn (1855) and Heidelberg (1858). He became Professor of Physics in Berlin in 1871. Perhaps his most celebrated contribution to physics is his work on the conservation of energy (1847). His work on visual perception is remarkable (there is no other way of putting it) for its width and its depth. He invented two instruments that remain essential: the ophthalmometer for measuring the curvatures of the lens of the eye, and the ophthalmoscope for observing the living retina (though he was anticipated by two years in the latter by Charles Babbage, who did not fully appreciate it). Helmholtz measured the reaction-time of the nervous system in 1850, which had most important consequences. He combined a thoroughgoing Natural Science approach with such philosophical sophistication that he could think of physiological processes (themselves depending on the principles of physics) as mediating procedures of inference – which he regarded as necessary for perception.

These 'Unconscious Inferences', for making sense out of stimuli, were generally unpopular at the time, as they demoted consciousness – and so also self-knowledge and moral responsibility.[3] They were disliked by many physiologists, also, because this theory of perception precludes any simple stimulus–response accounts, and so limits the significance of what can be derived from recordings from the nervous system, at least before function has been understood in data-processing terms. Only now, with electronic computers, can such theories of perception be outlined at all adequately. (One may compare this situation with trying to make sense of signals recorded from a computer, without knowing its programs.)

Helmholtz stressed the importance of experience, and held that there is little or no innate knowledge. He said also that Mind is not an independent causal entity. He thus represents a position very different from Freud, who wished to give conscious and unconscious Mind the role of controlling behaviour, partly from deep-seated inheritance of ancient knowledge. Helmholtz represents the purest Empiricism: he strove hard to bring Mind into Natural Science.

One of Helmholtz's most important ideas, as already mentioned, was that perceptions are derived by inference – Unconscious Inference from sensory signals: [4]

The psychic activities that lead us to infer that there in front of us at a certain place there is a certain object of a certain character, are generally not conscious activities, but unconscious ones. In their result they are equivalent to a *conclusion*, to the extent that the observed action on our senses enables us to form an idea as to the possible cause of this action; although, as a matter of fact, it is invariably simply the nervous stimulations that are perceived directly, that is, the actions, but never the external objects themselves. . . . And while it is true that there has been, and probably always will be, a measure of doubt as to the similarity of the psychic activity in the two cases, there can be no doubt as to the similarity between the results of such unconscious inferences and those of conscious conclusions.

A particularly illuminating passage follows:

These unconscious conclusions derived from sensation are equivalent in their consequences to the so called *conclusions from analogy*. Inasmuch as in an overwhelming majority of cases, whenever the parts of the retina in the outer corner of the eye are stimulated, it has been found to be due to external light coming into the eye from the direction of the bridge of the nose, the inference we make is that it is so in every new case whenever this part of the retina is stimulated; just as we assert that every single individual now living will die, because all previous experience has shown that all men who were formerly alive have died. . . .

But, moreover, just because they are not free of conscious thought, these unconscious conclusions from analogy are irresistible, and the effect of them cannot be overcome by a better understanding of the real relations. It may be ever so clear how we get an idea of a luminous phenomenon in the field of vision when pressure is exerted on the eye; and yet we cannot get rid of the conviction that this appearance of light is actually there at the given place in the visual field; and we cannot seem to comprehend that there is a luminous phenomenon at the place where the retina is stimulated. It is the same way in the case of all the images that we see in optical instruments.

There are many cases of such immutable links, which are clearly the result of experience and cannot possibly be given by inherited physiological structures or connections, and yet many physiologists are still, as in Helmholtz's day, reluctant to ascribe these – or perception of objects from retinal images – to learning. An example of where learning must give the links is the sound or sight of a word and its significance. The word 'bread' (written or spoken) is only conventionally related to the edible object; yet it is as strongly linked for English-speaking people as are the shapes of loaves to their colour, consistency or taste. Indeed, Aristotle was fundamentally misled by this apparently immutable relation between words and things (since he lacked the check of alternative languages), into thinking that names are of the essence of things. For Aristotle the word 'sun' was as much part of the nature of the sun as its heat or light: this is the psychological basis of Nominalism – a kind of unconscious inference, or Inductive association.

Helmholtz's notion of Unconscious Inference was badly received; because at that time consciousness seemed necessary for inference. He tried to rename it 'inductive conclusion' but, as Richard and Roselyn Warren (1968) say, 'It was too late, his position was too well known, and the old name persists even today.'[5] They point out that the Unconscious Inference notion was accepted, in a rather strange form, by Pavlov. It was conveyed to Russia by I.M. Sechenov, who worked with Helmholtz, and it seems that Pavlov learned of it from Sechenov and gave it a Conditioned-Response interpretation. Pavlov wrote:[6]

Evidently, what the genius of Helmholtz referred to as 'unconscious conclusion' corresponds to the mechanism of the conditioned reflex. When, for example, the physiologist says that for the formation of the conception of the actual size of an object there is necessary a certain length of the image on the retina and a certain action of the internal and external muscles of the eye, he is stating the mechanism of the conditioned reflex. When a certain combination of stimuli, arising from the retina and ocular muscles, coincides several times with the tactile stimulus of a body of certain size, this combination comes to play the role of a *signal*, and becomes the conditioned stimulation for the real size of the object. From this hardly contestable point of view, the fundamental facts of the psychological part of physiological optics is physiologically nothing else than a series of conditioned reflexes, i.e., a series of elementary facts concerning the complicated activity of the eye analyser.

Pavlov was suggesting that his S–R (stimulus–response) links explain 'Unconscious Inference' in terms of physiological mechanisms. Helmholtz was, however, more concerned with *procedures* (although he was a physiologist), leaving aside for the moment how logical procedures are carried out by physiological mechanisms. In this, he anticipated and was a forerunner of the most recent work on Artificial Intelligence, where procedures are more important than details of the mechanisms by which they are carried out. This is not to minimize the importance of physiology; it does, however, emphasize the need to discover the *functional* significance of physiological mechanisms (see chapter 3).

Helmholtz proposed a thoroughgoing Empiricist philosophy of perception with a strong emphasis on the importance of learning: We see this clearly in the following passage from one of his *Popular Lectures*, 'The Recent Progress of the

Theory of Vision' (1868), where he discussed whether perception of space is given from physiological innately given structures (as held by 'intuitionists') or whether basic characteristics of the external space of objects have to be learned. This led him to consider how spatial relations, and other characteristics, are represented: [7]

What, then, is it which comes to help the Anatomical distinction in locality between the different sensitive nerves, and . . . produces the notion of separation in space? In attempting to answer this question, we cannot avoid a controversy which has not yet been decided.

Some physiologists, following the lead of Johannes Müller, would answer that the retina or skin, being itself an organ which is extended in space, receives impressions which carry with them this quality of extension in space; that is conception of locality is innate; and that impressions derived from external objects are transmitted of themselves to corresponding local positions in the image produced in the sensitive organ. We may describe this as the *Innate* or *Intuitive* Theory of conceptions of Space.

Helmholtz then suggests evidence to settle the philosophical controversy: [8]

It obviously cuts short all further inquiry into the origin of these conceptions, since it regards them as something original, inborn, and incapable of further explanation.

The opposing [innate] view was put forth in a more general form by the early English philosophers of the sensational school. . . . Its application to special physiological problems has only become possible in very modern times, particularly since we have gained more accurate knowledge of the movements of the eye. The invention of the stereoscope by Wheatstone made the difficulties and imperfections of the Innate theory of sight more obvious than before, and led to another solution which approached much nearer to the older view, and which we will call the *Empirical* Theory of Vision. This assumes that none of our sensations give us anything more than 'signs' for external objects and movements, and that we can only learn how to interpret these signs by means of experience and practice. For example, the conception of differences of locality can only be attained by means of movement and, in the field of vision, depends upon our experience of the movements of the eye.

According to Helmholtz, all that the sense organs have to do is to transmit differences, corresponding to important differences in the external world, for perception to develop. The 'intuitionist' physiologists, on the other hand, emphasized the importance of 'local signs' – that the kinds and positions of physiological signals correspond innately to features of the external world. Helmholtz summarizes this as follows: [9]

The Empirical Theory regards the local signs (whatever they really may be) as signs the signification of which must be learnt, and is actually learnt, in order to arrive at a knowledge of the external world. It is not at all necessary to suppose any kind of correspondence between these local signs and the actual differences of locality which they signify. The Innate Theory, on the other hand, supposes that the local signs are nothing else than direct conceptions of differences in space as such, both in their nature and their magnitude.

Just what is inbuilt and what learned remains a most difficult question to answer. Ernst Mach (1885) essentially follows Helmholtz's emphasis on learning; but suggests that, for example, straight lines are seen by innate physiological characteristics, and also by the following of various rules for handling visual and other information: [10]

It is to be noted that every point of a straight line in space marks the mean of the depth-sensations of the neighbouring points . . . the assumption forthwith presents itself that the straight line is seen with the least effort. The visual sense acts therefore in conformity with the principle of economy, and, at the same time, in conformity with the principle of probability, when it exhibits a preference for straight lines.

There is also a long-standing tendency to think that apparently simple or primary sensations are given by simple or unitary physiological mechanisms. A particularly clear example of this (mentioned briefly earlier) is the colour yellow. The chief proponent of 'local signs' for space perception, Ewald Hering, supposed that yellow is signalled by a single class of colour receptors tuned to this frequency of light, and not by mixture from other colour-tuned receptors. The argument was simply that yellow appears primary, and simple, so it must be signalled by a correspondingly simple physiological mechanism. The sensation given by monochromatic yellow light is, however, indistinguishable from the sensation given by a mixture of a red and a green light of suitable intensities. Further, many different experiments show that yellow is not signalled by a 'special yellow receptor' even for monochromatic light that appears yellow. Physiologically, yellow is always a mixture of colours, although it appears and is said often to be 'primary'. So there is no straightforward relation between physiology and sensation or perception.

Now that we have evidence, from electrophysiology, of the retina and visual cortex, of 'feature detectors' tuned to orientation, movement or other 'simple' features [11] it may be tempting to suppose that they give the corresponding perception, or sensation to which they are tuned. This is a temptation to neuro-physiologists, because they could then identify detectable physical brain processes with perceptual features: which would indeed be psychophysics. But, from such cases as the perception of yellow, we can hardly expect often to find simple relations.

It must be confessed that Helmholtz's position does not give any account of consciousness; for if Unconscious Inference gives perception and behaviour, why should we have any awareness? However, Helmholtz's account leads directly to Artificial Intelligence. By removing the problem of meaning from consciousness and emphasizing functional logical processes it allows meaning to be considered in terms of digital computation, with programs selected for situations or needs. This use of the notion of programs for describing brain function has recently been described with great erudition and elegance by J.Z. Young (1978). The Gestalt psychologists also attempted physical explanations but using analogue analogies. We shall look at this attempt now.

Brain analogues: Gestalt psychology[12]

Pictures can represent things; or at least they can represent things to us – to creatures with brains. Does the brain itself function by pictures? Does it picture reality? This is the central doctrine of the Gestalt school. The notion is that the brain produces electrical fields of the same shape as observed objects. This supposed identity of shape from objects to brain fields was called 'isomorphism'. The brain fields were also supposed to adopt certain preferred forms – to have tendencies towards 'good form'. 'Good forms' were simple, convex rather than concave; and should have minimal potential energy, as soap bubbles adopt nearly spherical stable forms. The supposed tendencies for the brain fields to adopt certain forms in preference to others, were classified into the Gestalt 'Laws of Organization'. These 'Laws of Organization' became explanatory concepts for Gestalt psychology. They were invoked to explain why features appear grouped together in ways generally corresponding to objects. This at least raises a good question, to which we should return, even if the Gestalt theories are rejected.

The founders of the Gestalt school, especially Kurt Koffka (1886–1941), Wolfgang Kohler (1887–1967) and Max Wertheimer (1880–1943), were concerned with how the mosaic of stimulation of separate rods and cones on the retina, with parallel neural channels to the brain – and G.D. Müller's 'Law of Specific Energies' that each nerve fibre can only transmit one kind of sensation – could give this organized grouping to perception. They each had somewhat different answers. Wolfgang Kohler was most specific, suggesting that grouping (or the forming of 'Gestalten') obeys physical principles extending beyond organisms. In his paper *Physical Gestalten* (1920), he drew analogies (or pointed to supposed identities) between interactions and groupings in perception with electrical fields in physics. Section 25 of *Physical Gestalten* reads: [13]

A. Nervous fields excited by pressure, pain, or temperature stimuli display electromotive forces when circumscribed parts of the sensory surface are stimulated or when the type of stimulation is different at different places on this surface.

B. If the nerve regions corresponding to the two ear labyrinths are unequally stimulated, an osmotic communication of this dissimilarity will result, and a difference of electrical potential will arise between the two.

C. Most important of all is the nervous field corresponding to the retinas, for these are not peripheral sense organs like the cochlea, but parts of the brain itself. Two colours simultaneously seen have each a particular excitation-equivalent in the nervous system, and the contour between them is the equivalent of an electromotive force in the nervous system between the two areas. The amount and direction of the potential difference are thus determined by the nature of the stimulating colours. If the two colours are gradually made to resemble one another, both the electromotive force and the contour diminish and disappear also. So long, however, as there is differentiation in the visual field – e.g. that of figure and ground – there will be a corresponding electromotive force in the nervous field. Figure perception is represented in the optic field by differences of potential along the entire outline or the border of the figure.

There is much of interest in this quotation. We find a certain amount of physiological mythology, but the attempt to show how perception is given by stimuli and yet is more than the stimuli is important. It raises a good question but does Kohler provide an adequate answer? We may note that he ascribes perceived differences between objects to physical contours and even the figure–ground distinction seems to be attributed to 'differentiation in the visual field'. But are objects seen as separate by simple geometrical characteristics of their contours? This is the assumption behind Gestalt psychology; but it does not hold. As the words in a speech spectrogram displaying the energy patterns of speech run into each other without breaks, though we hear words as distinct 'things', so objects are seen as distinct from each other though there may be no clear physical breaks or discontinuities between them. The theory has it both ways, regarding perceptions as pictures (isomorphic representations), and also as constructing perceptions of objects by the 'Laws of Organization'.

The 'physical Gestalten' of the brain were supposed to be copies of reality and to be what we experience. To return to Theophrastus' objection to copies for perception, 'How should we hear this internal sound itself?' We might add, 'How do we see the brain trace of the circle, or the house?' Can we say that *they* are conscious? Kurt Koffka, in *Principles of Gestalt Psychology* (1935), would have said in answer, 'Yes.' This seems clear from his enthusiastic espousal of Wertheimer's notion of isomorphism, which we see in Koffka's *Principles*: [14]

Wertheimer's Solution. Isomorphism. And now the reader can understand Wertheimer's contribution: now we will see why his physiological hypothesis impressed me more than anything else. In two words, what he said amounted to this: let us think of the physiological processes not as molecular, but as molar phenomena. If we do that, all the difficulties of the old theory disappear. For if they are molar, their molar properties will be the same as those of the conscious processes which they are supposed to underlie. And if that is so, our two realms, instead of being separated by an impassable gulf, are brought as closely together as possible. . . .

There is a deep logical problem here, concerned with identity and analogy. If we say that 'Their molar properties will be the same as those of the conscious processes' and also hold with Wertheimer that the 'molar properties' are those of typical physical processes outside brains, then we should say that these extra-cerebral processes are conscious. If, on the other hand, only an analogy is intended, then Koffka's 'impassable gulf' remains.

Koffka does, however, make an interesting and important point in this section of *Principles*. He distinguishes between two kinds of 'environments': on the one hand, the environment of the physical world of objects, and on the other, the internal environment of the supposed brain traces. Considering behaviour in relation to his two environments he opts for the more interesting alternative. He calls these two 'environments', 'geographical' and 'behavioural' respectively. (His use of 'behavioural' may have been a sop to the dominant Behaviourist school in America at that time, but he clearly means the states of the brain related to experience.) He writes: [15]

So far the behavioural environment has been introduced as a mediating link between geographical environment and behaviour, between stimulus and response. These two terms denote objects which seem to have a very definite place in our system of knowledge; they both belong to the external world. But what is the locus of [i.e. where is] the behavioural environment?

Koffka then describes an experiment by G. Révész showing that chicks respond to an illusion of size as we do:

Why do the animals choose one of the two equal figures, when they have been trained to choose the smaller one? Described in these *geographical* terms their behaviour seems unintelligible, and neither stimulus properties nor experience can supply even the semblance of a satisfactory answer. But everything becomes perfectly plain and simple if we answer our question as every unbiased layman would answer it, by saying: The animals chose one of the two equal figures because it *looked* to them smaller, just as it looks smaller to us.

This equating of distortion – a phenomenon of perception – with the 'inner' environment of brain processes, and/or experience, will be discussed later (chapter 18). Meanwhile, we may note Koffka's remark that these brain processes (and also experience?) are a 'mediating link' between stimulus and response.

We should note the emphasis that the Gestalt school place on their 'brain traces' having similar shapes to objects perceived; and shapes identical to what is experienced – hence, indeed, their term 'isomorphism'. But how can 'traces' in the brain representing external objects have the same shape as objects if they have their own 'Laws of Organization'? This is indeed having it both ways. It is a deep conflict and inconsistency in Gestalt theory.

Suppose now we drop the 'brain picture' notion. Consider, rather, a form of representation more like writing. If we write the number 7, or the word 'seven', this bears no isomorphic relation to seven things, or to the concept seven. If, again, I write or type the word 'house', this does not look at all like (and it is not at all geographically, architecturally or topologically like) a house which one could live in. Why, then, did the Gestalt psychologists choose isomorphic representation? This is only one of an infinite set of kinds of representation; but of course it is special – for it is like pictures. It could have been a simple analogy with pictures that suggested to the Gestalt psychologists representation by isomorphic brain traces – but this allowed a special kind of explanation for characteristics of perceptions.[16]

The Gestalt writers emphasized the way that dots group themselves perceptually to form simple closed patterns. Once a circle, for example, is believed to be represented by a circular brain trace, they thought the physics of such supposed traces could be invoked to explain perceptual phenomena, such as sets of dots tending to form circles; for a circular trace can be supposed to have minimal potential energy and so be stable. Then nearly circular objects should tend towards perfect circularity. If, on the other hand, objects are represented by quite different shapes – as houses are represented by the very differently shaped word 'house' in the English language – then no such inference could be made. So the Gestalt psychologists needed isomorphism to bridge in their terms physics to perception, and to consciousness.

Is this representing by isomorphic brain 'pictures' plausible? We may consider the problem more generally. The Gestalt writers went beyond visual shapes to think of musical tunes as showing 'good Gestalten' or 'closure' – and that tunes remain when transposed into a different key because their isomorphic principles for visual perception applied to hearing. But how could music – sound – be isomorphically represented? It is absurd to think that there are tunes playing in the brain. Now consider colour: it is just as absurd to suppose that when a traffic light turns from red to green, we see this because part of our brain changes colour from red to green. It is almost equally absurd to think of a complex three-dimensional object such as a house as represented by a three-dimensional model house in the brain. So the account cannot be general enough.

Lashley carried out the ingenious experiment of embedding gold wires into the striate region of a monkey's brain, after training to discriminate straight from bent lines. With the introduction of the shorting wires which should have distorted such 'brain fields', the monkeys continued to separate straight from bent lines – as though they remained undistorted. There is no physiological evidence for isomorphism.

The Gestalt notion of a physical basis of perception in terms of the tendency of isomorphic brain traces toward forms of lowest potential energy demands that important characteristics of perception are innate. These features of 'closure', 'common fate' and so on may apply to *all* human observers, which might suggest that they are innate; but this generality might come about by learning, for there are common properties of objects and situations encountered by all babies in all cultures. Most objects, everywhere, are of rather closed shapes, so a perceptual tendency to 'closure' could be developed in each individual separately by experience rather than given by the physics of the brain as conceived by the Gestalt school.

Either of the first two alternatives are preferable to the 'physical Gestalten' notion, for there is no reason why such minimal potential energy of isomorphic brain traces should be appropriate for the shapes of objects. Further, we experience strong 'grouping' effects in situations that are definitely learned. Letters grouping to form words as units shows this very clearly. To take one example, we see the word THERAPIST as a word, without realizing that introducing a gap changes it to the very different THE RAPIST. Another example is NOSMO KING and NO SMOKING. This grouping must be learned, for it depends on the conventions of our language.

Just as pictures or maps are only one kind of description of a scene or a country, isomorphism – or pictures in the brain – is only one kind of representation for perception. A more sophisticated alternative is that the brain develops functional models of external situations. These may have quite different shapes from what they represent, though they may be analogues of objects. This notion was developed by K.J.W.Craik, at Cambridge, in his important book *The Nature of Explanation* (1943). A related notion was developed in the influential book *Organisation of Behaviour* (1949) by the Canadian psychologist D.O.Hebb, which introduced his notion of 'Phase Sequences', thought of as analogues developed inductively to match situations, to make behaviour generally appropriate. We shall now consider these 'analogue' brain traces, which are not supposed to be pictures but, rather, more abstract representations of objects and situations which are seen as being built up from many individual experiences, presumably by some kinds of Inductive generalizations carried out by brain mechanisms. So here many of our previous considerations come together: the relation of computer programs to the physics of machines, and the distinction between analogue and digital computers (pp. 57–8, 91); what Induction is, and whether learning is or can be Inductive (the second half of chapter 8 and much of chapter 9); the notion of hypothesis-testing for learning, as investigated by Krechevsky (p. 284); learning in simple analogue machines, such as Grey Walter's mechanical animal (which was devised in about 1950), *M. Speculatrix* (pp. 86–92); and the difficult problems of what it is to *represent* which rise up to meet us in many places (especially chapter 12).

Looking ahead, we shall consider possible relations linking brain and Mind, or whether they are different aspects of the same thing (chapter 16); and questions of consciousness, and the nature or status of the Self, on all of which an adequate theory should at least say something of interest (chapters 16 and 17).

Brain models: Kenneth Craik

The writer who first clearly stated the notion that perception and thinking may be given by neural models of the external world was the brilliant Cambridge psychologist K.J.W.Craik. Kenneth Craik was tragically killed in an accident when aged only thirty-one. In *The Nature of Explanation* (1943) he says: [17]

> By a model we thus mean any physical or chemical system which has a similar relation-structure to that of the process it imitates. By relation-structure I do not mean some obscure non-physical entity which attends the model, but the fact that it is a physical working model which works in the same way as the process it parallels, in the aspects under consideration at any moment.

This emphasis on dynamic features of the world which the organism has to reflect is a significant advance over the 'brain pictures' (as we have called them) of the Gestalt psychologists. Craik was writing while analogue computing was the vogue, before digital computers took over. Analogue computers represent functions by mechanical or electrical devices that display functions (transfer functions) between their inputs and outputs, without representing the mathematics, or logical processes, required to go through the steps needed for deriving the answer by mathematical or logical procedures. The implication is that the brain may work logically or mathematically without using logic or mathematics.

Craik considers how numbers may be represented and used by his essentially analogue Internal Models, by which he considers the brain functions. But here he seems to be falling into the trap which Wittgenstein warns us against. Introducing his position, Craik writes:

> Our question . . . is not to ask what kind of thing a number is, but to think what kind of mechanism could represent so many physically possible or impossible, and yet self-consistent processes as number does. . . . It is likely then that the nervous system is in a fortunate position, as far as modelling physical processes is concerned, in that it has only to produce combinations of excited arcs, not physical objects; its 'answer' need only be a combination of consistent patterns of excitation – not a new object which is physically and chemically stable. . . .

He continues with this interesting passage:

> I see no great difficulty in understanding how anything so 'different' from physical objects and concepts and reasoning can tell us something more about those physical objects; for I see no reason to suppose that the processes of reasoning *are* fundamentally different from the mechanism of physical nature. On our model theory neural or other mechanisms can imitate or parallel the behaviour and interaction of physical objects and so supply us

with information on physical processes which are not directly available to us. Our thought, then, has objective validity because it is not fundamentally different from objective reality but is specially suited for imitating it – that is our suggested answer.

Craik is suggesting that human inference has its validity and logical power because and only when it shares the structure of reality. This is pushing the concept of analogue computing to its limit. But how can thinking be like the objects of thought? Craik is surely slipping back here into a 'picture' account, with all its attendant difficulties. There may be some similarities, but surely we should look for formal similarities for representations of generalities of the world and for logic and number. To say that generalities and abstractions are in the world is to take up Platonism, though this does not seem to be what Craik intended. His theory of brain processes as put forward here states a theory of number, and by implication of all symbols. He regards symbols as parts of the world rather than as objects standing for parts or aspects of the world. This has such difficulties as: how could the brain (or any symbolic system) represent abstractions? Craik is forced to say that abstractions such as numbers exist as other objects exist; for his brain states are non-representing objects. He thus goes back on Gauss's insights, which rescued us from the classical difficulties of the status of mathematics, logic and language. This is surely a step backwards, which we should regard with extreme caution. Craik's Internal Model notion can, however, be modified to avoid such difficulties, as very likely he would have gone on to do had he lived to appreciate the power of digital computers, whose states are very different physically from what they represent with their codes and formal languages.

Cell Assemblies and Phase Sequences: Donald Hebb

The Canadian psychologist Donald O. Hebb published his highly influential theory of brain function in *Organisation of Behaviour: A Neuropsychological Theory* (1949), suggesting a working paradigm which has inspired informative experiments and useful criticism. Hebb's *Organisation of Behaviour* has entered the minds of psychologists and drawn them to neurophysiology over the last thirty years. Equally important is Kenneth Craik's *The Nature of Explanation* (1943), already mentioned; but somehow this never made the same impact: only now do we see how good it was. These two books have much in common, for they both suggest that the brain models the world with analogue representations – Craik's Internal Models and Hebb's Cell Assemblies and Phase Sequences. They are also both related to Sir Frederic Bartlett's Schema, which Bartlett put forward in *Remembering* (1932). In all three accounts, emphasis is placed on the dynamics of brain processes for filling gaps and extrapolating from data with some kind of analogue-type representation of brain traces. This is most fully worked out by

Hebb, who pays close attention to clinical findings from local brain damage, and the evidence of early learning for perception. He looks for specific neural processes to explain the phenomena; while Craik argues, more generally, from engineering-type analogies, and Bartlett from systematic changes in memory of pictures and stories with successive recall and with social factors such as familiarity.

Hebb aims to show 'how a rapprochement can be made between (1) perceptual generalisation, (2) the permanence of learning, and (3) attention, determining tendency or the like'. This he does as follows: [18]

It is proposed first that a repeated stimulation of specific receptors will lead slowly to the formation of an 'assembly' of association-area cells which can act briefly as a closed system after stimulation has ceased; this prolongs the time during which the structural changes of learning can occur and constitutes the simplest instance of a representative process (image or idea).

The physical basis of learning is, for Donald Hebb, structural changes of the synapse with related signal transmission, but the situation is not simple because very many synapses in the assembly are the unit of memory rather than single synapses. These concepts were worked out at a fundamental level by Warren McCulloch and his collaborators, at the Massachusetts Institute of Technology. This approach led to several related mathematical analyses of the dynamic properties of richly interconnected neural nets, and how they could be stable and at the same time adaptive, and provide both individual and generalized memory (the former appearing in consciousness, the latter not, but necessary for skills). In its most general form, this kind of account becomes, perhaps, the holographic type of model proposed by Karl Pribram (1966), though here the details of the physiology command less attention. G. S. Brindley (1969) has proposed a new nerve net model.

Hebb's attempt was distinguished by its clear references to specific physiological processes and neuropsychological data. It at once suggested tests, and where new physiological and behavioural data should be found. The emphasis on specific physiological cells and circuits implied that there are limitations set by performance characteristics of neural components. This is, however, extremely difficult, if possible, to show without full 'circuit diagrams'. To see this, one has only to think of electronic circuits, or of what people can attain when working together, which one would hardly guess from what can be done from knowledge of individual abilities. Certainly, component characteristics can limit performance of electronic devices; but it is also true that very often the immediate limitations can be avoided, with sufficient cunning. So this kind of argument from physiology is by no means straightforward or clear-cut, because component limitations can often be overcome by appropriate circuit design.

The power of digital computers comes from their ability to perform very different tasks with the same components. There are very many components of only a few kinds performing many kinds of function, and what they do is far removed from the function of the machine as a whole, and functions change according to control instructions of its prevailing programs. This flexibility, and relative independence of component characteristics, together with the much greater emphasis on the logic of programs, is the 'New Look' of neurological models of brain function. So the next generation of theorists may well come from computer science rather than from physiology. There is, though, little hope of advancing this understanding without active deep cooperation between those concerned with the brain's 'hardware' characteristics and limitations, and those concerned to see how it is logically possible to achieve learning, perception and behaviour by possible components. What AI has done – and this is almost sufficient in itself – is to show the inadequacy of our present concepts for explaining even the simplest behaviour. Whether the next advances will come mainly by seeing how the brain works or by inventing intelligent machines, is a question for the future. Personally I believe that AI is vitally important for developing concepts necessary for interpreting the findings of physiologists: I doubt whether we can see how the brain works without understanding how it *could* work.

Seeing without brain links? J.J.Gibson

The distinguished psychologist J.J.Gibson attempted to describe perception without reference to brain processes (and indeed he tended to deny even retinal images).

The late James J.Gibson (1904–79), affectionately known as Jimmy Gibson, worked for many years at Cornell University and made several notable experimental contributions, by stressing the importance for perception of features of what he called the 'ambient array', which he supposed is selected and picked up to give visual perception according to principles of 'ecological optics'. He hoped to find simple mathematical relations between features of the ambient array of light and corresponding features of perception. In this quest he found evidence of the importance of what most people would call stimulus features, such as gradients of the density of textures, and 'optical flow' and dynamic parallax with observer motion. Gibson rightly criticized the emphasis of much perceptual research on pictures, viewed by a static observer, with a single viewing eye, while neglecting observer motion through the world of normal objects lying in three-dimensional space. Gibson is an Empiricist because he lays emphasis on sensory experience, but he does not hold any form of representational Empiricism, which we shall be considering shortly.

Gibson's philosophy of perception – and here I have no intention of being rude, for there are many excellent insights from the past – returns essentially to the view of perception held until the tenth century AD, before Alkindi and Alhazen appreciated light rays. Gibson goes back, in effect, to the notion that information of objects in three dimensions is given by light without need to infer distances, or the three-dimensional forms of objects. For him, all the information for visual perception is in the light, without reference even to retinal images.

Indeed it has recently become possible to extract three-dimensional information from light reflected by objects with holography; but this is not at all Gibson's idea. Although holography has occasionally been suggested as a model of vision, there are insuperable objections, because the eye does not have phase information, and I shall not further consider it here.

What Gibson does is to go back to the ancient notion of 'shells' or 'simulacra' of objects supposed, before Alkindi, to structure light in space. Gibson moved toward this position from his first book *The Perception of the Visual World* (1950) to the second book *The Senses Considered as Perceptual Systems* (1966). In *The Visual World,* he tolerates light rays and retinal images, but he explicitly attacks notions of perception as inferences from cues provided by retinal patterns. He lays every emphasis on stimulus patterns. What it amounts to is that for him stimuli are correlates giving immediate knowledge of objects. However, they are not, for him, 'pictures' of objects, and they are not 'read', as pictures are read.

Gibson emphasizes the question: just what are patterns of stimulation that give rise to perceptions? This has proved useful and has inspired much valuable research. Our concern is with the extreme notion that perception can be considered simply as a passive 'picking up of information' from stimulus patterns. (By 'passive' I do not mean observers stationary in space. I mean that the observer does not need to make active inferences from the stimulus pattern to the object producing the patterned retinal image. Unfortunately, Gibson uses the word 'active' for motions of the observer, which for him is highly important.[19] It is a pity he did not call this 'observer motion', as 'active' generates a disastrous ambiguity from the use of normal language in this context.)

In the later book (Gibson, 1966), we read:[20]

. . . the function of the brain when looped with its perceptual organs is not to decode signals, nor to interpret messages, nor to accept images. . . . The function of the brain is not even to organize the sensory input or to *process* the data, in modern terminology. The perceptual systems, including the nerve centres at various levels up to the brain, are ways of seeking and extracting information about the environment from the flowing array of ambient energy.

Consider the plight of the newborn child: according to Gibson, 'He does learn to perceive but he does *not* have to learn to convert sense data into perception.' What the baby has to do, on this view, is to discover stimulus invariants. The stability of the visual world with observer motion is supposed to be given by invariants of the 'optical array', which are discovered in infancy, especially by moving around in space to discover which features remain invariant. (So presumably, an animal or child incapable of motion would be effectively blind. There is, however, evidence that this is not so.) Gibson calls this approach 'ecological optics'. Ray optics (geometrical image formation) is dispensed with, in favour of already structured ambient light reflected from objects. By holding that the light is already structured, he holds a pre-Alkindi theory, which dispenses with light rays and retinal images for perception. This may strike one as bizarre (if only because it is possible to see retinal images in other people's eyes) but Gibson has drawn attention to the importance of certain kinds of stimulus structure for the recognition of edges, surfaces and so on: features of 'ecological' importance with his account.

In objecting to retinal images, he rightly complains of how 'slippery' a word is 'image'. But surely this goes much too far: 'The cerebral image in the brain, the physiological image in the nerve, and the retinal image in the eye are all fictions.'[21] If these are fictions, what is not fictional?

Who speaks of 'images', rather than, say, 'maps' in the brain (except perhaps for 'mental images' which may indeed be misleading)? Whoever spoke of 'physiological images' in optic nerves? It is absurd to dismiss retinal images as 'fictions': they are just as real as images in cameras, though of course the brain does not 'see' the images in its eyes as a photograph is seen. For Gibson the 'image' is optical structure in the ambient light; so he asks: 'Could it be that the retinal image is a sample of a *permanent* image?'[22] I think by 'permanent' he must mean structured ambient light, very like the ancient 'shells' or simulacra of theories of perception held before optical images were discovered, and made use of in the camera obscura and later in photographic cameras and other optical instruments whose lens-focused images can hardly be denied. It is possible that Gibson was so worried about the conceptual pitfalls of an 'inner eye' seeing the retinal image – requiring another 'inner eye' to see this eye's image, and so on (see below) – that he came to deny retinal images for perception and reverted to a primitive view which, from what we know of optics, is strictly untenable.[23] However this may be, his account of perception has proved to be extremely useful. Gibson has shown how ignorant we were of just what specifies borders, surfaces, concavities, and so on.[24] Much new knowledge has been discovered by many disciples, inspired by Gibson's rejection of the active information-processing brain in favour of saying in effect that perceptions are in the light. This has, ironically, proved most useful for workers in Artificial Intelligence – who need to know just which features of optical images are significant for scene analysis and object recognition by compute programs – though the computer programs provide just the kinds of activities that Gibson rejects for human perception.

Although, as I see it, Gibson's general account of perception is philosophically incorrect, it has proved useful. This seems indeed to be a counter-example for the theory of knowledge, or truth, of Pragmatism, suggesting that truth is that which is useful; but perhaps Gibson's account is not useful enough. We should find a test for this and other accounts of perception, in attempts to design and make machines to see. Can Artificial Intelligence test brain links of seeing?

Seeing machines: programs for brains?

What is now known as Artificial Intelligence has ancient origins, back to Hellenistic Greece, but in this century it has grown from Cybernetics (control systems based on servo-mechanisms) and Bionics (producing engineering equivalents of biological design principles, developed through Natural Selection). AI differs from these, and also from pattern recognition devices, by relying upon stored information, which is categorized as knowledge. The assumption is that knowledge-based performance may be intelligent though mediated by machine. In practice AI devices or systems may be autonomous, or they may be intelligence-amplifiers, when used to enhance human performance in decision-making. However, there is no clear boundary here as such aids as slide rules, pocket calculators or even memo pads may be regarded as 'intelligence amplifiers' for humans though they are not intelligent.

Some AI systems are capable of recognizing objects or scenes, and some can interact with worlds (real or computer-simulated). Information for real-world interaction is generally provided by a single TV camera, augmented by touch probes. Such fully robotic systems have learning capacity: they may learn to extract useful 'visual' features, and they may calibrate their internally represented 'visual space' by touch exploration of objects in their world. Such robotic devices employ pattern recognition with stored knowledge in order to infer (from signals and the stored knowledge of their object world) the three-dimensional shapes and non-sensed properties of objects – from the limited and never strictly adequate sensed data. Their pattern perception *per se* is usually not as sophisticated as it could be, the emphasis being on making effective use of limited real-time sensed data by programs employing stored knowledge. This generally fails in atypical situations, for inferences to the world depend on appropriate assumptions.

It should be stressed that current AI research is aimed at developing programs rather than sophisticated hardware. The Robot devices can serve as test beds for suggesting and testing programs. There are typically rather few check procedures, so the devices are easily fooled, for example by atypical lighting or shadows.

There are however a few programs which actually make use of shadows and other contingent variables and typical restraints of positions.

For seeing machines, a computer is usually programmed with knowledge of object characteristics, enabling the system to go beyond given data to infer hidden or non-sensed features. Features of objects may be hidden, by masking by nearer objects, or through inadequacy of the available video signal. Inbuilt knowledge of objects makes it possible to recognize them from various points of view and over wide ranges, though available features change with orientation and range. Identification features may be pre-programmed, but machine-adaptive learning is possible. In particular, a robotic machine may have a mechanical arm with touch capability, for exploring surrounding space and handling objects whose positions and shapes are discovered by touch, to modify the visual processing. It is believed that calibration by contact and the discovery of features by touch are essential for sophisticated vision in animals and infants, and also for machines.

An essential concept for machine vision is processing 'from bottom-up', and processing 'from top-down'. An example of the former is line repair, in which gaps in signalled contours are filled in by interpolation procedures. However, it is always the case that *true* gaps may be filled by such low-level procedures, then fictional lines or other features may be produced and we see things that are not there. This could create ghosts, flying saucers and all manner of things. These errors may be avoided by 'top-down' procedures which judge the probability of there being gaps, corners and so on, from high-level stored knowledge of objects. Unusual objects can always beat even these combined bottom-up and top-down procedures: hence many of the common illusions of human vision have counterparts in machine vision (see pp. 399 *et seq.*).

The most successful existing object-recognition programs concentrate on edges and vertices. They are weak on surfaces, as signalled by texture, and they are more successful with straight-line objects than with objects having predominantly curved edges. They are capable of recognizing objects partly masked by nearer objects, and distinguishing and recognizing both the eclipsed and eclipsing objects. This is perhaps the most impressive power of current 'scene analysis' (analysing line pictures) and object-recognition programs. They tend however to be slow for real-time object recognition; but this serious limitation of such Robot devices may be transformed with the introduction of parallel processing, which waits upon adequate computer hardware.

It is found that conditional probability assessment by the machine is highly important for object recognition, especially where there is ambiguity as to the identity of individual shapes as parts of objects. This is 'higher-level' than line repair. The machine may select several possibilities for each shape signalled, and consider identification in terms of their relative positions and orientations. Thus for a ship, a rectangular shape is more likely to be a funnel if it is above, rather than below or to the side of, another rectangular shape which could be a hull. The hull hypothesis is also confirmed by the funnel assumption, so there is mutual support for the final object hypothesis: that this is a ship. There is suggestive evidence that complex interactive probability assessments and adjustments are important in human vision, whose efficiency is by no means rivalled by current machine systems, except for freedom from human fatigue, boredom and distraction.

The most celebrated program is still that of Terry Winograd (*Understanding Natural Language*, 1972). Its aim is set out at the beginning: [25]

When a person sees or hears a sentence, he makes full use of his knowledge to understand it. This includes not only grammar, but also his knowledge about words, the context of the sentence, and most important, his understanding of the subject matter. To model this language understanding in a computer, we need a program which combines grammar, semantics, and reasoning in an intimate way, concentrating on their interaction.

The Winograd program recognizes, describes and may even invent new names for arrangements of objects, such as cubes and pyramids, stored within the computer and visible on a display screen. [26] This is not a robotic object-handling system, though it could be adapted to be an effective Robot. It employs Systematic Grammar, which emphasizes the limited and highly structured sets of choices made in producing a syntactic structure, abstracting the features that are important for conveying meaning. This is based on the structure of a Natural language – English – though simplified. A parser (an interpreter accepting recognition grammars written in a procedural form called programmar) builds a syntactic logical tree, and creates and uses feature lists described in its sub-structures. There is also a dictionary, and there are problem-solving programs allowing objects to be moved, positioned or balanced appropriately to their shape or other characteristics. The computer language used is Lisp; with this, the computer can accept typed normal English and punctuation, looks up words in its dictionary and performs morphemic analysis, such as realizing that 'detecting' is the 'ing' form of the word 'detect'.

The system's knowledge of its object world is given in a collection of planner programs known as 'blocks'. This includes theorems and knowledge about achieving goals in its world, and making Deductions from it; thus it talks about its world, which in some sense it 'sees'. Here is a specimen of conversation. Human conversation is in italics, machine conversation in capitals. Comment follows each stage.

1   *Pick up the big red block*
OK
The system answers 'OK' when it carries out a command. In order to pick up the red block, it has to clear it off by finding a space for a green one and moving the green one away.

2   *Grasp the pyramid*
There are three pyramids in the scene, so the program does nothing.

3   *Find a block which is taller than the one you are holding and put it into the box*
The system has to figure out what is meant by 'one' and 'it', as well as using Deduction to decide which block it is holding.

BY 'IT' I ASSUME YOU MEAN THE BLOCK WHICH IS TALLER THAN THE ONE
I AM HOLDING

This is printed out to keep us informed.

OK

The hand has set down the red block to pick up the taller blue one.

4   *What does the box contain?*

THE BLUE PYRAMID AND THE BLUE BLOCK

Sometimes the machine fails: it may ask for help; or it may discover solutions.

It is an important point of methodology that the main purpose of academic, as opposed to industrial applications of robotics is not so much to produce the best result that might be attained, by allowing the device a great deal of sensed information, but rather to restrict the available information so that it has to adopt problem-solving procedures: in common parlance to be intelligent. (In chapter 10 we call this 'Kinetic Intelligence', rather than 'Potential Intelligence'.) For example, providing the Robot with two eyes, or with radar range-finding, might inhibit the development of programs for making use of perspective convergence at a single eye, and the typical shapes of corners of objects for inferring depth and distance. There is however some controversy here, if only because there are various theoretical and practical aims to the study of Artificial Intelligence. Current progress is well described by Margaret Boden in her book *Artificial Intelligence and Natural Man* (1977), which contains extensive references. Unfortunately, support for this research has been at a low ebb over the last decade since its peak in the 1960s when the main centres were the Massachusets Institute of Technology (MIT), Stanford and Edinburgh University. But there are signs of a revival. With Professor Donald Michie, and Professor Longuet-Higgins, F.R.S., I helped to set up and run the Department of Machine Intelligence and Perception at the University of Edinburgh. We built the Robot 'Freddy'. The best current account of techniques of computer vision, describing languages and programs, is by Patrick Winston, of MIT – *Artificial Intelligence* (1977).

The first clear statement that man is a thinking machine is by Julien Offray de la Mettrie, in *L'Homme Machine* (1748). De la Mettrie (1709–51) was born at Saint Malo in Brittany. After studying rhetoric and becoming a Jansenist, he studied and practised medicine. He was so struck by the dependence of mental states on bodily states such as hunger and disease that he attacked Descartes for his Dualism, and suggested that Mind is given by the organization of the Matter of organisms. For this he received the active hatred of priests of all denominations and of his physician colleagues, so that he had to leave his country to live for a time at Leyden where *L'Homme Machine* was published. It is however little more than a credo, lacking detailed arguments or suggestions as to how a machine could be intelligent; and he is far from clear how the structure or motion of matter could be intelligent or conscious.

The first paper to discuss in detail how machines may be intelligent, and especially how one may judge claims of machine intelligence, is Alan Turing's paper, which appeared in the philosophical journal *Mind,* in October 1950, entitled 'Computing machinery and intelligence'. Turing was a distinguished code-breaker in World War II, which led him to design computing machinery, which was then secret; and later the ACE computer at the National Physical Laboratory, and at Manchester where he did his last work. With Princeton and MIT these were the pioneering achievements in electronic digital computing.

Turing was, as logician and mathematician, a principal architect of the modern computer, and prophet of its powers. Asking 'Can machines think?' he asks for definitions of 'machine' and 'think'. It is interesting that a British philosophical journal of that time, and especially *Mind,* printed a paper which rejected the importance of the normal uses of these words, and asked instead for operational criteria. Turing suggests a form of the 'imitation game'. A man and a woman hidden behind a screen, and communicating to an interrogator (with a teletype to avoid recognition by tone of voice), are asked questions to establish which is the man and which the woman. They are allowed to lie. Now one of the people is replaced by a machine. If the machine cannot, in this way, be distinguished from the person, it is deemed intelligent, though it may be a machine. Turing points out that the game may be criticized because machines may solve problems with inhuman speed and accuracy; but it will be allowed to pretend fallibility in such matters. It may, also, have a random generator to initiate moves or questions (here there is an essential extension from Babbage's conception). As to whether, in Turing's view, such a machine will be or can be conscious – aware of pleasure, grief, sex, and so on – he replies that neither machine nor man can convince the interrogator of their awareness, and so the machine is no different from the man or the woman. This is a neat form of solipsism – or at least of Behaviourism. It is strange that Behaviourism as a philosophy of psychology is false, though Turing's test in principle fails to distinguish (conscious) people from (unconscious) machines!

We have only touched upon current work on Artificial Intelligence and its historical beginnings, which can indeed be traced back to the Greeks, though almost nothing could be done before the modern digital computers and even these are inadequate. This discussion may serve as an introduction to the question (which is important for AI, as for biology and philosophy): are perceptions – and indeed all knowledge – perhaps somewhat doubtful hypotheses?

Are perceptions hypotheses?

Since the time of Descartes, and largely because of his powerful doubting, belief in certain direct knowledge has slipped away. With this loss of certainty, has grown the significance and status of hypotheses. There is, however, considerable resistance to regarding perceptions as hypotheses; if only because this gives a weak base for science, and because it may seem implausible for untutored brains to have anything like the sophistication of science. Is this criticism just?

Russ Hanson's view

Norwood Russell Hanson (1924–67) made a significant contribution to the philosophy of science, and especially the role of the observer in science, especially in two books: *Patterns of Discovery* (1958) and *Observation and Explanation* (1972). I shall draw principally from the earlier book as this is of most direct relevance here.

Hanson agrees with Popper in thinking that hypotheses – which he generally calls 'theories', though we need not make a distinction at this point – are not, at least usually, mere aggregations of data, and neither are laws of physics. To take Hanson's example: [27]

> The reason for a bevelled mirror's showing a spectrum in the sunlight is not explained by saying that all bevelled mirrors do this. On the inductive account this latter generalization might count as a law . . . . But only when it is *explained* why bevelled mirrors show spectra in the sunlight will we have a law of the type suggested, in this case Newton's laws of refraction. So the inductive view rightly suggests that laws are got by inference from data. It wrongly suggests that the law is but a summary of these data, instead of being what it must be, an explanation of the data.

There are important implications (or implicit meanings) here; especially that generalizations, useful as laws, are more than summaries of data: are theory-loaded, and are explanatory.

Hanson goes on to suggest that hypotheses are not purely of 'psychological' origin, but that there is a logic to their discovery and formulation. Thus '. . . the initial suggestion of an hypothesis is very often a reasonable affair. It is not often affected by intuition, insight, hunches, or other imponderables as biographers of scientists suggest.' This passage continues: [28]

> If establishing an hypothesis through its predictions has a logic, so has the conceiving of an hypothesis. To form the idea of acceleration or of universal gravitation does require genius: nothing less than a Galileo or a Newton. But that cannot mean that the reflections leading to these ideas are unreasonable or a-reasonable. Here resides the continuity in physical explanations from the earliest to the present times.

Here Hanson and Popper part ways. For Popper, hypothesis generation is essentially random, as mutations of organic forms are supposed random: for Hanson, hypothesis generation is a function of directed intelligence – in extreme cases, genius. So, for Hanson, there could be psychological laws for hypothesis generation, much as there are physical laws of the object world. Data can, for Hanson, initiate hypotheses: [29]

Kepler did not *begin* with the hypothesis that Mars' orbit was elliptical and then deduce statements confirmed by Brahe's observations [the Danish astronomer Tycho Brahe, 1546–1601]. These . . . set the problem – they were Johannes Kepler's starting point. He struggled back from these, first to one hypothesis, then to another, then to another, and ultimately to the hypothesis of the elliptical orbit.

Hanson imagines Kepler standing on a hill with Tycho Brahe, watching the dawn. Kepler believed that the earth moves, annually, round the sun, but Tycho continued to accept the Ptolemaic conception of the sun moving round the stationary earth. Hanson uses the imaginary dialogue between them, to ask whether perceptions are related to scientific hypotheses: [30]

Kepler regarded the sun as fixed: it was the earth that moved. But Tycho followed Ptolemy and Aristotle in this much at least: the earth was fixed and all other celestial bodies moved around it. *Do Kepler and Tycho see the same thing in the east at dawn?*
We might think this an experimental or observational question. . . . Not so in the sixteenth and seventeenth centuries. Thus Galileo said to the Ptolemaist '. . . neither Aristotle nor you can prove that the earth is *de facto* the centre of the universe . . .'. 'Do Kepler and Tycho see the same thing in the east at dawn?' is perhaps not a *de facto* question either, but rather the beginning of an examination of the concepts of seeing and observation.

Here we may say that Hanson is considering the relation of 'conceptual' to 'perceptual' hypotheses, and asking whether the first affect the second. Kepler and Tycho return to the scene: [31]

Kepler and Tycho are to the sun as we are to [an ambiguous figure] when I see the bird and you see only the antelope. The elements of their experiences are identical; but their conceptual organization is vastly different. Can their visual fields have a different organization? Then they can see different things in the east at dawn.

Hanson's conclusion is: 'There is a sense, then, in which seeing is a "theory-laden" undertaking.'

One might assume that Russ Hanson will go on to say that perceptions are like hypotheses of science – but this is not quite so. Since I *do* wish to make this move, I must give Hanson's case in detail. This he states with the Tycho–Kepler confrontation, as follows: [32]

The physical processes involved when Kepler and Tycho watch the dawn are worth noting. Identical photons are emitted from the sun; these traverse solar space, and our atmosphere. The two astronomers have normal vision; hence these photons pass through the cornea, aqueous humour, iris, lens and vitreous body of their eyes in the same way. Finally their retinas are affected. Similar electro-chemical changes occur in their selenium cells. The same configuration is etched on Kepler's retina as on Tycho's. So they see the same thing.

Hanson goes on to say that many people, including experts on vision or neurology who ought to know better, are drawn to the retina as though its stimulation is almost all that matters or need concern us. But, as Hanson says: [33]

These writers speak carelessly: seeing the sun is not seeing retinal pictures of the sun. The retinal images which Kepler and Tycho have are four in number, inverted and quite tiny. Astronomers cannot be referring to these when they say they see the sun.

He continues: [34]

. . . . Seeing is an experience. A retinal reaction is only a physical state – a photochemical excitation. Physiologists have not always appreciated the differences between experiences and physical states. People, not their eyes, see. Cameras, and eye-balls, are blind. Attempts to locate within the organs of sight (or within the neurological reticulum behind the eyes) some namcable called 'seeing' may be dismissed. That Kepler and Tycho do, or do not, see the same thing cannot be supported by reference to the physical states of their retinas, optic nerves or visual cortices: there is more to seeing than meets the eye-ball.

It seems to me very odd that differences of 'visual cortices' (any brain state?) are not allowed to explain the perceptual differences.

Hanson then develops this by discussing visual ambiguity, such as the Necker cube reversing figure (fig. 35): [35]

Do we all see the same thing? Some will see a perspex cube viewed from below. Others will see it from above. Still others will see it as a kind of polygonally-cut gem. . . . It may be seen as a block of ice, an aquarium, a wire frame for a kite – or any of a number of other things.

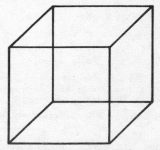

35   The first visually ambiguous figure to be described and investigated – the Necker cube. It reverses spontaneously in depth, even when its image is optically stabilized on the retina. (From R.L. Gregory and E.H. Gombrich (eds.), *Illusion in Nature and Art*, Gerald Duckworth and Co. Ltd, 1973.)

Do we, then, all see the same thing? If we do, how can these differences be accounted for?

Hanson goes on to suggest that to say that the changes of perception are, or are due to, changes of interpretation commits us to saying that there are two processes, perceiving and interpreting: [36]

This sounds as if I do two things, not one, when I see boxes and bicycles. Do I put different interpretations on fig. [35] when I see it now as a box from below, and now as a cube from above? I am aware of no such thing . . . [it] is simply seen now as a box from below, now as a cube from above; one does not first soak up an optical pattern and then clamp an interpretation on it. Kepler and Tycho just see the sun. That is all. That is the way the concept of seeing works in this connection.

Note that Hanson relies on introspection of our perceptual processes. But this is a dangerous approach, and is not supported here by experiments. This dependence on knowing by feeling whether we are thinking or interpreting applies to the rest of the discussion: [37]

The different ways in which these figures are seen are not due to different thoughts lying behind the visual reactions. What could 'spontaneous' mean if these reactions are not spontaneous? . . . One does not think of anything special; one does not think at all. Nor does one interpret. One just sees, now a [cube or] staircase as from above, now a [cube or] staircase as from below.

The sun, however, is not an entity with such variable perspective. What has all this to do with suggesting that Tycho and Kepler may see different things in the east at dawn? Certainly the cases are different. But these reversible perspective figures are examples of different things being seen in the same configuration, where this difference is due neither to differing visual pictures, nor to any 'interpretation' superimposed on the sensation.

Hanson goes on to consider a figure in which we seem to see (or rather understand) more than is 'given' (fig. 36). He writes of this: [38]

Your retinas and visual cortices are affected much as mine are; our sense-datum pictures would not differ. Surely we could all produce an accurate sketch of fig. [36]. Do we see the same thing? I see a bear climbing up the other side of a tree. Did the elements 'pull together'/ cohere/organize, when you learned this? You might even say with Wittgenstein it has not changed, and yet I see it differently . . . '.

One does not, however, see the bear as one sees the illusory contours and shape in fig. 42 (p.401). There seems to be a *gradation* from seeing visually to 'seeing' intellectually.

Considering 'organization', Hanson has a nice phrase, which I put into italics: [39]

Organization . . . is not an element in the visual field, but rather the way in which the elements are appreciated. Again, *the plot is not another detail in the story*. Nor is the tune just another note. Yet without plots and tunes details and notes would not hang together

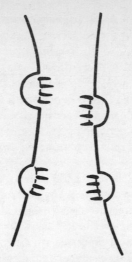

36 This may look like meaningless marks. But given the notion that it is a bear behind a tree, it looks very different. (From Norwood Russell Hanson, *Patterns of Discovery*, Cambridge University Press, 1958.)

. . . were this lacking we would be left with nothing but an unintelligible configuration of lines.

Finally, Hanson relates this 'perceptual organization' problem to observation in physics: [40]

> To say that Tycho and Kepler, Simplicius and Galileo, Hooke and Newton, Priestley and Lavoisier, Soddy and Einstein, De Broglie and Born, Heisenberg and Bohm all make the same observations but use them differently is too easy. It does not explain controversy in research science. Were there no sense in which they were different observations they could not be used differently . . . . It is important to realize, however, that sorting out differences about data, evidence, observation, may require more than simply gesturing at observable objects. It may require a comprehensive reappraisal of one's subject matter. This may be difficult, but it should not obscure the fact that nothing less than this may do.

Let us go back to Hanson's objection to saying that perceptions are both seeing and interpreting – that for him interpreting is not part of seeing. His reason is that interpreting is intellectual while seeing is not supposed to be intellectual, and that (according to him) we are aware of interpreting but not aware of processes of seeing. This seems to me an important mistake. We may attempt an analysis of Hanson's position here, which is probably a hang-over from Direct Realist theories of perception which are still seductive for philosophers seeking a secure basis for knowledge. It seems most odd to hold that animals such as squirrels lack perception. A great deal of perceptual research is carried out with animals as subjects (and infants, which Hanson also suggests do not have perception) so this is clearly contrary to the view of experimental psychologists and physiologists working on vision. If reversing figures such as Necker cubes demonstrate effects of the 'knowledge and theory', which Hanson denies to sub-human animals and infants, they should not be ambiguously affected by what to us are visually ambiguous figures or objects. So here we have, presumably, an experimental test. Unfortunately little if anything so far as I am aware is known of animal responses to Necker cubes; but animals do show appropriate behaviour to half-hidden objects such as the 'bear behind the tree'. Indeed, it seems very clear that animals do deploy knowledge and do not act merely from sensory signals. Hanson's view is here a hang-over from stimulus–response accounts of behaviour. The crucial point is that animals, as well as adult humans and scientists, predict from limited sensed data to situations which can be related only by kinds of inference. In fact we have every reason to believe that perceptions have their richness and integrity as well as their predictive power through inference. This is almost self-evident to the psychologist working on perceptual processes (though with exceptions), but it is anathema to philosophers seeking unadulterated, theory-free and assumption-free sensory data.

Hanson, it seems to me, gives no good reason for separating interpreting from seeing. It is clear that the processes of seeing (including object recognition by computers, developed since Russ Hanson was writing) involve many processes which could well be described as 'interpreting' – though we are not aware of these or any processes of perception. We come across this issue again in Wittgenstein's discussions of ambiguities of perception.

Ludwig Wittgenstein's view

In *Philosophical Investigations* (1953),[41] Ludwig Wittgenstein addressed himself, somewhat differently to the problem of ambiguity and whether perception includes interpretation. Thus of the Necker cube: 'But we can also *see* the illustration now as one thing now as another. – So we interpret it, and *see* it as we *interpret* it.'

Wittgenstein treats perceptual ambiguity very differently here from Hanson: Hanson rejects changes of interpretation, while Wittgenstein regards the visual changes as changes of interpretation. Later, Wittgenstein says:

But how is it possible to *see* an object according to an *interpretation?* – The question represents it as a queer fact; as if something were being forced into a form it did not really fit. But no squeezing, no forcing took place here.

Then:

And is it really a different impression – In order to answer this I should like to ask myself whether there is really something different there in me. But how can I find out? – I *describe* what I am seeing differently.

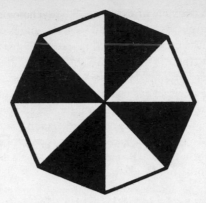

37   Rubin's double-cross figure. This example of visual ambiguity is not of depth but rather of form: it is seen as either of two alternative objects, as different regions become 'figure' or are relegated to 'ground' and virtually ignored.

Wittgenstein distinguishes between ambiguities such as the Rubin 'double-cross' (as Wittgenstein calls it),[42] and the ambiguity of pictures of objects that we must know as distinct kinds of objects for the ambiguity to occur:

Those two aspects of the double cross (I shall call them the aspects A) might be reported simply by pointing alternately to an isolated white and an isolated black cross [fig. 37].

One could quite well imagine this as a primitive reaction in a child even before it could talk.

Contrasting this with the Jastrow duck–rabbit figure, Wittgenstein adds: You only "see the duck and rabbit aspects" if you are already conversant with the shapes of those two animals. There is no analogous condition for seeing the aspects A.'

38   Jastrow's duck–rabbit, an example of visual ambiguity where alternative objects are duck or rabbit, with all features incorporated in both perceptions.

To put this distinction in the terminology of computer scene analysis, we may say that the ambiguity of the Rubin 'double-cross' figure is due to inadequacy, or conflicts of stimulus data used for 'bottom-up' processing. This is not so different from what Gibson said (pp. 375–7). But the ambiguity of the Jastrow duck–rabbit is due to features representing ducks and features representing rabbits being equally weighted given our knowledge of ducks and rabbits. It cannot be both a duck and a rabbit, so each is entertained in turn. This ambiguity is generated by 'top-down' processing; for it depends on stored data (of ducks and rabbits). This is the point that Gibson missed. He had to miss this point, because he denied that perception involves or requires knowledge of the world beyond strategies for selecting useful information from the 'ambient array'.

Wittgenstein brings out the importance of object knowledge with a subtle example: 'For how could I see that the posture [of an animal] was hesitant before I knew that it was a posture and not the anatomy of the animal?' Then considering a flourish (of, for example, an elaborately written letter): 'It could deviate from the correctly written letter in a variety of ways. – And I can see it in various aspects according to the fiction I surround it with. And here there is a close kinship with "experiencing the meaning of a word".' And two pages later: 'It is almost as if "seeing the sign in this context" were an echo of a thought. "The echo of a thought in sight" – one would like to say.' Wittgenstein distinguished between 'interpreting' and 'seeing' thus: 'Do I really see something different each time, or do I only interpret what I see in a different way? I am inclined to say the former. But why? – To interpret is to think, to do something; seeing is a state.' This is much closer to Hanson's position, in tending to reject 'interpretations'.

Wittgenstein considers also whether there could be human beings lacking the capacity to 'see something *as something*. . . . Would this defect be comparable to colour-blindness or to not having absolute pitch?' Wittgenstein calls it 'aspect blindness'. Thus:

> The aspect-blind man is supposed not to see the aspects A change. But is he also supposed not to recognise that the double cross contains both a black and a white cross? So if told 'show me figures containing a black cross among these examples' will he be unable to manage it? No, he should be able to do that; but he will not be supposed to say: 'Now it's a black cross on a white ground' [fig. 37].
>   . . . . Ought he to be unable to see the schematic cube as a cube? It would not follow from that that he could not recognize it as a representation (a working drawing for instance) of a cube. But for him it would not jump from one aspect to another.
>   . . . . The 'aspect-blind' will have an altogether different relationship to pictures from ours.[43]

This he compares with experiencing (and not experiencing) meanings of words. For Wittgenstein, pictures are like sentences.

Wittgenstein describes his discussion of ambiguity as a conceptual investigation, but it is possible to investigate these matters experimentally. Indeed, ambiguous figures are extremely useful tools for perceptual research, especially because they allow us to distinguish between what is given by stimulus patterns, 'upwards' from effects of stored knowledge working 'downwards'; for the spontaneous switches of perception must be neurally centrally determined.[44]

We shall look now at what Thomas Kuhn has to say about this problem of ambiguity, for 'paradigms' of science and for perception.

Thomas Kuhn's views on ambiguity

The notion of scientific hypotheses (though not so much perceptions) being theory-laden was developed with remarkable force in Thomas Kuhn's book *The Structure of Scientific Revolutions* (1962). Kuhn argues that normal science works by accepted 'paradigms', which are assumptions (often implicit) that dictate the practice and habits of thought of almost all scientists. Psychology could be an exception – non-normal science – for it lacks a controlling paradigm. Examples of paradigms are Ptolemaic and Copernican astronomy; Newton's mechanics; Relativity; Evolution by Natural Selection, and modern wave optics. A 'revolution' occurs when a paradigm succumbs to a rival, such as Kepler's paradigm overturning Ptolemy's. For Kuhn, the paradigm Revolutions are only clear-cut after the work of Newton (at least for optics). Before then, there were many competing paradigms, held by individuals or schools simultaneously: current psychology is like this now.

Perhaps curiously, Kuhn has very little to say about perception. He does not seem to think of perceptions as at all like hypotheses. It is almost as though Kuhn takes observations as given, as though he were a Direct Realist, in spite of his Relativist notions of scientific knowledge. This is not so for Hanson, as we have already seen. Kuhn discusses visually ambiguous figures in a context similar to Hanson's discussion. He has, however, rather different things to say about them:[45]

The subject of a Gestalt demonstration knows that his perception has shifted because he can make it shift back and forth repeatedly while he holds the same book or piece of paper in his hands . . . unless there were an external standard with respect to which a switch of vision could be demonstrated, no conclusion about alternative perceptual possibilities could be drawn.

Kuhn relates this to science:

With scientific observation, however, the situation is exactly reversed. The scientist can have no recourse above or beyond what he sees with his eyes and his instruments. If there were some higher authority by recourse to which his vision might be shown to have shifted, then that authority would itself become the source of his data, and the behaviour of his vision would become a source of problems.

What are we to make of this? Strictly speaking (and this is what matters in such issues) the psychologist must be assuming that the object of the Gestalt demonstration – the spontaneously changing figure – does not change, as objects may change physically. He must be assuming that the rabbit does not change into a duck, and back again, or the Necker cube transform so that its back and front occupy different positions, as they might in physical space. It would not be impossible to engineer such changes in these figures. One can indeed imagine an experiment in which physically changing figures, or objects, are presented, and the subject of the experiment is required to decide when the change is produced by the engineering and when by his own perception. Indeed, if we go back to Necker's original observation of the depth-switching of rhomboid crystals (1832), we find that he was puzzled as to whether it was the crystals or himself that was changing!

How would a Direct Realist deal with visual ambiguity? This is, of course, a most serious problem for Direct Realism; for how can even a selection of the object world be both part of this world and yet differ from it? The strongest held logical truth is that a thing is itself and not another thing ($p = p$). One way out is to say that illusions are essentially different from other perceptions, but this is extremely implausible, especially as perceptions may be but partly illusory.

This is a serious difficulty for J. J. Gibson, for he equates perception with the 'stimulus array'. How can he reconcile changing perceptions with unchanging objects, or stimulus arrays? He meets the difficulty bravely: [46]

In the goblet–faces display [fig. 39], the stimulus is the same for the two percepts but the stimulus *information* is not. In the absence of texture and parallax, the information for edge-depth or superposition has been arranged to specify two opposite directions of depth. There are two counter-balanced values of stimulus information in the same 'stimulus'. The perception is equivocal because what comes to the eye is equivocal. In such displays, the information for one-thing-in-front-of-another must come from variables of the mutual contour at the optical junction of the two things.

He thus revises his definition of 'stimulus information': *'The same stimulus array coming to the eye will always afford the same perceptual experience insofar as it carries the same variable of structural information.'* So 'if it also carries different or contradictory variables of information it will afford different or contradictory perceptual experiences'. I am far from understanding what this means, but I suspect that 'information' is being used in a way that he should not himself allow in this theory. If it is information, in any usual sense of the word, some kind

39   Rubin's vase–faces ambiguous figure. (From R.L.Gregory and E.H.Gombrich (eds.), *Illusion in Nature and Art*, Gerald Duckworth and Co. Ltd, 1973.)

of representation is implied; but this is just what Gibson attempts to deny for perception.

For the Direct Realist, there is only one reality; however, if we accept that perceptions are representations, there can always be discrepancies between what is and what seems to be. The perceptual psychologist cannot go behind the scenes of reality (though he may rig them) any more than a physicist can who is concerned with 'What is a change of things, what is a change of paradigm?' To the psychologist some perceptions are accepted as illusory just when and because they depart from his belief, or paradigm, of what is the case physically. His subject may indeed be unsure as to whether he has rigged the ambiguous figures, so that they might be changing by a hidden mechanism, but the psychologist himself must be very sure that there is no such mechanism for change if he is to claim that these are 'Gestalt switches', or whatever, in the observer.

We may generalize this to claims by psychologists of perceptual distortion. He always has to assume that the object being observed is not distorted, from some assumed standard. He may feel confidence in his control of such matters, while the physicist may be under grave doubt as to whether the queer observation is due to a queerness of perception, or whether it is what is being observed that is queer. We can find examples of uncertainty of whether something observed as queer is 'physical' or 'psychological'; for instance, the principal investigator of moiré patterns, Gerald Oster, has confided (in a private communication) that he has severe doubts over some of these phenomena, as to whether they 'belong' to physics or to perception. This is a particularly interesting case, for Gerald Oster has the know-how to solve experimental problems as a physicist and he also has the technical knowledge of perception, which might have allowed him to assign these phenomena confidently to physics or to perception if this were at all easy. It is surprisingly difficult to decide this case between physics and perception, though this may be rare. To my mind most surprisingly, Kuhn takes the classical Empiricist position for perception – that it is direct knowledge that may be trusted in a way that scientific knowledge cannot be trusted. But this example runs him into difficulties. He has in effect to accept that psychologists carrying out perceptual experiments have a certainty about things which not even a physicist would claim. But psychologists are, at best, scientists; so why should their science be so certain? Why should they be beyond the relativity of paradigms? It is common experience for perceptual psychologists to confuse a perceptual phenomenon with an artefact of their apparatus. They are indeed very far from having special knowledge, except for introspection, and this is not at all reliable for processes of perception. Their problem is lack of a consistent paradigm – which makes them especially vulnerable as scientists, as surely Kuhn would agree.

One might think that this possibility of the *object* changing during a 'Gestalt switch' is plain silly, for we know that it does not change (even though Necker himself became confused). But over the last thirty years, the architects of Quantum Theory have suggested that just this can, and does, occur – that even the act of observing sets what is, from might be. This we shall discuss later (chapter 19).

Philosophers defending the Empiricist position, that all knowledge comes from the senses, have a strong commitment to see perception as generally reliable and not too dependent on beliefs – which may change and may be incorrect. This dependence on the assumption of reliable and 'neutral' perception continues from the classical Empiricism of Locke, Hume and Berkeley, to this century's Logical Positivism. Dependence on unquestionable perception has indeed grown since Locke, for verification by observation has since been suggested as necessary for statements to have meaning. Perception is the basis of Empiricism and the basis of science, so it is surely important for philosophers to take a close look at the characteristics and limitations of how we perceive. Even in physics there is a surprising lack of emphasis on the characteristics of the observer, though he is essential in all experiments, and essential for selecting and interpreting all data.

I must confess to not understanding why Kuhn thinks of perception in the way that he evidently does. Is it a grasping for some certainty, when his insight of shifting scientific paradigms makes 'truth' uncomfortably relative? Absolute or certain truth is an obsession of almost all philosophers. Just what are the similarities and the differences between perception and science? I conclude that both Kuhn and Hanson draw unfounded distinctions between scientific observation and perception. This does not, of course, indicate that there are no distinctions to be made. It would indeed be exceedingly odd if the procedures and inferences employed explicitly by scientists were in every way identical to the brain processes of perception. The sense organs are not exactly like 'corresponding' scientific instruments. Eyes are not exactly like cameras, ears exactly like microphones – or brains exactly like any man-made computer. We look here, however, not for surface similarities (though there are many) but rather for *deep identities* of perception and science, especially in their procedures.

Once we see perception as depending on rules of inference and knowledge, it is hardly possible to hold that perception is directly related to the reality that we perceive. If the past as stored in memory is necessary for perception, and indeed constitutes much of what we experience as the present, then we cannot accept a Direct Realist account: we are forced to adopt some variation of the alternative – that perceptions represent reality with more or less validity. We shall explore the notion that the senses and their brain mechanisms represent the world perceptually very much as scientific hypotheses predict and explain the world conceptually.[47]

Chains of hypotheses

We have now reached the watershed – the crescendo – of our theme: to try to discover the relation of Science to Mind; to understand Mind in terms of the nature of Science; and perhaps to understand Science through understanding Mind, with hypothesis as the central linking notion. Of course, to expect to reach adequate understanding would be ridiculously optimistic; arrogant in the extreme as a hope, let alone as a claim that we had succeeded.

To suggest that perceptions are like scientific hypotheses is to suppose that the instruments of science parallel in essentials the sense organs; and that the procedures of science, in essentials, parallel processes carried out by the brain to generate perceptions.

There will clearly be differences – if only surface differences – between perceptions and hypotheses of science; but for the analogy, or suggested link, to be interesting there must be deep similarities and preferably identities. These we shall try to seek and test. I shall start by setting out three claims, which I may hope to justify: (1) that perceptions are essentially like predictive hypotheses in science; (2) that the procedures of science are a guide for discovering processes of perception which are knowledge-based; and (3) that many perceptual illusions correspond to, and may receive explanations from, understanding systematic errors occurring in science – through loss of calibration, and misplaced assumptions or knowledge.

It is hoped that by exploring this analogy – or perhaps deep identity – between science and perception we may develop an effective epistemology related to how the brain works. I shall attempt to indicate concepts and processes that seem important for the opening moves towards the end play of this understanding. The approach is based on regarding perception and science as constructing hypotheses by 'fantasy-generators' – which may hit upon truth by producing symbolic structures matching physical reality.

Three steps to perceptual hypotheses

In the first place, we may regard the sense organs (eyes, ears, touch receptors, and so on), as transducers, essentially like photocells, microphones and strain gauges. The important similarity, indeed identity, is that the sense organs and detecting instruments convert patterns of received energy into signals, which may be read according to a code. As signal, the neural activity is fully described in physical terms and measurable in physical units; but the code must be known in order to use or appreciate it as data. We suppose that the coded data are – in perception and science – used for generating hypotheses. For perception we may call them 'perceptual hypotheses'. These are what are usually called visual and other sensory 'perceptions'. Here are the three stages of perception, in these terms:

1   Signals

Patterns of neural events, related to input stimulus patterns according to the transducer characteristics of sense organs. For the eye, for example, there is a roughly logarithmic relation between intensity of light and the firing rate of the action potentials at the initial stage of the visual channel. Colour is coded by the proportion of rates of firing from the three spectrally distinct kinds of cone receptor cell, and so on. These transducer characteristics must be understood before the physiologist can appreciate what is going on. He can then (with other knowledge or assumptions) describe the neural signals as data representing states of affairs.

2   Data

Neural events are accepted as representing variables or states according to a code, which must be known for signals to be read or appreciated as data. This necessity of knowing the code is surely clear from examples such as signals conveying data in Morse code. The dots and dashes have no significance, and may not even be recognized as signals conveying data when the code is not known. The same holds for the words on this page; we must know the rules of the English language, and a great deal more, to see them as more than patterns of ink on paper.

It is generally true that a lot more than the code must be known before signals can be read as data. For a detecting instrument (such as a radio-telescope, magnetometer or voltmeter) it is essential to know something of the source – whether it is a star, a given region of the earth, or just which part of the circuit a voltmeter is connected to. The outputs of some instruments (such as optical telescopes, microscopes, and X-ray machines as used in medicine) may give sufficient structure for the source to be identified without extra information. This is especially so when the structure of the output matches our normal perceptual inputs. This is, however, somewhat rare for instruments. A voltmeter provides no such structured output by which we can recognize its source of signals without collateral knowledge. Some sense organs (especially the eye) provide highly structured signals allowing identification of the source; but visual and any other sensory data can be ambiguous (including touch, hence the game of trying to identify by touch objects in a bag), and indeed all sensory and instrumental data are, strictly speaking, ambiguous. The fact that vision is usually sufficient for immediate object identification distracts us from realizing the immense importance of contextual knowledge for reading data from signals. Scientific data from instruments are almost always presented with explicit collateral information on how the instrument was used, what source it was directed to, its calibration corrections and scale settings. The gain-setting of oscilloscopes and the magnification scale of photographs and optical instruments of all kinds are essential for scientific use: if the scale is given incorrectly, serious misinterpretation can result, even to confusing the surface of a planet with microscopic structures. So, not only must the signal code be known, but also a great deal of context knowledge for signals to be read as data. This holds for perception as it does for the use of scientific instruments, though for perception the context knowledge is generally implicit and so its contribution may not be recognized.

It is important to note that signals can be fully described and measured with physical concepts and physical units, but this is not so for data. Data are highly peculiar, being (it is not too fanciful to say) in this way outside the physical world, though essential for describing the physical world.

The codes necessary for reading signals as data are not laws of physics. They are, rather, essentially arbitrary and held conventionally. Some may be more convenient or efficient than others, but in no case are they part of the physical world as laws of physics are, or reflect, 'deep structures' of reality. Further, data are used to select between hypothetical possibilities, only one of which (if indeed any) exists. The greater the number of alternatives available for the selection (or rather the greater their combined probability) the greater the quantity of information in the data (Shannon and Weaver, 1949). The information content thus depends not only on what is but on the hypothetical stored alternatives of what may be. But these are not in the physical reality of the situation, so data cannot be equated with what physically is; neither can they be equated with signals, for data are read from signals by following the conventional rules (which are not physical laws) of a code.

3 Hypotheses
There is, unfortunately, no general agreement as to just what hypotheses are, or what characterizes them. This, it must be confessed, is a weakness in our position.

40   Adaptation, or re-calibration, takes place when the oblique lines are viewed. Move the eyes backwards and forwards along the horizontal bars between the slanting stripes (to avoid after-images) for a minute or so, and then fixate the vertical stripes. They will appear slanted, as the eye has adapted to the obliques. See also p. 515. (From Colin Blakemore, 'The Baffled Brain', in R.L. Gregory and E.H. Gombrich (eds.), *Illusion in Nature and Art*, Gerald Duckworth and Co. Ltd, 1973.)

If there is no agreement on what are hypotheses, how can we argue cogently that perceptions are hypotheses? Just what is being claimed? With the present lack of agreement, one must either be vague or stick out for a particular account, which may be arbitrary, of the nature of hypotheses. Current accounts range from Popper's view that they have no prior probabilities and no predictive power (and that they cannot be confirmed but only disconfirmed) to extremely different accounts, such as that they can be in part predicted; that they can be used for prediction; and that they can be confirmed (though not with certainty) as well as disconfirmed by observations. I shall not entirely follow Popper's account of hypotheses (1972), but hold, rather, an alternative account: that they have predictive power, and that they can be suggested by observation and Induction, and can be confirmed or refuted though not in either case with logical certainty.

   It may be objected here that if perceptions are themselves hypotheses, they cannot be invoked to confirm or refute the explicit hypotheses of science. This is, however, no objection, for it is common experience that a perception can confirm or refute other perceptions. And one scientific hypothesis may (it is usually held) confirm or refute other hypotheses in science. So there is no clear distinction between hypotheses and perception here to make my argument invalid.

There is, however, this problem of the lack of agreement of what constitutes hypotheses. The notion of hypotheses has grown in importance with the rejection of hopes of certainty in science such as hopes in the supposed certainty of geometrical knowledge, before non-Euclidean geometries; and with Kuhn's paradigms (1962). Perhaps all scientific knowledge is now regarded as hypotheses. But if Popper is right, would we have any wish to associate perception with hypotheses? For in his view they have none of the powers that we attribute to perception. What, then, are hypotheses?

I suggest that hypotheses are selections of signalled and postulated data organized to be effective in typical (and some novel) situations. Hypotheses are effective in having powers to predict future events, unsensed characteristics, and further hypotheses. They may also predict what is not true. I shall assume that we accept that these are important characteristics of scientific hypotheses and perception.

To amplify this, we may now consider in some detail similarities – and also differences, for there clearly are differences – between hypotheses and perceptions. We shall look first at similarities when perceptions and scientific hypotheses are appropriate. We shall then go on to compare them when they are inappropriate, or 'false'; and finally we shall consider ways in which perceptions clearly differ from hypotheses of science.

Perceptions compared with scientific hypotheses

Appropriate uses of perception and science
Interpolation across gaps in signals or data. This allows continuous behaviour and control with only intermittent signals, which is typical of organisms and important in science, though rare in machines.

Interpolations may be little more than inertial, or may be highly sophisticated and daring constructs. Let us first consider interpolation in a graph such as fig. 41. The curve is derived according to two very different kinds of processes. It is generated from the readings by following procedures, which are easy to state and to carry out automatically, without particular external considerations. The most common procedure here is fitting by least squares. The curve may not touch any of the points representing the readings and yet it is accepted as the 'best' curve. It is an idealization – a hypothesis of what should occur in the absence of irrelevant disturbances and an infinite set of readings with no gaps.

Any graph of experimental or observational data has some scatter in the readings through which the curve is drawn. The scatter may be due to random disturbances of the measuring device; or to variation in what is being measured, such as quantal fluctuations. These kinds of scatter have very different statuses, though for some purposes they may be treated alike and there may be a mixture of the two. In any

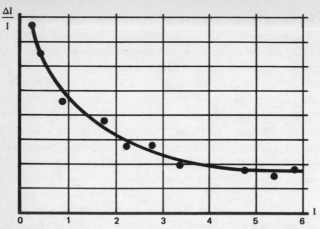

41 The Weber function, relating discrimination of differences of intensity ($\Delta I/I$) to intensity (I). The ΔI increases proportionally with I except for these low intensities, then the logarithmic relation breaks down, as $\Delta I/I$ increases. Note that the experimental points by no means all lie on the curve. The status of the curve is that it is a hypothesis: it predicts the observations that would be made from a larger (even infinite) number of readings.

case the curve may not touch any of the points indicating the readings. So it is a kind of fiction, accepted as the fact of the situation, and it may be true.

The second kind of procedure for obtaining the curve and for gap-filling is selecting a preferred curve, on theoretical or other general grounds, which may be aesthetic. The first kind of procedure is 'bottom-upwards' from the readings, by following procedures without reference to contextual considerations; the second is 'top-downwards', from stored knowledge or assumptions suggesting what is a likely curve. This may be set by a general preference, or prejudice, for example for a linear, or a logarithmic, or some other favoured type of function; or it may be set by particular considerations. Both have their dangers: the first biases towards the accepted, and the second tends to perpetuate false theories by bending the data in their direction.

The example of a graph illustrates that hypotheses – for the accepted curve or function may be a predictive hypothesis – can be non-propositional. Perhaps hypotheses are generally thought of as sets of propositions, but there seems no reason to restrict hypotheses to propositions as expressed in language. An equation such as $\mathbf{E} = \frac{1}{2}\mathbf{mv}^2$ is a hypothesis, in this case concerning quite abstract concepts (energy, mass and velocity), believed to represent something of the deep structure of physics, but it is not propositional in form. It could be written in language as a set of sentences expressing propositions, but this would be relatively clumsy. It could also be expressed as a graph, and this could be adequate for some purposes. Analogue computers, indeed, work from this kind of non-propositional representation (see pp. 57, 91).

There seems no reason to hold that 'perceptual hypotheses' require a propositional brain language, underlying spoken and written language, though this might be so. The merits of this notion need not be considered here, as we are free to regard hypotheses as not necessarily being in propositional form.

Perhaps interpolations are generally regarded as gap-filling in situations for which we do not have complete or strictly adequate readings, but interpolation can be far more elaborate than this: for example, postulating unknown species to fill gaps in evolutionary sequences. For a visual example, consider fig. 42. Perhaps 'illusory contours' are edges of objects postulated to account for gaps in available sensory signals or data. They take more or less ideal forms, and they are fictions (generally useful) joining data. This indeed defines interpolation in perception and science.

Extrapolation from signals and data, to future states and unsensed features. Extrapolation allows hypotheses to take off from what is given or accepted into the unknown. Going beyond accepted data is not very different from filling gaps, except that interpolations are limited to the next accepted data point; however, extrapolations have no end-point in what is known or assumed, so extrapolations may be infinitely daring – and so may be dramatically wrong!

Extrapolation beyond the ends of graphs of functions supported by data is sometimes essential (as for determining 'absolute zero' temperature by extrapolating beyond the range through which measures can, even in principle, be made). Extrapolations can leap from spectral lines to stars, and from past to future. With interpolation and extrapolation, data become stepping-stones and springboards for science and perception. This is to say that perceptions are not confined to stimuli: just as science is not limited to signals or available data. Neither, of course, is confined to fact: this is what gives them their extraordinary power (Gregory, 1968).

42   Illusory – cognitive? – contours. These may be postulates of nearer, eclipsing, objects, in which case the surprising *absence* of sensory signals can evoke object perception. (From R. L. Gregory and E. H. Gombrich (eds.), *Illusion in Nature and Art,* Gerald Duckworth and Co. Ltd, 1973.)

Discovery and creation of objects, in perceptual and conceptual space. The perceptual selection of sensed characteristics may create objects. There is also strong evidence for generating visual characteristics from what are accepted as objects.

We know too little about the criteria for assigning data to objects, and creating object hypotheses from data. What science describes as an object may or may not correspond to what appears to the senses as an object; different instruments reveal the world as differently structured. Further, general theories change what are regarded as objects. There is evidently a complex multi-way traffic here by which the world is parcelled out into objects; it seems that here again we have 'upwards' and 'downwards' procedures operating. The various rules of 'closure', 'common fate' and so on, emphasized by the Gestalt psychologists (see pp. 367–71), reflect features typical of the vast majority of objects as we see them. Most objects are closed in form, and their parts move together. These common object characteristics become identifying principles – but they may structure random patterns to create object forms.

More recently, the work in AI on object recognition makes use of typical features – especially the intersections of lines at corners of various kinds – to describe objects and their forms in depth (Guzman, 1968, 1971).[48] The effectiveness of these classifications and rules depends on what objects are generally like (see Marr, 1975, 1979). For exceptional objects, the rules mislead as we see in the Ames demonstrations (Ittelson, 1952) and in distortion figures (Gregory, 1968, 1970). The object-recognizing and object-creating rules are applied upwards to filter and structure the input. (It is, however, interesting to note that they may have been developed downwards, by generalized experience of what are, through the development of perception, taken as objects. For the Gestalt writers this is largely innate; but our similar experience and needs might well generate common object criteria through experience.)

Knowledge can work downwards to parcel signals and data into objects; as knowledge changes, the parcelling into objects may change, both for science and perception. We see this most clearly when examining machines: the criteria for recognizing and naming the various features as separate depend very much on our knowledge of functions. Thus the pallets on the anchor of a clock escapement are seen and described as objects in their own right once the mechanism is understood, though they are but shapes in one piece of metal, which happens also to look like an anchor. So we see here again the importance of 'upward' and 'downward' processing in perception and science – the complex interplay of signals, data and hypotheses. Unravelling this is surely essential for understanding the strategies and procedures of perception and science. It is also important for appreciating the status of objects. How far are objects recognized, and how much are they created by perception and science? This is a deep question at the heart of Empiricism.

What may be a profound difference between perceptual and conceptual objects, is that perceptual objects are always, as Frege put it concrete objects (Dummett, 1973), while the conceptual objects of science may be abstract objects. The point is that objects as perceived have spatial extension, and may change in time, while conceptual objects (such as numbers, the centre of gravity of concrete objects, and the deep structure of the world as described by laws of physics) cannot be sensed, may be unchanging and spaceless, and yet have the status of objects in that they are public though not sensed. We all agree that the number 13 is a prime number, that it is greater than 12 and less than 14, that it is odd and not even, and that all prime numbers except 2 are odd numbers. This kind of agreement is characteristic of the agreement and public ownership of objects as known by the senses (tables, stones, and so on); yet numbers cannot be sensed, though they are as 'public' as tables and stones – and so in this sense are objects.

This situation is rendered even more difficult by the consideration that, clearly, concrete objects have some features that are abstract: as we believe especially from scientific knowledge. Take, as an example, centre of gravity. Stones have centres of gravity, which is useful as a scientific concept, and are indeed what Newton took to be the 'objects' of the solar system for his astronomy. Centres of gravity may indeed lie not in but between concrete objects, such as between binary stars. Does the centre of gravity exist, as stones and stars exist? Or is it a useful fiction created as a tool for scientific description?

Even within what is clearly perception, we can be uncertain of what is 'concrete' and what is 'abstract'. We see that a triangle has three sides, and yet number is regarded, at least by Frege, as 'abstract'. Are shadows 'concrete objects'? The trouble here is that they are known by only one sense (if we except differences of sensed temperature), and they have few causal properties. Also, they are always attached to what is clearly a concrete object (which may be the ground), and by contrast they seem far less concrete, almost abstract, though we see them.

These are exceedingly difficult issues, which can hardly be resolved without deeper understanding of hypothesis generation, and further analysis of the similarities and differences between perceptual and conceptual hypotheses.

If we consider such 'objects' as electrons, which are clearly inferred indirectly from observational evidence, how do they compare with concrete objects of perception? If we believe that normal perception of concrete objects such as tables and stones requires a great deal of inference ('Unconscious Inference', to use Helmholtz's term), then the difference may not be great. The more perception depends on inference, the more similar we may suppose is the status of perceptual and conceptual objects.

We shall now consider inappropriate or 'false' hypotheses and perceptions. Here I describe certain phenomena of perception, such as various kinds of illusions, as our actually *seeing* what are *described* when they occur in science as errors – various kinds of ambiguities, distortions, paradoxes and fictitious features. The claim is that these categories, which are normally applied to arguments and descriptions, appear in perceptions, as experiences of recognizable kinds which can be investigated much as the phenomena of physics can be investigated; though some of these perceptual phenomena (illusions, which we shall classify) require rather different explanations from errors in physics – most are essentially identical.

Inappropriate uses of perception and science

Ambiguity, sometimes with spontaneous alterations and disagreements. The point here is that alternative hypotheses can be elicited by the same signals. There are many examples of visual ambiguity in which a figure (or sometimes an object) is seen to switch from one orientation to another, or transform into another design or object (L. A. Necker, 1832). This has been attributed to bi-stable (or multi-stable) brain circuits (Attneave, 1971), and, very differently, to putative hypotheses in rivalry for acceptance when their probabilities on the available evidence are nearly equal. The first would be an account in terms of signals, the second in terms of data. Inspection suggests strongly that the second is what is

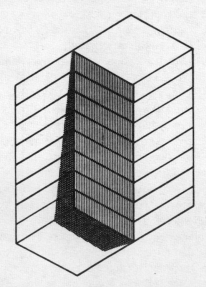

43   This visually depth-ambiguous figure has a dark rectangle, which may be a mark or a shadow. When the corner sticks in, it is more likely to be a shadow – and is then generally seen as lighter. Shadows tend to be rejected as objects, or as features of objects, and are largely ignored or suppressed. (From R. L. Gregory and E. H. Gombrich (eds.), *Illusion in Nature and Art*, Gerald Duckworth and Co. Ltd, 1973; after Ernst Mach (1959).)

going on in most cases – for the stimulus pattern can be greatly varied but what it represents matters a great deal. There are, however, many examples of the first kind: retinal rivalry, from different colours presented to the eyes producing spontaneous alternations, and lines of different orientation presented binocularly, producing rivalry. Here it is purely the stimulus characteristics that matter. On the other hand, figures such as the Necker cube, the Schroeder staircase or the Boring wife–mistress figure present equal evidence, for example, for two very different faces, which gives the ambiguity. For the Necker cube there is no evidence favouring either of two or more orientations. For both figures the ambiguity, no doubt, depends on our knowledge of faces and cubes (see note 11, p. 591 for the case of 'S.B.'). It is likely that different experience might change the bias of ambiguous alternations, in cases such as the Boring figure.

That science can be ambiguous is shown by the frequent changes of opinion and the occasional disputes which give it light and heat. For a current example of scientific ambiguity: are quasars astronomically near objects with abnormal red shifts, perhaps due to their powerful gravitational fields (and thus not obeying the Hubble law of increasing red shift with distance), or are they very distant, but of enormous intrinsic brightness? Here is a clear case of an important ambiguity that

44 The Boring wife–mistress figure. Devised by the American psychologist and historian of psychology, E. G. Boring, this is a dramatic example of object-perception ambiguity. Some features favour the young girl perception, or hypothesis, and others favour the old woman. It is possible to switch from one to the other by fixating with the eyes features favouring the young girl or the old woman. The alternations can however occur without changes of fixation. (From Norwood Russell Hanson, *Patterns of Discovery*, Cambridge University Press, 1958.)

is not yet quite resolved. It might be resolved by further data derived by instrumental signals, or by a change in the general theoretical position, for which this is a central question. In short, the change that resolves the paradox might be 'upwards' or 'downwards' – both in science and in perception.

Distortions, especially spatial distortions. Distortions can occur at the signal level by loss of calibration (as by sensory adaptation); by inappropriate calibration corrections; or by mis-match of the instrument of sense organ 'transducer' to the input (or affecting the input, as by loading with a voltmeter of low internal resistance, or detecting temperature by touch of thin metal which rapidly adopts the skin temperature). Distortions may also occur in the data and stored knowledge level, as when knowledge is transferred inappropriately to the current situation, so that signals are misread. Signal errors are to be understood through physics and physiology; data errors (which are cognitive errors) are understood by appreciating what knowledge or strategies are being brought to bear, and in what ways they are inappropriate to the current problem or situation.

Visual distortions can occur with (1) mirrors, mirages, sticks bent in water, or astigmatic lenses giving optical distortion of the input; (2) astigmatic lenses of the eyes (physiological optical distortion); (3) inappropriate neural *correction* of optical astigmatism (a calibration correction error); (4) neural signal distortion (which may be pathological or may be due to other signals interfering by crosstalk, or neural lateral inhibition, or some such); and (5) signals being misread as data (especially by 'negative transfer' of knowledge: generally from typical to atypical situations). Each has fairly obvious corresponding errors in physics.

I shall not expand on these, except the last, and that only briefly. Here again we find the distinction between processing 'upwards' and 'downwards' important. To take an example of misreading data that has received a great deal of attention, though explanations are still controversial, we may consider visual distortion illusions. Since the perceived size of things is ambiguously represented by retinal image size – size must be scaled. Visual scale is set by what I have called 'constancy scaling' (Gregory, 1962, 1968, 1970). It seems that scaling can be set upwards, from stimulus patterns normally accepted as data for distances, especially converging lines and corners normally indicating depth by perspective. When these stimulus shapes occur without their normal depth – as when perspective is presented on a picture plane – they may be accepted as though they normally represent depth – now to set the scaling inappropriately. Features represented as distant on a picture plane are perceptually expanded, for normally expansion with increased object distance is required to compensate for the shrinking of retinal images with object distance; but this is not appropriate for the plane-perspective drawings. Scale-setting is essential for maintaining perceived size independently of object distance (giving 'size constancy'), but when scale-setting by perspective features occurs other than by the retinal projection of parallel lines, etc. actually lying in the three dimensions of normal space – then the scale is set inappropriately to generate distortion 'upwards' from these stimulus patterns.

'Downwards' distortions occur when an incorrect depth hypothesis is adopted. This is clear from depth-ambiguous objects, such as wire cubes, which change shape with each depth reversal, though the retinal input and neural signals from the eye remain unchanged. Ambiguous objects and figures are extremely useful in this way, for separating 'upward' from 'downward' perceptual processes (Gregory, 1968, 1970), as the stimulus patterns remain constant.

Astronomy is rich in examples of scales set upwards from instrumental readings (with fewest assumptions by heliocentric parallax) and also downwards from considerations such as the mass–luminosity relation applied to a certain class of variable star, so that their observed periodicity can be used, together with their apparent luminosity, to infer distance. This involves a great deal of stored knowledge and associated assumptions. When these change, the Universe may be re-scaled.

There seems to be a remarkable similarity in the setting of scale for perceptual and for scientific hypotheses. Perceptual space is not, however, Euclidean, except for near objects. Consider the perception of an engine-driver: the rails appear parallel only for a few hundred metres, then they converge alarmingly. The driver can use his perceptual Euclidean near-space, in which parallel lines never meet, with confidence; but for greater distances he must reject his non-Euclidean *perceptual* space in favour of his Euclidean *conceptual* space, to drive his train further without a certainty of disaster. If, now, the driver reads Einstein in his spare time, he will adopt still another space. (Then what he relies upon professionally will become for him a parochialism, adequate for the job but not for fuller understanding.) Each view – perceptual or conceptual – will appear distorted from the spaces of his other views: relative spaces indeed! (See chapter 19.)

Paradoxes, especially spatial paradoxes. Paradoxes can be generated by conflicting inputs, or by generating hypotheses from false or inappropriate assumptions. A well known conflicting-input perceptual paradox is given by adapting one hand to hot water and the other to cold, and then placing both hands in a dish of warm water. To one hand this will be cold and to the other hot. The adaptation has produced (or rather is) mis-calibration, which gives incompatible signals to produce a paradox, since we do not allow that an object can be both hot and cold at the same time. Mis-calibrated instruments give the same conflicts and paradoxes.

In recent science there has been a relaxing of the strictures of paradox, such that what now seems paradoxical to common sense may, sometimes, be accepted as

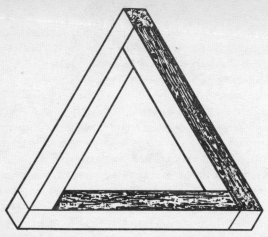

45 The impossible triangle figure. This figure was devised by the British geneticist Lionel Penrose and his son, the cosmologist, Roger Penrose. It is the origin of many later impossible figures and pictures, including the Swiss artist Maurits Escher's powerful studies of visual space. It is unfortunate that the Penroses called this an 'impossible object' rather than an 'impossible figure'; for three-dimensional objects can be made that look impossible, from restricted points of view (see fig. 46), and these introduce somewhat different considerations. (From R.L.Gregory and E.H.Gombrich (eds.), *Illusion in Nature and Art,* Gerald Duckworth and Co. Ltd, 1973.)

scientifically true. An example is light accepted as both waves and particles. Also, what appears paradoxical may be understood as non-paradoxical – as indeed for the 'impossible triangle' drawing or object (figs. 45 and 46).[49] Here we discover that conceptual understanding is sometimes powerless to correct or modify even clearly bizarre perceptions. We can, at the same time, hold incompatible perceptual and conceptual hypotheses: so we can see a paradox.

The 'impossible triangle' is clearly a cognitive illusion, for there is nothing special about this as a stimulus to disturb the physiology or signals of the visual channel. By making a model (fig. 46), it may be seen and understood that this occurs with a special view of a normal object. When viewed from the critical position, the perceptual system assumes that two ends of what appear to be sides of a triangle, are joined and lie at the same distance though they are separated in distance. Even when we know this, we still experience the visual paradox. It is very interesting that this false visual assumption – that the ends are at the same distance though they are separated – can generate a perception that is clearly extremely unlikely, and recognized as unlikely or even impossible. This shows convincingly that perceptions are built up by following rules from assumptions. Since perceptions can be extremely improbable and even impossible, it follows that perceiving is not merely a matter of accepting the most likely hypotheses. Figures and objects of this kind present useful opportunities for discovering

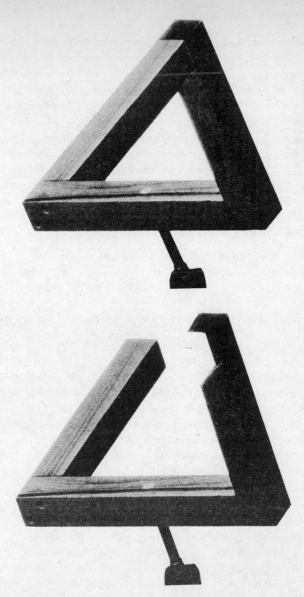

46 The impossible triangle object. This is based on the impossible triangle figure of Lionel and Roger Penrose (fig. 45). It only looks impossible when features in fact lying at different distances line up, when they are seen as lying at the same distance. This false assumption generates a paradoxical perception. It is important to note that this paradoxical perception continues, although we know conceptually the answer which resolves the paradox – and what is seen from a different point of view. (From R.L. Gregory and E.H. Gombrich (eds.), *Illusion in Nature and Art*, Gerald Duckworth and Co. Ltd, 1973.)

perceptual assumptions and rules by which perceptual hypotheses, which may conflict with high-level knowledge, are generated 'upwards' from assumptions by rule-following.

Our ability to generate and accept extremely unlikely perceptions must be important for survival, for occasionally highly unlikely events and situations do occur and need to be appreciated. Indeed, perceptual learning would be impossible if only the probable were accepted. At the same time, though, there is marked probability-biasing in favour of the likely against the unlikely – as in the difficulty, indeed the impossibility, of seeing a hollow mask as hollow, without full stereoscopic vision (fig. 47). So there, again, are the two opposed principles – processing upwards and downwards; the first generating hypotheses, which may be highly unlikely and even clearly impossible, the second offering checks 'downwards' from stored knowledge, and filling gaps which may be fictional and false.

Fictions, sometimes to fit and sometimes to depart from fact. We can see perceptual fictions in phenomena such as illusory contours (fig. 42, p.401). These were described by Schumann (1912) and have recently become well known with the

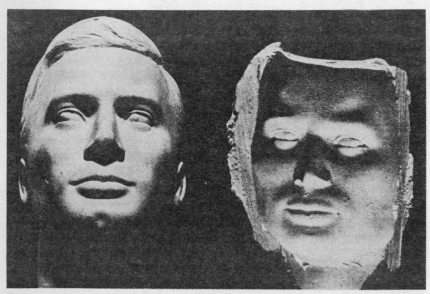

47 A hollow face. The right-hand face is the hollow mould of the left-hand face, and yet it looks not hollow, as it is, but with its nose sticking out as with a normal face. This occurs with typical lighting, and even with considerable stereoscopic information. So sensory data are rejected to favour the likely normal face against the unlikely hollow face. (From R.L. Gregory and E.H. Gombrich (eds.), *Illusion in Nature and Art*, Gerald Duckworth and Co. Ltd, 1973.)

beautiful examples due to Kanizsa (1955).[50] If we are right in thinking that they are postulated masking objects to 'explain' the surprising gaps in these figures (Gregory, 1972) we at once assign to them a cognitive status. Related examples are shadows of writing: letters which would cast such shadows are seen, though there is no stimulus pattern of letters. The letters are evidently fictions to 'explain' what are accepted as shadows by postulating letter-shaped objects (Gombrich, 1960). It is reasonable to suppose that a very great deal of perception is in this sense fictional: generally useful but occasionally clearly wrong, when it can be an extremely powerful deception. No doubt this holds also for science.

It is particularly interesting that the absence of signals can be acceptable as data. (This was so for the discovery of the Van Allen radiation belts, when the space probes' sensors were overloaded so that, surprisingly, they failed to provide any signals.) The gaps of the illusory contour figures provide data for an eclipsing (though non-existent) nearer object which is perceptually postulated (Gregory, 1972). These examples indicate ways by which the hypotheses of science and perception become richer than signalled data. They show also that we cannot equate neural signals with experience; for *no* signals can provide experience and perception, when their absence is evidence of objects.

Causes and inferences as links between hypotheses and perception and the world. Hypotheses of science and perceptions are, I believe, linked to reality very indirectly. What kind of relations do they have? There are two important questions here: (1) are they causally linked? and (2) are they linked by inference? We should allow the possibilities that either or both may be true, or false. Let us consider these.

That hypotheses are causally linked to reality would be held if there are what we have called signal links between reality and hypotheses. We accept this for transducers and signal channels, but what of the signals when read as data? There can be gaps in signals. These are often filled by interpolation processes (pp. 399–401), so here science and perception clearly maintain hypotheses (and maintain continuous control from hypotheses) through signal gaps. Nevertheless, we do not want to say that this gap-filling requires processes outside causal explanation. Part of the aim of theories of perception and accounts of science is indeed to explain gap-filling and these explanations, if they are to be like most explanations, should preferably be causal. We may, however, expect to find some deep conceptual difficulties over data, though not signals, as causes.[51]

What of data distortions? Are they breaks in causal sequences? One can see cases of data errors (rather than signal errors) in which there clearly are no causal breaks in the signals, and it is possible to understand why the signals are read as misleading data. This occurs when signals are read normally as though occurring in a typical situation, when in fact the situation is atypical. Particular perceptual examples are, on this account, many of the distortion illusions (as we have suggested: p.406). Here the scaling is supposed to be set quite normally by signalled features, but these features do not have their usual significance in these figures. For example, converging lines on a picture plane are read as perspective, as though the convergence were produced as in normal three-dimensional space when parallel lines lying in depth are imaged on the retina. Picture perspective misleads not by distorting neural signals, but by providing signals that are read as depth data although the picture is flat. There is no break in the usual causation of perception; but there is marked distortion, and the distortion is of data, not of signals. To understand why this happens, we always need to know what knowledge has been transferred to the current situation (in this case that convergence of lines is associated with depth), and why in this situation it is inappropriate. Superb demonstrations, some dynamic, such as the rotating trapezoid (a perspective-shaped window), were invented by the Transactionalist psychologist who started as a painter and worked at Princeton, Adelbert Ames Jr. Ames wrote very little, except for the jottings of *The Morning Notes* (1960); but designs for the demonstrations are fully described (W. H. Ittelson, 1952).[52]

What of paradoxical hypotheses? Since we do not accept that reality can be paradoxical, we cannot accept that paradoxical hypotheses or perceptions can match or represent reality. There is therefore some kind of gap – but is this a causal gap? We can think of signal distortion paradoxes where there clearly is no causal gap, for example the hands sensing hot and cold for the same bowl of water, when one hand has been adapted to hot and the other to cold. Here we have incompatible signals owing to the adaptation of one or more channels, combining to form a paradox – but without any causal break.

Figures such as the Penrose 'impossible triangle' drawing (fig. 45) or our impossible model (fig. 46) are very different from signal distortions. We attribute these perceptual paradoxes not to signal errors, but to false assumptions. So these are top-down errors. The question is: do top-down injections of data or assumptions produce causal gaps in signal processing? They certainly introduce considerations that may be very far removed from the current situation, and for quite displaced assumptions it may be very difficult indeed to see how they have come into play. If we give them a 'mental' status, then we may be tempted to say that they are caused mentally; it seems better, however, to say that the situation is something like a filing-card index, in that references are sought, by criteria of relevance and so on, as formulated within a physical search system. This may go wrong by misreading of signals or by indexing errors, and it may also produce misleading data because what is generally relevant is not appropriate in the particular situation. These mistakes are very different, but they can be explained in terms of the logic of the procedures plus the mechanical steps used to carry out the procedures. The first may be inappropriate: the second can break down mechanically.

To come to the second question, if probabilities can affect reasoning and acceptance or rejection of hypotheses, then just how can we hold that signals though not data are causal? To maintain a causal account we must allow that assessed probabilities have causal effects. We may, however, translate this into: signals have causal effects, according to the significance in the situation of the data that they convey.

If the distinction between signals and data is seen as a dualism, at least this dualism does not apply uniquely to Mind and brain. As argued earlier, it applies with equal force to instruments supplying signals and data for Science. So the activity of science becomes a test-tube – indeed a laboratory – for appreciating the Mind–brain problem.

We may now consider some *differences* between scientific and perceptual hypotheses.

Differences between scientific and perceptual hypotheses – when they are appropriate or inappropriate

1 Perceptions are from one vantage-point, and run in real time: science is not based on a particular viewpoint

Perceptions differ from conceptions by being related to events in real time from a local region of space, while conceptions have no locale and are essentially timeless. They not only lack any locale in the three-dimensional space of the physical world; but they may express variables and relations in all manner of conceptual spaces, which may not be claimed to exist though they are useful fictions for descriptive purposes.

So perception is far more limited in range and application than conception. The basis of Empiricism is that all conception depends upon perception. But conception can break away from perception, to create new worlds – though perhaps always using as building-blocks the objects of perception.

2 Perceptions are of instances: science is of generalizations

We perceive individual objects, but we can conceive, also, generalizations and abstractions. Thus we can see a triangle, but we can conceive general properties of all triangles – triangularity. Is this difference absolute, or rather, a matter of degree? A chess player may claim that he sees the situation rather than the pieces; and when reading, one is more aware of the meaning of the words than their form – and this can hold when the words express generalizations. This is a tricky issue requiring investigation. I incline to think that there is not a sharp distinction here between perceiving and conceiving (except, perhaps, for awareness?).

3 Perceptions are limited to 'concrete objects': science also has 'abstract objects'

This distinction, due to the logician Frege, raises a tricky issue closely bound up with the deepest problems of perception and epistemology. The problem is that the distinction is not clear-cut (M. Dummett, 1973).

Concrete objects are what are, or are believed to be, sensed. They may be simple or complex. Thus a magnetic field may be simple and a table is complex. It is not, however, at all clear that sensing is ever free from inference; for example, perceiving a table is far more than sensing various parts, and sensing a magnetic field requires all manner of inferences about the transducer and how it is placed and used. The contribution of inferences and assumptions to sensing even simple objects makes the distinction between concrete and abstract objects difficult and perhaps impossible to make clearly, for abstract objects – such as numbers and centres of gravity – are, or at least may be known via sense experience, and perhaps nothing is sensed 'directly'. If nothing is sensed or perceived directly – if all perception and all scientific observation, however instrumented, involve inference – then it seems that there are no purely concrete objects, as we know them. This is indeed a major conclusion from the thesis that perceptions are hypotheses. This conclusion applies equally to perception and to science.

4 Perceptions are not explanations, but conceptions can be explanatory

Scientific hypotheses are closely linked to explanation (it is an *explanation* that the tides are caused by the pull of the moon). Perceptions certainly have far less explanatory power, but perhaps they do have some. One understands social situations, or mechanisms, through looking: is this understanding part of the perception? I incline to think that it is. This difference is rather of degree than of kind.

5 Perception includes awareness: the Physical Sciences exclude awareness

This is by far the most striking difference between hypotheses of science and perceptions: sensations are involved in perception (though not all perceptions), but awareness, or consciousness, has no place in the hypotheses of physics. The scientist may be aware that he is working on a hypothesis, but the hypothesis is not itself aware – or so we assume! On the other hand, I do want to say that awareness is an integral part of many perceptual hypotheses, so here we have a clear distinction between scientific hypotheses and some perceptions.

Returning to our distinctions between signals and data: a traditional view was that perceptions are made up of sensations, but this I have rejected. It must, however, be confessed that the role, if any, of awareness or consciousness in perception is totally mysterious. Much of human behaviour controlled by perception can occur without awareness: consciousness is seldom if ever necessary. Perhaps consciousness is particularly associated with mis-match between expectation and signalled events, but if this is so, its purpose remains obscure – because it is not clearly causal.

Popper and Eccles (1977) argue from phenomena of visual ambiguity – especially maintaining or changing visual orientations by will – that Mind, as associated with consciousness, has some control over brain. This argument was also suggested by William James (1890), but – as William James points out – it could be other brain processes affecting reversal rates, or whatever. There seems no good reason to suppose that consciousness in this situation is causal (see p. 470).[53]

Is consciousness so difficult to understand and describe, just because it is not part of scientific hypotheses about the physical world? Is it these, only, that provide conceptual understanding? If so, we must be careful with our suggestion that perceptions are hypotheses; for if all we can know are hypotheses of physics, then perceptions are found to look like hypotheses of physics. This is an impasse for which I have no ready answer. I can only hope that further consideration will unravel or cut these Gordian knots of knowing.

We may conclude that, all in all, there are marked similarities and important identities between hypotheses of science and perceptions. It is these that justify calling perceptions 'hypotheses'. The differences are, however, extremely interesting, and I fear that I have not done them justice. This is not through any desire to minimize them, but rather that I do not know what to add. Possibly this is because we think in terms of the hypotheses of science, so that when something crops up that departs from them drastically, we are lost. We are lost for consciousness. It is very curious that we can think conceptually and see perceptually with such effect 'outwards' but not 'inwards'. It may be that developments in Artificial Intelligence will provide concepts by which we shall see ourselves.[54]

13 Meaning to say

'Sir, what is poetry?'
*'Why, sir, it is much easier to say what it is not. We all know what light is; but it
is not easy to tell what it is'.*

Samuel Johnson (1709–84) [1]

The most striking – and a unique – feature of Mind is the acceptance and use of
things as symbols standing for other things. Symbols may stand for, refer to, or
mean other things which may or may not lie within the world of physics. In either
case symbolic power, which is immense, is outside the causal net of the physical
world, and is inside Mind. This is so even if Mind *is* brain processes. And it is so
even though computers (clearly obeying physical laws) work by handling symbols
according to symbolic rules. In this sense we find Mind in computing machines.
Here we find Mind in the power of symbols, though we may not find consciousness.

Ethology stresses the importance of cries and songs, and all manner of gestures
in animal behaviour for conveying states and needs – and even intentions – in
animals who may or may not be conscious or aware as *we* appreciate sensations,
and what we take to be reality as perceived by our senses. For us language,
especially poetry, and music evoke conscious states without which they would
be nothing. But perhaps most symbols affect our behaviour without awareness:
hence the subtle powers of persuasion of leaders, teachers, advertisers and indeed
seducers. Recent studies of 'body language' and 'non-verbal communication'
spell this out in experimental detail. The germinal work was, however, Charles
Darwin's remarkable book *Expression of the Emotions in Man and Animals*
(1872), which contains a wealth of insights into the symbolic power of gestures,
cries, scents and facial expressions. This book remains unrivalled.

Darwin explored the notion that symbolic gestures (and strictly speaking all
gestures are symbolic) originated from functional features of behaviour. We might

say that this move from function to symbol is the move from physics to Mind – which occurred in Evolution with the development of intelligence. Intelligence requires symbols: it is through symbols that behaviour is different from all other events in the Universe. Symbols allow events to represent other events, possibilities and abstractions, which do not exist as objects of sense exist, though some may be hidden deep structures of reality.

Symbols for sexual recognition, for warning of danger, for signalling the location of nectar, for threat, enticement, territorial ownership and the fun of play, have existed for hundreds of millions of years before man – and yet we alone have structured grammatical language with which questions and abstractions can be phrased. Perhaps this is the single most amazing thing about human Mind.

The origin of language

How spoken language developed in human societies is unknown, but we may guess that it derived from gesture and onomatopoeia. What is so special about human language is its grammatical structure, which, if Noam Chomsky is correct, is in all Natural languages closely related to what he calls the Deep Structure of language, which is inherited not learned, and which, we may suppose, is derived from and indeed may in large part be perceptual classification. A biological problem for Chomsky's theory is the rapidity, on a biological time-scale, of the development of grammatical language; possibly this has been so rapid because there has been a take-over of perceptual Deep Structure having biologically ancient – indeed pre-human – origins (Gregory, 1971).

Language in animals

Until a few years ago the gap between human and animal language seemed unbridgeably immense but recent experiments may suggest that chimpanzees have considerably more language ability than they normally show, or than was guessed that they could have.[2] The first chimpanzee to impress in this way was Washoe. She was trained by B.T. and R.A.Gardner (1969). The Gardners, realizing the chimpanzee's physical inability to vocalize, used American sign language for the deaf (Ameslan), which consists of 55 basic elements or gestures ('cheremes'). Washoe learned to combine these, so that she developed a repertoire of 132 signs. She could produce sentences of up to five signs, including nouns, adjectives, verbs, imperatives and negatives. She often requested food or drink, or to be taken for a walk, but did not seem to ask questions. It is debatable whether she employed grammatical structure.

Using a different language, D.A.Premack (1970, 1976) taught another chimpanzee, Sarah, to build up sentences with plastic chips on a magnetic board. She can study these for as long as she likes, so short-term memory limitations are not a handicap with this language as they are for the sign language. Sarah can form sentences such as 'All the crackers are round' or 'None of the crackers are round.' It has been pointed out by Fodor *et al.* (1974) that although Sarah is impressive, there is no generation of new kinds of sentence forms. The overall conclusion so far is that chimpanzees show generalization, but not human language or intellectual ability, especially for asking questions.[3]

It has been found by Norman Geschwind (1964) that there are anatomical differences in the neonate brains of man and chimpanzee which show that we have far greater richness of association pathways between the senses, and to the limbic system which is associated with the emotions, than the apes. Geschwind suggests that the uniquely human richness of associations is necessary for accepting objects as symbols, and developing and using human-like language.

It is an unsolved puzzle how far it is that we have such powerful language because our brains are social, or whether and how far our mental superiority is due to language as a tool. Communication allows sharing and so individual gaining of knowledge, and it also seems to provide symbolic counters and tallies by which our kind of thinking is achieved. Just how important 'internal language' is, is very difficult to say, but probably it does structure (literally, 'articulate') thought.

Spoken language is almost certainly two to four million years older than written language, which as far as we know is not more than ten thousand years old. The first written languages were pictures, which gradually came to take on abstract significance by becoming conventional signs – which had to be learned, as it was no longer self-evident what they meant (see pp. 48–57).

The question of how language has meaning is a philosophical one, or at least it generates philosophical questions – especially when we ask what the difference is between true and false statements, and what kinds of truth statements may have. A deep question is how language is related to the world (see chapter 8).[4]

Theories of meaning

There have been many theories of meaning since the Greeks; but because they were only familiar with a single language, it is hardly surprising that they thought of the meaning of words as being attributes of objects. For example, they would think of the sun as having the attribute of being hot and also as having the attribute of its name – 'sun'. The word 'sun' (or rather its equivalent, $\mathring{\eta}\lambda\iota os$, in Greek) was part of the sun just as its heat and brilliance belong to, and are, the sun. This is 'Nominalism'.

Nominalism

This theory of meaning parallels the Direct Realist theories of perception, which held that perceptions are parts of perceived objects. Nominalism was the generally accepted theory of meaning until challenged in the twelfth century, especially by Pierre Abélard (1079–1142), who argued that names stood for concepts in the Mind.

The Nominalist notion that names are properties of objects may be in part a curious anthropomorphism from our personal names. This is picked up by Lewis Carroll in *Alice Through the Looking-Glass* (1872), where Alice says, touching a tree: 'What *does* it call itself, I wonder? I do believe it's got no name – why, to be sure it hasn't.'

The crucial objection to Nominalism is that names can refer to things which do not, or indeed could not, exist, such as glass mountains, or a sea of ink. More important are such examples as 'infinity' or 'zero'. These cannot be said to exist as physical objects exist, yet they are highly important. When they are supposed to exist, absurdities are generated, as also pointed out in *Alice:*

'Who did you pass on the road?' the King went on, holding out his hand to the Messenger for some hay.

'Nobody,' said the messenger.

'Quite right,' said the King: 'this young lady saw him too. So of course Nobody walks slower than you.

'I do my best,' the messenger said in a sullen tone. 'I'm sure nobody walks much faster than I do!'

'He can't do that,' said the King, 'Or else he'd have been here first. . . .'

Such confusions (here amusing but sometimes seriously misleading) are generated by Nominalism's need for objects to give meaning to names and statements. For words such as 'nobody', 'zero', 'infinity' or 'perfection' – which clearly have meaning – existing objects are postulated to give them meaning. So Nominalism plants the Universe with all manner of objects unknown to sense or reason. This is indeed the basis of Plato's Ideal Forms, including the notion that there are perfect geometrical forms, of which we see in the physical world only imperfect examples, for which geometry does not quite apply. To see how easy it is for us to slip into Nominalism, consider: 'Unicorns do not exist because they have better things to do!' Russell's Theory of Descriptions has done much to free philosophy of the ills of Nominalism.

Nominalism (and the incipient Nominalist tendencies are in all of us) not only generates objects unknown to sense or reason (which is the sin of metaphysics) but it also distorts how we think of the world as parcelled up into separate objects. What is an object? Is a book an object – or is it a collection of pages? Is a tree an object? Sometimes we think of each leaf as an object and sometimes we think of the entire tree as an object. The same holds for books and pages. We find this situation for objects that are sometimes used as a whole (such as carrying a book) and at other times used in parts (turning each page, for reading). Names are generally attached according to what we need to identify and name for uses.

We categorize the world into separate objects in perception, and we describe the world as being made up of separate objects by the words of language. It is an interesting question how far perceptual and verbal classifications into objects are the same. They are certainly similar, but there seem to be hardly enough names for the objects into which the world is divided perceptually. During perceptual learning – such as when learning to see biological cells with a microscope – new objects appear from initially random or meaningless patterns. When given names, such as 'nucleus' and 'mitochondrion', the student sees these patterns as objects. What is seen and accepted as objects also depends upon whether they are regarded as functional units. A hand, or an arm, or the pages of a book are functional units, though they are complex structures. In microscopy the criteria for what is a functional unit may be highly theory-laden, and so may change as theoretical descriptions change. It may then look different. In an extreme form, this gives rise to a theory of meaning and truth: Pragmatism.

Pragmatism

Considerations such as these do lead to thinking about meaning in terms of how symbols are used. This is very different from the Nominalist notion that the meaning of a name is a property of the object it denotes. With this change of what we mean by 'meaning', there is a related change in what we mean by 'true' and 'false'. For Nominalists, a false statement which is nevertheless meaningful must have an object corresponding to the false statement. So for them the Universe is filled with non-objects, giving meaning to false statements. This generation of ghostly entities is removed at a stroke by saying that meaning, and perception, are within the observer.

Is, truth, then, some kind of correspondence between states of Mind and the physical world? Then false statements and perceptual illusions would be described as mental departures from the state of affairs described or perceived. There is, however, a difficulty, or at least an inadequacy in this account. The snag is raised by the question: how could we ever know that a statement is true or a perception veridical? – for we cannot stand outside and see the correspondence, or lack of correspondence, with reality or whatever it is that is being described or perceived. This is the difficulty with the correspondence theory. If it is true, how can it be applied? It is such considerations that lead to considering meaning and truth of symbols in terms of their use.

This was first expounded clearly by the highly original (and personally eccentric) American philosopher, Charles Sanders Peirce (1839–1914).[5] The basic idea first appeared in an article by Peirce (published in the January issue of *Popular*

Science Monthly, 1878), entitled 'How to Make Our Ideas Clear'. The basic idea is that beliefs are rules for action. Pragmatism was taken up as a psychological theory by William James (1842–1910), especially in his *Pragmatism* (1907). James supported Peirce financially, and defended him in the intellectual world, though not to Peirce's liking. He so disagreed with James's version of Pragmatism that he renamed his own version with the clumsy word 'Pragmaticism'.

Pragmatism may have developed from the Positivism of the French philosopher and sociologist Auguste Comte. Comte (see pp. 250, 586) suggested that knowledge passes through the stage of theology, to metaphysics, into a positive stage where beliefs are tested by experience. Pragmatism (and later Logical Positivism) suggested that truth must be established by observation or experiment; and that meaning depends on the possibility of verification and is set by the manner of verification. This was not quite Peirce's position, however. He did not believe that individual statements or propositions could, or need be, verified to have meaning. He concluded that truth is what is inevitably found in the long run by scientific enquiry. He rejected the notion that every statement must be individually verified to be meaningful. So he was not a 'Logical Atomist'.

William James gives an example of the use of the pragmatic method from an episode at a camping party in the mountains, where his friends discussed a squirrel moving rapidly round the trunk of a tree, with a man running around the opposite side of the tree, following its every movement so that it is always the other side of the tree, from the observer. The question was: 'Does the man go round the squirrel or not?'[6] James settled the dispute between his friends (half of whom thought the man went round the squirrel while the other half thought he did not, though they all agreed that the man went round the tree) by saying:

Which party is right depends on what you *practically* mean by 'going round' the squirrel. If you mean passing from the north of him to the east, then to the south, then to the west, and then to the north of him again, obviously the man does go round him . . . but if on the contrary you mean being first in front of him, then on the right of him, then behind him, then on his left, and finally in front again, it is quite as obvious that the man fails to go round him. . . . You are both right and both wrong according as you conceive the verb 'to go round' in one practical fashion or the other.

James calls this an example of the pragmatic method for settling metaphysical disputes, claiming that with it, 'It is astonishing how many philosophical disputes collapse into insignificance the moment you subject them to this simple test of tracing a concrete consequence. There can *be* no difference in abstract truth that doesn't express itself in a difference in concrete fact.'

But do we normally discover what we mean by considering practical outcomes? Or is this a recommendation by these philosophers? Outcomes are often distant, in science and in life, from what we mean, or at least from what we may think we mean. Verification is far from simple – and it itself requires assumptions, not all of which can be verified. This move of Empiricism towards regarding verification as necessary for meaning and truth is, however, profoundly important. It is, at the least, useful for disposing of crude metaphysics. Where it misleads is in supposing (as William James supposes with his form of Pragmatism) that meaning is given directly by use. Concepts are not individually subjected to verification, and may be useful and meaningful even though we have only hazy notions of how they might be verified. There can be enormous gaps between possible verifications – and yet meaning continues. This is like perception, in that we continue to perceive although objects are only partially and intermittently available to the senses. So, for both perception and meaning of language, we have to give far more weight to processes – thinking and perceiving processes – continuing though signals for verification are not continuously available. This puts the emphasis for perception, thinking, and language on continuing internal processes. Perception, thinking and language do, however, fall into error when the gaps between sensory signals giving opportunities for verification are too great. Then fantasy, hallucination, illusion and metaphysics take over.

The problem is to discover just what these internal processes of language and of perception may be. This is an endeavour shared by philosophy, cognitive psychology and linguistics.

Logical Positivism

An extreme form of 'Operationalism' – Logical Positivism – gives as a theory of meaning that the meaning of a statement (the proposition conveyed) is the manner of its verification. This approach is highly compatible with Behaviourism: it may also suggest that some kind of 'objective knowledge' is possible if, indeed, meaning can be described in terms of specifiable operations on the world of objects without reference to observers.

This extreme Operationalism was developed in the Vienna Circle school in the 1920s. The most distinguished champion is A.J. (Sir Alfred) Ayer. In his *Language, Truth and Logic* (1936), Ayer set out as a young man to demolish metaphysics – but without demolishing too much. (In this he is careful to avoid Bradley's stricture in *Appearance and Reality* (1893)[7] – that a man who is ready to prove that metaphysics is impossible is a brother metaphysician with a rival theory of his own.) Ayer's attack is at the very root of any statements. He charges metaphysicians as uttering merely meaningless sentences, sentences that are not propositions because they cannot be verified. What, though, of statements which seem meaningful (such as statements about the past or distant future) but which cannot be verified? There are deep problems here.

The basic position of Logical Positivism is that for sentences to have meaning it must be possible, at least in principle, to demonstrate truth or falsity – for each individual proposition. This became a theory of meaning: that the meaning of a proposition is bound up with the manner of its verification. But what is verification?

Ayer puts sense experience as primary. He argues that there can be no meaning without reference to actual or possible sense experience. Here, Ayer distinguishes himself from Kant's rejection of 'transcendental metaphysics'. He describes Kant's rejection as based on the alleged fact that the human understanding is so constituted that it loses itself in contradictions when venturing beyond the limits of possible experience. Kant's argument is weaker and less radical than Ayer's: one might indeed suppose that human beings might in time learn to be effective 'transcendental metaphysicians' on Kant's argument, but never on Ayer's. Kant is merely saying that non-sensory knowledge is too difficult for us: Ayer is saying that it is in principle impossible, for what he calls logical reasons.

Ayer has to be careful that he is not 'overstepping the barrier' into metaphysics in his arguments denying the possibility of metaphysics. This is a criticism that has often been levelled against Logical Positivism. Ayer's defence should be presented in full: [8]

It cannot here be said that the author [Ayer] is himself overstepping the barrier he maintains is impassible. For the fruitlessness of attempting to transcend the limits of possible sense experience will be deduced, not from psychological hypothesis concerning the actual constitution of the human mind, but from the rule which determines the literal significance of language.

This passage continues: 'Our charge against the metaphysician is not that he attempts to employ the understanding in a field where it cannot profitably venture, but that he produces sentences which fail to conform to the conditions under which alone a sentence can be literally significant.'

Ayer goes on to describe what are his 'conditions under which alone a sentence can be literally significant'. He then objects to metaphysics on the grounds that metaphysical sentences are not meaningful (are not propositions) because they do not have even in-principle verifiability: [9]

We say that a sentence is factually significant to any given person, if, and only if, he knows how to verify the proposition which it purports to express – that is, if he knows what observations would lead him, under certain conditions, to accept the proposition as being true, or reject it as being false. . . .

If, on the other hand, the putative proposition is of such a character that the assumption of its truth, or falsehood, is consistent with any assumption whatsoever concerning the nature of his future experience, then, as far as he is concerned, it is, if not a tautology, a mere pseudo-proposition. The sentence expressing it may be emotionally significant to him: but it is not literally significant.

Ayer continues by saying that the same applies for questions as for alleged statements: questions are only regarded as meaningful if there is some way envisaged for answering them by reference to sense experience. If there is no link seen to experience, then it is not a 'genuine question'.

It will be clear from all this that a lot of weight is attached to sense experience. This is exactly in the tradition of classical Empiricism, and Realist theories of perception. We may say, though, that the Logical Positivists make 'sensory experience' even more central and important than did Locke, Hume and the rest of the earlier Empiricists, for they at least allowed meaning to non-sensory statements. Ayer puts tremendous weight upon sensory experience, perception; but will it bear the weight? I consider that it does not. Meanwhile, however, we should look more closely at Ayer's notion of the meaningfulness of allegedly 'empirical' statements as depending upon verification, or a possibility of verification by perception.

Ayer distinguishes between 'practical verifiability' and 'verifiability in principle'. His example of the latter is the now historically rather startling 'That there are mountains on the farther side of the moon.' Lack of verification of this statement, necessary to allow this as a proposition, is that: [10] 'No rocket has yet been invented which would enable me to go and look at the farther side of the moon, so I am unable to decide the matter by actual observation. But I do know what observations would decide it. . . .'

The first 'observations' of the farther side of the moon (which showed some, though surprisingly few, mountains) were not made by a man, but by an Orbiter automatic space probe. This indeed transmitted pictures to earth which men interpreted as mountains on the back of the moon, but we may imagine that the probe was entirely automatic, to signal back not pictures but rather the machine's own assessment. A machine might receive signals from T V cameras, process them, and report back to earth in statements, perhaps on teletype: 'There are mountains.' Would Ayer accept such non-human 'perception' as adequate for verification – to give meaning to such statements? This is surely an important question for the philosophy of science.

There have been such 'observations'. The Van Allen belts of intense radiation around the earth were discovered by an automatic space probe. This case is even more interesting, for there was total absence of signals – which inspired the hypothesis that there must be intense radiation blocking the instruments by overloading them. So absence of signals provided data (see pp. 401, 411).

This is like inferring that a person is 'shocked into silence'. Silence, or non-signals, may be accepted as data or as a message. The question here, however, is: would Ayer extend 'sensory experience' to automatic instruments? Or is sensory experience quite special? In Galileo's time there was a resistance to accepting telescopic observations for revealing the Universe, or for suggesting or testing hypotheses. Does this historical resistance to instrumental data extend, philosophically, to machines displaying powers of artificial intelligence and perception?

We should look more closely at what Ayer says about verification by human sensory experience. He distinguishes between 'strong' and 'weak' verification: [11]

A proposition is said to be verifiable, in the strong sense of the term, if and only if its truth could be conclusively established in experience, but it is verifiable, in the weak sense, if it is possible for experience to render it probable.

He then gives reasons for deciding that it is only the weak sense which is needed to give meaning to statements of alleged fact. Ayer, however, revised the original 1936 edition, in 1946, and added an important Introduction in which some of the original position is considerably modified. There are now three variants of the definition of weak verification. In the Introduction to the second edition, Ayer points out that his original distinction between 'weak' and 'strong' verification is suspect, because 'All empirical propositions are hypotheses which are continually subject to the test of further experience; and from this it would follow not merely that the truth of any such proposition never was conclusively established but that it never could be. . . .'

Interestingly enough, though, Ayer does not altogether give up his 'strong' verification. In the 1946 Introduction, he maintains that there is a class of indubitably true empirical propositions, and that these can give 'strong' verification: [12]

I have come to think that there is a class of empirical propositions of which it is permissible to say that they can be verified conclusively. It is characteristic of these propositions, which I have elsewhere called 'basic propositions' that they refer solely to the content of a single experience, and what may be said to verify them conclusively is the occurrence of the experience to which they uniquely refer.

He goes on to say that 'it is impossible to be mistaken about them except in the verbal sense'. So the same great weight is placed on perception.

Strong (direct) verification is given in 1946 in these terms: 'A statement is directly verifiable if it is either itself an observation-statement, or is such that in conjunction with one or more observation-statements it entails at least one observation-statement which is not deducible from these other premises alone.' [13]

Weak (indirect) verification was originally (in 1936) accepted as sufficient to allow meaningful statements: 'it is possible for experience to render it probable'.

In the revised form (1946), we have: 'If some possible sense experience would be relevant to its truth or falsehood.' This is amended and developed in the text of 1946: [14]

Let us call a proposition which records an actual or possible observation an experiential proposition. Then we may say that it is the mark of a genuine factual proposition, not that it should be equivalent to an experiential proposition, or any finite number of experiential propositions, but simply that some experiential propositions can be deduced from it in conjunction with certain other premises without being deducible from those other premises alone.

And this is amended, in the Introduction to the 1946 edition: [15]

A statement is directly verifiable if it is either itself an observation-statement, or is such that in conjunction with one or more observation-statements it entails at least one observation-statement which is not deducible from these other premises alone; and secondly, that these other premises do not include any statement that is not either analytic, or directly verifiable, or capable of being independently established as indirectly verifiable.

Ayer explains that he has put in 'other premises', which may be analytic (non-empirical, or non-observational), to allow for scientific theories in which a 'dictionary' may be provided for transforming non-verifiable statements into verifiable statements by steps of reasoning. He emphasizes the importance of this to distinguish between 'metaphysical' statements, which not only 'do not describe anything that is capable, even in principle, of being observed, but also that no dictionary is provided by means of which they can be transformed into statements that are directly or indirectly verifiable'.

A move that is sometimes made is from verification as necessary for meaning to a radical Behaviourism, linked to Pragmatism, in which the manner of verification gives and may define the particular meaning of propositions. This is an attempt to make meaning 'objective' – by removing it from the understanding and cognitive processes of intelligent beings such as ourselves. However this may be, it may well be said that the call for verification has had an excellent effect on philosophical discussion, by demanding clear examples and by linking philosophy more closely to the practice of science. I strongly suspect, however, that it does put too much weight on perception. It requires a theory of perception that is not tenable: a Direct Realism. If perception could be supposed to be direct knowledge of external objects and events, then it might take the weight demanded; however, as we have found, Realism has objections far too serious for it to be tenable.

It may be noted that Pragmatism, Logical Positivism, and also Popper's position all deny active Mind and active observers, in favour of operational criteria for meaning which they hope may apply and can be used without reference to stored knowledge. This has a parallel in Behaviourism. They are all attempts to avoid Mind. Very different is Wittgenstein's position – or rather two positions which he held at different times – to which we now turn.

Wittgenstein's picture theory of meaning

In the *Tractatus*, Wittgenstein starts by considering the nature of pictures, which to him include any kind of representation, including statues, maps and photographs. It seems that he was led to the notion of pictures representing with more or less accuracy (and so being true or false) from hearing that French law courts use models of cars and buses to represent road accidents. The models stand for the cars or lorries in an accident, their arrangement standing for the situation at that time. This led him to think that words, and their arrangements in sentences, stand for facts and situations of the world. Words, like the arrangements of the toys, may more or less accurately correspond to situations. For Wittgenstein there must be identities of form between the picture and what it represents for them to represent. This he calls the 'pictorial form'; however, pictures can represent not only facts but also possibilities – as for plans for work to be carried out.

The crucial issue is how pictures can connect with situations. Wittgenstein imagines pictures put up to 'feel' reality, and to match it more or less exactly. But there are many characteristics of pictures (which are themselves facts of the world) that are irrelevant. It may be, for example, their colour or their three-dimensional form which is relevant or irrelevant: it is for the observer to select what is important for reading the picture. However, there must be, in common with the picture and what it represents, common logical form.

How do we know whether a picture is true or false? This, to Wittgenstein, cannot be known by any examination of the picture alone. It is a matter of whether the picture corresponds to reality. Whether it does so or not is generally an empirical question, requiring observation or tests of some kind.

Pictures are facts of the world, as are other objects; but they are special because they represent other objects or situations. The picture may be like objects that it represents: this similarity, or limited identity, he calls its 'pictorial form'. This is identity of shape or colour. Wittgenstein does not, however, attribute the power of pictures to their spatial or colour similarities with what they represent: he speaks also of what has been translated as their 'representational form'. This allows pictures to represent, even though they are not replicas, or copies, of what they represent. For example, they may be and usually are far too small, and two-dimensional, even though they represent or convey full-sized objects in three dimensions.

What is the minimum required for a picture to represent? What is necessary for the observer to read other objects or situations from pictures? Pictures can to some degree be abstract, and they can be ambiguous, as we have seen (pp. 383–415). Wittgenstein holds that the structure of pictures has, or needs to have, logical form applying to what is represented. This is expressed in aphorism 2.18 of the *Tractatus:* 'What every picture, of whatever form, must have in common with reality in order to be able to represent at all – rightly or falsely – is logical form, that is, the form of reality.' The picture is 'true' insofar as its logical form (which 'represents a possible state of affairs in logical space') agrees with reality. By simply looking at the picture we cannot tell how far it is true. 'In order to discover whether the picture is true or false we must compare it with reality.'

Wittgenstein departs from what Kant would say at this point, by declaring (2.225): 'There is no picture which is *a priori* true.' (Or would Kant say only that what is possible may be known before experience?) At this point in Wittgenstein's discussion we find a perhaps rather surprising agreement with Popper's Principle of Transference: 'What is true in logic is true in psychology.' (By 'psychology' Popper means not the study or science of Mind, but Mind itself.) Popper suggests (see pp. 251 *et seq.*) that we cannot think of anything outside logic. Wittgenstein seems to say much the same in the *Tractatus* (3.03): 'We cannot think of anything unlogical, for otherwise we should have to think unlogically.'

This is, however, perhaps a little different. Wittgenstein is not quite equating psychology (Mind) with reality but is rather setting limits to what we can picture or say (3.031): 'we could not say of an "unlogical" world how it would look.' And (3.032): 'To present in language anything which "contradicts logic" is as impossible as in geometry to present by its co-ordinates a figure which contradicts the laws of space. . . .' Unfortunately, Wittgenstein was unaware of paradoxical pictures such as the Penrose 'impossible triangle' (fig. 45, p. 408), which was not invented until 1958.

Wittgenstein moves from pictures to propositions, using his notion of pictures to give a theory for the meaning of language. After pointing out that sentences do not look like pictures, he seeks a deep similarity. He suggests (3.32) that the sign (for example a word) is 'the part of the symbol perceptible by the senses', but the choice of signs (unlike pictures) is arbitrary. The deep similarity with pictures is in the structures of propositions, which to Wittgenstein mirror the logical structure of reality. This logical form of propositions is, however, by no means clearly visible in sentence structures. It is a major task of philosophy to reveal these hidden structures of propositions which, when appreciated, allow us to draw lines (though they are complex) between propositions and the world.

Wittgenstein's position is a subtle form of Empiricism in which truth is discovered by putting up propositions against reality – but yet we can never step outside perception, or language, to see how well they match the world. His analogy here is placing a ruler or scale against an object. But this description is, surely, hardly satisfactory: when using a ruler we can see both the ruler and the objects that it is laid against. Considering perception itself (or the propositions of language) we are, so to say, trapped inside: we cannot step outside to compare our perceptions or descriptions with what we picture.

Wittgenstein came to develop his 'language games', to explore the power, the limitations and the sometimes misleading characteristics of language. This might be analogous to testing and calibrating instruments. By trying out instruments – or statistical or other techniques for handling scientific data – one can come to judge their power and limitations and thus to see when to rely on and when to distrust instruments and techniques for gaining knowledge, though direct comparison in particular cases is impossible. This inquiry is the activity of the later Wittgenstein, who refined, and was critical of, his earlier picture theory, though he never quite abandoned it.

He came to warn against thinking of perception as an 'inner picture', in any crude sense. In the *Philosophical Investigations* (1953) he writes: 'The concept of the "inner picture" is misleading, for this concept uses the "*outer* picture" as a model. . . .'[16] This important point was made by Theophrastus. The notion of perception as a copy is also criticized by Wittgenstein from considerations of spontaneously changing perceptions of ambiguous figures such as Necker cubes and the Jastrow duck–rabbit. He writes of these, and of puzzle pictures with hidden objects, in *Investigations*:[17]

> I suddenly see the solution of a puzzle-picture. Before, there were branches there; now a human shape. My visual impression has changed and now I recognise that it has not only shape and colour but also a quite particular 'organisation'. – My visual impression has changed; – what was it like before and what is it like now? – If I represent it by means of an exact copy – and isn't that a good representation of it? – no change is shewn.

This is indeed a serious objection to the Gestalt school's representation by 'isomorphic' brain traces (see pp. 367 *et seq.*).

Wittgenstein started by dismissing psychological issues and theories as unimportant for his main thesis, but later came to reverse this view. In a celebrated letter to Bertrand Russell, written in 1919, he says, 'I don't know *what* the constituents of thought are but I know *that* it must have such constituents which correspond to the words of language.' Russell then asked, 'Does a thought consist of words?' 'No,' Wittgenstein replied, 'but of psychical constituents that have the same sort of relation to reality as words. What those constituents are I don't know.' He also dismissed psychology as of no more importance for his studies of the theory of meaning and perception than any other science. But later in the *Investigations,* he comes to consider phenomena of perception, especially ambiguous figures, as important for these philosophical issues (see p. 388).

What differences are there between pictures and propositions? Two differences stand out. In the first case, pictures may represent but they do not assert. One may draw a picture of Jack falling down a hill; but this is not to assert that Jack *is* falling down the hill, or indeed that he ever did, or will, fall down the hill. Propositions, we may note here, have tenses, though pictures do not. Another most important difference concerns 'not'. It is impossible to draw a picture of a negative fact (except perhaps by leaving out something expected) but we can always *say* that so-and-so is not the case.[18] We may similarly say that 'not' applies to language though we never perceive negative facts. We may perceive that something is missing by not being able to see it; but this is different from seeing that something is not the case, as we may intellectually infer or appreciate conceptually that something is not the case.

Wittgenstein tries to deal with 'negative facts' in picture terms with his analogy of applying a scale, or ruler, to reality. Suppose one holds a ruler up to a window when measuring for a curtain, and the window ends at the six-foot mark. We at once know that the window is not seven feet, nor nine feet, nor a mile high. All other lengths but six feet (outside the margin of error) are quite literally ruled out. Considering now a complex picture, the same consideration may be said to apply to all its features – when alternatives are ruled out by what is accepted as matching or fitting the picture. So in this sense, pictures *can* tell us what is not the case. There are, however, some curious puzzles. We may accept that nothing can be both six and seven feet in length at the same time; so by knowing it is six feet, we at once and without further evidence know that it is *not* seven feet. And consider another example: colour. Suppose the picture shows a rose to be red. We know that if it is red it is not yellow – for objects cannot simultaneously be more than one colour all over. But how do we know this? Why should it be so? Is it a generalization from what we have observed? Is it an accidental property of the visual colour system of the eyes? If so, could there not be another species, or even exceptional human beings, for which this is not true? Or is there something about the 'logic' of colour and space which makes this inevitably true – perhaps as nothing can be six and seven feet in length at the same time? Wittgenstein would argue for this position, that objects cannot be of two colours all over at once for logical reasons, but perhaps he never showed convincingly why or how this can be a logical rather than an empirical matter. And we *can* hear several tones at once. (See Wittgenstein's *Remarks on Colour* (1977), written 1950–1.)

A danger of this kind of argument is that what look like restraints on possibilities set by logic may be set merely by our lack of imagination. It is possible to find what seem to be clear examples of this, even in the writings of Wittgenstein. For example, *On Certainty* discusses and rejects the possibility of getting to the moon. This is discussed most fully in the following passage:[19]

But is there then no objective truth? Isn't it true, or false, that someone has been on the moon? If we are thinking within our system, then it is certain no one has ever been on the moon. Not merely is nothing of the sort seriously reported to us by reasonable people, but our whole system of physics forbids us to believe it. For this demands answers to the questions, 'How did he overcome the force of gravity?' 'How could he live without an atmosphere?' And a thousand others which could not be answered.

Wittgenstein adds that we would feel intellectually very distant from someone who claimed to have been on the moon but could not say how it had been achieved.

Wittgenstein is particularly interesting on the question of how far we can push doubt. He rejects the extremely general doubts of Descartes on the grounds – and surely this is an exciting and powerful argument – that the language that we have to use to express such general doubts cannot stand the weight imposed on it. To accept that language continues to work when it is called upon to express doubt of what we normally accept as surely true, is irrational, for at some point we have to question whether we can express or justify the doubt. He would not allow that doubt could be expressed adequately to reject other people's thought or consciousness, or that there is a world of facts. He hints, though, of what he feels may be true but cannot say. We see this in the final words of the *Tractatus:* '7. Whereof one cannot speak, thereof one must be silent.' Where does Wittgenstein think that our 'pictures' of the world come from? He would say that some come from perception and some from science – but none from philosophy. For him, Ethics and aesthetics have no propositions, and the 'propositions' of philosophy are ultimately empty. Philosophy is, however, not a waste of time for him, for it may show how to think and speak, and how we can be misled. He looks, though, to the sciences for propositions applicable to the world. Wittgenstein has little to say about the methods of science: of how science is or should be carried out. He does not tell us how propositions should be placed like rulers against the world to test for truth, though he does make it clear that what is true cannot be known by examining the internal structures of propositions – it is their match with reality that counts. But reality cannot be known directly, so how can we have reason to believe empirical propositions? Presumably this comes from knowing that appropriate scientific methods of enquiry have been carried out. Or are there other ways of checking scientific knowledge? But this is not quite the kind of question that concerned Wittgenstein.

Before returning to these issues (chapter 20) we shall look briefly at dreams, and at consciousness.

14   Meaning to dream

We are such stuff
As dreams are made on, and our little life
Is rounded with a sleep.

William Shakespeare (1564–1616)[1]

Ancient dreams

Dreams have been described and explained in many ways ever since the first 'dream books' from Mesopotamia, as early as 3000 BC. Dreams were important also for the Egyptians, for the Greeks, and for India and China. Indeed, there has been universal interest in dreams throughout recorded history. Particular dreams have had historical effects: dreams such as Pharaoh's dream in the Bible, of seven fat and seven lean kine as interpreted by Joseph; and Jacob's dream of the ladder linking levels of reality. Dreams were from the earliest times regarded as divinely inspired. The Egyptians regarded dreams as rehearsals for death. The Egyptians and the Greeks had special sacred places and, later, temples for 'incubating' dreams. Here people would go to dream answers to their problems.

The earliest recorded dreams are written in cuneiform script, on clay tablets from Nineveh. Here, 1,500 years before the Old Testament, are accounts of a great flood. Most interesting for us is the *Epic of Gilgamesh*, King of Uruk. Gilgamesh, who is now thought to be a historical figure, goes on an adventurous journey with his friend Enkidu, in Mesopotamia (now Iraq), before the third millenium BC. Gilgamesh finds the plant of eternal life; but it is seized by a serpent and lost to him. This itself sounds like a dream sequence; but in the Epic this is the reality, and Gilgamesh's and Enkidu's dreams are equally important experiences. On the day Enkidu is stricken with sickness, to die, he dreams:[2]

'There is the house whose people sit in darkness; dust is their food and clay their meat. They are clothed like birds with wings for covering, they see no light, they sit in darkness.

I entered the house of dust and I saw the Kings of the earth, their crowns put away for ever; rulers and princes, all those who once wore Kingly crowns and ruled the world in the days of old. They who had stood in the place of the gods like Anu and Enlil, stood now like servants to fetch baked meats in the house of dust, to carry cooked meat and cold water from the water-skin. In the house of dust which I entered were high priests and acolytes, priests of the incantation and of ecstasy; there were servers of the temple, and there was Etana, that King of Kish whom the eagle carried to heaven in the days of old. I saw also Samuqan, god of cattle, and there was Ereshkigal the Queen of the underworld; and Belit-Sheri squatted in front of her, she who is recorder of the gods and keeps the book of death. She held a tablet from which she read. She raised her head, she saw me and spoke: ''Who has brought this one here?'' Then I awoke like a man drained of blood who wanders alone in a waste of rushes; like one whom the bailiff has seized and his heart pounds with terror.' Gilgamesh . . . wept quick tears. . . . He opened his mouth and spoke to Enkidu: 'Who is there in strong-walled Uruk who has wisdom like this? Strange things have been spoken, why does your heart speak strangely? The dream was marvellous but the terror was great; we must treasure the dream whatever the terror; for the dream has shown that misery comes at last to the healthy man, the end of life is sorrow.' And Gilgamesh lamented, 'Now I will pray to the great gods, for my friend had an ominous dream.'

This conveys the ancient force and the mystery of dreams, which we still feel.

The Egyptians and early Greeks developed accounts of 'dream worlds' as detailed as their accounts of the physical world. The Greek God Hermes is the translation of the Egyptian Thoth: God of learning, writing and magic. Hermes leads the souls of the dead to the Other World. He passes the village of dreams, which in Homer's *Odyssey* (c. 850–800 BC?) is on the shore of Okeanos on the far edge of the waking world. The God of dreams, Morpheus, was son of Hypnos, the God of sleep. The God of death was his twin brother Thanatos. Dreams were divided by Homer into those that arrived through the Gates of Ivory and those that arrived through the Gates of Horn. The first were true and the second false. Dreams were immensely important as revelations and for prediction and warnings. It was indeed often considered that men are wiser when asleep than when awake.[3]

Several Greek philosophers saw dreams differently: possibly they had professional pride in being wise while awake! Plato saw unconscious and potentially dangerous desires in dreams. In the *Republic*, Plato writes:[4]

What kind of desires do you mean? Those which bestir themselves in dreams, when the gentler part of the soul slumbers and the control of reason is withdrawn; then the wild beast in us, full-fed with meat or drink, becomes rampant and shakes off sleep to go in quest of what will gratify its own instincts. As you know, it will cast away all shame and prudence at such moments and stick at nothing. In fantasy it will not shrink from intercourse with a mother or anyone else, man, god, brute, or from forbidden food or any deed of blood. In a word, it will go to any length of shamelessness and folly.

in every one of us, even those who seem most respectable, there exist desires, terrible in their untamed lawlessness, which reveal themselves in dreams.

This is a very different approach from dreams being visits: Plato is thinking of dreams as coming from within the individual, and expressing his hidden desires, fears and so on. This implies active psychological processes, of which dreams are manifestations. All this is remarkably close to Freud's accounts of dreams, in recent times. Indeed this passage from Plato might well have been written by Freud.

Plato's pupil Aristotle held a view different both from dreams as divinely inspired, and also from Plato's psychological theory. Aristotle tried to explain dreams in terms of his notions of physiology. Aristotle did not believe that dreams are visits from, or to, the Gods. His main argument against a divine origin for dreams is that stupid men and animals have dreams. Aristotle's discussion of dreaming gives us a clear account of much of his physiology and of his notions on visual and other illusions.[5]

Sleep is explained as necessary to restore fatigue of the 'primary organ' of common sense, which links the senses. (This for Aristotle is the heart, but the notion of a common-sense organ, which combines the evidence of the various special senses, is found in much later writers including Descartes, who attributes this function to the pineal gland.) Aristotle ascribes sleep to evaporation of food, causing liquid and solid substances to pass to the head which (with the eyelids) becomes heavy, and may nod. Dreams are mirrored on the turbulent fluids, as images may be reflected on a lake. When the turbulence is great, nothing can be discerned, and generally there are distortions. This notion is combined with the idea that in sleep the dreamer is uncritical and more than usually sensitive to small stimuli. For this reason dreams may be used by physicians for diagnosing the small beginnings of disease.

Dreams he explains as due to increasing sensitivity during sleep; described as follows:[6]

Stimuli occurring in the daytime, if they are not very great and powerful, pass unnoticed because of greater waking impulses. But in the time of sleep the opposite takes place: for then small stimuli seem to be great. This is clear from what often happens in sleep; men think it is lightening and thundering, when there are only faint echoes in their ears. . . .

Aristotle suggests that this is useful for medical diagnosis, for 'Since the beginnings of all things are small, obviously the beginnings of disease and other distempers, which are about to visit the body, must be small. Clearly then these must be more evident in sleep than in the waking state.' Considering whether dreams can prophesy, Aristotle gives a cautious rejection, after making the following important distinctions:

Now dreams must be either causes or signs of events which occur, or else coincidences; either all or some of these, or one only. I use the word 'cause' in the sense in which the moon is the cause of an eclipse of the sun, or fatigue is the cause of fever; the fact that a star comes into view I call a 'sign' of the eclipse, and the roughness of the tongue a 'sign' of fever; but the fact that someone is walking when the sun is eclipsed is a coincidence. For this is neither a sign nor a cause of the eclipse, any more than the eclipse is a cause or a sign of a man's walking. So no coincidence occurs invariably or even commonly. Is it true that some dreams are causes and other signs – of what happens in the body for instance? At any rate even accomplished physicians say that close attention should be paid to dreams. . . .

Aristotle writes on dreams with his usual perspicacity and common sense. By contrast he makes most later writers look wild and woolly!

A celebrated book, classifying dreams in terms of their supposed significance, was written in the second century A D, at the time of Hadrian, by Artemidorus of Ephesus. (He called himself Daldianus, from his mother's birthplace, Daldis in Lydia.)[7] Three thousand dreams are discussed analytically in the five books of his *Oneirocritica*. This classification of dreams has continued in use to the present day, though it does not seem particularly helpful.

The classes of dreams are:

1   Dreams influenced by the past but having no future significance. This kind of dream could give direct knowledge; and

2   Dreams that were supposed to determine the future. These are divided into: (a) direct prophecies; (b) previsions of future events; and (c) symbolic dreams, which needed interpretation to be understood.

The interest and the difficulties in considering dreams are over their interpretation. If dreams are indeed meaningful, why can we not see their significance immediately – as we see the significance of perceptions immediately in waking life? True, some perceptions have to be interpreted, for example some scientific observations or pictures having high symbolic content; but this is hardly typical of normal waking perception – though it is typical, if they have meaning, of dreams. This takes us almost at one bound to Freud. Before Freud gave to dreams immense significance (which is not, however, always accepted), a theory came into prominence more like Aristotle's, associated with the French philosopher Henri Bergson (1859–1941): that dreams are triggered by bodily sensations and made of memories. For Bergson, everything ever experienced is remembered, but memories actively struggle against each other to reach consciousness. This they succeed in doing most easily in sleep when they are not held by waking interests. Hence, dreams are composed of memories triggered by small sensory stimuli. Dreams that are clearly related to surrounding events are quite frequently reported, so there is evidence for this kind of theory.[8]

Other such commonsense notions may be added, such as rehearsing, or worrying over problems, thinking in rather disorganized ways during sleep, and acting out alternative scenarios for action. On these views, dreaming is regarded as similar to 'day-dreaming', and as somewhat similar to the 'free association' which became so important as a diagnostic tool for Freud and Jung.

It is worth pointing out here that the content of dreams is indeed all from our normal experience – for it seems safe to say that we never experience anything in dreams that are not combinations (though sometimes novel combinations) of normal waking experience. It is worth pointing out also, as possibly in favour of a 'physiological' theory, that dream-like states clearly related to memories from past experience, though jumbled up, can be elicited by brain stimulation in conscious patients during brain operations.[9] Freud's account sees dream experiences as symbolic hidden inner realities.

Sigmund Freud on dreams

Perhaps Freud's most fundamental idea is that dreams are infantile wish fulfilments. While writing *The Interpretation of Dreams* (1900), he was sometimes haunted by the thought (in a letter written on 9 June 1898) that 'The whole matter resolves itself into a platitude. Dreams all seek to fulfil *one* wish, which has got transformed into many others. It is the wish to sleep. One dreams so as not to wake, because one wants to sleep.' This is not at all Freud's theory of dreams: he hoped to be able to uncover the structure and the initiating energies of Mind in dreams.

For Freud, dreams are related to typical characteristics of neuroses. The strange, difficult-to-understand symbolism of dreams he attributes to their neurosis-like structure. Rather curiously, he started with a physiological explanation for why they are so indirectly related to experience, or to wishes.

As his biographer, Ernest Jones, puts it: [10]

He [Freud] makes the momentous distinction between two fundamentally different mental processes, which he called primary [unconscious] and secondary [conscious] respectively. He notes that the primary process dominates dream life, and he explains this by the relative quiescence in the activity of the ego (which at other times inhibits the primary process) and the almost total immobility; if the cathexis of the ego were reduced to nothing, then sleep would be dreamless. He also states that the hallucinatory character of dreams, which is accepted by the dream consciousness so that the dreamer believes in what is happening, is a 'regression' back to the processes of perception which he relates to the motor block in the usual direction of discharge. The mechanisms found during the analysis of a dream display a striking resemblance to those with which he had become familiar in analysing psychoneurotic symptoms.

Jones points out that Freud's initial explanation of why many dreams appear senseless in terms of physiological blocking (which Freud admitted he found unsatisfactory) is odd – for he makes no use of the process of 'repression' already familiar to him in the field of psychopathology. What is odd is that Freud chose at this point (especially as he had just started his self analysis) a Natural type of explanation – in terms of physiology – rather than an Unnatural explanation, in terms of concepts of Mind. The great English neurologist John Hughlings Jackson (1835–1911) had earlier said: 'Find out about dreams, and you will find out about insanity.' Freud always had a healthy respect for physiology, and indeed his first work was on the action of cocaine as an analgesic. He started with this kind of explanation. In later editions of *The Interpretation of Dreams,* increasing emphasis was placed on symbolism, with still greater emphasis in the later *Introductory Lectures.* (1916–17).

Freud was deeply concerned with why dreams are often apparently nonsensical, if they are indeed windows to the deep nature of man and the individual. This led him to distinguish between the 'manifest' and the 'latent' content of dreams. The principal task of dream interpretation by the analyst is to discern the latent content, which is the deep meaning, from the manifest content, which is the often trivial-looking experienced contents of the dream. It might be suggested that the task of the analyst is quite like ancient attempts to read underlying significance in Nature – from eclipses, occultations and the positions of the horns of the Moon. This now seems to us inappropriate for the Natural world, but it may be acceptable for discovering Mind. Is it that our notions of psychology are still primitive? Or is it indeed appropriate to look for hidden content and structures of Mind in dreams? We may look now at Freud's account in more detail, especially at his thoughts on regression.

This discussion in *The Interpretation of Dreams* is acutely interesting.[11] Freud starts by asking whether we can explain why 'a thought, and as a rule a thought of something wished, as objectified in the dream, is represented as a scene, or, as it seems to us, is experienced'. He points out that in dreams the thought is 'transformed into visual images and speech' of an immediate situation – often with the 'perhaps' omitted.

The notion is developed that

> The only way in which we can describe what happens in hallucinatory dreams is by saying that the excitation moves in a *retrogressive* direction. Instead of being transmitted towards the *motor* end of the apparatus it moves towards the *sensory* end and finally reaches the perceptual system. If we describe as 'progressive' the direction taken by psychical processes during waking life, then we may speak of dreams as having a 'regressive' character.

This notion of 'regression' is not original to Freud. The first hint is in Albertus Magnus (thirteenth century), and Thomas Hobbes' *Leviathan* (1651): 'In sum, our dreams are the reverse of our waking imaginations, the motion, when we are awake, beginning at one end, and when we dream at another.'[12] For Freud this regression (or 'retrogression') is a justification for using dreams as direct evidence – one might say direct perceptions of – the contents of the unconscious, for he holds that '*In regression the fabric of the dream-thoughts is resolved into its raw material.*'

The physiological state of sleep may produce the changes of excitation supposed to produce regression, but Freud points out that regression also occurs in waking patients suffering hallucinations as in hysteria. They are also regressions '– that is thoughts that undergo this transformation into images – but the only thoughts that undergo this transformation are those which are intimately linked with memories that have been suppressed or have remained unconscious'.

Freud emphasized the importance of infant experience and fantasies in dreams, to conclude that in dreams we may retrace our own experience and even the experience of the human race: [13]

dreaming is on the whole an example of regression to the dreamer's earliest condition, a revival of his childhood, of the instinctual impulses which dominated it and of the methods of expression which were then available to him. Behind this childhood of the individual we are promised a picture of a phylogenetic childhood – a picture of the development of the human race, of which the individual's development is in fact an abbreviated recapitulation influenced by the chance circumstances of life. . . .

Dreams and neuroses seem to have preserved more mental antiquities than we could have imagined possible; so that psycho-analysis may claim a high place among the sciences which are concerned with the reconstruction of·the earliest and most obscure periods of the beginnings of the human race.

Freud speaks of 'regression' both for the reversal of the normal waking order of the brain processes in dreaming and also for the return to infant experience and fantasies. This seems confusing, but perhaps he saw a deep connection between the two. An important fact that emerges is that at this stage of his thinking he adopted a largely naturalistic physiological account of this phenomenon of dreaming – to develop a psychological account in terms of the power of symbols and of censorship. Freud, therefore, moved in his thinking in the reverse direction from the history of the Natural Sciences.[14]

Freudian symbols

In the *Introductory Lectures,* Freud expresses his committed view that dreams are symbolic accounts of unconscious mental activity having deep significance. We may look especially at the lecture 'Symbolism in Dreams':

Symbolism is perhaps the most remarkable chapter of the theory of dreams. In the first place, since symbols are stable translations, they realize to some extent the ideal of the ancient as well as the popular interpretation of dreams, from which . . . we had departed widely. They allow us in certain circumstances to interpret a dream without questioning the dreamer, who indeed would in any case have nothing to tell us about the symbol. If we are acquainted with the ordinary dream-symbols, and in addition with the dreamer's personality, the circumstances in which he lives and the impressions which preceded the occurrence of the dream, we are often in a position to interpret a dream straightaway – to translate it at sight, as it were. . . .

You will now want to hear something of the nature of dream-symbolism and to be given some examples of it. I will gladly tell you what I know, though I must confess that our understanding of it does not go as far as we should like.

The essence of this symbolic relation is that it is a comparison, though not a comparison of *any* sort. Special limitations seem to be attached to the comparison but it is hard to say what they are. Not everything with which we can compare an object or a process appears in dreams as a symbol of it. And on the other hand a dream does not symbolize every possible element of the latent dream-thoughts but only certain definite ones. So there are restrictions here in both directions. We must admit, too, that the concept of a symbol cannot at present be sharply delimited: it shades off into such notions as those of a replacement or representation, and even approaches that of an allusion. With a number of symbols the comparison which underlies them is obvious. But again there are other symbols in regard to which we must ask ourselves where we are to look for the common element. . . . It is strange . . . that the dreamer should not be acquainted with it but should make use of it without knowing about it: more than that, indeed, that the dreamer feels no inclination to acknowledge the comparison even after it has been pointed out to him. . . .

The range of things which are given symbolic representation in dreams is not wide: the human body as a whole, parents, children, brothers and sisters, birth, death, nakedness – and something else besides. The one typical – that is regular – representation of the human figure as a whole is a *house*. . . . It may happen in a dream that one finds oneself climbing down the facade of a house, enjoying it at one moment, frightened at another. The houses with smooth walls are men, the ones with projections and balconies that one can hold on to are women. One's parents appear in dreams as the *Emperor* and *Empress*, and *King* and *Queen* or other honoured personages; so here dreams are displaying much filial piety. They treat children and brothers and sisters less tenderly: these are symbolized as *small animals* and *vermin*. Birth is almost invariably represented by something which has a connection with *water:* one either falls into the water or climbs out of it. . . . Dying is replaced in dreams of *departure*, by a *train journey*, being dead by various obscure and, as it were, timid hints. . . . You see how indistinct the boundaries are here between symbolic and allusive representation.

Freud now introduces sexual symbolism and its importance: 'The male genitals, then, are represented in dreams in a number of ways that must be called symbolic, where the common element in the comparison is mostly very obvious. To begin with, for the male genitals as a whole the sacred number 3 is of symbolic significance. . . .'

Freud goes on to list obviously similar-shaped objects – umbrellas, posts, trees, weapons and so on – to continue.

Among the less easily understandable male sexual symbols are certain *reptiles* and *fishes*, and above all the famous symbol of the *snake*. It is certainly not easy to guess why *hats* and *overcoats* or *cloaks* are employed in the same way, but their symbolic significance is quite unquestionable.

Considering the female, some of the more unlikely symbols are ships, cupboards, stoves, and more especially rooms. Here room symbolism touches on house symbolism. The breasts are represented by 'apples, peaches and fruit in general'; and 'Jewels and treasure are used in dreams as well as in waking life to describe someone who is loved.' Here Freud warns again that dream symbolism is not straightforward or obvious: 'You must not picture the use of the translation of these symbols as something quite simple. . . . It seems almost incredible, for instance, that in these symbolic representations the differences between the sexes are often not clearly observed.'

I suppose the question for us is whether such surprises are *too* great to allow belief in Freudian symbolism. It is exciting and brilliantly worked out by a man of genius; but how much is it Freud's psychology – his own Mind?

Carl Jung on dreams

The account of dreaming, symbolism and myth due to Carl Gustav Jung (1875–1961) – perhaps equally celebrated – is very different. It is indeed hardly surprising that a serious rift developed between Freud and Jung, so that they went their separate ways. Jung suggests that dreams reflect the 'collective unconscious', which to him is a shared pool of archaic experience represented by symbols that may be found developed independently in very different regions and by children. These he calls 'archetypes'. (He minimizes, though, the extent of ancient trade routes and the spread of artefacts, so his anthropological evidence for spontaneous universal archetypes can be questioned.) By supposing that the 'unconscious' in some sense exists as an entity between individuals and this is seen in dreams, Jung is suggesting in effect that dreams have a kind of 'objectivity'. He was very much concerned with why some objects, or peculiarly related events, should be significant, and others quite insignificant. This he explains with his odd concept of synchronicity. For Jung, when coincidences occur as by chance but seem meaningful, they occur because they 'like' to occur together. This is not a causal explanation but is much more like alchemy and astrology. It harks back to omens, to soothsayers and to the ancient Unnatural notion that the Universe is alive and aware of man. We may look at Jung's last comments on his ideas: [15]

Just as a biologist needs the science of comparative anatomy, however, the psychologist cannot do without a 'comparative anatomy of the psyche'. In practice, to put it differently, the psychologist must have a sufficient experience of dreams and other products of unconscious activity, but also of mythology in its widest sense. Without this equipment, no-body can spot the important analogies; it is not possible, for instance, to see the analogy between a case of compulsion neurosis and that of a classical demonic possession without a working knowledge of both.

In the following paragraph, Jung gives what he finally sees as the status of the archetype concept:

My views about the 'archaic remnants', which I call 'archetypes' or 'primordial images', have been criticised by people who lack a sufficient knowledge of psychology of dreams and of mythology. The term 'archetype' is often misunderstood as meaning certain definite mythological images or motifs. But these are nothing more than conscious representations; it would be absurd to assume that such variable representations could be inherited.

The archetype is a tendency to form such representations of a motif – representations that can vary a great deal in detail without losing their basic pattern. . . . My critics have incorrectly assumed that I am dealing with 'inherited representations', and on that ground they have dismissed the idea of the archetype as mere superstition. They have failed to take into account the fact that if archetypes were representations that originated in our consciousness (or were acquired by consciousness), we should surely understand them, and not be bewildered and astonished when they present themselves in our consciousness. They are, indeed, an instinctive trend, as marked as the impulse of birds to build nests, or ants to form organised colonies.

Here I must clarify my relation between instincts and archetype: what we properly call instincts are physiological urges, and are perceived by the senses. But at the same time, they also manifest themselves in fantasies and often reveal their presence only by symbolic images. These manifestations are what I call the archetypes. They are without known origin; and they reproduce themselves in any time or in any part of the world – even where transmission by direct descent or 'cross fertilisation' through migration must be ruled out.

The notion of synchronicity is primitive indeed. Can we really believe that some combinations of events of the Natural world have special symbolic significance for us? Certainly the artist uses Nature for his symbols, and highly effective they can be; but Jung's view is essentially saying that Nature is the artist, and the writer of novels, to give our lives significance. This is an antique dream, linked with astrology and alchemy. It is difficult to believe that this dream still lives on in waking life, and that is after all our deep reality.

Can meaning be read from dreams?

To suggest that dreams can be interpreted implies that they have meaning – though this must be discovered by learning how to interpret or read them.

We may add to this discussion with a comment on interpreting dreams. They have to be interpreted because they are apparently nonsensical. One might at least in principle object that 'interpretations' are little more than projections of the analyst's predilections – rather as we see forms in ink-blots. But, then, the most 'objective' experiments in physics also need to be interpreted by experts or they are meaningless. There is here a special problem: it would be quite surprising if the philosophy, or pre-suppositions, of the analyst as understood by the patient did not affect his dreams. We might expect that patients of a Freudian analyst will have typical Freudian dreams (at least while he is in sympathy with his analyst) while Jungian patients may have 'Jungian' dreams – filled with ancient archetypal symbols, as he has got to know them through the Master's writings, or books of ancient art. Perhaps, rather as the physicist Werner Heisenberg pointed out, observation can and in some cases must affect what is observed (see p. 542), so what we know or believe may affect what we dream. If we are affected by theories of perception and theories of dreaming, psychological tests of theories, especially of dreaming, are difficult to apply.

How can we be reasonably sure that dreams are not 'ink-blots' for the analyst to project his own predilections upon – to see in them evidence for his own assumptions which may look ever more justified through denying counter cases? This is indeed a tremendous danger. Consider the wish fulfilment notion: surely wish fulfilments must all be pleasant? Freud admits this, following data which he cites from Florence and Hallam Weed (1896), who find that: 'their own dreams are . . . 57.2 per cent "disagreeable" and only 28.6 per cent positively "pleasant". And apart from these dreams, which carry over into sleep the various distressing emotions of life, there are anxiety-dreams, in which that most dreadful of all unpleasurable feelings holds us in its grasp until we awaken.'[16] Freud answers these and related objections by saying:[17]

We must make a contrast between the *manifest* and the *latent* content of dreams. There is no question that there are dreams whose manifest content is of the most distressing kind. But has anyone tried to interpret such dreams? To reveal the latent thoughts behind them? . . . It still remains possible that distressing dreams and anxiety-dreams, when they have been interpreted, may turn out to be fulfilments of wishes.

Freud admits that dreams can be distorted. He admits also that dream symbols can be taken as positive or negative: very often he states that what is represented should be interpreted as the opposite of how it appears. He also admits that dreams are truncated and have important bits missing; that they may have been missing in the dream or that the waking memory in recall may have been faulty – to lose essential items or to fill in gaps with features that never were in the dream. This we see in recall of normal events (Bartlett, 1932) and in many perceptual gap-filling phenomena. Freud admits these and other difficulties or weaknesses of his position. It is for the reader to judge whether his defences are adequate.

I shall end this discussion of Freud with one further point, which will touch on what it was that Freud thought he saw in dreams, and which is central to psychoanalysis. As we saw earlier, Freud identified mental entities, such as the Id, the Ego, and the Superego. Now what kind of entities are these? Consider the Superego: this is seen as a check on unbridled instinctive behaviour. It is a censor. Now the Superego and the rest are sometimes thought of as being themselves personalities, or part-personalities, as it were inhabiting the Mind. But if personality, or Mind, is to be explained, can we use examples of what we want to explain as themselves explanations? It is characteristic of physical ('Natural') explanations that what is to be explained is not contained in the explanation. This at least is so since Plato; and Plato was not in this sense a scientist. (Plato was content to 'explain' the cutting power of knives in terms of knife-like cutting particles constituting their blades. How 'cutting' occurs is not explained in a way satisfactory to the Natural Sciences.) Now are Freud's mental entities, which he sees revealed in dreams, any better than Plato's physical particle entities, which he supposes have the properties he wants to explain? If the emphasis were on interactive effects of these entities, then it might be an acceptable kind of explanation. But Freud invokes the kinds of interactions or relations between his mental entities that human beings experience; they are, for example, at war with each other in highly human ways. This does make such explanations look 'humanistic' (and as such acceptable), but it makes them very different from the successful explanations of the Physical Sciences.

Consider the Superego notion further. In order to act as a censor, it must appreciate the meaning of what it censors. It is thus a kind of filter; but it is operating according to meaning. It is thus set apart from engineer's filters. This might be simply because engineers have not yet succeeded in making 'meaning filters'. The trouble is that meaning remains almost outside Information Theory (p. 156). Suppose engineers come to design 'meaning filters': then censorship could be given by functional brain components; rather as blood flow or nerve action potential rates may be regulated or controlled by 'Natural' mechanisms – or rather we would see this as possible.

It seems entirely possible that something like 'meaning filters' will be made as Artificial Intelligence develops. If so, Freud's accounts may come to look very different, and far more acceptable to Natural Science. At the same time they might look less 'humanistic'. But very possibly it is a logical mistake to describe components of people as humanoid part-people: this is Plato's mistake yet again. As it stands now, this is how Freud's account looks; but it might be transformed by translations of his mental-entity terms into processes of Natural neurological mechanisms. This is a 'dream' of physiology and Artificial Intelligence.

Dream experiments

There was a resurgence of interest in dreams some twenty years ago when new experimental techniques were brought to bear, and since then several dream laboratories have been set up. W. Dement and N. Kleitman (1957) related rapid eye movement 'REM', and also non-rapid eye movement, 'NREM', to dreams during sleep, by gently waking the subject during various kinds of sleep as indicated by eye movements and brain activity. Dement suggests that dreams are needed to prevent insanity, as Freud thought, that if we understood dreams we would understand insanity. Dement's suggestion is, however, very difficult to test. It has been found that dreams may be punning allusions to words spoken to the sleeper. Some people are wittier asleep than awake. Indeed, Wilhelm Flies, when reading the proofs of Freud's *Interpretation of Dreams* (in the autumn of 1899), complained that the dreams were too full of jokes, which inspired Freud to develop a theory of humour, in *Jokes and their Relation to the Unconscious* (1905): this is surely the most unfunny book on humour. It turns out that the Mind is still active and very likely highly active in sleep, and that the world of the senses is not entirely cut off, but may be interpreted in 'mad' ways. Dreaming confirms the notion that brain activity is spontaneous and is normally controlled by sensory signals, and what we might call project goals, rather than being merely passive and dependent on signals from the world. Dreams do thus give us – and psychotherapists – a window, if a very strange and perhaps distorting window, into the dynamics of the Mind. It turns out, though, that the dream world is (in normal terms) mad; though (in normal terms) we are sane – so dreams cannot be taken at face value and their symbols are certainly difficult and perhaps impossible to read reliably.

Do dreams have a biological function?

It is remarkable that almost all animals have periods of inactivity, when they are vulnerable to attack. There must surely be biological necessity for sleep, and yet the necessity is not understood beyond vague references to 'repair' of tissues. If so, how do sharks (which are exceptional, because having negative buoyancy they have to swim continuously) manage without sleep? Are sharks indeed the key to this question? Much the same mystery applies to dreams: are they useful for keeping neural processing active to prevent atrophy through disuse, and perhaps to aid development? (REM sleep is greatest in neonates.) Do dreams test out, repair gaps, and mend inconsistencies in brain 'programs'? There is no clear evidence. It has, however, been shown that animals deprived of visual stimulation do suffer atrophy of the visual system; so dreams are not adequate to keep the visual brain in physiological order.

The usual way to find out whether something is essential, is to remove it and note the consequences. This has been done for dreaming by William Dement (1969), who systematically deprived animals of REM sleep over many consecutive days, and found that, after deprival, the proportion of REM to other kinds of sleep increased, almost to make up the deprivation. This suggests that it has some kind of function. This impression is strengthened by the remarkable observation that the intensity of eye movements became greater and greater, until (Snyder, 1966): 'As one observes these phenomena the impression is inescapable that an irresistible force is steadily building and that if allowed to continue its pressure would soon become uncontainable.' Snyder surely rightly criticizes Dement's suggestion that dreams serve to prevent the build-up of toxic substances in the brain; for surely this is an absurdly complex manifestation for such a simple operation as 'detoxication'. But conceivably there is build-up of toxic substances which produce dreams, rather as many known hallucinogenic drugs such as LSD produce vivid and mad imagery.

Ian Oswald found that for humans, barbiturates suppress REM sleep, and also there is 'a profusion of eye movements' (Oswald, 1969); and, as for the animals, there is also a 'rebound increase of REM sleep above normal'. Oswald and his colleagues at Edinburgh have carried out heroic experiments on humans, which accord with the animal sleep experiments in the recordings of eye movements and electrical brain activity (EEG) as well as showing that there can be, as it were, translations into dreams from words or other sounds presented during sleep.

It is interesting that anaesthetics produce selective losses of function, in a sequence, generally with pain to go first and hearing last. There is indeed some evidence that patients under surgery can sometimes hear (perhaps while in a semi- or complete-dream state) conversations around them, which may produce lasting nightmares or even psychiatric disturbance. It is important to note that the degree of sleep or anaesthesia is difficult to assess (hence the necessary skills and monitoring aids of the anaesthetist), and in general there is more reception of sensory information than may appear. (I have tried this out and confirmed it myself.)

The REM sleep we associate with dreaming is found throughout the mammals, but hardly, if at all, before mammals in the evolutionary sequence. The animal of most interest here is the opossum, which has been called a living fossil of the first mammals. The opossum sleeps 75–88 per cent of its life, and its REM sleep occupies about one-third of sleep. So its REM sleep, when we might suppose it is dreaming, occupies more time than its waking life. Perhaps, then, the Mind of man has evolved from pre-human dreams. [18] Perhaps we are even now awakening from the dream-myth world of our origins.

Part 4 Regarding consciousness

When asked the question, what is consciousness? we become conscious of consciousness. And most of us take this consciousness to be what consciousness is. This is not true.

Julian Jaynes (1976) [1]

15 What is consciousness?

For there was never yet philosopher
That could endure the toothache patiently.

William Shakespeare (1564–1616) [2]

How should we regard consciousness? Why is our own awareness so difficult to
think about, and why is it so hard to reconcile consciousness of the world and of
ourselves with physics? There seems to be just one special difficulty: consciousness is private, while physical objects are public. Consciousness is difficult to
discuss because it is uniquely private; so we do not have analogies from our shared
concepts of the physical world at all adequate to describe our experience of it, or
of what we take to be ourselves. There is, however, a kind of bridge, in perception.

John Locke described consciousness as 'the perception of what passes in a
man's own mind'. On this account we should expect theories of consciousness to
be related to theories of perception. It is perhaps the general view of Empiricists
that consciousness is perception of mental events, and is somewhat similar to
perception of events of the external world; we must remember, however, that
Empiricists generally hold a Realist account of the perception of physical objects.
Locke may have thought of introspection as direct knowledge of Mind, very much
as he held perception of Secondary Characteristics to be direct knowledge of
physical objects in external space. If we reject Realist accounts of perception, as
we have to do, then we can hardly accept without question that introspection is
direct knowledge of Mind. If we take Locke's view that it is perception of what
goes on in our Minds, then we may suspect that our knowledge of ourselves is a
construction: perhaps very much as perceptions are hypotheses of the physical
world (see chapter 12).

Consciousness is sometimes regarded as a kind of moving light illuminating
facets of Mind. Imagine consciousness as the beam of light of a torch or search-
light, directed around the dark Universe of the Mind. We are the beam of light, or

we are an eye looking along the beam, and see nothing outside it. But there is a great deal around it, and much that is never illuminated. All this is Mind. Present consciousness is what is lit by the patch of light of the beam. One could easily be fooled into thinking that everything is continuously alight – that consciousness is the whole of Mind – for we only experience what is in the beam.[3]

Let us compare this situation with the movements of the eyes while we search, visually, the physical world. Behind us there is nothing – not blackness – nothing. This is the nothingness of blindness. In foveal vision, on this analogy, we have the centre of the torch beam (and indeed this is how vision was conceived in the ancient world) while beyond the visual field there is in experience nothing. We can, though, turn the eyes and extract something – many things – from the nothingness all around. This is probing and sampling physical reality with the eyes.

Now let us compare this with introspection, as looking at the Mind with the 'inner eye' of consciousness, with this supposed torch beam that lights upon and illuminates facets of Mind. We know, though of course this has been questioned by many philosophers, that the world (the room that we are in, or whatever) is there to be probed and sampled by the moving eyes, and that it is still there when the eyes are shut. We are not, however, at all sure just what this world, when the eyes are shut, is like – apart from how it appears – though it is this that science tells us, for science does not claim to tell us what the world *looks* like. We are, however, led to science's conscious-less accounts from how the world appears. It is a primary claim of Empiricism that appearances are the road towards understanding what the world is like, though appearances drop out of the final description – and that they are a minor part of reality, though scientific accounts are tested by observations depending on appearances. So inferences are made from, and to, appearances to suggest and test scientific accounts. There seems to be a fall-off in consciousness with the increasing abstractness of accounts – ending in the nonconscious knowledge of science.

What happens if we transfer this idea of 'inner perception' (i.e. the notion of perceptions of the external world as being hypotheses requiring inference) to introspective knowledge of Mind given by 'inner perception'? Is this, also, indirect and dependent on inference? It seems an interesting notion that the consciousness of the moving inner 'patch of light' may in this way be analogous to moving eyes giving knowledge of the world in the centre of vision (though largely augmented by inferences of many kinds) and unconsciousness with the *nothing* outside the visual field – when we rely entirely on assumptions that there is a world around us in space and time. Although we experience nothing behind us, we do act on assumptions that there are things, particular things, though unsensed. It is always possible that something will attack us from behind, or fall on us, and one sometimes takes a step backwards without checking that there is not a hole in the ground. Sometimes, of course, there is a hole in the ground, or a fearful obstacle which may prove disastrous. One may knock something over or step on someone's toe, or perhaps injure a child. Knowledge of what is behind one is entirely assumptive, and is more liable to error than perception in the visual field. It is assumptive because it is not signalled and it is outside consciousness. We cannot, though, equate increased knowledge given by perception with consciousness, for increased knowledge may come from instruments, and from the senses without consciousness. It is most important to note that gaining knowledge does not depend on consciousness. It may often be associated with consciousness; but it is a very real question whether consciousness is ever necessary for gaining or for applying knowledge.

This is part of the more general question, 'Does consciousness do anything?' If it does, we may be tempted to include it in descriptions of the nervous system, as we include nerves and action potentials and other physical events that mediate behaviour causally. This would bring consciousness essentially within physical or mechanistic neurological accounts, even though it is unique in being private. If it is deemed by its privacy to be too odd for inclusion as causal, we shall have to consider why it arose, and presumably developed in organic Evolution.

The so-called Mind–Matter gap is particularly difficult to bridge conceptually because it is uniquely a private–public gap. Because of this we lack adequate language (in spite of the efforts of many philosophers), and we lack concepts and data (in spite of the efforts of psychologists), to link private with public – Mind with Matter. We can hardly conceive cause across this gap.

The second problem here, is how consciousness developed if it has no survival value. Could it have survival value if it has no causal effects? A possible answer is that it has not evolved with the Evolution of species, but is unique to man and is perhaps a result of social living and language. But we do speak of, and act upon belief in, animal consciousness; otherwise we would have no worries about cruelty to animals. In fact this gives us great concern.

A further difficulty in bridging Mind and Matter is that Matter has extension in space while Mind has not. It is difficult to conceive that spaceless Mind can affect Matter. Or is this but a prejudice? (After all, we do think of meanings of messages having effects on men, or computers, though meanings do not have extension in space.)

Although philosophers have no accepted language for talking about consciousness, I shall take 'consciousness' to be much the same as 'awareness', and roughly the same as 'sensation', though there may be preferences for one or another term depending on context. You or I may be said to be 'conscious of red', or 'aware of red'; but I shall not allow such sentences as 'I am aware of my sensation of red,' for this would suggest a kind of internal screen of consciousness, which I experience or perceive. This step can generate a regress without end. We may, however, do well to hold that we are not aware of the whole (or even of more than a very small part) of Mind: I shall take 'Mind' to be a much wider term, not necessarily implying consciousness.

How do we know that other people are aware – conscious? Presumably we reason (if unconsciously!) by analogy with ourselves. Another human being is more like oneself than an ape . . . a dog . . . a fish . . . an insect. As we go along this scale, to creatures very different from us, consciousness looks less and less likely (unless we make the assumption that everything is conscious). It is this dependence on analogy from oneself, and other normal human beings, that makes it so difficult for neurologists to assess states of consciousness in cases of severe neurological disturbance. This is especially so when speech is not available, as in states of coma. We rely very much in normal life on speech to indicate states of consciousness. How reliable is speech as a guide? The success of human relations depends very much on the answer.

Can we be sure from speech that another person (or a speaking animal) is conscious? I suspect that for speech to work there must be shared knowledge or assumptions; and so there is a limit to what language can convince us of. Can it convince us that other beings are conscious? If not, and this we shall discuss in a moment, what could convince us of consciousness in other people, animals, or computers? We should have to look for observable abilities which clearly depend on consciousness to be carried out. This implies, though, that consciousness has causal effects; but has it?

Is it possible to show that certain abilities are always associated with awareness, or consciousness? If this could be shown it might be a step towards determining that consciousness has causal effects. Consider, for example, perception: if it could be shown behaviourally that we cannot recognize objects except when we are aware of them, then there might be a strong case for saying that consciousness is necessary, causally, for object recognition (see pp. 457–8).

This argument that consciousness, or Mind, appears causal when physics seems inadequate, is in effect extremely ancient: it is the basis of primitive Animism. If the world cannot be explained by current technology, then magic and extra-human intelligences are invoked as causes. Here, in the computer case, we are supposing that the theoretical basis of current technology is inadequate for accounting for technological results. Should this occur, would we now revert to a version of ancient Animism? If so, would we do this *only* for computers – because they appear somewhat Mind-like?

This raises the question of how alleged paranormal phenomena, such as telekinesis, are described or explained. There is a tendency to postulate the action of an intelligence, in this clearly non-computer case of objects flung around a room, as by a wilful child. So we should consider more than computers. We have a progression, from active Mind supposed to move objects without neural connections; to active Mind (we imagine here) supposed to increase the power of a computer; to active Mind supposed to affect and increase the power of brain mechanisms, beyond what can (we suppose) be explained by physiology and computer technology – or any imaginable computer theory.

There are plenty of examples of things working better than they should (as well as the converse!); for example, storage of electricity in Leyden jars (see p. 145). Perhaps most dramatic is Marconi's demonstration of radio across the Atlantic, when there was no reason (as this was before the discovery of reflecting layers in the stratosphere) for greater than line-of-sight communication. Marconi's optimism was supreme and triumphed over the physics of his day. It looked like a miracle. But he did not postulate some Mind-like entity, or some kind of telepathy, or whatever, to explain his as-it-seemed impossible reception of the letter 'M' in Morse, across three thousand miles of ocean – with no physical explanation.

It is, however, true that Marconi found that he had to optimize his aerial array, according to more or less known physical principles, to achieve the amazing result. Suppose that Marconi had to make his aerials in the form of pentagons; that his messages only came through when magic-like rather than physics-like features were optimized. Would he – and would we now – then postulate the action of Mind, some inner consciousness of the circuits, if such magic symbol shapes were needed for successful transmission? I very much suspect that he might have taken this step, at least for a time, and we might have followed.

Returning to computers: they do work with symbols! Or at least, the *software* is symbolic. On the other hand the *hardware* is designed around physical principles; but we do accept the symbolic significance of physical states of computers. Similarly, the letter 'M' is significant as a signal; but we do not suppose that this symbolic power affects the Marconi apparatus in ways we cannot understand from physics. How then does it affect us? This is surely the heart of the puzzle.

The letter 'M' (and later much more complicated messages) can be read by brains, and by computers; but we do not suppose a computer consciousness that is affected or in any way involved, though we do think of brain consciousness as affected by (and affecting) symbols. Why do we make this distinction?

To return to our original question, but in a somewhat different form as we may now see it in the light of the Marconi example: what *kind* of computer surprise would tend to make us postulate active Mind or consciousness in a computer? Freud postulated mental causes, not only for conferring power to problem-solving but also to *rejection* of especially embarassing things. Suppose a computer behaved with such psychological quirks as Freud used as evidence of unconscious mental activity (because they could not be introspected). Would such quirks be accepted as evidence of causal Mind for a computer? I suppose that this step might be taken, if the quirks were not put into the machine's program, either by its makers, by its 'teachers', or in the case of computers that can learn, by its own experience. In short, we might suppose Mind in *irrational* computers, and especially when the irrationality has no rational explanation.

This comes back to saying that we tend to postulate active Mind in situations beyond understanding. This principle applies, indeed, to how we see human Mind. So a thoroughly rational – and so in physical and logical terms *understandable* – individual might not be credited with active Mind. He would be written off as a 'mere Robot'. Most curiously, then, it is unreason which is evidence for Mind. But we see 'unreason' when we fail to understand. It would surely be sad if increased psychological understanding destroyed the appearance of Mind; though this is merely a value judgement. It does, however, make one wonder whether psychology can be a science in its own right – if Mind is destroyed or disappears, with reason and understanding.

We turn now to problems of how we describe consciousness.

The language of consciousness

Suppose someone of our acquaintance utters the sentence 'I am not conscious.' He might be asleep. Then perhaps indeed he is not conscious when he utters this sentence. If so, what he utters will indeed be true. It will be true also from a tape-recorder, or from a synthetic speech machine that is programmed to utter the syllables 'I am not conscious;' for indeed it will not be. Now suppose our acquaintance says, 'I am conscious.' This might be said with meaning by someone awakening from an anaesthetic, to inform people that he has emerged from his operation safely. The machines (the tape-recorder or the speech synthesizer) might, however, utter this same sentence: 'I am conscious.' But would we not accept this same sentence, from such a source, as false?

We seem to require that a man has consciousness in order to assert or deny awareness of particulars, such as seeing red, and also of particulars such as that he has hands, or owns a house, or that he lacks hands or a house. So he can assert that he is conscious; but in asserting that he is not conscious, he denies what we deem is needed (consciousness) to allow the assertion meaning. So denying consciousness is different from denying that one has hands, and houses, and everything of these kinds.

If a man says 'I am conscious,' this is an information-bearing statement only if it could conceivably be false. It might be false if he were under an anaesthetic, and his brain was being stimulated to make the utterance. But then would he be a man speaking? Would we allow the use of the word 'I' for the unconscious man – or for the tape-recorder? It sounds odd to accept a tape-recorder as an 'I'. Is this because we believe that it cannot truthfully say it is conscious?

Is this what Descartes was getting at with his 'I think, therefore I am'? Surely not quite, for we could (and presumably Descartes could also) conceive a machine thinking, without predicating consciousness to the machine. The machine might solve all manner of problems difficult for humans and yet we might well deny that 'I am conscious' has meaning when spoken by the machine.

Has the statement 'I am conscious' meaning when spoken normally by a human? This is the central question. We give it meaning by accepting that other people are importantly different from non-conscious machines capable of making utterances. A man might say, 'I am cold.' We light a fire for him. After a time, he might say, 'Thanks, I am no longer cold.' The same might occur also with a conceivable talking machine. It might utter words to inform us, truthfully, that it is too cold for it to work efficiently, just as warning lights might flash on its control panel to call for help.

If the man says 'I *feel* cold,' and the machine utters the same words, with the same stress, 'I *feel* cold,' can these have the same meaning? Not if we suppose that only for the man can 'I feel' refer to consciousness — to awareness of being cold. 'I feel cold' only has meaning if we accept that the source of the statement can be conscious. We must therefore say that it is meaningless when spoken by a man if we are sceptical that he is (or ever can be) conscious, just as it is meaningless for the machine source. Whenever we accept this and similar statements as in this way meaningful we must be accepting that the source is conscious. So we have a kind of test for discovering what we accept as conscious. We accept anything as conscious that can utter 'I am conscious' meaningfully.

But this is not a test for what is *in fact* conscious; it can test only our belief of what is or is not conscious. It has, however, some specific implications. Consider an example, such as someone saying : 'I see red.' It is commonly said that 'Your red may be different from my red.' This is said on the grounds that there is no operational test for comparing your 'red' with my 'red'. Now consider again not some specific sensation such as seeing red, but the general statement 'I am conscious.' If I cannot assert that your red is like my red, why should I believe that your consciousness is at all like my consciousness? If I consider your consciousness to be different from mine, I cannot conceive your consciousness, just as I cannot conceive your particular red. But then you might as well be a machine if I cannot conceive your consciousness. If indeed I cannot conceive any consciousness different from mine this is the position I am in: to me you might as well have no consciousness. I suppose I am a queer machine to you: impenetrable protoplasm! (See D.Dennett, 1978.)

Let us return to your red being different from my red. I might conceive your red as being my blue, and your blue as my red. Then there is no special difficulty, except that I have nothing to indicate the truth of this translation. I can conceive (though I have no evidence for) translations of sensations, but I cannot conceive novelties of sensation. So how can I conceive your different consciousness? There might be conceivable new combinations or arrangements of sensations, but new individual sensations are not conceivable. Perhaps this is a criterion for 'simple', 'atomic' sensations. (But we must be careful, for combinations may be highly surprising, as in colour mixture; and sounds mix very differently from colours.)

An abstract painter aims to produce new sensations or new combinations of sensations. Does he ever know what he does, except to himself? If he paints a 'red' patch he claims that he 'sees red', and he may assume (though he may have no reason that he could justify) that we have the same red sensation. Suppose that the painter adds a separate 'blue' patch. He describes this as a 'colder colour, appearing more distant'. We agree. Does this add evidence that my colour sensations are similar to his colour sensations? Can common shared associations with colours, or other sensations, resolve this problem?

If we explain that red appears warm and near as due, say, to associations with blood and fire being warm, and generally seen at close quarters; and that blue appears cold and distant by association with distant haze and snow, then our *verbal* agreements would not be good evidence that our colour *sensations* are the same. For if it is cross-associations that give verbal agreement, then verbal agreement is not evidence that our individual primary sensations are the same, for now we have a strong alternative account. Suppose, though, it should turn out that there is no such adequate explanation for verbal agreement. Would lack of alternative explanations suggest that indeed there is similarity of sensations between the owners of sensory systems having verbal agreement? In other words, would lack of evidence of associations to common situations giving verbal agreement suggest that our *sensations* are indeed similar?

There seems to be something of a paradox here in Empiricist philosophy. Locke argued strongly for direct knowledge by sensory experience of Secondary Characteristics, assuming that we each have similar sensations of, for example, colour, because Secondary Characteristics were supposed to be parts of the surfaces of objects which we share. Shared objects in the physical world were supposed to give the same sense data; and so to link Minds by perception. But Locke held also that we know features of objects by associations from sensations. His emphasis on associations would seem to weaken just the premise that he needed for sensations as the basis of shared, common experience. It weakens the claim that words such as 'red' refer to common, shared, sensations.

But if we lack evidence for common sensations such as colour, how can we believe in shared sensations (or indeed shared knowledge) of primary features of the world such as the structures of things? It seems that we depend on assumed shared knowledge of the behaviour of things (their mechanical properties), as a function of structures. If a triangular wheel ran smoothly, or was reported as running smoothly, we would be too surprised to accept the observation without suspicion of illusion or lying.

So here we return to the privacy of consciousness. Locke's Secondary Characteristics were an attempt to make some features of consciousness public. But the attempt fails. It fails unless we put consciousness into the common world of objects. This is what Leibniz and Spinoza did; but few now would follow them. Consciousness for us is somehow linked to our own brains and not to shared objects. And it is linked in some very odd way; for if I see your brain (in an operation) I do not share your consciousness, even if you are not anaesthetized and say that you are aware.

What, then, is the relation between Mind and brain? Are they different but causally linked (somewhat as parts of mechanisms are linked), or are they causally independent yet somehow synchronized? Or are they not separate at all but somehow aspects of the same thing? [4]

The commonsense and traditional view is that Mind and Matter each exist; that they are very different; and that they causally interact. This is commonsense Dualism, which we shall discuss in the next chapter.

Consciousness in computers?

Imagine that we designed and built a computer with an 'eye' capable of recognizing objects. This has indeed, to some extent, already been achieved. Now suppose that the machine has a performance significantly superior to the designer's expectation, or his ability to explain. Could he be forced to the conclusion that the machine has awareness, consciousness, and that this gives an ability greater than he can explain from the electronics and software-programmed procedures? This seems a a possibility well worth considering. I suspect, indeed, that this is the only evidence we could ever get (apart from paranormal phenomena) of causal consciousness. (See Gregory, 1977.)

Blind sight

It has been discovered recently (Weiskrantz et al., 1974) [5] that a small lesion in the right occipital lobe can produce blindness, in the sense that the patient reports that vision is entirely lacking in part of the visual field – though he can detect objects placed in this 'blind' region, and even discriminate features such as the number of fingers, or bars, presented. Thus Weiskrantz et al. write:

> The present results indicate that visual stimuli can be localised with considerable accuracy and that certain aspects of their orientation and spatial distribution can be differentiated if they are larger than a critical size when they are presented in the 'blind' region . . . although there was no acknowledged awareness of this capacity. . . . Throughout the

experiments he insisted that he saw nothing except in his intact field . . . when shown his results he expressed great surprise, and reiterated that he was only guessing.

This raises in dramatic form the question, 'What does consciousness do?' Vision is impaired from normal in this 'blind' region, but by studying just what is available, evidence can be found for which visual capacities are related to consciousness and which are not, in human subjects. It is still, however, a question that might be impossible to answer: 'Is consciousness *causally* necessary?'

16 How are Mind and brain linked?

I don't think, therefore I am not.

New joke

We feel the prick of a pin (Matter affecting Mind); and we will to command our arm to move (Mind affecting brain). This is how it seems. It also seems that we have an internal private stream of mental events, running along in time, which sometimes corresponds to physical events (perceptions), and which are sometimes very different (imagination, dreams, hopes, fears, and so on). This is the way that things seem; but is it incorrect? Perhaps conceptually incorrect, somewhat as Aristotle was wrong in thinking the sky rotates around a fixed earth, although it appears to do so; or as the Greeks thought and saw the planet Venus to be two stars though it is but one?

Perhaps it seems so clear from 'observation' that Matter and Mind are essentially different, and interact causally, that we do not recognize counter-evidence. The longstanding difficulty of demonstrating the earth moving and rotating is a warning of how strong evidence has to be to shift conceptual insight away from appearances. This has, however, happened many times in physics, but perhaps never with success in psychology. Mind and Matter are usually seen as very different interacting substances, related as interaction between public and private and as interaction between extended solid Matter with spaceless unextended ghostly Mind. On this view, Minds are active things, and exist in some way in their own right, and may even be not at all dependent upon Matter. This is the view of theology, a basis for survival after death and a hope of immortality. With this as a gift it is hardly surprising to find it the most popular account; but is it true?

We shall look now at alternatives, and return to interacting Dualism having considered other relations, and the possibilities of *no* Mind and that Mind is identical with physical features of brain. Here I shall develop an analogy suggested by Leibniz, who considered possible (causal and non-causal) relations between a pair of synchronized clocks. Leibniz wrote in *The Principles of Nature and Grace* (1714): [1]

> Now, this may take place in *three* ways. The *first* consists in mutual influence; the *second* is to have a skilful workman attached to them who regulates them and keeps them always in accord; the *third* is to construct these two clocks with so much art and accuracy as to ensure their future harmony. Put now the soul and the body in the place of these two clocks; their accordance may be brought about by one of these three ways. The way of influence is that of common philosophy, but as we cannot conceive of any material particles which may pass from one of these into the other, this view must be abandoned. The way of continual assistance of the creator is that of the system of occasional cases; but I hold that this is to make a *Deus ex Machina* intervene in a natural and ordinary matter, in which, according to reason, he ought not to co-operate except in the way in which he does in all natural things. Thus there remains only my hypothesis – that is, the way of harmony. From the beginning God has made each of these substances of such a nature that merely by following its own peculiar laws, received with its being, it nevertheless accords with the other, just as there were a mutual influence or as if God always put his hand thereto in addition to his general co-operation.

This is a useful analogy, for Dualism supposes that there are events, mental and physical, running together in time. A question for Dualism is how they remain synchronized. So let us develop Leibniz's analogy, using modern knowledge of how clocks may be synchronized electrically. This may serve at least as an *aide-memoire*.

The clock analogy

We shall consider three kinds of synchronized clocks, which are in fact used:

1   Master clocks, which are autonomous timekeepers.

2   Slave clocks, which are autonomous but labile timekeepers, which receive frequent corrections from the master.

3   Repeater dials (including domestic electric clocks run off the mains) which depend on uninterrupted master signals to keep going: they have no autonomy. (They may have 'catch-up' mechanisms to re-start them correctly following a break in master signals.)

We may now consider a rather different case: clocks driven directly by 'real time', from worldly events. An obvious example is sundials. Pairs of sundials will keep the same time, though they are merely passive 'repeaters' (not autonomous clocks), by following 'real time' as from the sun. Other examples would be Foucault pendulums, or gyroscopes moving together as the earth rotates under them.

Applying the clocks to brain—Mind parallelism

Suppose for a moment that the brain and Mind are separate entities, with physical and mental events running parallel in time, like our pairs of clocks. Suppose also that there is some kind of causal link (analogous to electrical clock-pulses) to synchronize them. Which system would, given our horological technology, work best?

Master–master

A pair of master clocks keeping exactly the same time though without causal links of any kind. This is the most unlikely horologically, as in practice independent clocks always differ in their rate.

This is, however, Leibniz's choice. It is known as Epiphenomenalism.

Master–slave

The autonomous master provides occasional corrections to the slave which is an autonomous clock, generally set to run either slightly fast or slow. So if the correction signals are lost for a long period the slave will continue running, though with steadily increasing error. It will catch up (or down) to correct the error gradually when the master signals are restored. In normal running, these two clocks do not run exactly together, but have short-term variations with long-term agreement.

This would be interactive Dualism, with Mind and brain each allowed some autonomy though with overall control of one by the other. Either Mind or brain might be master or slave. Generally Mind is conceived as master by philosophers, though physiologists might accord the honour to brain.

Master–repeater

A master provides uninterrupted signals to a repeater dial. If the signals are interrupted, when restarted the repeater dial will always be 'slow' by the lost time, until it is re-set. Re-setting may be automatic (in practice rapidly, at the hour). This is the most common and the simplest synchronized clock system.

Here either Mind or brain is regarded as entirely passive and entirely dependent on the other. An extreme reflex psychology would be of this kind, with Mind passive (though perhaps aware).

Slave–slave

There might be two (or better, very many) interacting slaves, each correcting the others so that there is a pooled average time. There would be small short-term varying discrepancies. On long-term average they would agree with each other.

This would be symmetrical interactive Dualism, or a kind of democracy with equal voting rights and no leader. It may be the view of strong upholders of psychosomatic medicine.

The other Dualist possibilities remain empty, for neither slave nor repeater can drive a repeater or a master clock; and one master driving another would deny both 'masters' the autonomous status of master. There are, however, some non-Dualist possibilities to consider.

Real world–repeater

A sundial is a kind of repeater, following time as given by the motion of the sun from its shadow, or by its image cast by a lens. This is a kind of analogy for extreme Direct Realism, or Idealism where the world is the Mind.

This applies to Locke's Secondary Characteristics and to Realist theories of perception in general. These theorists usually, however, allow some mental events or processes to occur free of the world; though philosophers and psychologists holding strong intuitionist theories of knowledge and truth may hold this Realist kind of view in extreme form.

One clock–one world

The clock is not a timekeeper, but a time-maker; this is solipsism.

One clock appearing as two clocks

This might occur with a clock reflected in a mirror, but this is not what I have in mind here. There are very many other ways in which a clock (or any other object) might appear, or be conceived, as two or more objects. This can happen with an appearance given, say, visually, and a description in words. Consider the clock and its photograph, or the clock and a verbal description. Are these two clocks or one clock? Once we know that the photograph is a representation of the clock we would say there is one clock and a representation. When the representation or description is very different in appearance from the clock, as perceived, then there is a danger of not realizing that there is a clock and its representation: they may seem two very different things. For example, a description of a clock as 'a seconds-compensated pendulum, with dead-beat escapement and equation of time dial' looks very different from the clock as we see it or touch it, and from its photograph. If we did not know that the description referred to the clock, as we know it in experience, we might assume that two things are involved; and they could be accepted as very different things, for it may not be at all clear that the description refers to any kind of clock. (This could also be so of the photograph, for people not acquainted with pictures or photographs as representations.)

Another description might be in terms of the molecular structure. Would this allow us to identify a clock that we know by experience? This seems highly doubtful. And if the description changed in synchronism with the clock, we might realize that they are somehow related, though appearing exceedingly different. The identity theory of Mind and brain suggests that Mind *is* brain states, appearing different rather as two kinds of descriptions of a single thing appear different.

Before discussing this further, we may look at Dualism in more detail, as this is the view held very generally, not only by the man in the street but also by the man in the study.

Defenders of Dualism
René Descartes' Dualism

The principal modern philosopher to propose and defend Dualism was of course Descartes. René Descartes (1596–1650) was born at La Haye, near Tours, and was educated in a Jesuit college. He was perhaps the greatest Rationalist philosopher. His Mind–brain Dualism set the paradigm for almost all later psychology. His main works are *Le Discours de la Méthode* (1637), *Meditationes de Prima Philosophia* (1641) and *Principia Philosophiae* (1644).

Descartes suffered poor health as a child, and developed the life-long habit of contemplating while lying in bed. His contemplations led to doubts that permeated the entire Western world. In his late boyhood he studied classics and philosophy at the Jesuit College of La Flèche, and retained a respect for his teachers, though he grew to reject their doctrines with a passion that kindled his own philosophy. He pinpointed the moment of commitment, to undertaking a reappraisal of the whole of human knowledge, to 10 November 1619, while he was sitting beside a stove, absorbed in the meditations which appear in the second part of *Discourse on Method*. That night he had what he regarded as three prophetic dreams, described in a paper entitled 'Olympia', now lost, which inspired him to develop rules of thinking for attaining knowledge. He set out twenty-four Rules (the last three are now lost) which set out the power of Mind and the role of consciousness in understanding.[2] In this early work, which led to Descartes' *Discourse on Method*, we see why he separated Mind from Matter, and gave Mind some autonomy. The point is that he saw Mind as accepting sensations, and departing from sensations to generate imagination – and follow rules that are not laws of physics. He suggested that to understand a problem we must contemplate it, and gain understanding by an inner intuitive awareness which is conscious. He thus stressed the importance of non-physical rules and non-corporeal links for action, and awareness, which may be very different from physical laws and yet essential for discovering and appreciating truth. He also conceived projective geometry, and stressed the importance of visualizing forms for understanding Nature. Here are some of Descartes' 'Rules for the Direction of the Mind':[3]

Rule 1   The end of study should be to direct the mind towards the enunciation of sound and correct judgments on all matters that come before it.

Rule 2   Only those objects should engage our attention, to the sure and indubitable knowledge of which our mental powers seem adequate.

Rule 4   There is need of a method for finding out the truth.

What Descartes means by 'method' is, 'rules so clear and simple that anyone who uses them carefully will never mistake the false for the true, and will waste no mental effort; but by gradually and regularly increasing his knowledge, will arrive at the true knowledge of those things which he is capable of knowing'.

Rule 5   Method consists entirely in the order and disposition of the objects towards which our mental vision must be directed if we would find out any truth. We shall comply with it exactly if we reduce involved and obscure propositions step by step to those that are simpler, and then starting with the intuitive apprehension of all those that are absolutely simple, attempt to ascend to the knowledge of all others by precisely similar steps.

For Descartes, 'in this one rule is contained the sum of all human endeavour'.

Rule 9   We ought to give the whole of our attention to the most insignificant and most easily mastered facts, and remain a long time in contemplation of them until we are accustomed to behold the truth clearly and distinctly.

In other Rules, Descartes emphasizes attention to instances, seeing connections between instances, and so on; so (like Bacon) he believes in rules of Mind that transcend Nature, and which may be developed by practice. Descartes believed (Rule 5) in moving to the simplest possible to understand the rest.

Perhaps it was following this rule that led him to doubt all that he could, until he came to his own thinking Mind. His argument is interesting: he starts by doubting all around him including his own body, imagining that he is being deceived by some baleful influence, so that all that he senses is illusion. But, he argues, if this were so, there is still himself that is being deceived, so he – his thinking Mind – must exist though all around is illusion. He concludes that he is more sure of the existence of his Mind than of his body:[4]

I suppose, then, that all the things I see are false; I persuade myself that nothing has ever existed of all that my fallacious memory represents to me. I consider that I possess no senses; I imagine that my body, figure, extension, movement and place are but the fictions of my mind. What, then, can be esteemed as true? Perhaps nothing at all, unless there is nothing in the world that is certain. . . . Am I so dependent upon body and senses that I cannot exist without them? . . . But is there some deceiver or other, very powerful and very cunning, who ever employs his ingenuity in deceiving me? Then without doubt I exist if he deceives me, and let him deceive me as much as he will, he can never cause me to be nothing so long as I think that I am something. . . . We must come to the definite conclusion that this proposition: I am, I exist, is necessarily true each time I pronounce it, or that I mentally conceive it.

. . . I do not now admit anything which is not necessarily true: to speak accurately I am not more than a thing which thinks, that is to say a mind or a soul, or an understanding, or a reason, which are terms whose significance was formally unknown to me. I am, however, a real thing and really exist; but what thing? I have answered: a thing which thinks.

Having got down, as he sees it, to bedrock, Descartes builds up his physics and his philosophy of Mind. In the *Principles* he writes: 'We must know, therefore, that although the mind of man informs the whole body, it yet has its principal seat in the brain, and there it not only understands and imagines, but also perceives. . . .'[5] This he justifies with evidence such as that sleep, and diseases that only affect the brain, upset the senses; and cutting the nerves from the senses interrupts sensation. He also describes, in several places, the fascinating phantom limb phenomena, where patients after amputation experience limbs that are no more, and argues that losing parts of the body does not produce losses of Mind or consciousness. 'It is the soul that sees, and not the eye; and only by means of the brain does the immediate act of seeing take place.'

As Anthony Kenny points out: [6]

Descartes arrived at his position by taking seriously optical illusions and errors of sense. Consider the familiar case of the straight stick that looks bent when half immersed in water. In such a case, Descartes believed three things happened: (1) light rays reflected from the stick excited motions in the animal spirit in the optic nerve and thus affected the brain; (2) an idea of colour and light results in the mind; and (3) the will makes a judgement that the stick is bent. It is the second that is really called sensation; the first could occur purely mechanically in animals, and the third does not occur when the cautious man refrains from misjudging the stick. The result of this threefold apparatus is that in what is strictly called sensation – the idea in the mind – there is no error. There can be mistaken judgement and there can be mechanical breakdown, but the pure sensation cannot be mistaken, and so the interests of epistemology and theodicy are safeguarded.

Descartes' famous link between brain and Mind – the pineal gland – he described in *The Passions of the Soul:* [7]

That there is a small gland in the brain in which the soul exercises its functions more particularly than in the other parts. It is likewise necessary to know that although the soul is joined to the body, there is yet a certain part in which it exercises its function more particularly than in all the others; and it is usually believed that this part is the brain, or possibly the heart: the brain because it is with it that the organs of sense are connected, and the heart because it is apparently in it that we experience the passions. But in examining the matter with care, it seems as though I had clearly ascertained that the part of the body in which the soul exercises its functions immediately is in nowise the heart, nor the whole of the brain, but merely the most inward of all its parts, to wit, a certain very small gland which is situated in the middle of its substance and so suspended above the duct whereby the animal spirits in its anterior cavities have communication with those in the posterior, that the slightest movements which take place in it may alter very greatly the course of these spirits, and reciprocally that the smallest changes which occur of the spirits may do much to change the movements of this gland.

Descartes justifies his choice of the pineal gland as the seat of the soul in the next passage: [8]

The reason which persuades me that the soul cannot have any other seat in all the body than this gland wherein to exercise its functions immediately, is that I reflect that the other parts of our brain are all of them double, just as we have two eyes, two hands, two ears, and finally all the organs of our outside senses are double; and inasmuch as we have but one solitary and simple thought of one particular thing at one and the same moment, it must necessarily be the case that there must somewhere be a place where the two images which come to us by the two eyes, where the two other impressions which proceed from an object by means of the double organs of the other senses, can unite before arriving at the soul, in order that they may not apprehend how these images or other impressions might unite in this gland by the intermission of the spirits which fill the cavities of the brain; but there is no other place in the body where they can be thus united unless they are so in this gland.

Descartes regarded the body as a machine without feeling or purpose, but able to keep going and be responsive to events (to be called 'stimuli') by reflex action. The Mind (or 'soul') he at first thought permeated the entire body, on the grounds that one can feel to one's fingertips. This notion he abandoned, mainly because of the evidence of people with amputated limbs reporting sensations in 'phantom limbs' that they had lost. This forced him to narrow down the seat of the Mind, to the brain and finally to the curious brain structure, the pineal gland. (This he chose because, of all the brain structures, it is not duplicated in the two hemispheres.) Descartes thought that the optic nerves run from the retina to the pineal gland, there to represent reality. He was perhaps the first in modern times to consider how perceptions might be representations in the brain, rather than pictures or copies, or selections of objects, as held by the Direct Realists.

Descartes is saying in his Dualism that Mind and body are made of different substances. He denies identity of Mind with physical brain states, and so he is not a Materialist; and he does accept that his body exists independently of Mind, and so he is not an Idealist. His notion of Dualism is based firmly on the notion of mental and physical substances as being essentially different. But what is substance? We have failed to find any kind of answer, for substances are not supposed to have any properties other than given by form. This has been held universally since the alchemists. This change of view is indeed the great divide between alchemy and chemistry (see pp. 96–118).

It might, however, be argued – and Descartes does suggest this kind of argument – that the 'substances' are essentially different because the kinds of language that we use for describing mental and physical phenomena are not merely different, but logically different. The power of this argument depends entirely on what is to be meant by 'logically different'. And whatever this is taken to mean, it would seem that a logical difference is being used to establish a contingent difference. This takes the argument out of Empiricism.

For an Empiricist, logical distinctions are in language, but they cannot be used to justify empirical or contingent distinctions unless we know that what is being said is true. For example, if squares and circles are logically different, this arises from our language (or mathematics) for describing these kinds of objects. It may be conceivable that, on a different classification, they would not be described as logically different, though they would still look very different. Descartes does suggest empirical-type evidence for not equating Mind with the body, for example that the body may decay though the brain is not impaired.

For him, understanding is not at all related to brain states. Thus, in Article 17 from *The Passions of the Soul:*

> After having thus considered all the functions which pertain to the body alone, it is easy to recognize that there is nothing in us which ought to attribute to the soul excepting our thoughts, which are mainly of two sorts, the one being the actions of the soul and the other its passions.

Descartes seems to hold this view because he does not accept that a machine could be made that would understand. Thus he writes in *Discourse on Method:* [9]

> If there were machines which bore a resemblance to our body and imitated our actions as far as it was morally possible to do so, we should always have two very certain tests by which to recognise that, for all that, they were not real men. The first is, that they could never use speech or other signs as we do when placing our thoughts on record for the benefit of others. For we can easily understand a machine's being constituted so that it can utter words, and even emit some responses to action on it of a corporeal kind, which brings about a change in its organs; for instance, if it is touched in a particular part it may ask what we wish to say to it; if in another part it may exclaim that it is being hurt, and so on. But it never happens that it arranges its speech in various ways, in order to reply appropriately to everything that may be said in its presence, as even the lowest type of man can do. And the second difference is, that although machines can perform certain things as well or perhaps better than any of us can do, they infallibly fall short in others, by which we may discover that they did not act from knowledge, but only from the disposition of their organs. For while reason is a universal instrument which can serve for all contingencies, these organs have need of some special adaptation for every particular action.

So here again we find, as we found for the Greek difficulties with self-initiated movement, which we attributed to their lack of prime movers (see pp. 47–8), an argument based on technological inconceivability. (Ultimately, too, Popper's notion of 'refutation' from observations or experiments is the same: see pp. 249–62.)

Is it true that machines can never be made to understand speech? As soon as this was achieved, Descartes' conclusion would fail. Current failure to achieve it may be temporary and so cannot demonstrate that he is correct, for his argument depends on supposing it to be forever impossible. It seems fair to add that advances in technology since his time make us far less sure that understanding speech is impossible for machines. It is indeed now possible, in rudimentary form, and there is active research on speech recognition by Artificial Intelligence.

We may, however, suppose that a machine that understands speech convincingly (and this is technically extremely difficult to accomplish) may have a separate Mind. How would we know that the machine is not, at this point, Dualistic – with a Mind of its own? Perhaps we would conclude just this if its performance was surprisingly good, from our knowledge of engineering. This would be another kind of surprise. Would this surprise be explained by our inadequate understanding of engineering, computing or whatever? These may be questions for future computers to answer (see p. 457).

Descartes' Dualism was largely based on his inability to imagine thought without a thinker – his own Mind – though he could imagine objects, including his body, as illusory. Technology pushes the limits of conceivability into new possibilities; so, as science advances, philosophy becomes more uncertain of its conclusions as possibilities multiply. Descartes was explicitly aware of the current technology of his time, and was himself active and unusually successful in science. It is a warning to us that so outstanding an intellect could be led so far astray by taking analogies from technology; for example, Descartes' physiology was disastrously wrong, as he took the heart to be the source of heat for digestion and the brain, and he entirely misunderstood the role of respiration. He was no nearer to current knowledge than Aristotle in these matters, and was perhaps led even further astray by the greater variety of physical systems which he could use as models. Where William Harvey (1578–1657)[10] adopted the appropriate model of a pump for the heart Descartes took as his model a furnace – to be fundamentally misled. We tend now to take computers (or rather programs) for our models of brain function, and Mind, and this raises for us interesting suggestions for considering Mind–brain Dualism. But is the new computer technology appropriate for models of Mind and brain function – or is it our seducer?

William James's Dualism

James was writing at a time when stimulus–response accounts of behaviour looked a lot more adequate than they do now. They were then generally taken as the basis of behaviour: now they are seen as far less important than control of behaviour by internally stored knowledge, without continual need of external triggering stimuli to maintain behaviour. When James does discuss behaviour not triggered by a stimulus, it is for him something special and requiring a mental explanation – Mind affecting brain. At least, this is how I read him.

William James discusses Mind affecting brain in *The Principles of Psychology* (1890) especially under the heading 'The Intimate Nature of the Attentive Process', where he discusses attention and anticipation as caused by Mind. He gives as examples: (1) the accommodation or adjustment of the sensory organs, and (2) the anticipatory preparation from within of ideational centres concerned with the object of which the attention is paid.[11] The point that he makes here is that the eyes are adjusted before, in anticipation of, what visual signals will be needed; so their movements are not always controlled by stimuli, but may occur in darkness, or otherwise in the absence of any visual stimuli – according, he suggests, to purely internal processes of mental images. He says there can be 'the feeling of an actual rolling outwards and upwards of the eyeballs, such as occurs in sleep, and is the exact opposite of their behaviour when we look at a physical thing'. It is 'the exact opposite' because it originates from inner (mental) imagination of the world, and not from the outer present (physical) world. The question is: do we know that it is internal mental events which are moving the eyes, or could it be internal physical events? This question must be asked, for it is clear that there are physical (physiological) processes capable of moving the eyes from within. If everything inside the skull could be said to be mental, this whole situation would be much simpler!

William James goes on to consider various perceptual examples of what he sees as Mind controlling brain. He cites an interesting observation of Helmholtz's, which I am not sure has been investigated since. Helmholtz found that simple stereograms could be made to fuse by will, when presented as after-images from the flash of a single spark, so that eye movements are ineffective. Helmholtz says: '*if I chance to gain a lively mental image of the represented solid form (a thing that often occurs by lucky chance) I then move my two eyes with perfect certainty over the figure without the picture separating again*'. This must be a 'central' effect because the after-images are stuck to the eyes.

William James is careful to point out, however, that it could be that other physical brain processes are affecting the fusion of the images (or Necker cube reversals or whatever), in which case such effects would be no evidence for Mind affecting brain. I shall quote James in full here, where he considers this Physicalist alternative (though he comes to reject it):[12]

When . . . I symbolized the Ideational preparation element in attention by a brain-cell played upon from within, I added 'by other brain-cells, or by some spiritual force' without deciding which. The question 'which?' is one of those central psychologic mysteries which part the schools. When we reflect that the turnings of our attention form the nucleus of our inner self; when we see that volition is nothing but attention; when we believe that our autonomy in the midst of nature depends on our not being pure effect, but a cause –

> *Principium quoddam quod fati foedera rumpat,*
> *Ex infinito ne causam causa sequatur –*

we must admit that the question whether attention involves such a principle of spiritual activity or not is metaphysical as well as psychological, and is well worthy of all the pains we can bestow on its solution. It is in fact the pivotal question of metaphysics, the very hinge on which our picture of the world shall swing from materialism, fatalism, monism, towards spiritualism, freedom, pluralism, – or else the other way.

James urges that thinking, attention and will have significant effects, if only that will 'would deepen and prolong the stay in consciousness of innumerable ideas which else would fade more quickly away'.[13]

James ends this discussion by objecting to the Materialist analogy that the sense of will, occurring during difficulty, is merely physical – like the 'turbulence of rivers in constricted regions'. He waxes eloquent:[14]

Meanwhile, in view of the strange arrogance with which the wildest materialistic speculations persist in calling themselves 'science', it is well to recall just what the reasoning is, by which the effect-theory of attention is confirmed. It is an argument from analogy, drawn from rivers, reflex actions and other material phenomena where no consciousness *appears* to exist at all, and extended to cases where consciousness seems the phenomenon's essential feature. The *consciousness doesn't count,* these reasoners say; it doesn't exist for science, it is nil; you mustn't think about it at all. The intensely reckless character of all this needs no comment. . . . For the sake of that theory we make inductions from phenomena to others that are startlingly *un*like them; and we assume that a complication which Nature has introduced (the presence of feeling and of effort, namely) is not worthy of scientific recognition at all. Such conduct may conceivably be *wise,* though I doubt it; but scientific, as contrasted with metaphysical, it cannot seriously be called.

Whatever the reader makes of William James's impassioned prose in favour of mental images of Mind controlling brain, he states very clearly this traditional 'commonsense' view that consciousness has significant, if small, causal effects on perceiving, thinking and behaviour. On the other hand, he presents no convincing evidence.

Karl Popper's and John Eccles's Dualism

This view, which William James advocated, has recently been defended in similar ways by Sir Karl Popper and Sir John Eccles, in *The Self and Its Brain* (1977). They take two cases of what they regard as strong evidence for Mind affecting brain.

The first involves the fascinating reports of people who, having their brains electrically stimulated while undergoing brain surgery (Penfield and Roberts, 1959), experience streams of memories or other mental images, and at the same time are aware that they are in the operating theatre. Popper and Eccles suppose that because the externally signalled theatre, and the memories internally remembered though elicited by electrodes, occur simultaneously, they cannot both be physical. So one must be mental. This assumes, though, that different brain regions cannot operate simultaneously. But why not? Many activities are normally carried out simultaneously. I see no evidence for independent Mind here. These experiments are of the greatest interest; but surely they do not demonstrate causal Mind – only causal electrical stimuli, which may be internal or external. The second case involves the speeding up or slowing down of Necker cube reversals, by act of will. This they take as evidence of Mind affecting brain without explicit reference to the alternative – which William James does – of allowing that other brain processes may produce or inhibit the perceptual reversals. Why should this particular example illustrate, or demonstrate, Mind affecting brain? Surely there need not be two essentially different kinds of processes – mental and physical – and after all, computer programs work with just such interactions.

However this may be, for Popper and Eccles, Mind and brain are separate, with weak slow-acting interaction (like master–slave, or perhaps slave–slave clocks: see p. 461). Thus Popper describes illusions: [15]

. . . there are two kinds of illusions – illusions delivered to us or imposed upon us by the brain, and illusions which have a mental origin, let us say, wish-fulfilment. It is apparently built into our organism and into the whole 'mechanism of interaction' between the brain and the mind that the mind should be in many respects dependent on the brain, in order not to fall too easily into that kind of illusion which we experience in fantasy.

I would say that this whole field can be used to show at the same time a kind of gulf and also a kind of dependence between the self-conscious mind and the brain.

But do any illusions give us the right to make such statements? Granted that we can perceive one thing and know at the same time that this perception is distorted or in some other way false, surely it by no means follows that one of these (the perception or the knowledge) is mental and the other physical. Indeed, computers could not be given check procedures that involve recognizing discrepancies if this were so. All that we can infer is that the brain can process more than one thing at a time and that discrepancies can be noted, as when we recognize illusion. Of course this could be rewritten 'The *Mind* can process more than one thing at a time,' but this again gives no reason for saying that there is evidence here of Mind and brain as separate, interacting entities. I conclude that James's alternative (which he disliked) is in no way discounted by any such evidence. So none of these examples seems to me to provide any evidence for interactive Mind–brain parallelism; neither does the double awareness of the patient with his brain stimulated in the operating theatre. This could as well (and with less assumption) be two brain (or two Mind) processes as one a brain and the other a Mind process as Popper and Eccles suppose.

There seems no strictly neurological evidence for interactive Dualism. There does, however, seem to be some evidence against it, as we shall now consider.

Attacks on Dualism
1   Awareness comes too late

Do we act because we are aware of a situation, or are we aware of a situation because we have acted? One normally thinks of a cause preceding its effects, so this question of a temporal sequence is clearly important. In the case of a startle reaction to a sudden event, one frequently finds that one has acted, and *then* experiences shock and realizes what has happened. In the case of accidentally putting one's hand under a too-hot tap, one feels the pain a second or more *after* rapid withdrawal of the hand. Only then does one know one is hurt.

The James–Lange Theory of Emotion is closely related. This supposes that we react to, for example, a fear-provoking stimulus – and that we then become aware of visceral body changes elicited by the increase in adrenalin and other physiological reactions to the stimulus. This removes the sensation as causally producing 'flight' or 'fright' behaviour, for it comes too late – after the behaviour – and so it cannot be the cause.

It is, no doubt, not always possible to be sure that this is the temporal sequence; but if, indeed, awareness often comes after the response to an event, it is hardly possible to suppose that the awareness is necessary, at least for most behaviour. This leaves the way free for descriptions of behaviour in physiological rather than mentalistic terms. We are, though, still left with the question of why we have consciousness if it is not causally significant, in the way that neural action potentials are clearly causally significant for behaviour and perception.

In interacting systems (such as some of the clock systems that we have considered), there can be short-term variations and alternations of leading and lagging in time, though they remain synchronized in the long run. Could this consideration be applicable here? It seems unlikely that it could apply to the sequence of events initiated by input stimuli, but it may be worth considering. In the meantime, the delay in awareness so often found may seem to be strong evidence against the notion that awareness is causally important. I shall, however, now suggest a possible reason for delays of awareness.

2 Prediction comes too soon

It should surely be possible to apply Bacon's and Mill's Methods of Inductive enquiry to establish the importance (if any) of consciousness, by asking in which situations we are most aware. Behaviour falls to bare maintenance of life processes under anaesthesia, so evidently awareness is not necessary for remaining alive. This is itself interesting, and important to note. When the physiology of anaesthetics is better understood, they may well prove to be extremely powerful tools for investigating consciousness and brain function.

It might be generally agreed that one is scarcely, if at all, aware of the particular movements, and certainly not the particular muscles, used in familiar skills such as walking or driving. This changes dramatically, though, when something unusual happens, or when a muscle has cramp or otherwise fails to function properly.

One is, I think, most aware of *failed predictions* during skilled performance. Is awareness at its greatest when there is a mis-match between (as Craik would put it) the prevailing Internal Model, and the signalled success of performance, as guided predictively? When the prevailing Internal Model (or 'hypothesis', as we might prefer to call it) is signalled as inadequate, or inappropriate, then we become aware. This at least is how it seems to me.

This has a rather interesting implication for how we may think about 'perceptual hypotheses' (pp. 395–415). We do not need to say that they are always associated with awareness – so we can without inconsistency reject the notion that perceptions are made up of sense data. At the same time, though, we may be aware of perceiving, say, a tree. Is this only when the tree is surprising – giving a mismatch with our general expectations? Are we aware only of surprising features of a very familiar tree?

If consciousness is primarily (if not entirely) associated with predictive inadequacy or failure, as seems to me to be likely, we have some sort of an answer (p. 401) to why awareness often comes too late to be causally important, in the way that action potentials are causally important. The point is that mis-matches between prediction and signalled result must come after the result is signalled. This applies to the hand burnt by hot water, as it does to a sudden emergency, say when driving a car. It is suggested by the expert on pain, Pat Wall, that pain which may be delayed after serious injury for a day or more serves as a continual reminder to be cautious and to remain inactive. This is a somewhat different notion: both could be true.

We still have to consider what consciousness does in correcting predictions, or selecting new hypotheses, or whatever. It would seem that for it to serve such a purpose it must have causal power of some kind. Or is it an aspect of something else, which is causal? The suggested evidence for causal Mind I find unconvincing (except that if it has evolved it must surely do something), and the fairly clear evidence against it seems to be that awareness often occurs after the events or the actions that Mind might be supposed to control. On the other hand, the suggestion that awareness is mainly found, and so is associated with, failed predictions from inadequate Internal Models or hypotheses, may get round this difficulty. There is indeed a sad shortage of evidence for deciding the issue.

One may suspect that at our most creative we are not particularly aware, and indeed may be quite unaware of what is going on. As someone said: 'I don't know what I think until I have said it.' How much is one aware of thought processes when talking philosophy? I suspect that awareness here is after the event, even for our 'highest' mental processes.

Should it be possible to time, with high accuracy, 'mental' events and corresponding 'physical' brain events, it should be possible to settle the causal argument, for causes are supposed always to precede what they cause. If there is no time difference, this could be evidence for an identity theory.

A most interesting kind of psychophysical experiment is to record electrical evoked potentials from the human cortex and relate this activity in time to awareness. Experiments of this kind have been carried out by Benjamin Libet (1965, and especially 1977). Libet claims that the sensation of touch from a weak stimulus is earlier than the related brain activity produced by the stimulus. There are however severe difficulties over the interpretation of these experiments, as pointed out by the philosopher Patricia Churchland (unpublished). It is difficult to time awareness accurately; and it may not be clear that the relevant activity is being recorded. However this may be, this is a line of enquiry well worth following up, if only to establish just what are the neural correlates of consciousness. We end, though, with a curious point: we would surely not accept as immediately related recorded cortical activity which does not occur simultaneously with awareness. We are still far from conclusive experimental evidence, and there seems to be a lack of an adequate paradigm for interpreting the available evidence; or for designing experiments to answer these questions, of extreme neurological and philosophical interest. We need further evidence along these lines.

Perhaps the most dramatic as well as the most direct experiments relating brain with Mind are electrical stimulation with microelectrodes of the brains of conscious human subjects. The subjects are of course patients undergoing brain surgery – usually, if not always, suffering from epilepsy; so they are by no means normal. We must recognize also that they are likely to be under very considerable stress. Nevertheless, the reports of these patients are of the greatest interest. I shall quote from the important book by Wilder Penfield and L. Roberts, *Speech and Brain Mechanisms* (1959). Also referred to here is W. Penfield and H. H. Jasper, *Epilepsy and the Functional Anatomy of the Human Brain* (1954).

Penfield and Roberts describe the experience of a patient receiving stimulation of the temporal lobe, as generally recognizing a play-back or 'flash-back', as they call it, from their past. Here is a detailed report of such a case: [16]

Some patients call an experiential response a dream. Others state that it is a 'flash-back' from their own life history. All agree that it is more vivid than anything that they could recollect voluntarily.

A patient (Penfield and Jasper, 1954, p. 137) was caused to hear her small son, Frank, speaking in the yard outside her own kitchen, and she heard the 'neighbouring sounds' as well. Ten days after the operation she was asked if this was a memory. 'Oh, no' she replied. 'It seemed more real than that.' Then she added 'Of course, I have heard Frankie like that many, many times – thousands of times.'

This response to stimulation was a single experience. Her memory of such occasions was a generalization. Without the aid of the electrode she could not recall any one of the specific instances nor hear the honking of automobiles that might mean danger to Frankie, or cries of other children or the barking of dogs that would have made up the 'neighbourhood sounds' on each occasion.

The patients have never looked upon an experiential response as a remembering. Instead of that it is a hearing-again and seeing-again – a living-through moments of past time.

Penfield and Roberts give the following general comments, or description of what they find: [17]

Time's strip of film runs forward, never backward, even when resurrected from the past. It seems to proceed again at time's own unchanged pace. It would seem, once one section of the strip has come alive, that the response is protected by a functional all-or-nothing principle. A regulating inhibitory mechanism must guard against activation of other portions of the film as long as the electrode is held in place, the experience of a former day goes forward. There is no holding still, no turning back, no crossing with other periods. When an electrode is withdrawn, it stops as suddenly as it began.

A particular strip can sometimes be repeated by interrupting the stimulation and then shortly re-applying it at the same or a nearby point. In that case it begins at the same moment of time on each occasion.

They conclude:

Every individual forms a neuronal record of his own stream of consciousness. Since artificial re-activation of the record, later in life, seems to re-create all those things formally included within the focus of his attention, one must assume that the re-activated recording pattern of neuronal activity may be considered much more than a record, for it was once used as the final stage of integration to make consciousness what it was.

They go on, however, to indicate a Dualist account of consciousness: [18]

One might suppose that originally, like a strip of film, its meaning was projected on the screen of man's awareness, and somehow it was held in place there for a brief time of consideration before it was replaced by subsequent experience and consequent neuronal patterns. . . . Consciousness, 'forever flowing' past us, makes no record of itself, and yet the recording of its counterpart within the brain is astonishingly complete.

What are we to make of this? Are they justified in introducing this 'screen' of awareness – or is this an elementary philosophical mistake, leading to an infinite regress of 'screens' and 'viewmasters'? It is not clear how such a regress can be avoided: surely this account is deeply confusing. Can the obvious difficulties of 'screen' accounts be removed, by supposing that Mind and brain are identical?

Are Mind and brain identical?

I shall introduce Mind–brain identity with the words not of a philosopher but rather of a nineteenth-century naturalist, who was a personal friend of Charles Darwin: George John Romanes (1848–94). In *Mind and Motion* (1885), Romanes writes: [19]

> We have only to suppose that the antithesis between mind and motion – subject and object – is itself phenomenal or apparent: not absolute or real. We have only to suppose that the seeming duality is relative to our modes of apprehension: and, therefore, that any change taking place in the mind, and any corresponding change taking place in the brain, are really not two changes but one change.

This is a remarkably clear statement of the brain–Mind identity notion. Romanes continues:

> When a violin is played upon we hear musical sound, and at the same time we see a vibration of the strings. Relatively to our consciousness, therefore, we have here two sets of changes, which appear to be very different in kind; yet we know that in an absolute sense they are one and the same: we know that the diversity in consciousness is created only by the difference in our modes of perceiving the same event – whether we see or whether we hear the vibration of the strings. Similarly we may suppose that a vibration of nerve-strings and a process of thought is really one and the same event, which is dual or diverse to our modes of perceiving it.

> The great advantage of this theory is that it supposes only one stream of causation, in which both mind and motion are simultaneously concerned.

Romanes gives another analogy, which we might also include:

> In an Edison lamp the light which is emitted from the burner may be said to be caused by the number of vibrations per second going on in the carbon, or by the temperature of the carbon; for this rate of vibration could not take place in the carbon without constituting that degree of temperature, which affects our eyes as luminous. Similarly a train of thought may be said indifferently to be caused by brain-action or by mind-action – for *ex hypothesi*, the one could not take place without the other . . . volition is produced by mind in brain, just as light is produced by temperature in the carbon. And just as we may correctly speak of

light as, say, the cause of a photograph, so, we may correctly speak of volition as the cause of bodily movement. . . . It thus becomes a mere matter of phraseology whether we speak of the will determining, or being determined by, changes going on in the external world; just as it is but a matter of phraseology whether we speak of temperature determining, or being determined by, molecular vibration.

Romanes regards this identity as a hypothesis for which there is no evidence; but for him it is 'the only one which is logically possible, and at the time competent to satisfy all the facts alike of the outer and the inner world'.

This is in many ways a highly attractive account. It does, however, raise questions that are far from resolved. In the first place, clearly much hinges here on what we mean by 'identity'. It cannot be taken to mean that everything we say of brain states, we can say of consciousness. The problem, indeed, is to find anything in common! So this is no 'surface' identity. What criteria for identity should, then, be satisfied for brain states to be accepted as identical with conscious states? If we have to accept criteria for 'identity' that would be accepted in other cases (such as electron flow and lightning, or electromagnetic radiation and light), just how like such cases does the Mind–consciousness relation have to be, to be accepted as an 'identity'?

If it is a unique case – and this is the trouble about consciousness – is it helpful to apply criteria taken by analogies? At least two criteria for 'identity' between two things (A and B) would normally be demanded. First, that there are precisely related time relations between the occurrences, or changes, in A and in B. Secondly, that A and B are coincident in space. Whether there are exact time relations between brain states and consciousness seems to be an empirical question that might be answered by experiment (as discussed on p.474). Spatial coincidence of brain states and consciousness poses a deeper problem, for it seems misleading to say that consciousness occupies space. How, then, can brain states and consciousness be identical if one occupies space and the other does not? Could this, however, be a case where we take over criteria of 'identity' from common physical object examples that are inappropriate for this special case?

This problem for the identity theory is considered by J.J.C. Smart in a paper, *Sensations and Brain Processes* (1959). He dismisses the objection as an *ignoratio elenchi* (disproving a statement different from that advanced by an opponent). Smart says that he is 'not arguing that an after-image is a brain process, but that the experience of having an after-image is a brain process. It is the *experience* which is reported in the introspective report.' His point is that though an after-image appears to be in outer space, the experience is not.

We often speak ambiguously about after-images. The ambiguity is that 'after-image' covers both physical changes at the retina and the experience that may be reported. Normal perception of objects is given by retinal signals that are virtually the same as when the retina is signalling after-images: we then speak of perceiving objects, not perceiving retinal activity.

We do not ever say that the experience is located in the retina. Is it less odd to say that the experience *is* located in the brain? The trouble is that the experience is not spatial, while the brain is a physical object in physical space. The identity theory supposes that (both for after-images and for normal perception) non-spatial experience can be identical with an object (the brain) lying in space. The identity theory is asserting identity when one term is not in space though the other is in space: the theory therefore flouts the usual requirement for identity of spatial coincidence. Since the brain exists in space but experiences do not, for the identity theory to be acceptable we must relax the usual spatial requirement for identity.

We do allow that two spatial things can be in some ways identical though they have spatial differences; for example, a lump of lead and a sack of feathers may be identical in weight. Even though lead and a feather are very different in most respects, they do have this identity. But the physical brain is a spatial object while Mind is non-spatial, and has no properties of objects such as mass, solidity and opacity to some wavelengths of the electromagnetic spectrum. It might however be said to change in time. The principal objection to the identity theory is simply that Mind and brain do seem so extremely different that it is hard to see resemblances, let alone anything much that could be identical.[20]

Can we think of some other relation that might hold between brain and Mind such that Mind is intrinsically part of brain (or at least brain function), without our having to say against all appearances that they are identical? I suggest that we should, as a candidate, consider *meaning*.

Meaning and Mind

There is a form of identity that allows extremely large physical differences and that does not demand spatial congruence: *identity of meaning*.

If we write '$6 \times 9 = 54$', the marks either side of the identity sign '$=$' are very different. They are physically, spatially, different; and yet they have identical meaning. Perhaps even more physically different and yet with identical meaning, is 'Six times nine = 54'. This notion may be extended to instruments, including slide rules and other analogue devices, and to mechanical and electronic digital computers. They represent, from one physical state, to a very different physical state or abstraction, according to some kind of code relation. To see the identity we must appreciate the code. The suggestion is, then, that certain physical brain states represent ideas and experience so that these are formally identical though physically very different: rather as '$2 + 3$' is identical with '5' though they are spatially different.

This relation is clearly broader than what is needed for consciousness, for much is represented by states of the nervous system of which we are not conscious. So we still have to explain why we are conscious of only some – representing – brain states. Similarly, we do not want to say that computers are conscious because they represent in this kind of way. Clearly, then, this kind of identity can only be claimed as a necessary and not a sufficient condition for consciousness.

To ask why only some and not all representing is associated with consciousness might be regarded as an empirical question, to which we could find an answer by experiments. This at least is where I would look for the key to consciousness. Here there are some suggestive clues in various well-known phenomena, especially of perception. Considering ambiguous figures such as fig. 39 and paradoxical figures such as fig. 45, there are obvious similarities to ambiguous and paradoxical sentences. It would be very interesting to discover just what brain changes occur (and do not occur) when perceptions shift from one alternative to another in ambiguous perceptual situations. Possibly the representing brain states remain unchanged with the changes of perception, as the physical sentence remains unchanged although its meaning shifts. For example: 'He has fallen down on the job' has alternative meanings with no change in the marks (or sounds, if spoken) of the sentence. The same sentence without change carries alternative propositions. We might look, then, for unchanging representing brain states; but with changes in the reading mechanism, with shifts of perception with ambiguous figures or objects. On this view, ambiguities and puns are critically important phenomena, for which we should seek neural correlates.

Paradoxical perceptions are interesting in a rather different way. Physical obects cannot be paradoxical, and yet perceptions can be paradoxical. This shows (as pointed out on p.407) that paradoxical perceptions cannot be said to be directly related to at least some perceived objects, which is an argument against Direct Realism. If we now shift this argument away from perceived objects and towards brain states representing perceptions, we see that they also can hardly be directly related to (or themselves be) perceptions. They could, however, have a *meaning* relation.

We can write paradoxes very easily with physical marks accepted in a formal code. For example, '$6 \times 8 = 54$' is a paradox, given acceptance of our normal meanings of these marks (symbols) and the rules of multiplication indicated by '$\times$'. So, on a meaning account, the problem of how brain states give paradoxical perception fades away, for paradoxes can be represented by, though not embodied in physical objects.

We may soon be technically equipped with brain recording techniques adequate to discover the features of representing associated with consciousness. Should this kind of account turn out to be correct, how will we answer the question: 'Is consciousness mental or physical?' The answer should be based on how we regard meaning. Is meaning mental or physical? Given that computers (consciousness apart) accept meaning, and computers are physical machines, we may wish to say that meaning can be physical. But this should not be said without realizing that it is only the procedures carried out by the machine which are relevant. The physics of the machine is not important, except when it fails to be appropriate. Are procedures 'physical' or 'mental'? Curiously, it doesn't seem to matter how we answer! Once a Dualist account is abandoned, the Mind–brain distinction similarly disappears.

So, to return to the clock analogy, we have a new relation to suggest: the position of the hands to the time they tell. These are identical, although the hands are physical, and have spatial position, while time is not spatial and has no extension or position. I suggest that there are not two clocks, as Leibniz supposed, but rather, on the same analogy, there is one clock (the brain) that tells the time and much else by its coded meaning – which we might be able to discover experimentally. We might then read brains as we read clocks, or books, from knowledge of the code or language by which their states have meaning. Consciousness is private, on this view, because there is only one reader for each 'clock-book' brain. Perhaps we shall learn to read other people's brains; consciousness will then be public as books and clocks are public – though each time they must be read privately.

The Cartesian Dualism of Mind and Matter becomes a Dualism of mechanism and procedures carried out by mechanisms, which indeed we see in all machines. If consciousness is associated with states of brains representing states of affairs or whatever, according to a code that is read and which we might discover; then by finding out just what is special about the brain's states, or the code, during consciousness (and anaesthetics might be useful research tools here) we might come to understand consciousness as well as we understand Matter. It is important to appreciate that Matter is ultimately mysterious, although we understand a great deal in physics. Perhaps consciousness and Matter are ultimately mysterious because for neither can we point to anything more general – or anything in our technology – to provide analogies for Matter or Mind.

Perhaps I have been trying here to explain one mystery with another; for *meaning* is almost as mysterious (see pp. 161–2) as the Mind–brain relation and consciousness. But considering the mystery of the fundamental concepts of physics – matter, time, energy, gravity – together with the immense success of physics at explanation, perhaps this is typical of science!

Part 5   Mind in us

In attempting to understand the elements out of which mental phenomena are compounded, it is of the greatest importance to remember that from the protozoa to man there is nowhere a very wide gap either in structure or in behaviour. From this fact it is a highly probable inference that there is also nowhere a very wide mental gap.

Bertrand Russell (1872–1970) [1]

17 The Self

'I can't explain myself, I'm afraid, sir,' said Alice, 'because I'm not myself, you see.' 'I don't see,' said the caterpillar.

Lewis Carroll (1832–98) [2]

What are we? What is our self-identity? Concepts of self-identity are expressed in various forms in religions, and in most religions something of Self is supposed to survive the body. This 'something' may be a kind of immortal inhabitant of the body: the soul. On such accounts it is not adequate to say that the body is the Self. This is so for theological accounts of Self; but there are others, and not all are concerned with survival.

Modern discussions start from the arguments of Descartes and Locke. As is very well known, Descartes started considering the Self through doubting everything – until he decided that he could not doubt his own thinking existence. In the course of this, he developed rules of thought, which for him are structures of Mind: structures and processes of Self. His Dualism – which implies that the Self is separate from the body because it is a different substance – reflects (or is a hangover from) ancient mythology, and has influenced philosophy perhaps more than any other statement in modern times. It may, however, be profoundly incorrect, or even meaningless; for what is to be meant by 'Mind substance', or (for that matter) 'Matter substance'?

Where is Mind?

On a Dualist account, Mind may be anywhere in the Universe. Minds may be hovering around, connected in some totally mysterious way to physical brains. Descartes saw this difficulty, hence his pineal gland notion. Alternatively, Mind may be spaceless, though not, it would seem, timeless. If we reject Dualism in favour of some kind of Mind–brain identity, then we must find Mind in the brain, and not outside it. Or at least, this must be so insofar as the brain is a closed system (and we may not wish to say that Mind is in space).

However, machines are not, strictly speaking, closed systems. Engines draw energy from their surroundings (hence their running costs), and they affect their surroundings (otherwise they would be of no use). It is somewhat arbitrary to legislate what is an object, of any kind, and for machines where there is energy or information transfer this is even more vague because they are closed systems.

Identity theories of the Mind generally hold that Mind is to be regarded as, or at least described in terms of, functions of the brain regarded as a machine.[3] So, to locate Mind, we must locate brain functions. We may then assign brain functions to various aspects of Mind. This might be done either from knowledge of brain function (physiology) or from (perhaps introspective) psychology. How, then, do we localize functions in machines?

Localization of function is surprisingly difficult to comprehend or establish, even in simple machines. As has been pointed out (see p.83 and Gregory, 1961) there are logical difficulties in localizing function by removing, or disturbing, parts of a mechanism. The trouble is that disturbing a mechanism or an electronic circuit, by removing or changing characteristics of components, generally affects the mechanism or circuit in strange ways, which themselves need explaining. It is now a different system, with different properties: to understand these the entire new system may have to be redescribed. For example, a small change in an amplifier can turn it into an oscillator; a speck of dust in the carburettor needle valve may turn a car engine into a coughing monster. To understand such effects we need a theoretical, conceptual, knowledge of the normal working of the system (circuit, engine or whatever) and physical–conceptual mapping, to say where functions are located for the normal or disturbed machine. This is why it is so difficult to make improvements, especially to highly designed machines such as jet engines or racing cars. A small change that looks like an improvement can have devastating effects on other parts of the system, which then in turn have to be corrected. There are further problems when there is redundancy, for then nothing may happen when parts are removed. In my view these are considerations that are still not fully incorporated into neurological thinking. If it is argued that the nervous system is a 'special case' to which such considerations do not apply, then the onus is surely on those who hold this to show that our general knowledge of mechanical and electronic systems does not apply to the nervous system. If this is urged, then how can we think of it? Is it implicitly held that it is so different because it is being controlled by Mind? But if so (and it is usually admitted that Mind's effect on brain is at the least mysterious) then, surely, to *localize brain function* is an even more daunting task; for who holds that we know anything about, or that it is meaningful to speak of, spatial relations of Mind? It is bad enough trying to localize functions of machines, let alone when some kind of controlling Mind is supposed to be involved.

There is strong evidence that conscious states can be elicited by physical stimulation of the human brain. Stimulation of the surface of the brain in man, especially of the occipital and parietal cortex, evokes sensations of vision, and memories. The 'pleasure centres' of the mid-brain (Olds and Milner, 1958) may evoke such intense pleasure that, when self-administered, the patient (only special patients are of course used) can hardly stop pressing the key administering the brain stimulation. If we assume that the term brain 'centres' is not entirely conceptually misleading, this may be evidence that aspects of consciousness are in some sense located in specific brain regions, and that we can discover where these are; but when we stimulate the nose with a flower (or an electrical stimulus) and experience a smell, we do not regard the nose as the 'centre' of the sensation. Just why and when should we believe this of a brain region, but not the nose?

There is a temptation to imagine that activity of cells recorded by microelectrodes gives immediate knowledge of what function is taking place in the cells, or even that sensations are somehow 'in' the cells. But this is a naïve view of localization even for simple devices, without the difficulties of sensation. For example, although we may be tempted to do so, it is misleading to say that a sewing-machine's ability to stitch is 'in the needle', or indeed in any specific part of the machine. All that we can say is that some regions, or components, have more or less crucial importance for the machine's performance, or the organism's behaviour.

For example, consider the components of a clock: the mainspring or the driving weights, the escape wheel and pallets, the pendulum, the hands and the dial. When we know how a clock works, we see the functions of these components, and then we can map the functions as seen conceptually with the structures seen perceptually lying with controlled motions in space.

We may be able to map the functionally conceived processes (which may be very different from anything in the machine's output, or in the animal's or human's behaviour) with the physical structure of the components and the way that they are connected. To do this, a first necessity is to establish what functions are going on – to appreciate how the machine, or the brain, or whatever, works – for obviously it is impossible to map functions before we know what the functions are that have to be mapped.

It is worth considering that once separate parts of a machine are put together to make the working machine, the parts lose their self-identity. Consider Meccano parts made up into a crane, a clock or whatever. The fact that they were separate before being bolted together is irrelevant, although we may be able to recognize the parts as they were in the box, or as they were incorporated in other models and perhaps performing very different functions. We can easily be misled by describing mechanisms in terms of the appearance of separate parts, for it is irrelevant to a machine's functioning how it may be taken apart, or how it was originally put together, or whether the same parts can be used in the same or in different ways in other machines (see chapter 3).

Sigmund Freud's view

There are machines in which the essential action takes place not in the physical structure but in spaces within the machine, or even in space outside it, for example ovens, rockets and cyclotrons. Freud makes this point in connection with how he thinks that consciousness may be related to physical brain function. Freud cites the founder of psychophysics, Gustav Fechner (1801–87) on the location of dreams. Fechner says: *'the scene of action of dreams is different from that of waking ideational life'*. Freud comments that 'this is the only hypothesis that makes dream-life intelligible', and 'what is presented in these words is the idea of psychical locality'. However, Freud rejects the notion that we can give detailed anatomical loci to mental processes, or that they need to correspond to any physical structure: [4]

I shall entirely disregard the fact that the mental apparatus with which we are here concerned is also known to us in the form of an anatomical preparation, and I shall carefully avoid the temptation to determine psychical locality in any anatomical fashion . . . I propose simply to follow the suggestion that we should picture the instrument which carries out our mental functions as resembling a compound microscope or a photographic apparatus, or something of the kind. On that basis, psychical locality will correspond to a point inside the apparatus at which one of the preliminary stages of an image comes into being. In the microscope and telescope, as we know, these occur in part at ideal points, regions in which no tangible component of the apparatus is situated. . . . So far as I know, the experiment has not hitherto been made of using this method of dissection in order to investigate the way in which the mental instrument is put together, and I can see no harm in it.

After disclaiming hope of adequate physical analogies for Mind, Freud continues: 'Strictly speaking, there is no need for the hypothesis that the psychical systems are actually arranged in a *spatial* order. It would be sufficient if a fixed order were established by the fact that in a given psychical process the excitation passes through the systems in a particular *temporal* sequence.' He calls this the ψ system. The temporal sequence from receptors to motor activity and perception 'does no more than fill a requirement with which we have long been familiar, namely that the psychical apparatus must be constructed like a reflex apparatus. Reflex processes remain the model of every psychical function.' This, surely, looks odd to us now.

Freud next considers memory, saying that memory traces are laid down beyond the apparatus receiving perceptual stimuli. The perceptual system, for him, has no memory whatever. Memory is regarded similarly to Donald Hebb's Phase Sequences (see pp. 373–5). Freud says:

> We must therefore assume the basis of association lies in the mnemic systems. Association would thus consist in the fact that, as a result of a diminution in resistances and of the laying down of facilitating paths, an excitation is transmitted from a given mnem. element more readily to one mnem. element than to another.

Considering now the perceptual system which precedes the memory systems:

> . . . our memories – not excepting those which are most deeply stamped on our minds – are in themselves unconscious. They can be made conscious; but there can be no doubt that they can produce all their effects while in an unconscious condition . . . moreover, the impressions which have had the greatest effect on us – those of our earliest youth – are precisely the ones which scarcely ever become conscious. But if memories become conscious once more, they exhibit no sensory quality or a very slight one in comparison with perceptions. A most promising light would be thrown on the conditions governing the excitation of neurons if it could be confirmed that *in the ψ-systems memory and the quality that characterises consciousness are mutually exclusive.*

Freud returns to dreams at this point, as helping us to understand this 'apparatus' of Mind:

> We have seen that we were only able to explain the formation of dreams by venturing upon the hypothesis of there being two psychical agencies, one of which submitted the activity of the other to criticism which involved its exclusion from consciousness. . . . Further, we found reasons for identifying the critical agency with the agency which directs our waking life and determines our voluntary, conscious actions. . . . We will describe the last of the systems at the motor end as 'the preconscious'. . . . We will describe the system that lies behind it as 'the unconscious', because it has no access to consciousness except *via the preconscious*, in passing through which its excitatory process is obliged to submit to modifications.

Freud's account here may appear physiological with the references to 'spatial order' of his systems and so on; but is it? In what sense can the 'two physical entities', one of which 'submits the activity of the other to criticism', be physiological? The point that I am getting at is that because meaning is involved, the censoring filter is not an engineer's signal filter, tuned to reject signals of certain physical characteristics: it is a meaning filter which hardly exists in engineering or physiology. If it is a meaning filter, then its criteria for rejection must be so subtle that even Freud has problems specifying what they are. Is this feasible for anything less than almost a complete person – a complete psyche with the full range of human experience and knowledge? One suspects that Freud's unit processes are far too subtle – and too knowledge-based – to be simple parts located within the 'system'.

Given that what I have called 'meaning filters' can hardly be realized even now, let alone when Freud was writing, we must assume that he was adopting some kind of brain–Mind parallelism, with Mind affecting brain systems. This is not so different from Descartes' Dualism (with the pineal gland serving as link between Mind and Brain), though Freud does not discuss how interaction may 'take place' or occur in physical, physiological, psychological or other terms.

Perhaps all the great philosophers consider questions of self-identity, but it is among the least discussed of the acknowledged problems of philosophy. This is strange, for upon one's view of self-identity depends one's justifications for regard and punishment of others, and appraisal of one's own actions. Thus Locke asks whether a man while drunk is the same man when sober – and whether an ill-deed performed while drunk should be punished, if when sober he is no longer the same self.

John Locke's view

John Locke discussed self-identity in the second book of the *Essay Concerning Human Understanding* (1690). Locke based self-identity on consciousness, as we see in the following passages: [5]

. . . to find wherein personal identity consists, we must consider what *person* stands for; which, I think, is a thinking intelligent being, that has reason and reflection, and can consider itself as itself, the same thinking thing, in different times and places; which it does only by that consciousness which is inseparable from thinking, and, as it seems to me, essential to it: it being impossible for anything to perceive without *perceiving* that he does perceive.

We can now hardly accept Locke's stricture that consciousness is inseparable from thinking, for this was effectively challenged by Helmholtz with his Unconscious Inference in perception and by Freud's emphasis on unconscious mental processes, and it is currently challenged by computer programs which, it may be claimed, think. However, this is clearly Locke's view, for he adds:

For, since consciousness always accompanies thinking, and it is that which makes everyone to be what he calls self, and therefore distinguishes himself from all other thinking things, in this alone consists personal identity . . . and as far as this consciousness can be extended backwards to any past action or thought, so reaches the identity of that person. . . .

Locke does consider enduring substance of the body as giving the continuity of the individual, but he rejects this in favour of consciousness as the basis of the

Self, though he acknowledges that consciousness has gaps. He uses the following example, imagining a person's consciousness all in his little finger, which is cut off: 'if the consciousness went along with the little finger when it was cut off, that would be the same self which it was concerned for the whole body yesterday'. From this he concludes: 'This may show wherein personal identity consists: not in the identity of substance, but, as I have said, in identity of consciousness.' And later:

Nothing but consciousness can unite remote existences into the same person: for whatever substance there is . . . without consciousness there is no person: and a carcass may be a person, as well as any sort of substance be so, without consciousness.

In spite of our deep sense of puzzlement over consciousness, Locke's account is probably the 'commonsense' view. We may, however, question whether we are the same Self in extreme moods of depression, anger or excitement. We are conscious in all these states, but we almost question whether we are the same Self. Very extreme changes of mood – especially when linked selectively to memories (multiple personalities) – may suggest that it is not only consciousness that is generally accepted as the criterion for Self. Perhaps memory and consistent characteristic behaviour are at least as important. These questions were discussed most fully by David Hume, who pondered long over these problems but was never satisfied by his conclusions – confessing that here he found himself in a 'labyrinth'. In his *Treatise of Human Nature* (1739–40), written when he was less than thirty years of age, Hume denies that we have direct knowledge (or, as Russell would put it, 'knowledge by acquaintance') of ourselves. But the arguments of Book 2 of Hume's *Treatise* depend on assuming just the opposite: that we do experience Self. Hume describes his puzzlement in the Appendix, where he writes: 'Upon a more strict review of the section concerning *personal identity:* I find myself involved in a labyrinth, that, I must confess, I neither know how to correct my former opinions nor how to render them consistent.'

David Hume's view

By 'identity', Hume meant existing through time without change or interruption. It is this that is for Hume the problem (if it is a problem) of how we are essentially the same person through life, although we change. The problem is in part to establish what, if anything, continues without change or interruption although our ideas and feelings frequently change, and are interrupted nightly by sleep. It is this continuous something, if it exists, which is taken to be the Self. It may be regarded as a kind of substance whose form may change although the self-substance remains unchanged as our moods or activities change.

Taking a substance to be an unchanging substratum, Hume asks whether the

Self might be a kind of substance somewhat analogous to matter, which also is classically supposed to remain unchanged although material forms change. However, he comes to reject this classical notion of underlying unchanging substance. He rejects it both for the material world and for the Self. He prefers to speak of a 'collection' of characteristics. As Norman Kemp Smith (1941) puts Hume's view: 'Bodies and selves, he insists, are by their very nature both unchangeable and complex. Each, therefore, is a collection in the sense that in its successive moments it is variable, and that at any one moment it is manifold.' [6]

Hume describes the Self as a 'bundle or collection of different perceptions', and he considers the case of the continuity of Self to be similar to that of plants, which also change and yet remain the same object. The 'internal' continuity of the Self through change is thus compared with what we accept as 'external' continuity through change of objects.

There is another aspect of Self, an aspect that is often regarded as unique to humans, and that may be invoked for establishing self-identity: the will. The will is classically the physiological *anima* which is the motive force of animals and the psychological source of the motives of human behaviour. Hume does not, however, take this 'Aristotelian' view of will: 'By the *Will*, I mean nothing but *the internal impression that we feel, and are conscious of, when we knowingly give rise to any new motion of our body, or new perception of our mind.*' [7] For Hume, then, the will is not a cause of action but is merely an event in a chain: and this event like any other may be regarded as caused by past events or as causing later ones. The will is thus not for Hume a separate causal agency which may be the Self. Perhaps we do generally think of the Self as our unique will controlling our voluntary actions, though, as Hume points out, this has immense difficulties.

Hume looks for links uniting the separate events of perceptions, feelings, memories and so on. He considers three kinds of possible links: similarity, contiguity, and cause. He soon rejects similarity and contiguity because for him there is not sufficient similarity or sufficient continuity or contiguity to serve as adequate links uniting experiences into an entity – the Self. He finally plumps for causation, though he holds a watered-down account of cause.

To ascribe mental causes is say to that there are mental events, rather as there are physical events, and that these 'mental events' are not, on this account of the Self, entirely dependent upon the physical events of the nervous system. In other words, Minds are at least in part separate from brains – the unity of the Self is not to be explained by physical continuity of the brain as an object. This, of course, depends on accepting a Dualist account, such that Mind is accepted as separate from, essentially independent of, the physical brain. If we resist this we must reject this account of Self. On the other hand, does physical continuity of the brain help? Here discussions of what would happen to Self in the case of a brain transplant are interesting, if not conclusive. Like Locke's notion of the consciousness being in the little finger, which is then removed – the Self for him going with the little finger – these are thought experiments useful for explicating hidden assumptions, though they do not give us new knowledge.

If A's brain is put into B's body, would A's Self move into B's body? Clearly, if the bodies were different (A might be a man, and B a woman) then the Self could hardly be the same – for our notion of Self is surely bound up with our potentialities and our behaviour. At most there could be but a kind of inner core of Self (which might be memories?) remaining after a radical change of body. Suppose, though, that A and B are identical twins and equally fit. Would they swap Selves with a brain transplant? Suppose memory were transferred, and the senses and the motor systems gave the same inputs: it is perhaps difficult to imagine that Self would not be swapped. This shows that we believe Self to be intimately associated with the brain. We might go further, to ask which parts of the brain are critical to Self: to establish by transplanting parts of the brain in which part or parts does the Self lie. The interesting question is, how could we decide? If my brain, or part of my brain, were transplanted into your body, I should need to remember myself now and compare this memory with my experience in your body. This, at least, would be appropriate on Locke's account of Self as dependent on consciousness. But if I listen to music, or take a psychedelic drug, I experience a marked change of consciousness, and I am often able to recognize that it is unusual, without a brain transplant. It is not clear whether Locke meant consciousness *per se*, or the typical consciousness or usual 'feeling tone' of the individual. If the former, I might argue that my consciousness has been transferred to your body with the brain transplant, and so I remain myself in a different body even though (because your body is different from mine) I feel very different in your body. If the latter, I do not need to do anything as drastic as a brain transplant to change my Self: listening to music or a joke may do the trick. On Hume's account we may say that the Self is not known directly, any more than any object is perceived directly, and the Self may be regarded as a hypothesis based on bits and pieces of experience. It may be a hypothesis like perceptual hypotheses of external objects.

In many ways we know ourselves no better than we know close friends. Indeed, I am struck by how little we know of ourselves by comparison with the immense richness of perception and conceptual understanding of the physical world. Further, it seems to me that what understanding I have is by analogy with my perception and concepts of the external world: I don't believe I know myself apart from interactions with the world of external things.

Returning to our discussion of consciousness, I suggested (p. 473) that what we are conscious of is at least mainly mis-matches between expectation and what is signalled via the senses as occurring. We may also be conscious of looking for matches and mis-matches between predictions of perceptual hypotheses and changes of current hypotheses – which are the most immediate updates that we ever get of reality.

Now if we are right in thinking that we are virtually only aware of discrepancies, or mis-matches, can we explain why we know so little of ourselves because there are but few mis-matches *within* compared with the interface mis-matches *between* ourselves and our expectations of external events as signalled? To put this another way: there are frequent mis-matches between perceptions, predictions, and our conceptions of the world. These seem to evoke awareness, which is absent with extreme familiarity and complacent understanding. One is only aware of the Self (I suggest) when there is emotional conflict or conceptual dissonance: hence indeed Descartes' feeling (which at least should apply to philosophers) that he knew himself when he was thinking.

Consider in this light one's awareness of one's body. Surely one is most aware of one's body when it fails to perform as expected. When carrying out familiar skills, one is seldom aware of any bodily sensation or even that one has a body. And this cannot be due entirely to peripheral sensory adaptation because it holds even though the body is in violent motion, for example rowing in an eight, after long practice when one is fit and not tired. Now I suggest that we extend just this to explain why we have so little perception of ourselves.

To try another thought experiment: what would happen if we had a richer input of signals from another person's body than from our own? Is it possible that we would believe ourselves to be aware of their body rather than our own body? If so, would our identification of Self move to their body? This does not seem impossible. On this kind of view it should be possible to become more aware of Self by provoking emotional conflicts and conceptual dissonances, and very likely this is so; but of course what is then experienced may depend on the conflicts and dissonances that have been cultivated. So it may be an 'artificial' or self-created Self that one experiences. It may be, also, that contemplation makes us more sensitive to dissonances and mis-matches, to make us aware of Self though we are passive and not in conflict with or uncertainty about events as signalled by the senses.

How do we know – or why do we assume — that other people only have one Mind (though of course they may be 'in two Minds' about something)? Is it their continuity of memory; or that we associate one Mind with one body; or what? Discussions of criteria other than bodily identity are given with a wealth of illustration by the Cambridge philosopher Bernard Williams, in *Problems of the Self* (1973).[8] These are essentially 'thought experiments'. Of more empirical content are the 'split brain' experiments of Roger Sperry, for these have been attempted, and the results widely discussed. It is worth pointing out, however, that the concept and possible results of splitting the brain by cutting the nerve fibres of the corpus callosum connecting the left and right hemispheres was for many years a 'thought experiment', which served to focus attention onto certain issues and to set up a framework for later experimental enquiry. It could indeed be that this is typical of science – of the relation between philosophy and science throughout their history. Bernard Williams's question is: 'Is the identity of a mind given by the identity of the body?' The split brain question is: 'Would isolating the two cortical hemispheres produce two Minds in a single person?' This question was asked by Fechner, long before the experiment was attempted. The questions are essentially similar: they both have neurophysiological implications, which have not been missed by those concerned with theological issues, especially the distinguished British neurobiologist and information scientist Donald MacKay (1980). We may also consider here cases of 'multiple personality', which it is claimed may include five or more 'Selves' in a single body, for example the case of Sally Beauchamp, described in great detail by Morton Prince in *The Dissociation of a Personality* (1905, reprinted 1978).[9]

The interesting question of whether separating the cerebral hemispheres by sectioning the corpus callosum would produce two Minds, and also whether the curious cases of 'multiple personality' are indeed cases of multiple Minds, were discussed by the psychologist William McDougall (1871–1938). McDougall spent much of his life at Oxford, where he held the title of Wilde Reader of the Mind, and moved to Duke University, North Carolina, where his interest and influence led to J.B.Rhine's celebrated experiments in ESP, Rhine being McDougall's student. McDougall was himself highly interested in ESP, and he accepted as a fact telepathy between distinct organisms. This led him to accept a theory of the unity of the Self as given by, as it were, internal telepathic communication between various Monads of the individual. Cases of multiple personality he attributed to some kind of loss of internal telepathy between Monads. He also used this notion, with external telepathy, to explain social cohesion. In *An Outline of Abnormal Psychology* (1926) McDougall writes:[10]

I am only the dominant member of a society. . . . There are many purposive activities within my organism of which I am not aware, which are not my activities but are those of my associates . . . I consciously control and adjust only a few of the executive processes of my organism, those only which are of primary importance for my purposes. But I and my associates are all members of one body; and, as long as the whole organism is healthy we work harmoniously together. . . . But, when I relax my control, in states of sleep, hypnosis, relaxation, and abstraction, my subordinates, or some of them, may continue to work and then are apt to manifest their activities in the forms we have learned to call sensory and motor automatisms. . . . And in extreme cases such a revolted subordinate, escaped from the control of the dominant member or monad, may continue his career of insubordination indefinitely, acquiring increased influence over other members of the society and becoming a serious rival to the normal ruler or dominant. Such a rebellious member was the famous

Sally Beauchamp [whose dissociation is described by Morton Prince]. . . .

McDougall concludes that 'All the facts of this order point to the view that the communications between the Monads are direct or immediate, that is to say unmediated or, as we may say, telepathic.'[11] He goes on to suggest that dreams are temporary take-overs by normally subordinate Monads.

Hypnosis is described similarly. McDougall cites a statement of one of the patients of Pierre Janet (1859–1947), pioneer of dissociative hysteria:[12]

In hypnosis also I am passive and my subordinates work independently of my control. They may receive and understand, retain and execute, suggestions of which I remain unconscious. And, if I carry out these suggestions in the post-hypnotic period, I may be surprised to find myself performing actions of which I have no intention and no prevision; and if I attempt to inhibit or prevent such actions, I may be aware of a real difficulty in . . . controlling and subduing the efforts of my subordinates.

McDougall extends the telepathically linked Monads theory of the Self to society:[13]

I suggest that Ireland affords a parallel to Sally. The organisation which is the British Empire may be likened to the organisation of a normal personality: Ireland has never been brought in to willing subordination to the whole system; like Sally, she has remained rebellious from an early date in the development, when she was estranged by an act of violence . . . Sally was the Ireland of Miss Beauchamp's body politic.

McDougall takes his telepathically linked Monads theory of Self to explain phenomena of memory, and of partial loss of memory, and impairment of sensory capacity with local physical brain damage. Thus he interprets the findings of the neurologist Henry Head (1861–1940), that destruction of certain brain regions may lead to impairment rather than complete loss of function, as some remaining internal telepathic communication between Monads supposed to be components of the Self may allow memory and self-identity to survive.

Margaret Boden (1972) suggests that McDougall's view can be transposed into current computer terminology, regarding sub-routines as Monads.[14] This is an interesting idea, but of course the status of McDougall's 'telepathy' needs to be clarified. If we ignore magnetic coupling between neighbouring circuits, or other such field effects, then results of loss of computer circuits through damage should be explicable in terms of parallel paths and redundancies and so on. And the same could be so for interpreting localized brain damage, or effects of isolating the hemispheres by sectioning the corpus callosum.

The effect of sectioning the corpus callosum by Roger Sperry and Michael Gazzaniga, at the California Institute of Technology, are described by Gazzaniga (1967):[15]

From the beginning one of the most striking observations was that the operation produced no noticeable change in the patients' temperament, personality or general intelligence. . . . Close observation, however, soon revealed some changes in the patients' everyday behaviour. For example, it could be seen that in moving about and responding to sensory stimuli the patients favoured the right side of the body, which is controlled by the dominant left half of the brain . . . the left side of the body rarely showed spontaneous activity . . . when he brushed against something with his left side he did not notice that he had done so, and when an object was placed in his left hand he generally denied its presence.

Lights flashed onto the right visual field were reported as seen, but generally this was not so for the left visual field. The right hemisphere was effectively blind – except when the patients were directed to point to the flashed lights instead of giving a verbal response. This they were able to do, almost as well as the verbal reports from the left hemisphere. This was evidently due to the location of the speech mechanisms in the left hemisphere. Similar results were found for recognizing and reporting objects as felt by the right or left hands. When a picture was presented visually to the dominant left hemisphere, the patients were able to describe it normally; but when presented to the non-dominant right hemisphere the patients were at imbecile level verbally, but they could respond appropriately and very well, as for example by feeling around for an object to match the picture of a spoon. In general, it was found that picture perception and emotional responses were unimpaired for the right hemisphere, but that verbal and perhaps logical function was lost for the right hemisphere. This has led to the notion that the left hemisphere is 'analytic' and the right hemisphere primarily synthetic or 'analogue'. This notion has been expounded especially by Robert Ornstein (1972). In the view of Michael Gazzaniga:

All the evidence indicates that separation of the hemispheres creates two independent spheres of consciousness within a single cranium, that is to say, within a single organism. This conclusion is disturbing to some people who view consciousness as an indivisible property of the human brain. . . . It is entirely possible that if a human brain were divided in a very young person, both hemispheres could as a result separately and independently develop mental functions of a high order at the level attained only in the left hemisphere of normal individuals.

Caution is required. The patients are all serious epileptics; and, it may be suggested, perhaps the separations of abilities and of awarenesses are not sufficiently dramatic to be described as different Selves, associated with the hemispheres. This is also a great difficulty in interpreting the supposed 'multiple personality' cases, and the alternating Fugue states so graphically described by Morton Prince. Are these alternative mental states sufficiently separate and stable, and so on, to be separate Selves; or are they, rather, extreme moods? After all, we all change in personality, and in consciousness, and immediate access of memory content, according to what we are doing or what we are thinking about, or what emotional state we may be in. While playing a game one's state may be extremely different from one's state when discussing, writing, cooking or making love. But these different states of consciousness and performance are not normally described as different Selves (and McDougall would describe them as different Monads of a Mind, not Selves). So on what grounds can it be said that one brain can hold two or more Minds?

Donald MacKay (1980) suggests that the criterion should be the 'thou-ness' of another person, as in a dialogue: [16]

. . . we might have strong grounds for recognising two conscious individuals if each could be fully 'Thou' to the other in dialogue: but this is just what can never be, so long as the deep central supervisory systems in these human cases are undivided. No mutual interpenetration of independent evaluators is possible, unless at the most superficial levels, where any 'dialogue' that could be implemented would amount to little more than the debates with oneself that every normal individual conducts without serious danger to his individuality.

And yet one may be 'beside oneself with anger': could this not exhibit multiple Selves in extreme cases of emotion or concentration? Should we demand clinical cases for evidence of multiple Minds in a single body? The answer seems to be the length of time each 'Self' can persist. 'Moods' are short-lived only, and are not different Selves or Minds but rather 'states' of Mind, though they may be extreme. I would however suggest that this distinction is ultimately arbitrary, and may reflect the theological notion of Minds persisting even beyond death.

Or is MacKay's criterion of 'Thou-ness', as in conversations, the key concept for Self? The neuro-physiologist Horace Barlow (1980) goes so far as to suggest that consciousness [17] 'even for immediate sensation, is a form of imagined discourse. Or to put this another way, the portion of the stream of sensory information of which one becomes conscious corresponds to what one is selecting for potential communication to others.' Barlow concludes: 'An audience, as well as an actor, is necessary for consciousness.' This is clearly related to MacKay's view; but it is in a sense opposed to it, for Barlow holds that discussion with an internal 'audience' can give consciousness, while MacKay holds that this is just what cannot occur, but that if it did, there would be grounds for accepting the theory of two Minds in a single body, which he wishes to say is impossible.

An intriguing experiment, which one can easily perform on one's friends, is to write letters with the finger, or gently with a pointed instrument, on the forehead or the back of the neck, or other parts of the body. The question to ask is: 'Which way round are the touch-letters experienced, or "seen"?' If, say, an **E** is written on someone's forehead in this way, with the eyes closed, he may 'see' it as from in front of his forehead or from behind, as though from inside his head. Is this an indication of where one's 'Self' resides? Since for many people the touch-letters are 'seen' from a viewpoint in front of their faces, this is no evidence for identifying the Self with the brain; but it might indicate where, as a kind of hypothesis, we suppose ourselves to be located.

Using mirrors, evidence is found that only primates and man locate themselves in front of mirrors and not behind them, as they appear to be. The reason for this is optical: it is very interesting that we know we are in front though we see ourselves behind mirrors. There is more to this: why are we right–left reversed in mirrors? Just what happens when we look at objects or our faces in mirrors? We always see ourselves right- and left-reversed, and not upside-down; why? This is so whenever we face a mirror by rotating around a *vertical* axis; but if we choose to face the mirror by standing on our heads, then we are upside-down and *not* right–left reversed. This is easier to understand for an object other than ourselves. Consider looking at a book in a mirror, giving 'mirror writing'. The reversal depends simply on the book's being rotated around its vertical axis to face the mirror, so that we can see the front of the book although we are behind it. If the book is rotated to face the mirror by rotating it around its horizontal axis, then we do *not* get right–left-reversed mirror writing: it is now upside-down and not mirror-reversed. For looking at ourselves, the situation is complicated by our interesting ability to go on seeing things the right way up although we are upside-down, quite apart from mirrors. Also, except for photographs, we normally only see ourselves in mirrors; and so we cannot compare seeing ourselves directly with how we appear in a mirror, although we can for any other object, such as a book. It is surprisingly easy to miss the point that we only get laterally inverted mirror writing when the book (or other object) is rotated around its *vertical* axis to face the mirror. It is rotation of the object to face the mirror – not the mirror – that produces mirror writing, and makes us right–left-reversed in a mirror, unless we stand on our heads. (But seeing ourselves through the mirror is due to the increased light-path.)

There is indeed a long history of confused discussions (including, in my view, Martin Gardner's account in *The Ambidextrous Universe,* 1964) which postulates active mental rotation.[18] This is clearly incorrect. The 'effect' occurs unchanged, even when there is no indication that a mirror is involved; for example, with a large wall mirror with no frame, or a photograph of a book taken in a mirror. Here there is no information that a mirror is involved and yet the same reversal is seen. Even a photograph taken of a scene in a mirror is 'mirror writing' – because it is right–left reversed, though there is no indication of a mirror when the book, or whatever, was rotated to face the mirror when its picture was taken.

According to Kant, there is no physical difference between left and right except as the observer makes a difference. Kant uses this argument as evidence for his Categories of Space – and so implies that the Self is bound up with spatial relations. This position of Kant's has been criticized by the philosopher Jonathan Bennett (1970), who effectively demolishes his argument, though he does not give what I regard as the correct explanation of mirror reversals. Plato was similarly confused, in the *Timaeus*.

All this shows how difficult it can be to think simple things out. To the point, here, are the experiments of Gordon G. Gallup (1977), who watched the behaviour of infants, primates and lower animals, viewing themselves in a mirror. Gallup found that only primates and infants over about a year old correctly touched a mark previously placed on one side of their face while they were anaesthetized. This result implies that the mirror image is not being taken merely at its face value; but that the observer 'realizes' he is seeing himself, laterally inverted. Gallup suggests that this realization is evidence for awareness of Self. On this criterion, self-awareness only occurs in primates and in humans after the first year of life. Another interesting book in this connection is Adrian Desmond's *The Ape's Reflection* (1979).

To make this issue even more confusing, the 'mental rotation' described by Shepard and Metzler (1971), which was discussed on p. 37, is entirely different from the right–left reversal we normally experience in mirrors – *always,* indeed, experience for objects rotated around their vertical axis to face a mirror. Shepard's effect is a slow mental rotation, as we *imagine* an object being rotated. This is certainly a cognitive (and a very interesting) effect. Something like this was supposed by Martin Gardner, and also by Kant, to take place for mirrors, as we have seen; but in my view this is mistaken and misleading. The difficulty (if it is difficult) is to see what kind of a problem mirror rotation is. It is not optical; it is not cognitive; and it is not conventional or verbal, as has been suggested; for mirror writing looks entirely different from normal writing, and it may indeed be difficult to read. To repeat, what we forget is the procedure of rotating the object. If we forget this we have a mystery – for we attribute the rotation to the mirror, or to the observer's Mind, or brain, and not to the object itself. But if we remember the procedure, it is obvious. How much of philosophy and science of Mind is difficult because we tend to forget about procedures?

However this may be, we normally see ourselves as entities capable of observing and understanding things very different and almost entirely separate from ourselves: things linked only by tenuous threads of light and (generally unconscious) logic. Can we learn more of ourselves, by turning to modern physics? Can we see ourselves reflected in physics? Perhaps we can use the tools and paradigms of modern physics as ancient myths, magic and machines were invoked at the dawn of written thought, to reveal ourselves in the Universe – though these we have largely rejected. How, then, are we related to the world of physics?

We have now reached extremely difficult problems and there are no stable answers. I shall approach the question of where, or whether, the observer fits into current accounts of the physical world in Relativity Theory and Quantum Mechanics, by first considering the science of psychophysics: attempts to measure sensations in physical terms. Our discussion here is bound to be somewhat technical.

18 Science in Mind

Measurement is one of the notions which modern science has taken over from common sense. Measurement does not appear as part of common sense until a comparatively high stage of civilization is reached; and even the common-sense conception has changed and developed enormously in historic times. . . .

Norman Campbell (1880–1949) [1]

Measuring sensation
Weber and Fechner

Psychophysics may be seen as attempts to find mathematical functions relating body and Mind – or relating behaviour and consciousness – by mathematical descriptions. In practice this endeavour is virtually limited to relations between stimuli and sensation, and the measures are usually the smallest increments between stimuli that give just-noticeable differences. This may, however, be extended to the psychophysics of aesthetics; indeed, this was attempted by the founder of psychophysics. Gustav Fechner (1801–87), in *Elemente der Psychophysik* (1860). He was the first to try to measure sensations. This was based (as Fechner realized) on the earlier work of many physicists and mathematicians, especially Bernoulli. Daniel Bernoulli (1700–82), who was born at Basle in Switzerland, was an anatomist, botanist and mathematician. He came from a remarkable family of mathematicians. His work on applying probability, or what we would now call statistical notions, for explaining the pressure changes of gases with changes of volume, temperature and velocity, affected Fechner's thinking by showing that probability concepts can be explanatory: that even if we cannot say just what each causal event is, yet we may be able to discover useful predictive relations and achieve insight into underlying processes which may involve unobservables. (A prime example is the importance of random variation in Darwinian Evolution (see p. 166), and, in modern physics, Heisenberg's Uncertainty Principle (see p. 540). Fechner was also influenced, and perhaps more directly, by Bernoulli's work on applying the theory of probabilities to games of chance. This led to discussions of *fortune morale* and *fortune physique,* mental and physical values that he believed to be related so that 'mental fortune' varies with physical fortune, according to the ratio of gain or loss with the total fortune of the gambler.

Fechner made use of the normal distribution of errors, following Laplace, Poisson and Gauss. Realizing the variance of 'thresholds' or 'limens', he invented methods of psychophysical measurement based on statistical considerations which are the very foundation of present-day experimental psychology. He was aware also of physiological work on sensory discrimination, especially that of Ernst Heinrich Weber (1795–1878), who, when Professor of Physiology at Leipzig, devised methods for measuring the sensitivity of the skin, and established Weber's Law.

Weber found, experimentally, that for the just-noticeable difference (**jnd**) between a stimulus (**I**) and an increase or decrease of intensity of this stimulus (Δ**I**):

Δ**I/I** = constant, for the **jnd**s.

Fechner made the assumption that if this is true for the just-noticeable difference between stimulus intensities, then this law should apply also to any small increment of sensation, **S**. So following Weber's Law, Fechner wrote:

Δ**S** = **c**Δ**I/I**,

where **c** is a constant of proportionality. (The introduction of Δ**I** is the mathematical equivalent of Fechner's assumption that all Δ**I**s are equal and can be regarded as atoms of sensation, and so can be added as units.) Fechner then integrated this equation, to give:

S = **c** $\log_e$ **I** + **C**

where **C** is the constant of integration, and e the base of natural logarithms. This gives the magnitude of the sensation, **S**, for any magnitude of stimulus intensity, **I**. The formula has, however, two unknown constants. Fechner eliminated the constant **C** by letting the threshold value, **r**, equal the stimulus, **I**. At this value, the sensation **S** is by definition zero. So:

when **I** = **r**, **S** = 0.

Substituting these values of **S** and **I** in the equation **S** = **c** $\log_e$ **I** + **C**, Fechner obtained

0 = **c** $\log_e$ **r** + **C**;

C = − **c** $\log_e$ **r**.

Now by substituting for **C** in **S** = **c** $\log_e$ **I** + **C**,

$$S = c \log_e I - c \log_e r$$

$$= c \log_e I - \log_e r)$$

$$= c \log_e (I/r).$$

This can be written in common logarithms (base 10) by changing the constant from **c** to, say, **z**:

$$S = z \log_{10} (I/r)$$

On this, Fechner's measurement formula, the scale of sensation, **S**, is the number of **jnd**s that the sensation is above zero or, rather, the number of **jnd**s above the threshold of awareness.

Fechner took a final step to suggest that **I** might be measured by its liminal value. For this, he took **r** as the unit of **I**. If **r** is the unit of **I**, then:

$$S = z \log I.$$

Unfortunately, Fechner called this 'Weber's Law', but it is not equivalent to Weber's experimental finding because Fechner added important assumptions and implications. The assumptions are that the unit of **I** is the liminal value of the stimulus; that it is valid to integrate **S**; and that we can assume that $S = 0$ at the limen. These are the assumptions required to measure sensation from Weber's Law. Whether these assumptions are indeed justified remains controversial to this day. Later work on intensity scaling, in which observers are asked, for example, to say which stimulus intensity is twice another intensity (or, as another example, lies halfway between two intensities), is also controversial.

It should be added that Weber's law ($\Delta I/I = c$) from which all this started is not precise, and breaks down markedly at extreme- and especially low-stimulus intensities, when greater incremental intensities are needed for discrimination than his formula indicates.

Weber was aware of this breakdown at extreme-stimulus intensities, and it was soon suggested that the law is improved for low values by an added constant,

$$\Delta I/(I + k) = \text{constant},$$

which is the equivalent of a small 'internal' stimulus. For vision, this was sometimes called the 'dark light of the retina'. For discriminations between lifted weights, the weight of the arm was sometimes added; but the constant required is much less than the weight of the arm. It has more recently been suggested that the basis of this added constant, extending Weber's ratio law to low-stimulus values, may be background spontaneous random neural activity, or masking 'noise', from which neural signals from stimuli must be discriminated. This gives a new physical interpretation of 'limen', or 'threshold', which is compatible with and to be expected from current knowledge of discrimination by any physical detector.

It may be added that there is evidence that even such apparently simple and basic discriminations as weight discrimination are affected by expectation of the weight. Not only do large weights feel, and are judged to be, lighter than smaller weights of the same scale weight (the well known 'Size–Weight Illusion'), but, more subtle and less well known, the discrimination (Weber's constant, c) is affected by the estimated weight as indicated by size, such that c is smallest when the density of the weight is around 1. It seems that a surprisingly heavy or a surprisingly light weight gives a raised or impaired discrimination. (It has been suggested that the basis for this may be that discriminations are given by neural mechanisms like Wheatstone bridges, in which the external signal is balanced against a pre-set internally assumed value. When this assumed value is markedly incorrect, discrimination in such bridge circuits is impaired.) Measurement by detection of differences between signal and expectation has engineering-type advantages over direct signalling of input (or stimuli) values, especially for giving a large dynamic range though the components have limited dynamic range, and also for reducing drift with unstable components. These would seem good reasons for such a stimulus–expectation comparator system in the nervous system (Ross and Gregory, 1964, 1970).[2]

Fechner was, however, not concerned with how the physiology works: he was concerned to relate body and Mind. To be more specific, he tried to show by measurement, and by mathematics, that body and Mind are exactly the same. He did, however, hold that sensations can only be measured indirectly, from stimulus discriminations. This was perhaps a tactical mistake – for is any measure strictly direct? Fechner's emphasis on indirectness fuelled the flame of his critics, who objected that he claimed to measure sensation when in fact he was only measuring physical discriminations. But is this fair when there is no such criticism of volt-meters for example, that they give information only indirectly?

One might argue that estimates of similarities between sensations are as 'direct' – and indeed more direct – than anything else, and that they are needed for the 'objective' measures of physics to be possible. What Fechner did was to reverse the usual order from judgements of similarities and differences of sensation to statements about the physical world by using physical stimuli (as measured by the methods and units of physics) to explore sensation. His critics should, rather, have questioned whether the normal procedure of physics can be reversed, to measure and explore characteristics of sensation. There is surely no obvious reason for not reversing the normal procedure for physics, to investigate the sensations on which all physics seems ultimately to depend.

It is important to distinguish between judgements of equal from judgements of not different. For Fechner's purpose, the notion of the limen (which can be traced back to Leibniz, the limit of what is just or is just *not* discriminable) is vital. Fechner noted that the limen is affected by attention to the stimulus, and that it is very different between sleep and waking, so he had a rather good reason to attribute the limen to consciousness rather than excitation. He thought, for example, that the stars in the daytime do stimulate the eye but that this stimulation falls below the limen for consciousness because of a psychological barrier. We would now say that the signal is masked and lost in background neural noise, when it is too small to be significantly different from the noise. This pushes statistical considerations into the physiology of the observer. This is a major concept of signal detection theory. Attention, however, remains important, and is ultimately mysterious, though no doubt it has a physiological *modus operandi*.

It has often been objected that there is no justification in assuming that **jnd**s are equal throughout the range of sensed intensities. But this doubt applies also to all units of 'objective' physical measurement. These issues are now familiar in Relativity, but they apply also, though often unrecognized, in classical physics.

What it is that Fechner measured can only be answered within a paradigm of psychology. This holds, indeed, for any measurement in science.[3] He claimed that Mind and body are one, but this by no means follows from Weber's empirical Law, or even from its formulation with Fechner's additions. But this is no criticism of Weber's or Fechner's accomplishments. For at least some paradigms, they provide meaningful measures of Mind.

Weber did not formulate algebraically his principle that '*When noting a difference between things that have been compared, we do not perceive the differences between the things, but the ratio of the difference to their magnitude.*'[4] Its formulation as $\Delta I/I$ = constant, is due to Fechner. Fechner claimed that he did not get the ratio notion from Weber, but that on the morning of 22 October 1850, while he was lying in bed, he saw that to relate body to Mind the thing to do is to make 'the relative increase in bodily energy the measure of the increase of the corresponding mental intensity' and that he had, independently, sufficient knowledge to consider the constant-ratio relation. He was, however, over-generous to Weber, who failed either to formulate or to derive general conclusions from his discovery. Weber did, however, point out at least one interesting consequence:[5]

I have . . . by this observation . . . refuted something that could quite plausibly be believed concerning the method by which we compare two impressions on the senses.

For instance, it would not be an inherently incongruous idea that we compare the images of two lines pictured on the retina in the same way as two objects held by hand. We compare the length of such things most accurately when one of them is superimposed on the other, so that we recognize both their similarity and their difference. The same thing could also occur in the eyes. A line might make an impression in the centre of the retina that persisted for some time when the eye was turned away: if the eye was turned upon another line so that its image fell on the same place, the length of this image could be compared with the impression of the line seen earlier, and their degree of similarity could then be appreciated. My observation is, however, inconsistent with this hypothesis. If the above method were the case, we should be seeing the absolute difference.

By the same reasoning he rejects the hypothesis that length is given by a procedure such that 'the mind unconsciously counts the nerve endings in the retina that are touched by the impressions of both lines'. This seems to have contemporary interest for the problem of how shapes are represented. (Indeed he provides arguments against both analogue and digital representations!)

Fechner devoted much of his unusually active life to developing a philosophy intended to embrace the inorganic, the organic and Mind, including consciousness – though he argued for the reverse of this order. He saw Mind ('cosmogenic' Mind) as primary – given by order of atoms – which in lesser degree is life, and in still lower degree merely inorganic. For Fechner, consciousness is not part of, or contained in, organisms, but is a characteristic of their atomic organization. He was thus a Functionalist, but carried this to the extreme of Panpsychism, in which Mind permeates all Matter. Stars, for Fechner, had Mind and soul, insofar as they were organized structures. The threshold for sensation (or limen) represented a kind of cut-off from the universal Mind. Below the threshold were secret processes hidden or rejected according to the dictates of the level of attention.

Fechner started his career studying medicine, which he gave up as he was distressed by its inadequacy. He went on to study physics and mathematics, carrying out research in electricity. He was brought up in the post-Kantian German Idealism of his time, in which everything was seen as intimately linked to everything else in the Universe. It may well be that his experiments with electricity, around 1824–40, confirmed in a most striking way this view of the unity of Nature. Anyone who has played with electrical discharge tubes, activated by electricity from an influence machine, must be struck by the lifelike sensitivity of the miasmic glow which follows every movement of the hand near the tube and yet has an unpredictability, a sensitivity as though it has a life of its own while responding to its surroundings. Perhaps significantly his experiments in psychophysics were inspired by a correspondence with an electrician, Wilhelm Weber (not E.H. Weber of the Weber–Fechner Law), and of course notions of 'animal electricity' were at that time prominent (see pp. 142–9).

Fechner did not think of sensation as given by transfer of energies along nerves, but rather believed that the organism was part of the total structure of the Universe; though he did suppose that there were inner self-contained energies, protected, as it were, below the level of attention, giving what was for him the all-important threshold of awareness. It may be suggested that the accounts of observation of atomic events in current Quantum Mechanics, especially by Heisenberg, are not so very different from Fechner's Leibniz-like philosophy (see pp. 537–47).

We may conclude that even with the later discovery of the logarithmic relation between frequency of action potentials in nerves and intensity of stimuli, just why sensation is (if it is) logarithmically related to stimulus intensity is not explained by psychophysics or by physiology. But this is not such a special case as it may appear, for all measurements and laws in science need interpreting. Weber showed that there are physical bases for sensation; whether Fechner was correct to equate physical activity with sensation (via the ratio function) remains an open question, which is answered in the affirmative only by proponents of Mind–brain identity theories. Their grounds are invariably philosophical, rather than anything known from measurements of psychophysics. This is not to minimize the importance of Weber's and Fechner's work. They founded experimental psychology. It is just remarkably difficult to design convincing experiments concerning Mind.

The clearest and most useful analysis of kinds of sensory experiments is due to the physiologist Giles Brindley, in his excellent review of visual physiology *The Visual Pathway* (1970). Brindley distinguishes between 'class A' tasks, sets of stimuli which give the *same* sensation or report or response, which are *matching* experiments; and 'class B' experiments, in which the subject must describe what he experiences or perceives. This important discussion is in Chapter 5.

Brindley says of ordinary ('class B') observations: [6]

> For physiology, the terms used in stating the theoretical background are physico-chemical and anatomical; so it would seem that no physiological hypothesis that is also stated in physical, chemical and anatomical terms can ever predict the result of a sensory experiment, in which a report of sensations is concerned.

He accepts, then, an irreducible gap between physics and sensation which physiology cannot bridge. There are however the 'class A' experiments, which do give knowledge of physiological processes, depending on a hypothesis which is exceedingly hard to doubt: [7]

> The additional hypothesis required is that whenever two stimuli cause physically indistinguishable signals to be sent from the sense organs to the brain, the sensations produced by these stimuli, as reported by the subject in words, symbols or actions, must be indistinguishable.

Much has been learned (especially perhaps on mechanisms of colour vision) using class A matching experiments. For example, the fact that a monochromatic red mixed with a monochromatic green light can exactly match – be indistinguishable from – a monochromatic yellow light even after adaptation of the eye to red or green, is extremely strong evidence that there is no special yellow-sensitive receptor; but that yellow is always seen by neural mixture, from red and green receptor channels though the stimulus be a single monochromatic yellow light. This is surprising, because yellow appears to be a simple primary colour, and not a mixture. So here a class A matching experiment has yielded new and surprising knowledge.

Class B experiments are, however, far richer. Indeed, the class B situation is the basis of all normal observations – though their interpretation is always open to doubt, and their ultimate psycho-physiological basis is unknown.

Scales of sensation

Although various kinds of scales for measuring different physical dimensions and kinds of phenomena have been devised and adopted through the history of science, few philosophers have discussed theoretical problems of measurement, or such questions as when which scales are appropriate, and why. Lengths are usually represented on linear scale functions, though not always, as witness the Mercator projection of the spherical earth on maps. A similar (stereographic) projection was indeed used in classical times for the astrolabe (see p.58). Logarithmic scale functions and several others are in common scientific use. Also, scale functions may arise from the contingent behaviour of materials used in instruments, or from such situations as taking equal angular or length measures of the sun's shadow from a gnomon to indicate intervals of time, or indeed to take the earth's rotation as a clock which gives intervals different from the now-accepted linear time.

For measuring quantities, we need some kind of physical scale, such as rulers or clocks, and also a conceptual scale for exhibiting or describing the results. They are usually chosen for convenience – and there are many criteria of convenience. Very often data can be transformed, for reasons of convenience, from one scale function to another without loss of information.

On what scale function do sensations lie? Is this indeed a meaningful question? Weber discovered that sensory discriminations are roughly logarithmically related to stimulus intensity, and Fechner suggested that sensations are also logarithmically related. But Fechner's claim is based on the assumption that stimulus **jnd**s are linearly related to intensity of sensation. What justification can there be for this? Clearly, physiological measures of, for example, the frequency of action potentials in afferent nerves as a function of roughly logarithmic stimulus intensity will not give the answer; for how is sensation related to neural activity? Unfortunately we do not know. If, however, we could measure intensities of sensation, we might be nearer to finding this out (see pp.204–10).

Before discussing the new psychophysics – which has attempted to measure sensations as directly as possible and without assuming that discrimination is in any simple way related to sensation though it is related to stimulus intensity – we should look at the kinds of scales available to the Physical Sciences. Several authorities suggest that there are four; and some are far more informative, though not always available, than others. They are nominal, ordinal, interval and ratio scales.

Nominal

This is the most general class, indicating only that the members are in the same class (and letters are as good as numbers). Thus the letters of the alphabet are arranged in a nominal scale.

Ordinal

This involves rank according to greater or less, along some dimension; for example, the hardness of materials on the Mohl scale of 'pecking order', of which substances scratch and which are scratched.

It is worth pointing out that although we generally accept that if A is greater than B and B is greater than C, then A is greater than C, this is not necessarily so. For example, if A usually beats B at tennis, and B beats C, it does not follow that A will usually (or ever) beat C. If such situations were common perhaps we would not have developed scales, for they would be virtually useless.

It is also worth pointing out that two variables may be related, such that when one increases so does the other – but only over a limited range, beyond which the relation may reverse. For example, as a man receives more and more money he may spend more and more on his house – until at a certain point he takes to travelling round the world and reduces his house expenditure. Physical scales, such as fluids for thermometers, are chosen which do not have this property in typical conditions.

Interval

This involves equality of intervals or differences, for example temperature (Fahrenheit or Celsius). These are quantitative, though the zero point is a matter of convention. These scales remain invariant when a constant is added. Thus for the Celsius and Fahrenheit scales, both are given by noting equal volumes of expansion, and there is an arbitrary zero for each. They can be converted, one to the other, by equations of the form $x' = ax + b$. It is meaningless, for interval scales, to say that, for example, one value is twice another.

Ratio

These have a zero, though not necessarily obtainable in practice (as, for example, for absolute temperature). Length, weight and electrical resistance are examples of absolute ratio scales. There are also derived ratio scales, such as density and velocity. Most physical scales are fundamental or derived ratio scales.

It is in the last resort impossible to say that a numerical scale function is 'right', or 'correct'. Scale functions such as linear or logarithmic are chosen for convenience and overall simplicity of description. We take it that 'equal' measures of length add 'linearly' (unless there is 'distortion of space'), and that, given this, other things obey logarithmic or some other functions. But conceivably we might regard what is now 'logarithmic' as 'linear'; and if this gave an overall simpler description of things this would be a good move. It is interesting that information, in Shannon's Information Theory, adds logarithmically, though this is chosen purely for convenience.

When we say that loudness increases logarithmically with signal amplitude, we are first of all saying that sound amplitude (or pressure) is measured on what is accepted as linear, as for measuring lengths, and then that from this physical scale the sensation loudness increases logarithmically. How far this is empirical and how far conventional is extremely difficult to determine. The answer is fairly clear while we are talking about discrimination but very difficult when we are talking about the appearance of intervals. Fechner tried to measure intervals of sensation by saying that sensation intervals are made up of equal **jnd**s and that just-noticeable differences correspond to equal differences of sensation whatever the range of stimulus intensity; but this is an assumption without justification, even though it is very commonly assumed in experimental psychology.

At the same time, however, all our physical scales are ultimately derived from sensations. This is so despite the fact that even lengths do not appear the same though measured as the same under various conditions: a very distant ruler appears shrunk. This is why instruments and standards such as rulers are so important, and why they have very largely taken over from sensation for measuring the world. Sensations are for preference used in science for detecting differences, and positions of pointers on scales, as we are quite good null detectors but are poor as judges of physical values without recourse to external comparison standards.[8]

Comparing sensations

A useful way of estimating degree of pain, such as post-operative pain, is to set up another painful stimulus which can be adjusted (for example a heat source on the forehead), and set to match it. This may, however, have a different *quality*, making matching difficult.

Effects of analgesics cannot be estimated in this way, for both pain sources are affected, unless they affect different kinds of pain differently or the analgesia is local. The effects of analgesic doses can, however, be estimated by setting the test pain to the threshold of acceptability – which is of great practical importance.

Kinds of sensations: sensation space

There is very general verbal agreement that the sensations called green, yellow, pain, tickle, loud, hot, sick, fear and love are very different kinds of sensations and seldom if ever confused. (An exception would be confusion of red and green for people with colour anomaly, but even here it is not sensations that are confused, but rather different wavelengths of light that are not distinguished though for most observers they give different sensations. The same applies to poor spatial discrimination with loss of visual acuity.)

What is interesting here is that although we cannot know another's sensation (their red, green, pain and so on), yet we generally agree on kinds of dimensions of sensation, and also that intensity of sensation generally increases with increase of stimulus energy. It is interesting also that analgesias have similar selective effects, as judged with sensation language, and that sensation language is useful for medical diagnosis. This is so even though there are not always clearly identifiably different physiological channels. The different kinds of pain are particularly mysterious physiologically. It is also so though kinds of sensation do not correspond to different kinds of objects, or physical dimensions. John Locke's distinction between Primary and Secondary Characteristics of the physical and sensation worlds seems to hinge on this point: where they correspond we have Primary Characteristics; where not, only Secondary.

Measurement of drives

Behaviourally observable drives, such as those of hunger, thirst, sex and maternal or paternal protection of the young, can be ordered or ranked in two ways. Pairs of drives may be put into competition. Thus water and food may be offered as alternatives, and the animal's choice noted. The other way is to set up a variable barrier of some kind (usually an electric grid), and note how high or painful a barrier, such as voltage, the animal will accept, to satisfy the drive. Drives can be ordered in this way. A usual order is maternal protection, thirst, hunger and sex. Experiments and measures of this kind are of course central in Behaviourism, extreme Behaviourists holding that we say nothing whatever of sensations (or that they have no kind of existence) though we carry out, report and understand behavioural experiments. For the Behaviourist, these kinds of measures are equivalent to measuring specific heat or inertia (or perhaps more realistically the performance of engines under various conditions) without reference to 'internal' states which cannot be measured from outside.

The problem for psychologists who are not Behaviourists is to suggest how to measure these internal states from inside – if necessary with the help of external scales or references. Whenever we equate behaviour with sensation or emotion, it may be said that we obtain measures, or at least ranking orders, by noting choices. For example, if someone says 'I find that picture beautiful,' and pays so much money (or such and such a proportion of his wealth) for it, then we might be tempted to infer that the money he parts with is a measure of his sensation of beauty. But unfortunately this is simplistic. There may be some uni-dimensional drives (hunger? thirst? sex?), but surely the drive to buy paintings is complicated by social and all manner of other considerations. Indeed we have plenty of evidence for this; what we lack is evidence that any behaviour is simply related to sensation, or that there are single dimensions of sensation corresponding to behaviour.

This point comes up in human experiments as a problem of what instructions the subject should be given. Should subjects of psychophysical experiments be asked to judge or compare object characteristics, or their sensations of what they may take to be object characteristics? Almost all experimenters are ambiguous and shift their ground at this point. It is indeed exceedingly hard to know which to ask of subjects – or if subjects are capable of obeying the instructions. C. Wade Savage, in his important book *The Measurement of Sensation* (1970), quotes the psychologist E. H. Galanter in this regard: [9]

The experimenter can never decide whether the subject is right or wrong in a scaling experiment . . . in scaling experiments we are forced to assume the uncontrolled ability of the subject to accurately report his sensations . . . the reproducibility of the data upon repetition of the experiment lends some support to this assumption.

As Wade Savage points out, the experimenter can tell if the subject is wrong over a stimulus or object judgement (for example that one of two sticks is the longer), though he cannot say the subject is wrong in saying that one appears longer, or does not. He may, however, conclude that the subject has suffered an illusion (or even a hallucination) from such a discrepant answer. But we do normally experience considerable agreement with our sensations and our physical beliefs of lengths and weights, and so on, of objects. In short, perceptual illusions are limited; so surely correlation between sensation and the physical world must be possible. Also, if we discern that there is perceptual illusion, then we have surely, somehow, measured sensation – because we are saying that it differs in some degree from physics. Or, if not measured, at least we have estimated sensation in physical terms. This distinction between measuring and estimating is significant almost to the point of distinguishing Mind from physics.

Measurements and estimates

It seems most important – as emphasized by Wade Savage (1970), though this is often ignored – to distinguish between measuring and estimating. It could indeed turn out that sensations can be estimated though not measured, as we understand 'measurement' in the Physical Sciences. Wade Savage defines measurement as '*a procedure of assigning numerals to objects within a dimension by means of an empirical process of comparing these objects with a dimension unit or units*'.[10] What should we mean here by 'dimension' and 'object'? Dimensions are colours, pains and sensations of length, weight and other physically measurable features of physical objects. Some of these can be measured for physical objects with rulers, scales and so on, but we cannot match a ruler against the appearance of length, or weigh the sensation of heaviness.

Attempts have been made to avoid this crucial difficulty by relaxing the criteria for measurement, by allowing that measurement can be assigning numerals to magnitude ratios. This is the basis of S.S. Stevens's procedures for measuring sensations by establishing ratio scales for brightness, loudness, heaviness, length and so on, for other sensations. Stevens claims that this function is a power law, with a different slope for different sensory modalities or dimensions.

S.S. Stevens's Power Law

Stevens is essentially the founder of the new psychophysics, which claims to measure sensations on ratio scales; but it should be recognized that Stevens was himself an acknowledged Behaviourist, and so did not officially recognize sensations. He tended to accept data as valid if they showed consistency, especially where consistency was maintained though the experimental conditions were varied. Hardly surprisingly, given his Behaviouristic leanings though (unfortunately) he did not distinguish clearly between measures of sensation and physical attributes of objects or stimuli. (Here I shall use ψ for sensation and ϕ for stimulus or physical attribute.)

Stevens admits that sensation ψ cannot be measured directly, but holds that sensation is, like temperature, a construct – built from observed stimuli and reactions of organisms. This is the Behaviourist's position. But he also says that 'what we want, of course, is an unbiased method, one that on the average lets E [the experimenter] make an estimate that is neither too high nor too low. Since we do not know in advance what his estimate should be, we can apply no independent criterion of validity.' It is, however, highly unclear what 'validity' can mean here, for it is not clear what would be incorrect or correct.

What Stevens does is to get his subjects to assign numerals to felt weight, or brightness or loudness, and to relate these ψ values to given ϕ values, which are presented over a considerable range. This gives tables of values such as:

Psychological weight	Physical weight
10	100
9	88
8	76
7	62
6	50
5	41
4	31
3	22
2	15
1	7

Relating assigned numbers to physical magnitudes, Stevens finds that the best fit is given by a power law, of the form

$$\psi = \phi^n,$$

where ϕ is the stimulus in physical units and ψ is sensation magnitude or intensity, as indicated by the subject's assigning a numeral from a previously agreed range of numbers, and n is an exponent found empirically.

The value of the exponent n is found to be different for each modality. Here are some experimental values:

Psychological dimension	Physical dimension	Exponent	Unit
Loudness	Sound intensity	0.3	Sone
Brightness	Light intensity	0.3–0.5	Bril
Taste	Solution concentration	1.0	Gust
Seen length	Measured length	1.1	Mak
Felt weight	Measured mass	1.45	Veg
Subjective duration	Clock time	1.05–1.2	Chron

It has been objected that these results are highly dependent on the instructions given to the subjects, and to the range of numerals allowed; to which Stevens counters that all measures and scales depend upon assumptions, which may be quite difficult to justify, and may be arbitrary and may lead to circularity. He would thus say that the situation is very similar to that of measuring temperature. [11]

What are dimensions of sensation?
Physical dimensions such as length, mass and time may be described by operations of measurement that we perform under a great variety of conditions. These three

are examples of *absolute* dimensions, for they may be given by one set of measurements without reference (at least explicit reference) to others. Taking area, volume or speed, we have to infer these from two or more simple or absolute measures; so these are derived. Sometimes absolute values such as lengths have to be derived, for example the distance of a star.

Are sensory estimates absolute or derived? Clearly some, such as distance seen by stereoscopic vision, are derived (by computation from the convergence angle of the eyes and retinal disparity). Apparent length of, say, a rod, is derived from the size of the retinal images and the assessed distance of the object. We know that if either retinal size or signalled or assumed distance is changed, apparent size changes systematically, and can be clearly inappropriate, so that we cannot equate apparent with physical length. What, though, is apparent length? There is a temptation to say that if we have a length, even an apparent length, it must be a length of something. But what is this 'something'? It is most important to recognize that 'apparent length' should not imply the existence of some internal thing, which has length. There is a temptation to suppose the existence of internal things having properties – ghostly properties – like characteristics (which may or may not be sensed) of physical objects. This conceptual danger is largely averted by speaking rather of perceptions as 'descriptions' or 'hypotheses'. This helps, if only because we are familiar with fictional descriptions which have no referent objects. And of course stories, or pictures, may have some features corresponding to referent objects, while other features have no referents or even possible referents. No doubt hypotheses in physics are generally mixtures of this sort (and perceptions that include illusions are true insofar as they agree with physics).

Now we may suggest that our hypothesis account of perception (see p. 395) may be applied here to sensation. We may say that sensations are components of perceptual hypotheses, and that the dimensions of sensation are general features of the structures of perceptual hypotheses. As we describe and explain the world in science – with values along dimensions of length, voltage and so on – so do we see the world with dimensions and magnitudes that roughly correspond to the dimensions of physical descriptions.

Several perceptual phenomena show very clearly that few if any sensory dimensions are 'pure'. For example, one might have thought that sensed weight would be pure, for it seems a simple matter of signals from the skin, joints and muscles, but it is not pure, as we know from the Size–Weight Illusion: that large objects are experienced and are judged to be lighter than smaller objects of the same-scale weight. Possibly when we sense weight we are deriving density without realizing it; but however this may be, weight is not a pure simple sensory dimension but is affected by our appreciation of other object characteristics (see p. 503).

If we go on to ask where sensory dimensions come from, we may (with Kant) say that they are innate, or we may (with Helmholtz) say that they are derived from experience of handling objects. Both could be true, and in any case innate dimensions may well have been derived by ancestral experience. No doubt dramatically different sensory modalities such as hearing and vision are determined innately, by the different sense organs and especially by the regions of brain they feed their signals to; but more subtle dimensions of sensation and perception are most likely derived from individual experience. So there seems to be a deep interaction between how the world of physical objects as we handle them generates sensory dimensions, and scales them, and how sensory dimensions affect how we see the world, and how they parcel it into objects and dimensions along which we scale and measure the world in science.[12]

Finally, we may consider a concept that has become important in recent psychophysical experiments on the senses: the concept of channels. It is obvious that the ears and eyes are separate channels; and it is clearly useful to think of the (three) kinds of retinal cone colour receptor cells as giving individual channels for 'red', 'green', and 'blue' light. (We can then say that yellow is seen by neural mixture of signals from the red and green channels, though the stimulus in monochromatic yellow light.) But it is not, at least at present, possible to distinguish these colour channels anatomically. They look the same *structurally* although they are separate *functionally*. When both descriptions are available, they do not always agree. For example it is unclear how the sensations of touch, tickle, warmth, cold and pain are related to the anatomically distinguishable skin receptors and their neural pathways. On the other hand, the retinal rod and cone receptors do correspond to 'scotopic' and 'photopic' vision respectively, for night and day vision.

What the recent research aims at is to isolate channels for highly specific features (especially for vision), such as for orientation; texture size; retinally signalled movement; binocular disparity; and many more. These channels are mainly isolated functionally, by *selective adaptation*. For an example, see fig. 40, p.398. This experiment, by C.Blakemore, R.H.S.Carpenter and M.A.Georgeson (1970), shows very clearly, as the reader can check, that the apparent orientation of bars, and their spatial frequency, change with prolonged viewing – indicating adaptation of specific channels, signalling orientation and spatial frequency.

This functional dissection reveals channel characteristics of the sensory system which are its modalities through which the world is sampled for perception. It is an important question how far physiological channels match the psychological dimensions of perceptual experience.

Part 6   Mind in Science

It appears to be a law that you cannot have a deep sympathy with both man and nature.

Henry David Thoreau (1817–62) [1]

19 The nurture of physics

There is a hierarchy of facts. Some are without any positive bearing, and teach us nothing but themselves. The scientist who ascertains them learns nothing but facts, and becomes no better able to foresee new facts. Such facts, it seems, occur but once, and are not destined to be repeated.

There are, on the other hand, facts that give large return, each of which teaches us a new law. And since he is obliged to make a selection, it is to these latter facts that the scientist must devote himself.

Henri Poincaré (1854–1912) [2]

Physicists seek 'objectivity', and yet they generally accept that all their knowledge is based on perception – including sensation. It is very strange – and I hope that this book does a little to redress the balance – that although physicists are deeply concerned with the basis of scientific method they are not generally concerned at all with perception and cognitive processes of understanding. Although all Science is in Mind they are not at all concerned with Mind and its place in Science. This may be because science is explicitly restricted to soluble problems, as beautifully and profoundly expressed by Sir Peter Medawar in *The Art of the Soluble* (1967) but, of course, it cannot be clear what is not soluble, at least before questions have been formulated and discussed in detail. Have we the right to make *a priori* judgements that such questions as these cannot be answered, and so are outside science? Are they supposed to be forever beyond reach? And if not for ever, what good reasons can there be for not considering them now? Surely, we should at least bear such questions in mind to have any hope of finding answers.

In the two great paradigms of modern physics – Relativity Theory and Quantum Mechanics – observers are mentioned but they are highly mysterious. In classical physics, there is seldom a vantage-point, or explicit observers – though perceptions are always from the vantage-point of the observer. We should, however, qualify the statement that physics is not concerned with points of view, for as we have seen the shift from the Ptolemaic to the Copernican description of the solar system is in part a change of viewpoint; Ptolemy describes the motions of the planets as seen from earth, while Copernicus describes them from some imaginary point in Space outside the solar system. This allows a far simpler description of the same motions. (The highly complex description of elliptical planetary motions, as a set of epicyclic circular motions, is equivalent to the analysis of a Fourier series.) The great simplification given by the change of observer position away from how we see it, to an imaginary observer in Copernicus' account, allowed the planets to be described as moving in ellipses although we do not see the ellipses. The shift from the Ptolemaic to the Copernican view does not, however, involve rejection of common sense, as we are quite used to things appearing different with changes of observer position. Even young children have some (though limited) ability in visualizing what a scene would look like from a different place (see p. 36), and we can rotate 'mental images' as investigated and measured by Shepard and Metzler (1971). The concept of an observer situated far from earth is explicit in at least one description of a Greek astronomer, explaining eclipses of the Sun and Moon, so this is no recent skill. The conceptual paradigm shifts of Relativity and Quantum Physics are, however, different: here common sense *is* violated by accounts giving greater simplification of what is observed by moving further away from how things appear.

When Newton, in the *Principia* (1687), developed the Copernican account of the astronomical world into a fully determined mechanism without need of guiding intelligence or any external agents it seemed that the entire future of the Universe could be predicted, given a complete knowledge of the positions and accelerations of all the bodies of the Universe at the present or some past time. This was well put by Pierre Laplace (1749–1827), in *A Philosophical Essay on Probabilities* (1820): [3]

Given for one instant an intelligence which could comprehend all the forces by which nature is animated and the respective situation of the beings who compose it – an intelligence sufficiently vast to submit these data to analysis – it would embrace in the same formula the movements of the greatest bodies of the universe and those of the lightest atom; for it, nothing would be uncertain and the future, as the past, would be present to its eyes. The human mind offers, in the perfection which it has been able to give to astronomy, a feeble idea of this intelligence.

One might indeed say that once the data were stored within the intelligence its 'perception' could be purely internal and infallible by following the principles of Newton's account of the Universe. This remained unchallenged as a concept (though never likely or indeed possible to realize in practice, if only for reasons of limited computing) until the Quantum Mechanics of the first decades of the present century.

If complete prediction, as supposed in principle possible on the Newtonian account, could be realized the observer would disappear; for all observation could be replaced by prediction. This is the ideal of Rationalist philosophy. Arguing this the other way: since prediction cannot be complete in practice (even on the Newtonian account), the observer is essential. This is the justification of Empiricism.

I have argued that perceptions are inferences from sensory signals and that they are, in status, like scientific hypotheses (p. 395). What emerges here is that as scientific hypotheses give deeper insights and there is more effective computing power for prediction, perceptions give way to scientific hypotheses. Where there is conflict we nearly always say that the senses are misled, as shown by physics. We continue, however, to rely on perception: 'perceptual hypotheses'. This does not mean that there are two realities but rather two or more accounts, which may have different uses. Which of these is true is a question that we may approach by asking whether 'distortions' in Relativity Theory are in *space,* or in *observations*.

We shall look now at some of the concepts of Relativity Theory with its radical departures from the space and time of perception and common sense. Where and what is Einstein's observer? How different is he from Newton's?

The observer in space: Relativity Theory

Relativity Theory was born from the prevailing philosophical emphasis on operational accounts which held that to be meaningful, statements must be immediately testable by the manner in which the observations are made, and from the notion of extreme Logical Positivism (see p. 422) that their meaning *is* their manner of verification. The 'observer' is not much more than an agency for putting rulers end to end and noting the simultaneity of light flashes originating from different regions of space. He may also carry out very simple experiments such as noting apparent relative velocities of objects carried through space in boxes, or elevators, or on railway trains. This assumes that observers are essentially similar – indeed identical – and entirely neutral, so that *their* characteristics do not appear. Indeed, they do not have any characteristics, and there is no account of what a perception is. For Newton, human perceptions (the *sensorium*) are essentially like God's, and are perhaps regarded as part of God's *sensorium,* as also is space. But for Newton, and before him Galileo, who discusses the problem more explicitly, perception is a deep mystery and entirely outside physics. This must be so on an 'objective' account. And yet they do speak of 'discovering', 'observing' and so on. It is as though the verbs have meaning but not their nouns. (Can one have perceiving without a perceiver – or perceptions? Or running without a runner? Perhaps so.)

In the Special Theory (1905) Einstein described all observers as experiencing the laws of Nature as the same, however they are moving, provided they are not subject to forces. In the later General Theory, inertial and gravitational mass were equated, and physical laws were held to be the same for all observers whatever their velocity or acceleration: observers are therefore equivalent whatever the inertial forces may be. We shall look at the implications of this.

Einstein's Theory of Relativity starts from the 'Principle' of Relativity, which is that observers at uniform motion should experience the same physical laws whether they are in steady motion or at rest.[4] This was stated clearly, a year before Einstein's celebrated 1905 paper, by Jules Henri Poincaré (1854–1912), the French mathematician, in 'The Principles of Mathematical Physics' (1904):[5]

The principle of relativity [is that] according to which the laws of physical phenomena should be the same, whether for an observer fixed, or an observer carried along in uniform movement of translation; so that we have not and could not have any means of discerning whether or not we are carried along in such a motion.

As D. W. Sciama (1969) points out, the Principle of Relativity applies to physical laws but not necessarily to parameters, which may give the game away for travellers in a closed train, or a box in space.[6] Thus, particular values of parameters measured from inside the train or closed box can tell the occupants that they are in motion relative to some other reference without violating the principle. Indeed, our brains are in boxes but we can tell from our senses that we are moving relative to other things, though of course we may be uncertain, for example, whether it is a neighbouring train or ours that is moving. Sciama takes the example of measuring an electric component of the earth's field from inside the box. It is important to note that any measures, not just those of the unaided senses (which are arbitrary for this discussion of 'observer'), are allowed. The issue here turns on what we take as 'laws' and what as 'parameters'. There seems to be considerable agreement among physicists as to which is which, though one might imagine doubts creeping in.

Poincaré saw that something was needed to make it impossible for observers to travel faster than light, for the Principle of Relativity to be maintained. So having stated the principle, and some of its main requirements and implications, he came very close indeed to formulating the Theory of Relativity, which is Einstein's. Einstein's theory has its seeds in questions such as Bishop Berkeley's questioning of how a single sphere in space could be said to be (or not to be) rotating – and so how can we say what is fixed, what moving? Newton, greatly bothered about such questions, resorted to the notion that things move in relation to the absolute space of the consciousness, or perception of God. Newton's theory was so successful in astronomy that its deep assumptions were assumed to be valid and were seldom questioned.

Newton's Laws are restricted to non-accelerating observers, which limited the generality of his theory; this worried him. He had to introduce fictional 'forces' to account for discrepancies, especially for rotation. Rotation, such as rotation of the earth, is an especially interesting kind of acceleration because it is continuous without having to be maintained (in the absence of friction) by application of forces requiring energy drawn from another system. This is true also of a pendulum, though here the 'forces' are not continuous as for rotation, but cyclic. But the forces that seem to be required depend on where the observer is and how he is moving. The most striking example of the difficulty here is the flattening of the poles of the planet Jupiter. We say that this is because 'Jupiter is spinning on its axis,' and indeed this is what we see from successive observations with a telescope. But spinning by reference to what? For Newton, it would be spinning in relation to the absolute rest of the view of the Universe that God has. For Descartes, it would be spinning by reference to something like the ether which became a basis of physics in the nineteenth century, with the acceptance of light as waves rather than Newtonian particles. (Particles could move from place to place carrying energy or information through empty space, but waves, by analogy with water waves and sound waves, evidently needed a medium to carry them. So the problem of 'action at a distance' was avoided by postulating a medium for light waves.) Looking at Jupiter's flattened poles, it seems clear to us that it is rotating round its axis; but if we imagine revolving in a synchronous orbit round Jupiter we might not expect to see it flattened, for it would appear stationary; and if we imagine one observer looking from earth while another revolves round it in a synchronous orbit from which it appears stationary, surely it would not appear flattened to the earth observers and at the same time *not* flattened to the observers orbiting around it? So this would seem not to be a question of how things appear but rather of how things are, although what seems to require forces depends on how the observer is placed, or how he is moving.[7]

As we have seen (p. 125), Newton takes a bucket of water spinning on a rope to illustrate the issue. The surface of the water is only curved when the water spins in relation to (as he sees it) absolute space; for the bucket may be stationary or spinning with respect to the water, when it is curved or when it is flat. Newton was forced to postulate 'fictitious forces', such as centrifugal forces, to explain even the curvature of water spinning in a bucket.

The essential problem was how to describe inertial resistance to acceleration. Einstein's answer to this is surprisingly similar to Bishop Berkeley's, and to Ernst Mach's: that the 'inertial frame' by which acceleration can best be described and is experienced comes from the great mass of the distant stars. This is so much greater than the mass of the bucket, or of the observers of Jupiter, that they have no appreciable effect on what is observed. For Einstein, this becomes a deep relationship, or rather identity, between inertial mass and gravity.

Einstein's first theory, the Special Theory, was, like Newton's, limited to non-accelerated motion; but in his later work Einstein developed what is called Equivalence Theory. This is an extension of the Principle of Relativity, allowing the relation between the world and the observer to be described in conditions of acceleration. Einstein illustrates it by imagining people inside an elevator being pulled with uniform acceleration. Whatever observations they make, they cannot tell whether they are being pulled at constant force, to give uniform acceleration, or whether they are at rest but affected by gravity. It was from this Principle of Equivalence that Einstein equated gravitational with inertial mass, which appeared as a purely fortuitous connection to Newton. (See, though, p. 522.)

Einstein started with the observer. He asked how it is possible for an observer (though a very simple and neutral observer) to know that events at different distances occur simultaneously when light (supposing this their source of information) has finite velocity, so that although light signals arrive simultaneously, the events signalled would be separated in time as a function of their relative distances. Actually this followed his teacher at Zurich, Hermann Minkowsky (1864–1909), who proposed the notion of a space–time continuum, in which space and time could not strictly be thought of as separate. Action at a distance was avoided without postulating an ether, as the motions of celestial bodies were seen as following along (or falling 'down') the curves of space set up by gravitational forces. Matter was seen as nodules of condensed energy, distorting space and bending light. This was first measured by the solar eclipse expedition of 1919 of Sir Arthur Eddington (1882–1944), in which the light passing close to the sun from more distant stars was seen to be bent, to the predicted amount, by the sun's gravitational field. This prediction, when verified by the eclipse observation, was accepted as a remarkable vindication of what had seemed a daring and exciting set of speculations but without reasons for believing it. How did Einstein discover, or derive, his theory? There was little experimental evidence available that was not known to Newton (though Newton was unaware of the precession of the perihelion of Mercury). It seems that Einstein did not know of the negative result of the Michelson–Morley experiment (see p. 529). Einstein used vividly described 'thought experiments', often with little or no mathematics, and he demanded operational criteria for accepting that observations are meaningful. This at least is so for his earlier work; later he greatly relaxed his early Positivism, in which the 'observer' had little more to do than look at clocks and put rulers end to end.

It is interesting that not all of Einstein's thought experiments were amenable to observational tests. Einstein attacked the ether as a medium 'at rest' and the carrier of light, using imaginary thought situations impossible to realize and incapable of producing observable results. Thus he imagined travelling at the speed of light looking at himself in a mirror held in his hand, with a candle that he carried with him. On the ether account, the light would never reach the mirror – so he would not be able to see himself. He would then know that he must be travelling at the speed of light. But this would violate the principle that the Universe should be essentially alike for all observers – the Relativity Principle. Einstein saw this thought experiment as a kind of evidence – indeed as evidence so strong that it challenged Newton – though he could not set it up or say how it might be set up.

Both the ether of nineteenth-century physics, and its demise in our century, are the results of seeking consistency: in the nineteenth century, consistency with the common sense of sensory experience in which waves must travel in a medium; and in our century, consistency with highly abstract (one might say Platonic) concepts far removed from experience. It is most interesting that for twentieth-century physics to accommodate the notion of light as being both waves and particles required the acceptance of what seems on physical analogies from sense experience to be paradoxical. Einstein and the early Quantum Physicists were able – perhaps forced – to accept this perceptual paradox as a better alternative than still deeper paradoxes in the conceptual space generated by the Relativity considerations. It is this that makes *reductio ad absurdum* a weak kind of philosophical argument; yet without it, where is sanity? Without a sense of the ridiculous there can hardly be persuasion, other than by force, and we cannot begin to distinguish between illusions and realities or agree on what is true.

Mind in space

We have considered perceptual distortions, especially distortions of 'visual space', as having two origins, which are quite different in their status (see p. 407). This difference goes back to our account of mechanical processes and procedures carried out by machines of all kinds (see p. 73). Some perceptual distortions are due to signal distortions of the mechanism of the nervous system, and others to inappropriateness of procedures or assumptions by which neural signals are read, as data for deriving the perceptual hypotheses that are our most immediate reality. In speaking of perceptions as 'hypotheses', we are emphasizing their predictive power and their amazing richness compared with the paucity of available signals or data. Considering the term 'visual space', we see this as sensory dimensions and scales of perception by which we interact with the world. 'Perceptual space' is hypothetical, and it can be modified by experience. The limits of this modification are not known (see pp. 226–7).

Considering, now, 'physical space': this was accepted without question as Euclidean until the development of non-Euclidean geometries early in the nineteenth century; and more drastically in the present century with Relativity Theory, in which space is said to be 'curved', and time to slow down in strong gravitational fields or accelerations.

We normally think of perceptual distortions – illusions of various kinds – as discrepancies between how things seem and how they are. They are thus discrepancies between Mind and the physical world. But, as notions of space and time have changed with changes of physical paradigms, we may view this as discrepancies between perceptual and conceptual hypotheses, which are not fixed. Either may change, and changes in either may change to some degree the other. Are perceptual or conceptual hypotheses ever 'True'? I postpone this question to almost the very end (see chapter 20). Meanwhile, we are concerned with the relation between the observer and the physical world. This has been approached by looking at various 'links': links of light, of logic and so on. It is time now to try to bring some of this together, and to look at the place of the observer in the two great paradigms of modern physics: Relativity Theory and Quantum Mechanics.

The notion of distortion is useful, though like all useful tools it is dangerous when misapplied. We see distortions as discrepancies, which may be studied: (1) between observations; (2) between observations and what is believed to be true of the physical world; or (3) between accounts of the physical world. Perceptual distortions suggest significant measurements relating perception and physics, and they cry out for theories of their origin – which may lie in perceptual processes or in physics. Bending of light in mirages is a clear case of physical distortion producing perceptual errors. The Muller–Lyer Illusion is, I think, a clear case of illusion within perceptual processing (see p. 406).

It is important to note that strictly speaking we should not say that anything is distorted, except by reference to other things accepted as not distorted. Distortions are discrepancies from some standard, and it is meaningless to say that anything, including a perception, is distorted without at least hinting at what the reference standard is supposed to be from which there is a discrepancy. In mirages, we have plenty of evidence to say that light is bent by refraction, for there are plenty of examples that we accept of light not being bent. But, as physical theories become more general, reference cases become more scarce – until there are none at all. It is then not possible to speak of 'distortions' – at least as the word is normally used.

In Einstein's Theory of Relativity, light is sometimes said to be 'bent by gravitational masses', but it is also said that 'A straight line is the path that light takes' (or: 'A straight line is the shortest distance between two points – which is the path that light takes in space'). But these are contradictory; or at least, the first only has meaning by reference to another paradigm of physics: in this case Einstein's account compared with Newton's.

We have a similar situation for forces, especially for the 'force of gravity'. For Newton, this is a force much like muscular force (though acting at a distance, and the relation between inertia and mass is a strange coincidence); but for Einstein, gravity is not a force: gravity is the way that objects fall, through what is sometimes called 'curved' Space. It is curved by (or its curvature is changed by) the presence of the earth, or any massive body; but an object in free fall is moving in a 'straight' line in that it is taking the shortest path, and is not being deviated by an external force. So what appears as force depends on the general conceptual paradigm, and the particular observational reference adopted.

For this and other reasons, it is surprisingly difficult to distinguish between distortions of physics and distortions of Mind. Ultimately it is difficult because we do not know quite how the observer fits into, or is part of, the physical world. We do not know our place in Nature.

Einstein's is a physical theory, but he does make frequent references – which is highly unusual in physics – to what an observer at a certain place, or travelling at a certain velocity relative to another observer, or whatever, would see. Consider, for example, a source of light in a moving train. The train is travelling very fast, near the speed of light, and there are two observers, A and B. Observer A is in the train and observer B is on the embankment observing the train as it rushes by. The central light source sends a flash of light simultaneously up the train to the front end of the carriage, and also down to the back of the carriage. Back and front are equidistant from the source. Now for observer A, on the train, the light flashes reach the back and front at the same moment; but for observer B, on the embankment, the light reaches the back of the train before it reaches the front. Thus the observers report different events; yet on Relativity Theory both are true! Neither observer's perception, or account of what happened, is to be preferred to the other; and yet, for common sense or for classical physics they are incompatible. Nor mally we would regard such inconsistencies of observations as the clearest evi dence for illusion; but not here, on Einstein's paradigm. So we learn that what is a discrepancy depends on what is not allowed by the prevailing paradigm. Gener ally we take common sense as our guide but clearly this is not adequate for considering Mind in Science. And how we see Mind in Science obviously depends on which science we accept. So even paradoxes are relative!

As is well known, Relativity Theory is based on the rejection of the notion of some absolute or preferred reference for observing or describing physical phenomena, and in particular motion. It is also well known that the critical experiment was the Michelson–Morley observation, that the velocity of light is not affected by the observer's motion through the supposed ether, which had been postulated as the medium for carrying light. The ether was accepted for interpreting Maxwell's equations, relating electrical with magnetic phenomena and showing that light is electromagnetic. Most curiously, it turned out that Maxwell's equations were incompatible with Newton's, and that they were compatible with Einstein's – although Einstein rejected the ether which Maxwell accepted as necessary. I mention this to bring out again the complexity and non-obviousness of incompatibilities, discrepancies and illusions. Just why did physicists conclude that Newton could not live with Einstein, though Maxwell (who accepted the absolute reference of the ether) should henceforth and for ever live with Einstein?

I suspect that here we find another weakness in Popper's position *vis-à-vis* refutation. What it comes down to is that what is accepted as refutation is just as subtle a matter as what is accepted as discrepancy, or illusion. Surely discrepancy is the primary concept here; and just when discrepancies should be accepted as refutations has not yet been formulated in the philosophy of science. I suspect that this is a very deep problem, and I am incapable of suggesting where to seek answers, beyond pointing out that scientists of different persuasions or interests may hold different criteria for what are unacceptable discrepancies – and so for what are 'refutations'. The same holds for Ayer's verification criteria (see p. 422).

We may home in now on some specific issues of Relativity Theory. This may help us to see how the observer is related to what he observes, in the Relativity paradigm of physics. We shall start with curious expansions and contractions of time and space, especially those associated with motion.

The distortions of Doppler

It is well known that the frequency of sound changes for observers moving towards or away from the source, and also when the source moves towards or away from the observer. This Doppler Shift occurs because more individual waves pass the observer in unit time with approach, and fewer with recession. Christian Doppler (1803–53) was an Austrian who became Professor of Physics at Vienna in 1851. We see in Doppler Shifts a kind of distortion of observer space, since the observed frequency depends on velocities. The explanation is, however, purely in terms of physics.

As the observer approaches the source of a sound, or the source approaches the observer, the received frequency (number of waves received per second) increases, and the distance between the waves appears to decrease correspondingly as he meets more waves in each second, as approach speed is added to the velocity of the sound in the air. This Doppler Shift is generally too small to notice for the speeds of walking or running, but they are dramatic for the higher velocities of technology: train whistles, or the sounds of aircraft.

Take an observer O moving towards a sound source with a frequency f_s. As we have said, as he is approaching the source he will meet more waves in unit time (say per second) than if he were stationary (and if he were moving away from the source he would meet fewer, until at the speed of sound he would experience no waves). Consider motion towards the source: if the observer's velocity is v_o, in one second he will meet an extra v_o / λ_s waves, where λ_s is the wavelength of the waves emitted by the source. So

$$f_o = f_s + v_o / \lambda_s .$$

As the speed of sound $c = f_s \lambda_s$, this gives

$$f_o = f_s (1 + v_o / c).$$

Interestingly, the shift in observed frequency of sound is not the same when the source moves towards the observer as when the observer moves towards the source. It turns out – and this is central to Relativity Theory – that this does not hold for light, though sound and light should obey the same equations if light were waves carried in an air-like fluid. For light, the Doppler Shift is given by a relation which, though only slightly different numerically for normal speeds, is profoundly different conceptually from the Doppler Shift for sound in air.

Consider the velocity of sound in air: if the observer moves towards a sound source, the velocity of the sound increases as his speed through the air increases, and the velocity of sound is affected by the wind speed. If, however, the source is moving, the velocity of the sound reaching him is not affected; though it would be if sound were particles, like bullets. Now for light, most remarkably, the velocity remains constant for any source velocity and any observer velocity (limited to the speed of light). So the case of light is different from that of sound, though both manifest properties of waves. This was shown experimentally in America, in 1887, by Albert Michelson and Edward Morley. This may be the most celebrated experiment in physics and is certainly the most famous negative result ever obtained. The 'null' observation required the utmost precision to be significant.

We do not need to go into details of the experiment, but the method is based on what happens to two swimmers in a flowing river, who swim at equal speed with respect to the water. One swims up and then down the river, and the other back wards and forwards across it. What distance do they cover? The up–down swimmer is handicapped, because he spends longer swimming against the current than with it. He covers less ground (his average speed in relation to the ground is less) than the swimmer who swims across the river. Michelson and Morley used a cross-shaped interferometer, which on this argument should have shown a change of the time of arrival of light, from a source carried with it, according to which arm was moving along and which across the ether, as the earth moved around the sun. This was detected by looking for shifts of interference patterns from the two light beams when combined; but no changes of the interference fringes were observed.

When Michelson and Morley found, in this way, that the velocity of light from the source carried with their interferometer remained constant for any orientation of the measuring equipment, two principal explanations were suggested: (1) the ether was being carried round with the rotation of the earth and dragged with it in its motion round the sun; and (2) much more strangely, that the null result was due not to the light remaining at the same speed with the motion, but that the differences could not be *seen* because the length of the interferometer arm shrank in the direction of motion through the ether. On this view, it was as though there is a conspiracy in nature to prevent us from ever measuring absolute observer velocity. But as Poincaré pointed out: *'A complete conspiracy is itself a law of nature!* Poincaré then proposed that it is not possible to discover an Aether wind by any experiment; that is, there is no way to determine an absolute velocity.'[8]

This was explained by the Lorentz–FitzGerald equations. The Irish physicist George FitzGerald (1831–1901) saw that the Michelson–Morley experiment would give its null result if the arm of the interferometer shrank exactly to cancel the effects of its speed through the ether. This may seem a bizarre notion, but it saved the rest of physics: especially Maxwell's equations. FitzGerald showed that on the assumption that the velocity of light appears the same for any observer velocity because any body (ruler or interferometer arm or whatever) shrinks in the direction of motion, the shrinkage must follow the relation $\sqrt{(1 - v^2/c^2)}$. Where v is the velocity of the motion, c is the speed of light, L is the length that the body would be at rest, and L' is the length of the body in its direction of motion,

$$L' = L\sqrt{(1 - v^2/c^2)}.$$

At the speed of light, L' becomes zero: the interferometer (or any other object) would shrink to an infinitely thin wafer. (It would shrink to almost half its length L when $v = 0.87\ c$ which is about 160,000 miles per second.)

The velocity cannot quite reach the speed of light. This can be thought of as due to mass increasing with velocity so that infinite force would be required to accelerate a body of rest mass M to the speed of light, when its mass, M', would be infinite:

$$M/M' = \sqrt{(1 - v^2/c^2)}$$

or

$$M' = M/\sqrt{(1 - v^2/c^2)}.$$

The mass is just more than doubled when $v = 0.87c$ (about 580,000,000 miles per hour), and is infinite at the speed of light.

The Dutch physicist Hendrik Antoon Lorentz (1854–1928) suggested that it is a contingent property of matter to shrink as it moves through the ether. In this case, the amount that an object shrank might well depend on the substance of which it is made; interferometers made of various metals, wood and so on were therefore tried, but all with the same null result. So the result did not seem to be contingent on properties of materials, but rather to be a fundamental principle of physics applying to light. This at once suggested to Ernst Mach, among others, that space has no ether – no medium for the transmission of light – and so no absolute reference for motion.

Lorentz saw that how the world appears depends on the motion of the observer, and can only be described consistently from a given inertial reference frame. This may be taken to mean that if anything is absolute it is acceleration. An inertial frame is a set of coordinates within which there are no accelerations, and only from such a basis can consistent observations be made. Of course in practice the errors are generally small – the discrepancies are generally too small to be detected – but for a conceptually consistent account this is not the point. Conceptual consistency has nothing to do with the limits of observation and measurement. This is perhaps where Rationalism enters Empiricism: demands of conceptual consistency are decisive forces for correcting, or at least changing, accounts of the Universe based on observations of the utmost available precision, which are never precise enough for certainty.

Einstein, in 1905, took the unprecedented step of accepting inconsistencies between observations and accounts from different reference frames, and saying that although they are inconsistent they are all equally valid. This, it seems, is Einstein's key contribution. Its significance is perhaps still not fully appreciated in philosophical discussions of epistemology; for we still do not seem to be clear about what kinds of inconsistency we can live with, and which must be resolved for acceptable rational accounts of Matter and Mind. We certainly demand consistency between observer's reports in the normal business of life: reporting accidents, the weather, and so on, as well as in the Physical Sciences. Where there are maintained inconsistencies, we are apt to say that these are due to 'subjective' factors, such that observers are not equivalent. It is, indeed, these differences that are the evidence that observations are not merely selections of physical reality, but are rather individual constructs, and so are Mind-ful. Normally we would accept evident inconsistencies only by relegation to subjective, or in-Mind, factors, and say that various subjective accounts are equivalent with respect to truth. Einstein extends this universally accepted tolerance of observation differences to the physical world. It is, however, important to emphasize that Einstein is not reducing observations (including imaginary observations) to solipsistic dream-like accounts, where anything goes because cross-checks due to the privacy of experience are ruled out: Einstein is concerned with the various situations that the observers are in and not with their individual differences, *qua* observers. We might say that it is these differences which are the evidence of Mind in a physical Universe.

Mind in Matter?

Are the distortions associated with very high velocities distortions of space or of objects, or are they observer distortions?

Many descriptions in science are based on deliberate distortions, such as Mercator projections for maps of the earth, or the remarkably early (first-century AD) projections of the heavens as represented on the flat plates of astrolabes, when the still earlier globe models were flattened, for convenience as portable robust instruments. Somewhat similarly, functions are plotted on graphs whose axes may be linear or logarithmic, or whatever, according to convenience. Some of the resulting curves may appear 'distorted'; but there may be no uniquely 'undistorted' curve, or 'undistorting' axis. Often it is merely that some are easier to read than others, and descriptions may be simpler according to which axis functions are chosen (see p. 509).

The distortions of Mercator projections and the stereographic projections of astrolabes are explicit distortions of picture-like representations, but with no claim that space itself is 'distorted' from common sense, or from Euclidean space. It is worth emphasizing here that visual space is not Euclidean, except for near objects. As we pointed out (p. 407), railway lines or tracks appear parallel only for the first few hundred metres; for greater distances they appear to converge in a highly non-Euclidean way, and then the train-driver's faith must take over to countermand his perception of how things are, or he will surely stop in his tracks, with this sight of continuously impending disaster. This is a case where appearance is rejected by knowledge, or belief, in a familiar and common situation; though it may be that such situations are 'Unnatural', as they occur with technology rather than in pre-technological situations. The suggestion here is that perception has not quite come to terms with the biologically extremely new situations of technology such as superhuman speeds and parallel tracks. It is curious that commonsense concepts are seldom violated by these discrepancies between how things are seen and how they are believed to be. This is not, however, at all true for Einstein's account of the Universe: here the world *is* far removed from common sense, perception and understanding.

To measure the forms of objects, we may distinguish between 'intrinsic' and 'extrinsic' measures. A fly walking over a surface may discover intrinsic form and work out the geometry by measuring each angle and length as he journeys over the surface; or he might stand back and receive a perspective-projection view to give extrinsic measures. It may be possible to derive the intrinsic from the extrinsic account, or vice versa. Thus a cartographer with chain and theodolite builds up intrinsic descriptions, which can be related to the extrinsic views of earth from Space. There should be no discrepancies once the proper scale and other corrections have been applied.

Suppose a set of measures or observations shows that Space is distorted by comparison with an expectation based on theoretical grounds or from observations made in different conditions. Or suppose that the intrinsic account does not agree with the extrinsic, after what seem to be the appropriate corrections (for example for perspective) have been applied. How can we describe such discrepancies?

1   Observations may affect and change what is being measured (for example, a ruler laid on a cushion or a jelly may distort the cushion or the jelly by the act of measurement).

2   The measuring instrument may be distorted by what is being measured (for example, the ruler may become heated or cooled).

3   There may be some perhaps unknown modification of the energy pattern received from what is being measured, to the instruments or senses – for example, by a mirror or for a mirage.

Now it is clearly important to say which, if any, of these is involved when it is said that 'Space is distorted,' or that 'Space is non-Euclidean.'

When Einstein says that light is deviated by inertial masses, such as the stars, he may be saying that space has a changing refractive index: that light is refracted by gravitational fields. So measures are affected by gravitational masses, somewhat as mirages are seen by refraction of light by air. For the ruler that was heated and cooled, and so changed in length, we say that this is a trivial artefact; but when any conceivable instrument is affected, to give the same 'errors', then the situation appears very different. It then looks like a universal situation, to be described by laws.

Now let us return to the equations of FitzGerald and Lorentz (pp. 530–1). Do these represent what happens to measures, or, rather, to what is being measured? Is the increase in mass and reduction in length a distortion of the matter of the interferometer, or distortion of space?

For the Relativity 'distortions', any measuring instrument (and light itself) is distorted as space is said to be distorted. So it is not possible to speak of discrepancy between the measure and what is being measured, for everything including the measuring instruments are 'distorted' together, so there can be no such discrepancies. It is, however, true that what is observed depends greatly on the distribution of masses in the Universe. Most dramatically, this follows from the explanation of Mach's principle for rotation: the bulge of Jupiter depends on the mass of the 'fixed' stars. The bending of light by the sun (an observation that is not perhaps yet fully confirmed) is a deviation from 'straight' as defined by light in the absence of gravitational 'disturbance'. Newton's First Law, that bodies move in straight lines unless disturbed, is thus a tautology if 'straight' is defined as how objects move when undisturbed.

Since Newton's definition of 'straight' is tautologous, it is strictly meaningless to say that Einsteinian space is a distorted form of Newtonian space. It can, however, be useful to say this when 'Relativity corrections' are in mind. These are corrections from predictions made on Newtonian assumptions, and so then the Einsteinian space appears 'distorted' from Newtonian space. The Relativity corrections are applied much as corrections for calibration, or refractive bending of light by the atmosphere from stars, may be applied. But, strictly, these kinds of 'correction' are very different.

Perhaps another way of looking at this is to consider translation from one paradigm to another. Conceptually structured theories or paradigms do not exactly fit, or nest, within each other. We say that one paradigm (usually the later one) 'corrects' errors, and makes up for 'gaps' in the other. This means, roughly, that the second 'fits the facts' better than the first, and 'covers more facts' than the first. This is like discrepancies and inadequacies of measurement – and like these it implies that we know what the 'facts' are if we claim to spot the discrepancies or the inadequacies. The Einsteinian 'distortions' are not distortions of measurement, although it may be said that the measuring instruments are 'distorted'. This sounds like nonsense; but the statements here are so general that there is nothing against which to compare them, so they cannot have normal meanings. Indeed one is tempted to say that all language has meaning only from comparisons: that all meanings are relative. It is something of a paradox, then, that Relativity Theory is so general that the language for describing it loses meaning because language meanings are relative. When there is nothing beyond very general statements to serve as references, meaning is lost; this is surely why physicists are driven to express their most general ideas in mathematics, which represents nothing but relation. They seem inadequate; so they are enriched by being 'interpreted'.

Do we confer relations by acts of measurement; or are these relations there, in space, for us to measure? This is an ancient philosophical question, but in a somewhat new form. Its relevance now is for how we should think of the status of measurements and observations in a Relativity world, in which the instruments and observers are within the world and subject to its laws, where geometry is not necessarily Euclidean, and where (so far as possible) explicit procedures for making measurements and observations must be stated to make observations 'objective' and so acceptable for science. (One might say that this is supposed to distinguish 'observation' from 'introspection', and to distinguish physical from conceptual space.) There is controversy here among philosophers of science. Bridgman, a distinguished proponent of Operationalism, holds that (as Adolf Grünbaum puts it in his paper, 'Logical and Philosophical Foundations of the Special Theory of Relativity', 1955): 'we human beings are the ones who first confer properties and relations upon physical entities by our operations of measurement'. But Grünbaum gives his own view as very different:

I do not think that the theory of relativity can be validly adduced in support of this homocentric form of operationalism. Einstein's theory asks us to conceive the topology and metric of space–time as systems of relations between physical events and things. But both these things and their relations are independent of man's presence in the cosmos. Our operations of measurement merely discover or ascertain the structure of space–time but they do not generate it.

I shall leave this controversy here. It hinges on whether man is observer or creator. We have argued that perception is a highly creative, enriching business, in that it adds a great deal to what is given; but this is very different from saying that physical reality is created by perception, which ultimately is solipsism. Personally I do not take this view, holding that such useful words as 'perception', 'communication', 'discovery' and 'refutation and confirmation' are, by solipsism, rendered needlessly meaningless. At the same time, no doubt Eddington was right to stress how what we observe is filtered (as coarse nets only catch large fish) by the limitations of measurement; these limitations are ultimately set by characteristics of the world such as the finite velocity of light and the quantal nature of energy required to receive information.

Kant, we may be sure, was wrong in thinking that the human Mind can conceive of only one – Euclidean – space. I should prefer to follow Poincaré (1914): [9]

. . . though geometry is not an experimental science, it is a science born in connection with experience; that we have created the space it studies, but adapting it to the world in which we live. We have chosen the most convenient space, but experience guided our choice. As the choice was unconscious, it appears to be imposed upon us.

Perhaps we still cannot quite decide from perceptual adaptation experiments whether Poincaré is right, but probably he is. It is in any case worth pointing out that if our perceptions were extremely limited in adaptive range, the world would appear to have absolute forms and values – as Kant saw it. We can only appreciate relativity in physics, or of beliefs, through broad views that can adapt with circumstances.

Poincaré held that it is quite arbitrary which geometry is accepted: that all is relative to assumption. There is, however, a universal constant in Relativity physics: the speed of light *in vacuo*. And the laws of physics appear the same to all observers, whatever their inertial reference frames: the Principle of Relativity. So there remain a few anchors in physics. Taking the language analogy, it is as though a few things are known (as Russell put it) by acquaintance, and these are accepted as absolute. They provide the constants for the equations and are the ultimate links between experience and science – between perceptual and scientific hypotheses.

Yet here we must be careful, for what is the status of this knowledge by acquaintance? Is this not the last vestige of Direct Realism, which I have attacked throughout in detail and abandoned? How did Einstein 'see', and be the first to see, that the velocity of light is an absolute constant, in a Relativistic world? He did not see this by anything like Russell's last vestige of direct undeniable knowledge. Einstein somehow inferred or guessed that if c were accepted as an absolute, then order could be made of the rest. The constants of physics are postulates that serve as keystones of theories. Without constants the structure falls; without the structure, there is nothing to support them as anything special. Both Newton and Einstein made light the keystone of their theories, though as the theories are different, light looks different – and as light looks different, so do the theories.

We may say that Relativity Theory, which was born of seeking objective truth by carrying out real or imaginary procedures, confirms, or at least is no exception to, the notion that truth itself is relative: that knowledge depends on assumptions or postulates that are taken as absolute because without them we cannot, with meaning, say or see. Events appear to occur (or at least we invariably describe them as occurring) in an irreversible time sequence, though 'subjective' time may run fast or slow by reference to clocks. And in memory we can move forwards and backwards within the range of time of our stored experience.[10]

How do the celebrated 'paradoxes of time' in Relativity Theory affect this situation? Relativity Theory has changed basic notions of time, by pointing out that observers in relative motion do not experience simultaneity of a single event. Since Relativity, simultaneity can only be defined relative to observers in a given inertial frame. This implies that time – clocks of any kind – run slow or fast according to the relative speed of the inertial frame. This generates the famous so-called paradoxes, such as the 'twin paradox' of time. The paradoxes do not, however, reveal internal inconsistencies within Relativity Theory. They are paradoxes only in that they violate common sense, which is the pre-Relativity metaphysics of the common man, as suggested no doubt by astronomical cycles and the seasons, and later by the technology of mechanical clocks that suggested a strict linearity and 'constant rate' of time. But of course the very notion of constant rate for time is suspect, for what is time's rate supposed to be constant to? Here we see just the same problem as for absolute velocities in the absence of a neutral 'rest frame' which was traditionally given by the ether. Differences in clock time between inertial frames in relative motion imply that all causal effects must be velocity-related, for nothing (including gravity and any force) can travel faster than light. So observers and physically causal events are equally isolated, or interactive in time, according to relative velocities. In this sense at least observers are part of physics: they are limited by physical considerations such as the finite velocity of light (and we suppose that brain function is limited by physical laws). It is thus most important to establish what these laws are, and also to see just how functions of machines or brains, including the functional processes of perception, which may include machine perception, are related to causes as described in physics. We have seen this as a problem (p.411) when considering why we should accept the validity of a logical argument carried out by a machine or a brain, and just how one distinguishes between a mechanical (or a physiological) error from an error of logic. It is surely this kind of distinction that removes the observer from descriptions of physics. Does it also remove him from the world described by physics? For Einstein, in his earlier writings, as for Popper, it seems fair to say that the observer is reduced to physically definable operations, which do not include anything of what a psychologist would mean by 'perception'. On such operational accounts, discrepancies (and so experimental 'refutations') are alleged incompatibilities between events occurring naturally and the highly selected events of operational experiments. But surely these can only be incompatible on certain assumptions; but assumptions are not part of the physical world. It is part of the seductive power of Direct Realism to give an account that seems to avoid the need for assumption, as it denies the active observer. But having rejected Direct Realism, it seems that we cannot avoid a radical and extreme Relativism for knowledge. As in Einstein's Theory, we can only avoid a chaotic Relativism by postulating some absolutes as necessary for seeing consistency. But surely seeing consistency (and inconsistency) is not to be described within physics, for the rules for testing consistency – as all logical rules – are not laws of physics.

Einstein, especially in his later writings, saw the Universe as mathematically simple and its essentials unchanging. He was a Rationalist, in the Platonic tradition and the later tradition of Spinoza and Leibniz, though Relativity was inspired by Empiricist perceptual and conceptual observations, especially by Berkeley and Mach. The second great paradigm of modern physics – Quantum Mechanics – is frankly Empiricist in its origins and formulation, and it sees the world as a complex muddle. It is far more, and indeed deeply, concerned than is Relativity with the place of the observer in the Universe. This applies especially to the thinking of its founders, Max Planck, Niels Bohr (1958) and Werner Heisenberg (1930, 1958), and to such later writers as Eugene Wigner (1967), John Wheeler (1975, 1977, 1979, 1980) and Abner Shimony (1963, 1965), and J.F. Clauser and A. Shimony (1978), who have more recently considered its philosophical basis and implications for how the world can be observed. It is to the role of the observer in the current Empiricism of Quantum Mechanics that we now turn.

The observer in matter: Quantum Theory
Sir Arthur Eddington asks, in his Tarner Lectures (*The Philosophy of Physical Science* 1959): 'What do we really observe? Relativity theory has returned one answer – we only observe relations. Quantum theory returns another answer – we

only observe probabilities.'[11] Eddington proceeds to make illuminating remarks about probability, pointing out that probability is commonly taken as the antithesis of fact but that for Quantum Mechanics the 'hard facts of observation' are probabilities. Quantum Mechanics is essentially different from classical physics, and from Relativity, because prediction of the velocity and position of particles is limited by irreducible indeterminacy; and predicted positions, velocities and other characteristics of particles are less precise and curiously different from observations. A further difference is that Newtonian physics cannot explain why there are not continuous gradations and continuously changing orbital parameters of electrons in atoms. What especially interests us, however, is the irreversible character of probability, and the irreversible relation of observing in Quantum Mechanics – and the notion that observations are interactions. It turns out that classical Realism is challenged – that the observer is essentially separate from what he observes.

Quantum Mechanics is based on uncertainties – while Rationalist philosophies are concerned to establish simple certainties. The uncertainty here is due to the finding that energy (like matter) is not infinitely divisible but is transferred only in discrete steps, or 'quanta'. Since information is transmitted by discrete energy changes, there is a strict limit to the precision of observations, and so to knowledge of structure and processes of the micro-world. Observations are, further, held to upset the micro-events providing information: there is an essential interaction between the observation and what is observed, on the microscopic scale. It is this scale, rather than the cosmological scale of Relativity Theory, that primarily concerns Quantum Mechanics. Since, however, the observation of macroscopic objects (tables and chairs as well as stars) depends upon micro-energy exchanges, the essential uncertainty of Quantum Mechanics underlies the whole of physics and is central for epistemology – and so for considering the observer and perception.

The notion of object–observer interaction may be illustrated by the sense of touch. Consider moving the finger to touch a light object, such as a matchstick lying on a table. The matchstick will be moved by the touch necessary to detect it; and indeed it may be moved so that it remains undetected. Something fragile, such as a soap bubble, may be destroyed by the detecting touch – so that it never can be known, at all directly, by touch. The resolution limits imposed by the discreteness of energy transfer are clearly expressed by Eddington's celebrated analogy to the fisherman who never discovers fish smaller than the holes (or the spatial frequency) of his net.

The graininess of observations follows from the original theory, announced in 1900, of the German theoretical physicist Max Karl Ernst Planck (1858–1947), which started from the finding that when a 'black body' is heated through a range of temperatures, the total radiation it emits follows the same law (fourth power of the absolute temperature), whatever it is made of. Also, whatever the substance, the frequency of the radiation having the maximum energy increases with the temperature, so that glowing metal, for example, changes from red to blue with increasing temperature, before it vaporizes. Planck found that these results could be neatly described by supposing that the radiation is emitted in discrete units, or 'quanta'. Further, the energy of each quantum, he supposed, is related to the frequency of the radiation, such that the energy of each quantum doubles with each doubling of the frequency of the radiation. This he expressed with the celebrated equation

$$e = h\nu,$$

where **e** is the quantum energy, ν the frequency and **h** Planck's constant, which gives the proportionality relation between quantum energy and frequency of radiation, right through the electromagnetic spectrum. The discovery that light is electromagnetic is due to the astonishing theoretical work of the Scottish theoretical physicist James Clerk-Maxwell (1831–79), who predicted what came to be known as radio waves, travelling with the speed of light, some twenty years before they were detected by Heinrich Hertz in 1887. It was the combination of the evident wave nature of electromagnetic radiation, over a frequency range far greater than visible light, and the need to quantize the energy to describe the black-body radiation law, that started Quantum Mechanics. It developed into the notion that matter, too, has this dual characteristic of waves and quantized particles.

What may seem to be more direct evidence of the discreteness (quantization) of energy, and so of essential limits to information transfer, is given by the 'photoelectric effect'. This was discovered by the German physicist Philip Lenard, for which he received the Nobel Prize in 1905, the year of Einstein's first classical Relativity paper. Curiously, Einstein did not receive his Prize for the Theory of Relativity, but, in 1921, for his explanation of the photoelectric effect. This is that electrons are emitted from certain metals under the action of light, but that increasing the intensity of the light does not knock off more electrons, though increasing the frequency of the light does so. This is explicable if the energy is quantized, for what does the trick is the energy in each packet, which is determined by the frequency. This was a clear demonstration of the power of the quantum concept. At the same time, interference patterns, and the interference properties of gratings for producing spectra for discovering the substances, and the generally recessional velocities of stars, grew in importance: so the particle and wave properties of light gradually assumed equal importance, and both were accepted as how light is. Which predominates depends on the situation: a grating, or a thin film, or a hologram, reveals wave properties; the photoelectric effect reveals corpuscular properties. This looks odd, but is it so different from many other combinations of properties of the macroscopic world? When are combined properties incompatible – sufficiently 'impossible' to refute hypotheses? This is surely a deep question. As we have seen (p. 119), Zeno set up paradoxes by appealing to incompatibles, but the current world of physics is full of incompatibilities with common sense. What is incompatible within science depends (if logical rules are not violated) on the accepted paradigm, and its accepted assumptions; so 'objective' refutation is, surely, hardly possible, however bizarre or even paradoxical the claim may be.

What Planck's theory did was to make it clear that all energy transactions – and so all detections, records and observations – are uncertain by the unit of quantum action. The quantum steps may be smoothed at the macroscopic scale, but observation is essentially uncertain because what is being observed must change (or be changed, by quantized energy from the observer or instrument) for detection to be possible; and because these changes occur in finite steps, perfect resolution or complete information is impossible. Observation of macroscopic objects is indeed only possible because their structures are redundant, so that averaging gives useful information of macro-structure from micro-events.

It might be thought that the quantum steps are so small that they can be ignored for considering visual perception, but this is not so, as S. Hecht, S. Schlaer and M. H. Pirenne (1942) showed, in a celebrated experiment on detecting flashes of light visually.[12] Moment-to-moment variation in effective sensitivity of the eye for detecting small short-duration flashes of light varies according to the mathematical function of a Poisson distribution, attributable to the small numbers of individual quanta. The steepness of the frequency-of-seeing-curve, plotted against mean luminance, gives a measure of the near-perfection of the eye as a detector of radiation, in the frequency range of visible light. The results show that the dark-adapted 'rod' cells of the retina respond to a single quantum, which is the theoretical limit, though between five and eight quanta are needed (spaced close together in time and space on the retina) for the flash to be seen. This requirement is, no doubt, to reduce 'false positive' illusory flashes due to the residual random activity ('noise') of the neural channel. So the visual system has coincidence gates, as used with Geiger counters for detecting high-energy particles such as cosmic rays with minimal noise contamination.

It was Werner Karl Heisenberg (1901–76) who introduced the Uncertainty Principle, that the quantum of action (Planck's constant, h) limits the precision by which the position and velocity of a particle can be known, so that observational precision is limited for any argument, including the eye, by the graininess of energy. He argued also that causation at the atomic level is uncertain. Observation and prediction of macroscopic objects can, however, be made with high precision because they are made up of statistically large samples of randomly perturbed events, or atoms.[13]

Heisenberg finally gave up any attempt to describe the constituents of atoms as waves or as particles – or indeed anything that we can visualize from perceptual knowledge. He resorted to pure numbers for his description of the atomic world. So the atom itself disappeared as a 'thing' or an 'object'. This is, however, still controversial. Perhaps most physicists still cling to perceptual-object analogies, but of course, there is no intellectual reason to suppose that the atomic or sub-atomic scale bears any recognizable analogy to what we seem to know by the senses. But the bare equations seem inadequate even to theoretical physicists.

The rival conceptual descriptions of waves and particles became embedded in entrenched theories, which each developed its own lines of defence against the other pictures, accounts or descriptions of what light is. A particle, like an object of sense, exists at a given place in a given time, and can be in only one place at the same time. And particles, to be like objects, must obey laws of momentum, collision, and so on. Wave trains, on the other hand, as we experience them, can exist in the same place at the same time – and unlike objects or particles can move through each other without disturbance. When it became clear that light exhibits both of these kinds of features, Niels Bohr (1885–1962) developed what he called 'complementarity': that light and microphysical processes could be described with either wave or particle concepts as alternative descriptions, but that in some situations one is more appropriate than the other. This notion of logical incompat-ibility here seems, however, to be losing favour: it is simply said that these properties are not *logically* paradoxical, however odd light may seem. Logical incompatibility has dropped out with abandoning the attempt to make Quantum physics an extension of classical physics. It requires a new paradigm for the strange combination of wave and particle characteristics to be accepted without being dismissed as paradoxical. If we go on to ask what light is, or what micro-physical processes are, below the resolution allowed by the indeterminacy of Planck's constant, **h**, then on an operational philosophy one would have to say that we cannot know. But is our knowledge limited to the resolution of observa-tions? One might well argue that all sorts of things have been discovered in science that could not be observed, or are too small or too distant or whatever, to be observed in detail (for example Harvey's capillaries later seen by Malpighi).

Heisenberg describes the ultimate limits to precision of observations, in the following way:

Determination of the Position of a Free Particle – As a first example of the destruction of the knowledge of a particle's momentum by an apparatus determining its position, we consider the use of a microscope. Let the particle be moving at such a distance from the microscope that the cone of rays scattered from it through the objective has an angular opening ϵ. If λ is the wavelength of the light illuminating it, then the uncertainty in the measurement of the x- co-ordinate according to the laws of optics governing the resolving power of any instrument is:

$$\Delta x = \frac{\lambda}{\sin \epsilon}.$$

But, for any measurement to be possible at least one photon must be scattered from the electron and pass through the microscope to the observer. From this photon the electron receives a Compton recoil of order of magnitude h/λ. The recoil cannot be exactly known, since the direction of the scattered photon is undetermined within the bundle of rays entering the microscope. Thus there is an uncertainty of the recoil in the x-direction of amount

$$\Delta p_x \sim \frac{h}{\lambda} \sin \epsilon,$$

and it follows that for the motion after the experiment

$$\Delta p_x \Delta x \sim h.$$

There is thus an essential uncertainty which applies to all perception by detection of photons; and similar considerations apply to all other sources and kinds of observational information.

Heisenberg (1930) gives a beautifully clear account of Bohr's notion of complementarity between observations and what is observed.

With the advent of Einstein's relativity theory it was necessary for the first time to recognize that the physical world differed from the ideal world conceived in terms of everyday experience. It became apparent that ordinary concepts could only be applied to processes in which the velocity of light could be regarded as practically infinite . . . the resolution of the paradoxes of atomic physics can be accomplished only by further renunciation of old and cherished ideas. Most important of these is the idea that natural phenomena obey exact laws – the principle of causality. In fact, our ordinary description of nature . . . rests on the assumption that it is possible to observe the phenomena without appreciably influencing them. . . .

Second among the requirements traditionally imposed on a physical theory is that it must explain all phenomena as relations between objects existing in space and time. This requirement has suffered gradual relaxation in the course of the development of physics. . . . Now, as a geometric or kinematic description of a process implies observation, it follows that such a description of atomic processes necessarily precludes the exact validity of the law of causality – and conversely. Bohr has pointed out that it is therefore impossible to demand that both requirements be fulfilled by the quantum theory. They represent complementary and mutually exclusive aspects of atomic phenomena. . . . There exists a body of exact mathematical laws, but these cannot be interpreted as expressing simple relationships between objects existing in space and time. . . . This indeterminateness of the picture of the process is a direct result of the indeterminateness of the concept 'observation' – it is not possible to decide, other than arbitrarily, what objects are to be considered as part of the observed system and what as part of the observer's apparatus . . . the concept 'observation' belongs, strictly speaking, to the class of ideas borrowed from the experiences of everyday life. It can only be carried over to atomic phenomena when due regard is paid to the limitations placed on all space–time descriptions by the uncertainty principle.

A special feature of Quantum Mechanics is that the distinction between the measurement and what is measured becomes blurred because the act of observing affects what is observed. We are dealing here with interactions between the world and the observer, or the detecting instrument, such that the observer is not clearly on one side and the thing being detected or measured on the other. So to interpret experiments and observations it is necessary to devise a model of the total situation, in which it is somewhat arbitrary to say what is the observer and what is being observed. (The story is told by Russ Hanson of an Oxford undergraduate, who staggered his tutor by asking, 'What is the external world external *to?*' On this account, 'external' merges by quantal interactions with 'internal' detection and measurement.)

In detecting waves or particles, instruments do not exhibit in any direct sense the properties being detected or measured, though the matter of which they are composed acts according to these principles. Light detectors provide amplification, and essentially irreversible changes from quantal energy jumps. Thus a photographic plate is selectively blackened by quantal phenomena, to produce a permanent record of these events at macroscopic scale. Similarly for the retina, there are averaging and multiplying processes allowing quantal jumps to trigger neural action potentials, though the energy of a quantal jump is not adequate to initiate action potentials directly. The amplifying processes are extremely important, but of course they are themselves noisy. They degenerate information by adding randomness. It is possible to estimate by experimental means the contribution of 'neural noise' to the loss from ideal sensitivity given the quantal considerations for the eye.

It is important to point out that although visible light, radio waves, X-rays and so on are all identical except for their frequencies, they give very different observable and causal effects, because they interact with matter in very different ways. So detectors for instruments working in different frequency bands of the electromagnetic spectrum must be made of different materials to have appropriate properties for the detection of chosen frequencies. Similarly, the responses of the eye depend, of course, on the atomic and molecular scale interactions with quanta of various wavelengths, or wave packet energies as quantized in units of Planck's constant, **h.** This is where the 'empirical' complexity comes in, with little hope at present of fully adequate accounts of just why light of some frequencies produce some effects, other frequencies very different effects, except for 'black bodies', which all behave the same. The situation is not so bad for low-pressure gases; but even here the positions of absorption or emission spectral lines can only be related to the atomic scale approximately, and then only for the lighter atoms. So there is complexity here far surpassing detailed understanding well before we come to phenomena of perception. Is perception merely a stage more complex – or is perception entirely different from other quantum-causal effects?

This question divides into two questions: (1) is perceptually guided behaviour an extension of quantum interactions at receptors? And (2) is consciousness – are conscious states – just an extension of these interactive quantal interactions between energy and matter? After all, to consider consciousness, physical effects of light such as heating, lengthening, emitting, exploding and turning blue (in frequency terms) are fairly different from each other; so, are the conscious states of perception so different that they need a non-physical explanation? Can they not exist within *any* paradigm of physics?

We have discussed reasons for saying that the behavioural aspects of perception are different from other physical events. Because they are predictive, they are affected by alternative possibilities as stored in memory – to act on alternative possibilities from strictly inadequate 'inputs'. Perhaps curiously, given these considerations, the *behavioural* aspect of perception can appear more mysterious from the standpoint of physics than *consciousness!*

This is not, however, so if we extend physics to control engineering, allowing computers with memories and predictive capability. Here we can see at least something of the behavioural aspect of perception initiated by quantal energy jumps (as in the photo-cells of an artificial eye) producing some, if crude, predictive behaviour. We see this in the engineering but not in the pure physics context, though we do not believe that consciousness is involved for the machines displaying perceptual behaviour.

So far as we know, 'Artificial Intelligence' seeing machines could have human-like behavioural skills, though they are not conscious or aware. So one is, in the face of these considerations, almost tempted to say that consciousness looks more alien to engineering than it does to physics – though the behavioural aspect of perception requires engineering and not merely physics solutions or descriptions. To raise this afresh: is consciousness a result of energy–matter interactions, at the quantal level, or is it produced by AI-like procedures carried out by brain processes? If the former, any matter may be conscious, as Spinoza would suggest. If the latter, so-called intelligent machines with object recognition capability should – if consciousness is causal – perceive differently and presumably better than we could expect from engineering design considerations. This would be so if the conscious aspect of perception helps the behavioural performance. And we could expect this aiding from consciousness to be recognizable, by surprisingly effective performances in machines that mimic perceptual brain processes, as we understand them in physiological–engineering terms. We should not limit concepts of physiological–engineering to classical or macroscopic physics; indeed Quantum Mechanics is essential for understanding and sometimes for designing electronic devices, at least as far as solid state components are concerned. Relativity 'corrections' are also important in electronics where velocities of accelerated electrons or other particles approach that of light. So we may descend from these recondite considerations of the paradigms of current physics to ask how these specific effects should affect how we think of the physics and physiological engineering of brain function. Here Quantum Mechanics is almost certainly far more significant than Relativity considerations. What, then, is the significance of the randomness of quantal jumps – that they cannot from any knowledge be individually predicted, though the average of many can be predicted as for all macroscopic objects which appear lawful? Is this quantal uncertainty mere noise, or is it somehow used for brain function? Perhaps it is an important source of originality, and even the ultimate dynamic of Mind.

Some quantum phenomena are bizarre in the extreme, for example the phenomenon that optical interference patterns are produced by light passing through two holes in a screen, even though the photon rate is so low that seldom if ever during the exposure of the plate should there be photons passing simultaneously through the holes. How can photons interact, though displaced in time? Time, also, is very odd from the standpoint of common sense in Relativity Theory, as clocks slow with increasing velocity. This may occur right down to the quantum scale. There are many such considerations that make sense only within these new paradigms, which are surely important for understanding brain function – and so for seeing Mind in Science. For this, the notion of observation as the reducing, or 'collapsing', of the wave packet is most interesting. This notion arises from the difference between the blur (the Eigenfunction) of reality as described in Quantum Mechanics and the specificity of observations. The amplification and storage of detection seems to set precise values to the probability distribution of physical reality. This raises the question of how consciousness enters physics. Bernard d'Espagnat (1976) puts it in *Conceptual Foundations of Quantum Mechanics* in the following way: [14]

 . . . it is convenient to agree to call 'physical reality' everything that strictly obeys the linear laws of quantum mechanics as long as no consciousness comes into play . . . consciousness, therefore, is by definition a non-physical entity. However, this is not an outstanding peculiarity of the theory, since in most of the current philosophical views on the subject consciousness *is* considered already as a property of reality that cannot be fully reduced to the 'ordinary' physical properties. Indeed, these views radically distinguish consciousness from any other phenomena by asserting that it violates the otherwise universal law that there is no action without reaction. It can be acted upon but it never acts (the fact of being aware of some physical situation does not affect that situation).

Heisenberg and Eugene Wigner interpret the laws of Quantum Mechanics by invoking something close to a causal role for consciousness.[14] The collapsing of the wave packet, for observations to be specific, though reality at the atomic level is probability-blurred, is for them an effect on reality of observing by quanta. The observing instrument is not independent: measurements and observations are interactions. Observations are, however, peculiar and unlike other causal events because they produce irreversible changes in consciousness. So consciousness is somehow a result of interactions at the quantal atomic level, where the macroscopic laws of classical physics do not apply. And neither do the Laws of Relativity apply; or at least there is doubt in this, for there is as yet no unified field theory linking Relativity with Quantum Physics. Conceivably we must wait for this before we can understand consciousness.

Whether consciousness is the key to the collapsing of the wave packet – to making events from potential statistical potentialities – remains controversial among philosophers of Quantum Physics. Eugene Wigner (1967) holds that consciousness is the key; John Archibald Wheeler, on the other hand, holds that observations are amplified quantum effects which are irreversible, such as the blackening of photographic plates, even though there is no conscious observer. On this account presumably any irreversible change, such as a crater left by a meteorite on some dead world, far removed from conscious life, or a quantum-disturbed gene, or a track in a sliver of quartz buried deep in the earth, is an observation. This does seem an odd sense of 'observation' (and yet surely we observe without realizing it – without consciousness – most of the time).

At the edge of understanding, also, is time in relation to the observer, in Quantum Physics. By an extension of the 'two-hole' experiment, it appears that time is in a certain sense at our command. John Archibald Wheeler (1980) puts the situation beautifully:[15]

The photographic plate has been sliced to make a Venetian blind. It can be opened up or closed at will. If it is opened up, then the light comes through and is recorded at one counter or the other. In this way one knows through *which* hole the photon came. If it comes from the upper hole the lower counter goes off. If it comes from the lower hole the upper counter goes off. So far nothing paradoxical. Now, however, the new feature comes. We make the decision whether to open or close the blind, whether to make the photon go through both holes or only one hole, *after* the photon has *already* gone through this piece of metal with the holes in it! One here and now makes a decision which has an irretrievable effect on what one has the right to say about the past. There is an inescapable sense in which we in the here and now can decide what we have the right to say about what has already happened.

And although here we are talking about a couple of nanoseconds, there is nothing in principle that prevents the time from being billions of years.

Whether we shall find an answer to Einstein's thought— 'the most incomprehensible thing about the world is that it is comprehensible'— by finding that, after all, it is deeply incomprehensible; or whether we shall find ourselves comprehensible through Relativity or Quantum Physics, lies in the future, and there is nothing we can do about that. If David Bohm (1980) is right, Relativity and Quantum Physics are two aspects (like two viewpoints of a three-dimensional object) of a single, almost hidden, reality. Is this where consciousness lies? Or is it generated by brain function? I hope we find out.

Brains, we believe, construct predictive hypotheses of aspects of the world which are generally useful for survival. We now appreciate that most brain hypotheses, and especially perceptual hypotheses, are largely at variance with the realities of physics. Our perceptual and conceptual hypotheses float free, even from things that seem most immediately sensed and known, to create and journey into realms of fantasy, myth, poetry and illusion. Sometimes, the fantasy traveller returns to bring gifts back to our world. The creative Mind is Pandora's Box: sometimes the gifts are good and sometimes they bring disaster. Perceptions are, in any case, only approximations; there is always some error, which usually goes unnoticed. The mathematicians Christopher Zeeman and Peter Buneman, in a paper called 'Tolerance Spaces and the Brain' (1968), suggest that the notion of tolerance spaces is helpful here, as a mathematical tool for formalizing mental 'spaces', lying within detected or noticeable discrepancies. Tolerance is essential for any practical engineering device or machine to function, and as important for societies. Here we may, perhaps, see in tolerance spaces the necessary play for imaginative freedom. Perhaps where tolerance limits are set provides the basis of personality, as also of the structure of societies. If so, surely the setting of the limits within which we are free of the physical world and yet able to survive within it, distinguishes sanity from madness – and due tolerance is necessary for us to observe and understand – let alone to communicate.

20 The nature of knowledge

Socrates *I take it that men are wise in those things whereof they have knowledge?*
Theaitetos *Of course.*
Socrates *So knowledge and wisdom are identical?*
Theaitetos *Yes.*
Socrates *Well, that is just what puzzles me; I cannot satisfy myself as to what exactly knowledge is. Can we answer that question? What do you say? Which of us is going to speak first? Everyone who misses shall 'sit down and be donkey', as children say when they play ball; anyone who gets through without missing shall be our king and shall be entitled to make us answer any question he likes to ask. Why the silence? I hope, Theodorus, that my love of argument is not making me rude; I only want to start a conversation so that we shall all feel at home with one another like friends.*

Plato (*c.* 428–348/347 BC) [1]

Is truth in science?

I suggested almost at the beginning that generally accepted beliefs are required for societies to function effectively and survive, and that aesthetic agreement is needed for cooperative work – for decisions must be made though rational criteria from experience or theory are not generally available. As technology and science have developed, the range of rational grounds for decision has increased; but there are still many everyday decisions for which arbitrary choices must be made for cooperation to be possible. There is still great variety – indeed increasing variety – of houses, boats, tools and so on, which are artefacts with a history as long as civilization; but it is still necessary to agree how to build them and in what shapes. This is so when there are no clear functional reasons for particular designs. This, surely, is where aesthetics steps in. We somehow 'know' that a particular shape or proportion is 'right'. But what is 'right' this year may be very 'wrong' next year. In technologies general acceptance of prevailing forms and designs is essential for cooperative work; and typically only a few individuals – generally artists – have the power to change aesthetic standards.

There have been attempts to justify aesthetic preferences by relations in physics. Most famous and influential are Pythagoras' relations between musical intervals and the lengths of stretched strings. This gave a scientifically acceptable physical basis to music; but when mathematical ratios were used as aesthetic criteria of visual form, as in the relation of the Golden Mean, then the physical justification is no longer with us, and it looks like mumbo-jumbo. The musical intervals seem to be a very special case, where physics gives dependable rules; elsewhere we are on our own – the aesthetic rules are of Mind rather than physics.

The question to ask now is: are philosophical and theological beliefs, similarly, accepted for group survival without rational criteria? Are the bases of religions, of politics, of philosophy and even of science any more rational than aesthetics? I take it that aesthetics cannot be discussed apart from its technical considerations, such as how to lay on paint, draw in perspective or model drapery in stone; but the basis seems a-rational, and not open to persuasion structured by arguments. The same seems to hold for religions and politics: these are indeed almost perversely irrational. But perhaps they *need* to be irrational, and perhaps attempts to verbalize (or discover or invent) presuppositions render them worthless.

Turning to philosophy, we see a long history of accepted beliefs with occasional attempts to check, to clarify, and to invent new beliefs of the nature of the Universe and of man. While the Logical Atomism of seventeenth-century science and much of philosophy up to the Logical Positivists and Russell was tenable, it could be held that knowledge can be discovered by experiment without the need for underlying beliefs for interpreting observations. This now looks most doubtful, as science and the significance of accepted facts are seen to change with paradigm shifts of beliefs; which are not simply related to observations – if indeed there are any simple or direct observations. These we have rejected, having rejected Direct Realist accounts of perception, although as Thomas Kuhn (1962) has so cogently pointed out, shifts of underlying beliefs – changing paradigms of science – do affect how observations and experiments are interpreted. For example Natural History seen on a theological or on a Natural Selection paradigm for the origin and status of species looks very different. Perhaps, though, it does not follow that all observations and experiments are deeply suspect. It is extremely difficult to imagine that the observations, and the theoretical account of, for example, Harvey's circulation of the blood, will ever be destroyed by a change of paradigm. The meanings of 'circulation' and 'blood' may change with changes of knowledge or belief, but at least something seems to remain, in spite of the most drastic paradigm changes. In this sense science and technology seem to be ratchets, generating on the whole ever-increasing knowledge.

What, though, of the deepest belief statements of philosophy? They seem to ebb and flow with tides of opinion, casting flotsam and jetsam upon the edges of knowledge for our inspection. Philosophical beliefs or assumptions do not seem to grow as scientific knowledge and technology grow. For some, indeed, philosophy looks sterile and useless as the same old questions are endlessly discussed. If philosophy does make discoveries or important suggestions these tend to be snatched away into science – stolen, one might say – so that philosophy is left with imponderables and unanswerables. There is something in this, but it remains a fascinating game to pit the wits against Nature and discover at least something of the rules. If the rules that work are science, at the least it does no harm to try to formulate them and frame questions worth asking as clearly as we can.

Many people, perhaps indeed almost all people, do grow impatient with questions such as 'Do physical objects exist?' or 'Are other people conscious?' Most people are apt to say 'Of course they exist; of course people are conscious' and leave it at that. But when it is then asked just what kinds of objects exist (stones, clouds, numbers; past or future stones, numbers) then at once doubt does creep in and we do start to ask for criteria for 'object' and criteria for 'existence'. We find this also for the consciousness question: 'Are animals conscious?' produces doubts in everyone, especially concerning simple animals, though there seems no sharp divide between them and us, so the doubt may not be contained.

The trouble is, of course, that we have no way of proving beliefs of this kind, though we may all share them. When philosophy set out to find certainty, such lack of proofs seemed scandalous; now we are more content with uncertainty, perhaps in part because science itself is in continuous change, and especially in its deepest assumptions.

We live according to the strongest assumptions that there is an external world of objects, and that other people are conscious, with states of consciousness similar to our own. Thus we offer our friends wine, and try to please them according to our appreciation of their consciousness. All social life depends on it. Can we doubt these deepest beliefs? For Wittgenstein, as earlier for G.E. Moore, it is not rational to doubt physical objects, or consciousness in others. On the suggestion that all is a dream, Wittgenstein says (*On Certainty*, 1969): 'The argument "I may be dreaming" is senseless for this reason: if I am dreaming, this remark is being dreamed as well – and indeed it is also being dreamed that these words have any meaning.'[2] And there are limits to doubting set by the need for some beliefs in order for doubting to be possible; thus Wittgenstein:[3] 'If you tried to doubt everything you would not get as far as doubting anything. The game of doubting itself presupposes certainty.'

Putting this in a slightly different way, Wittgenstein argues very similarly to Thomas Kuhn, and also Russ Hanson, that questioning and testing require assumption and belief:[4]

All testing, all confirmation and disconfirmation of a hypothesis takes place already within a system. And this system is not a more or less arbitrary and doubtful point of departure for all our arguments: no, it belongs to the essence of what we call an argument. The system is not so much the point of departure, as the element in which arguments have their life.

This is closely similar to Thomas Kuhn's notion of 'Paradigms' and it is also closely similar to much of what Russ Hanson says in *Patterns of Discovery*. It means, in effect, that for some beliefs we simply cannot step outside the conceptual system in which we live. Outside it we are blind and stupid.

It may be thought that this, and Kuhn's paradigms, generate an uncomfortable relativity for belief, in which anything or nothing can be justified, according to the assumptions that one happens to be holding. Does Wittgenstein's argument that it is irrational to doubt what is almost universally held, help here? How can we know that strong counter-evidence will not appear or a change of assumptions or as Wittgenstein puts it, a change of the 'system of our beliefs'? Wittgenstein is curiously inconsistent at this point. He holds both that assumptions are necessary for accepting observations as evidence, and also that some beliefs (and not only very general or deep beliefs) must be accepted and counter-claims rejected. The examples that he has chosen are, in the light of later events, remarkably unfortunate for this claim. Returning to Wittgenstein's last writings, *On Certainty,* we find passages such as these:

117   Why is it not possible for me to doubt that I have ever been on the moon? And how could I try to doubt it?

First and foremost, the supposition that perhaps I have been there would strike me as *idle*. Nothing would follow from it, nothing be explained by it. It would not tie in with anything in my life.

When I say 'Nothing speaks for, everything against it', this presupposes a principle of speaking for and against. That is, I must be able to say what *would* speak for it.

Wittgenstein's certainty that getting to the moon is impossible reads very oddly now. It was not only impossible but inconceivable for him with his 1949 knowledge base, only a few years ago; and it seemed to him that he had the best knowledge. It is clear to us now that he was wrong to see this as impossible, as we believe that men *have* got to the moon.

Wittgenstein argues that although learning gives belief, some learning is better than other learning:

286   What we believe depends on what we learn. We all believe that it isn't possible to get to the moon; but there may be people who believe that it is possible and that it sometimes happens. We say: These people do not know a lot that we know. And, let them never be so sure of their belief – they are wrong and we know it.

If we compare our system of knowledge with theirs then theirs is evidently the poorer one by far.

It is very surprising for us that so imaginative and essentially clever a man as Wittgenstein would find landing on the moon impossible to accept only twenty years before (as we believe!) it happened. But, and this is a most serious question: if Wittgenstein can have been so certain and yet been proved incorrect so soon, why should we accept *any* belief statements of philosophers? If Wittgenstein is wrong in man-on-the-moon belief, why should we be persuaded by any of his, frankly persuasion-type, arguments? Why, indeed, should we accept that objects exist or that other people are conscious? Perhaps the only reason for holding these beliefs is that we cannot, at all readily, conceive or express any alternative. To say that they do not exist hardly seems an alternative that we can state with sense; though Wittgenstein would have no trouble in stating alternative beliefs about the possibility of men landing on the moon. The difference here seems to be the availability or non-availability of alternatives.

How, though, do we know that there are no alternatives to be found round the corner? Can we indeed say more than that beliefs are accepted as certain when imagination has run out of steam for generating alternatives? If this is all that there is to it, philosophical (or indeed scientific) 'certainty' should not be trusted any more than *not* seeing obstructions in a fog should be trusted.

The notion that all belief is relative and that no statement can be accepted without some doubt is disturbing both intellectually and morally. Men do sacrifice their own and other's lives for beliefs. This is indeed common in each generation. Unfortunately, the beliefs for which the greatest sacrifices are made – and what greater show of moral integrity can there be? – are often conflicting, so that clearly not all can be true. Also, we find that many kinds of belief are found associated with particular societies and educations. This at once suggests that these beliefs are relative. As we have seen, Wittgenstein wrote, 'What we believe depends on what we learn.' And then, 'If we compare our system of knowledge with theirs then theirs is evidently the poorer one by far.' But how? How can we know this without somehow stepping back to have a God's view of their knowledge, and ours, from some other more privileged vantage-point? The trouble is that there is no such vantage-point from which correspondence can be tested or seen.

This takes us back to the theories of truth that we discussed in chapter 13: correspondence, coherence, Pragmatism and the rest. We suggested that what is meant by truth is correspondence, between propositions and reality; though pragmatic usefulness and coherence of propositions are important as perhaps the only criteria for deciding what is true by correspondence. One might think that the largest set of coherent propositions is true, but this changes. Also, it is sometimes important to accept the truth of propositions that stand out as inconsistent, and so not cohering, as signs that all is not well with the coherent body. If exceptions were dismissed *ipso facto* as errors, or false, then exceptions could not be used in science. But striking exceptions have been of the greatest service to check and extend theories. Scientists are indeed magnetically attracted to exceptions.

Karl Popper goes so far as to suggest that it is only refutations that gain new knowledge. This implies that non-cohering propositions and observations have uniquely special importance. He supposes that these have the power to topple the most coherent theories, such as Newton's or Einstein's, and no doubt he is right in this. It implies that coherence may be a strong criterion of truth; however, it also implies that exceptions have extraordinary power, and so must sometimes be accepted as true against the coherent body – otherwise they would not have the power to topple coherent accounts but would simply be disregarded.

It seems, then, that both usefulness and coherence are criteria for seeking and establishing truth, but not what truth means. But if truth means correspondence – correspondence to what? The common view is correspondence to objects, to states of affairs, to laws, to structures of the Universe. Then the truth of observations, or perceptions, would be correspondence with observable features of the world – concrete objects – while more indirectly obtained knowledge may be correspondence with objects or features that we cannot perceive, or observe or detect with instruments. These are abstract objects. These might, as for Frege, include numbers. Numbers have most of the criteria of concrete objects. They have common features found by everyone, such as what divides into them, or whether they are prime. We 'handle' and use numbers almost as though they are physical objects.

'What, then, is truth?' The commonsense notion is that a statement – or a picture, a map, a theory or hypothesis, or a perception – is true when it corresponds with fact or reality. What we generally mean by 'true' is that there is such a correspondence. But how can we ever know that truth is attained without our being able to stand on the other side and see that there is correspondence? While it was considered that verification, or at least the possibility of verification, was necessary for meaning, this objection that correspondence could not be at least immediately verified seemed to dispose of correspondence as a theory of truth. But now that the requirement of verification for meaning has been relaxed, and has had to be relaxed, this is not such a strong objection. With the death of Logical Atomism we can now say that correspondence is what we mean by truth though it is not by observing correspondences that truth is verified. We may now look for other criteria for recognizing truth while holding that correspondence is what we seek and require and mean by 'truth'.

The theory of truth according to Pragmatism – that truth is what is useful – will not do as the meaning of 'truth', if only because what is useful depends on intention, and intention changes though truth does not change with changes of intention. (It is true that the moon is 240,000 miles distant, whether or not I intend to go there.) But Pragmatism's emphasis on what is useful is a guide to what is worth looking at or considering as possibly true. This surely is the powerful effect and rich contribution of technology to philosophy.

A related criterion of truth is prediction. This may be prediction in time, or prediction to unsensed characteristics. This last is most important for deciding what is an object. Prediction in time is most important for what is a mechanism, and for what are accepted as causes. Somewhat like usefulness, as in Pragmatism, predictions are not simply statements of the world but are rather statements of the observer's view of the world. (This, at least, is so on a 'subjective' account of probability, for which predictive power depends upon the accepted knowledge base.) One might say that successful predictions provide evidence of redundancy of structures, but failed predictions may show that the knowledge base is inadequate or ill-applied. This gives the opposite conclusion to Popper's emphasis on failed predictions, which he accepts as 'refutations' necessary for gaining knowledge. If probabilities are accepted as depending upon the current state of knowledge, then failures of prediction are ambiguous evidence. It is perhaps never clear that a hypothesis is ever strictly speaking refuted, for there can always be mistakes or misinterpretations even of what looks like devastating counter-evidence. Accepting coherence of propositions is not an infallible criterion that they correspond to concrete or abstract objects of reality and so are true, for large bodies of propositions may mutually cohere without discord in elaborately worked-out fictions that we accept as false.

To what, though, are true propositions supposed to correspond? For Frege, they may correspond to concrete or to abstract objects. We normally think of objects as concrete things, which we can touch and see (though it is not clear that we hear objects). It is not clear that a tune is concrete, as a table we can touch and see is a concrete object. This begins to suggest that the distinction is not as clear as we may at first believe, and yet we may hold that concrete objects exist in a way that abstract objects do not exist and so propositions concerning the one may be different (and perhaps differently true) from those of the other.

What makes *abstract* objects different from concrete objects is that they cannot be seen or touched and may have no location in space or time. But much of the accepted structure of what are accepted as *concrete* objects cannot be seen or touched, or even in any way observed or detected. Many things that we believe to be true of stones and tables cannot be perceived through the senses, or at all directly observed or detected by any instruments. They may be inferred from sometimes bizarre assumptions of physics (and the assumptions may change) to accept as true un-object-like things (such as sub-atomic particles) which are claimed by science from highly indirect observations. This is the power of science, to challenge, flout – and to destroy common sense and the Gods.

If we accept the essentially Helmholtzian position that I have been advocating here, this applies also to all perception and the objects of perception; for everyday perceptions are generated by inferences – from tacitly held and unconscious assumptions which, as many illusions show, can be inappropriate or false. On this account of perception (which I hold) the distinction made by Frege between concrete and abstract objects is extremely blurred. On this view, concrete and abstract merge and can hardly be distinguished. This merging is part of my claim that perceptions and hypotheses of science are essentially the same.

Is truth in Mind?

So far I have been considering knowledge of the external world. How far do these considerations apply to statements about Mind?

Can we accept correspondence to something else as the meaning of truth when speaking of experiences? If I say 'I feel a tickle,' does this require that the sensation corresponds to something else – some object like a tickle in me?

If Mind is related to brain function somewhat as, say, the power of an engine is related to processes carried out by its components, then there is no reason to suppose that the sensation of tickle corresponds to an internal tickle: it would rather correspond to processes that produce tickles. On one form of Mind–brain identity account, the brain state, or process, *is* a tickle, though described in physical terms; and on a functional account it *produces* tickles (as engines produce power). The correspondence relation is very different for these alternative accounts. To appreciate how an engine produces power, we must understand how physical laws have been applied to set up restraints to normally (or 'Naturally') occurring Degrees of Freedom to produce effects that do not occur in the Natural world. For machines, this requires that we understand the relevant physical laws and the design of the machine. To appreciate the effects of cams and push rods and valves and so on we need rather more than physics. On this view, for psychology, we similarly need to know rather more than physics and, it would seem, rather more than physiology. The question is: what extra do we need to know? If we knew this we should know how to understand Mind.

There seem to be two paths that we might now take. We might look for more and more physiological knowledge, which is like asking for more and more engineering know-how; or we might look for deeper insights from physics of the relation of the observer to the physical world. This is not to ask for more detailed theories of perception, as understood in physiology or psychology, but rather to ask how observation is possible in physics.

The trouble here is that what is accepted as possible depends on the basic physical concepts that are currently accepted. The picture looks very different in Newtonian physics, Relativity, and Quantum Physics. For Newton the *sensorium* was wholly mysterious and essentially outside his mechanics of the Universe. For

Einstein, the observer is a manipulator of rods and a clock watcher. Quantum Physics does, however, at least for some physicists, have an essential place and role of the observer: observations are interactions, giving irreversible effects in consciousness. Just how this happens is the deep mystery. It is fair to say, though, that Mind is no *more* mysterious than Matter.

We have come to accept in this book an extreme form of relativity of belief. This follows from the proposition that all observations depend on assumptions and that no assumptions can be independently justified. This situation extends even to formal proofs. The logician Kurt Gödel (1906–78) showed that the consistency of a formal system cannot be proved from within the system. Proof of consistency requires justification from meta systems. But this leads to an infinite regress for formal proof; for the meta system can ultimately only be proved consistent by referring to a further meta system, and so on *ad infinitum*. So, even consistency cannot finally be proved for any logical or mathematical system.

There may, indeed, always be surprises: it is always possible that a computer program of sufficient complexity might run into trouble in certain cases. We have also to consider the point that any computation, by computer or brain, depends on the reliability of the mechanisms carrying out the procedures. Strictly, it would require an infinite number of repetitions to check for random errors, and systematic errors would not show up as different answers to the same question or problem, so they could be missed.

The relativity of knowledge extends right down to the simplest and apparently most basic perceptions, for the signals of the nervous system must, to be accepted or read as data, be processed by procedures that may not be appropriate. We have argued that when the procedures for reading sensory signals as data are inappropriate, in a given situation, illusions are generated. These may go unnoticed. Perfectly reliable checks are not possible, for *they* require assumptions; and the neural mechanisms are not completely reliable.

On the other hand, one can get impatient with statements of extreme doubt. Wittgenstein, among other modern philosophers, has tried to save the situation by arguing that some doubts are irrational because we have no grounds strong enough to cast doubt. But it is striking how particular examples they have taken as undeniable have, and sometimes within a few years, been transformed so that they appear ridiculous. We saw this (p. 551) in Wittgenstein's example of men walking on the moon. From seeming impossible, this has become almost boring. The statements of the current physics of Relativity and Quantum Mechanics often appear so absurd or paradoxical that they would have been rejected a few years ago on the grounds of *reductio ad absurdum,* and are still exceedingly hard to accept. This raises the awkward consequence that apparent absurdity is hardly a criterion for belief. So is anything acceptable to test truth?

This is where one can become impatient with extreme Relativism of knowledge or belief, and it has dangerous social consequences. If we seriously question whether operations without anaesthetic produce pain, or that it is unpleasant to starve, or that any social or political system is as 'good' as any other, we run into trouble. It is indeed curious that philosophers both lead us into extreme doubts – following especially Descartes – and are guardians of Ethics which depends on subtle assumptions for the objectivity of 'good'; and what is good and bad, which are extremely hard to justify in general terms.

It seems that to maintain sanity we have to adjust our sights for 'truth' and what we demand as evidence over an enormous range, according to the situation that we are in (or believe ourselves to be in!). Perhaps much of madness is rigidity in this dimension. Most basic is the ever-growing and now vast discrepancy between the world as seen by the senses and by science. We do indeed seem to have two worlds: a world of hardness, colour and muscular pulls, and an incredibly different world of atoms reducing to probability distributions, and four kinds of forces, among them gravity which is not so much a force as curvatures of space. Even the distinction between space and time, which is so basic in experience, becomes blurred, to blend into another scheme of things with different conceptual categories. This is like a caterpillar going through its profound reorganization in the chrysalis, when even head and tail are reversed, to emerge as a quite different creature, perhaps even without memory of its past. We seem now to be in the chrysalis stage of our development.

It is, however, deeply misleading to call perception and science two worlds: they are alternative descriptions of what is surely one world. An adequate account of Mind in Science should unify these into a single description, in which we may see ourselves. This might just possibly be dangerous; even to the disaster of annihilation. We might with a unifying account disappear as objects of perception (much as 'object' used to mean an obstruction). Just possibly, as understanding grows we shall become transparent, invisible: so that we shall not know ourselves or exist, without returning to ignorance, and this may be impossible because the gaining of knowledge is virtually irreversible. One can imagine a Priestcraft of Deliberate Ignorance, organized to conceal truth, to preserve Minds as objects against the destroying light of the 'speculations' of science and philosophy!

However this may be, if we are right to think of perceptions as we have considered them here – as hypotheses essentially like the hypotheses which are the insights of science – then we should see Mind in Science with Science in Mind.

Postscript

I used to be indecisive, but now I'm not so sure.

Boscoe Pertwee [1]

Since the Babylonians, notions of the Universe created and run by quixotic intelligences of Gods has gradually given way to the super-miraculous power of tools in the hands of men, which has challenged the Gods by showing us that we have some control over our destiny – and that structures and functions of machines are keys to understanding the laws of the Universe. Machines throughout this time of three millennia have generally been seen as lacking volition or purposes – except as we direct them to our goals. So we enact the role of minor Gods. What, though, are we? Are we as demi-Gods outside the order of the mechanical reality? Our understanding seems far wider than our powers of action; but our powers have immeasurably increased as the tools and machines of technology have harnessed forces of nature, and made human skills and ideas cooperative. With this extension of possibilities, moral commitments have been extended: for there is no 'ought to do' when 'it can be done' is lacking. Now with current technology there is no clear limit to 'what can be done', so guilt may swamp us as we learn to swim.

Greek philosophy lost its concern with explanations of Nature (with physics) to become obsessed with the conduct of man, and structures of society, just when Greek technology expanded its power of 'can be done' beyond the compass of the individual ability to do, to command, or to be commanded. At this point spectators (including the spectators of the Olympic Games), the philosophers not the doers were for the Greeks the highest kind of men.

Around the time of Plato it seemed clear, from geometry, that perfect static solutions could be found; but mechanisms, though marvellous, were in practice subject to all kinds of errors and imperfections, so that their designs never worked perfectly, but lay in an unrealizable world of number and ideal geometrical form. It is possible that Plato saw this duality in technology as *reality* and *appearance*:

that what is realizable in the world is but an imperfect representation of underlying perfect design in the Universe. If so, did moral problems, as multiplied by technology, belong to the ideal timeless designs of the Universe, or to human realizations, working imperfectly in time? Plato gave the first answer, and Aristotle the second.

Possibly engineers are somewhat tainted by the imperfections of machines, compared to what machines would do if they could be made precisely and without friction or wear. The Greek wheeled computers of the heavens, of which we know at first hand the Antikythera mechanism of the first century BC, was an extreme case of this gap between design and performance. With the development of sophisticated wheeled clocks in the seventeenth century AD, almost frictionless machines came into being – and then came accounts of the motions of the stars based on the perceptual notion of ideal frictionless machines which never need winding up, or maintenance. Following Newton the Machine-Minder was dead. With the superb techniques of the clock-makers showing the way to Newton's perfect dynamic Universe – where, as for geometry, space was subsidiary to matter – Mind dropped out of the Universe of Science. True though, Newton himself thought of space as God's Mind: but this was for God the *idée fixe* of all time.

Throughout the history of philosophy, there has been a tendency to minimize situations of material gain, and of passions upon judgement, so that the philosopher is special. This has many consequences, not all happy. Socrates did not withdraw from passions: he imbibed almost to excess, perhaps to demonstrate that he was proof against temptations and weaknesses affecting his mind. In this tradition, the Mind must be above, and not influenced by the body. But of course this led to certain difficulties. The senses, giving evidence of states of affairs of the world, belong to the body, so they and observations were suspect. And how could a man free of normal frailty or temptation appreciate other people with sympathy, rather than cold moral judgement?

This branched two ways. Remaining a man, Socrates retained passion and temptation; but Diogenes in his tub, and others denying the life of man and living in protected celibate communities, removed themselves from viable living. It is remarkable how generally this second course has been practised and condoned, throughout the world, and indeed this is today sanctified by all the major Churches. This is the path of the Platonic world of static ideal reality beneath events. However, this can look like a profoundly immoral rejection of human life, and worse, it is to shut one's eyes to danger which is suicidal, and where others are involved this is murder.

There are, however, many different kinds of men and women who have very different needs. William James, in *Pragmatism* (1907), divided people into two kinds of mental make-up: 'the tender-minded' and 'the tough-minded'.[2] He lists their typical characteristics thus:

The tender-minded	The tough-minded
Rationalistic (going by 'principles')	Empiricist (going by 'facts')
Intellectualistic	Sensationalistic
Idealistic	Materialistic
Optimistic	Pessimistic
Religious	Irreligious
Free-willist	Fatalistic
Monistic	Pluralistic
Dogmatical	Sceptical

William James says that people who are markedly one have a low opinion of the other, and that 'their antagonism, whenever as individuals their temperament is intense, has formed in all ages a part of the "philosophical atmosphere" of the time'. He adds that, both in the past and now:

It forms a part of the philosophic atmosphere today. The tough think of the tender as sentimentalists and soft heads. The tender feel the tough to be unrefined, callous, or brutal. . . . Each type believes the other to be inferior to itself; but disdain in the one case is mingled with amusement, in the other it has a dash of fear.

James goes on to say that we have difficulty finding a philosophy to suit our needs. Now this emphasis on the individual psychology of philosophers is clearly significant. He sees personality traits as distorting thinking and observing, and setting up desires or needs to be fulfilled by a philosophy – which may thus be chosen to suit the individual rather than fitting the facts of the world.

In *Pragmatism* (1907), while discussing what James regards as the deepest philosophical problem, he writes, under the heading 'The one and the many': '. . . if you know whether a man is a decided monist or a decided pluralist, you perhaps know more about the rest of his opinions than if you give him any other name ending in *ist*'.[3] Since then, statistical methods have been designed to quantify structures of beliefs (G. A. Kelly, 1955). The question is, how far do philosophies and religions crystallize according to human needs, and how far are they structured by the results of technology and science? Peirce, James and the later Logical Positivists see operational procedures as tools for making us see and understand more clearly. To do this, we must break the accreted crystal structures of Mind, grown in social cultures which nurture myth (and we may suspect, art).

Logical Positivism is the philosophy based on the success of physics which sets up procedures of verification independent of human desires – and insofar as we grow away from our biological past, independent perhaps from our needs. Technology is not however inhuman in this way, for it provides tools, and shapes the environment to produce wonderful real-life fantasies, which are often ancient dreams come true. How else can one see flying, or listening to the radio, or speaking across the world, one's voice carried through space and bounced back by man-made satellites? How else can one see printing, trains, cars, and indeed cooking and clothes? These are all wonderful, and they are results of experiments and dangerous questions. It is not at all surprising that Socrates, Galileo and Bruno were feared, and became martyrs. They were seen to shatter crystal structures of Mind by questioning, and who could say what would result? Not all results of questioning and science are acceptable. Once on this path, there is no turning back: we can only go on to discover and invent solutions to the problems, and the actual and potential disasters generated by probing into the unknown.

But there are now few, if any, strong advocates of a radical Logical Positivism, in philosophy or science. After many years of careful thought, one of its chief proponents, Sir Alfred Ayer, has abandoned the attempt to formulate criteria of verification such that only bad metaphysics is rejected while important statements are allowed. He no longer believes that operational criteria can be found that will separate the wheat of facts from the weeds of subjective ethics and aesthetics. This implies that clear distinctions between sense and nonsense (or between objective and subjective) cannot be made by simple rule-following. These distinctions depend upon assumptions, which themselves change according to 'philosophic atmosphere'. Indeed, to expect to find infallible rules for discovering truth is itself a deeply metaphysical expectation.

The phrase 'philosophic atmosphere', used by William James in 1907, comes very close to Thomas Kuhn's 'paradigms' of 1962. James's emphasis on human strife, and individual needs for a philosophical view of the world to suit and justify oneself – 'The history of philosophy is to a great extent that of a certain clash of human temperaments' – is still perhaps not generally accepted (perhaps because of its very truth!), and Logical Positivism is essentially an attempt to remove altogether the individual, and so his wishes and needs, from the philosophical arena, and so from the Universe. William James found his own view of the world quite hard to reconcile with the Pragmatism he espoused, as he was more tender-than tough-minded; for it is the tender-minded who appreciate the variety of human belief, and see how powerful are hopes and fears for filtering and swaying evidence. This was appreciated by the classical scholar Francis Cornford, as beautifully expressed in his essay, 'The Unwritten Philosophy' (1967); and in his inaugural lecture at Cambridge in 1931:[4]

If we look beneath the surface of philosophical discussion, we find that its course is largely governed by assumptions that are seldom, or never, mentioned. I mean that groundwork of current conceptions shared by all the men of any given culture and never mentioned because it is taken for granted as obvious.

Slightly earlier, A.N. Whitehead wrote in *Science and the Modern World* (1925): [5]

In every age the common interpretation of the world of things is controlled by some scheme of unchallenged and unsuspected presupposition; and the mind of any individual, however little he may think himself to be in sympathy with his contemporaries, is not an insulated compartment, but more like a pool in one continuous medium – the circumambient atmosphere of his place and time.

It seems to me likely that general acceptance of how things are, and having goals to work for, are a vital necessity for group survival. This puts critical philosophy far removed from Natural Mind, if this term be allowed. Critical examination of assumptions is, however, the very basis of philosophy and science – which pays off when deep generalizations, Natural laws and design principles of machines are discovered beneath individual experience, for then we gain power over Nature. But to find deep structures requires arguments amounting sometimes to mortal strife, before Public Knowledge – to use John Ziman's phrase – can be attained, and with it more sophisticated and effective cooperation. Some ideas work, while others for various reasons do not. Here we see ideas put to trial somewhat as organisms are tested, by all manner of travails, to produce by an infinity of disasters new solutions, in some ways superior to the old. A major superiority of species which are higher up the biologically self-made ladder of Evolution, is increased knowledge – both stored in their structures and as tools for discovery.

What of the ethics of sacrificing animals, for dissection and experimentation? This is often seen as repugnant; but of course animals have from the start of life lived off lower links of the food chain, as we do; and one might suggest that gaining information is at least as worthy as gaining energy for one's survival. What characterizes civilization, though, is respecting the individual; and especially protecting the individual against the mainstream of the evolutionary competition of Natural Selection. Ethics and moral systems are concerned with rules, and so programs of the brain (J.Z. Young, 1978) which seem to be diametrically opposed to the biological Laws of Nature. Ethics, one supposes, may do biological damage to our species, as the unfit are protected to survive while the fit are subject to stress and sudden death in dangerous occupations and active combat. The security of mutual protection is, however, essential for civilization; and we do well to protect the agencies which protect our customs, and even our beliefs, though too much protection gives death by mental *rigor mortis,* so that we could become barely living fossils, adapted to a past world but not to the present. Risk-taking is necessary for survival; and with risk-taking, successful predictions rule. It is interesting how important were many kinds of prediction or attempts to predict in early societies: from omens, from the stars, by animals, by oracles, and by idiots, and dreams, and the thinking of the wisest men. We now see prediction as a vital element of intelligent behaviour, and indeed of Mind. We have suggested that intelligence may be defined as 'the generation of successful novelty'; and of course predictions that turn out right are successful and were novel.

The emphasis, at least in modern societies, is upon the individual; but in science the emphasis is on averages, on tendencies; and on laws applying to as many things as may be: the more general the law the better. When science is applied to individuals, in medicine or psychology, the individual decisions depend upon generalizations which have applied to many individuals. So, what we can do for each individual may be limited by knowledge which applies to very many, or even to all of us; which makes individualists difficult to appreciate, or protect by the methods of science. The individual can perhaps only be understood by the rapport of art and poetry and love; but these are almost outside society and science, and are far better outside society's dictates.

Each of our individual physiologies are, so far as we know, generally very similar: even subtle anatomical and biochemical analyses do not reveal differences related to the obvious differences of our characters and abilities. The brain of a mathematician looks just like the brain of a painter, or even of a moron. So, clearly, we have not yet tumbled upon the crucial physical features which give individualities of Mind. What, then, is the physical basis of human individuality? Presumably it lies in the detailed connections of the central nervous system. Perhaps if we could trace the circuits in sufficient detail, we could describe Minds by physical means. For this though (if computer analogies are at all appropriate), we should have to know the 'language', or the 'program' – the conventional rules – by which brains operate; and individual brains may not all be alike in this regard.

If our individual, internal brain languages are special to ourselves – requiring individual translation into our shared spoken languages – then psychology is faced with the scientific nightmare of unique cases. We do, though, have a host of reflexes, and instinctual drives and behaviour patterns, inherited from pre-human ancestors. When the eyes blink shut to a sudden noise, we re-enact a scene from the text of the ancient play of deaths through misadventure written and re-written from the origin of life through countless ancestral disasters.

If psychoanalysts are right we inherit something close to ancestral experience. Here, in a nut-shell of his, is C.G.Jung's view:[6]

In every individual, in addition to the personal memories, there are also . . . the great 'primordial images', the inherited potentialities of human imagination. They have always been potentially latent in the structure of the brain. The fact of this inheritance also explains the otherwise incredible phenomenon, that the matter and themes of certain legends are met with the whole world over in identical forms. Further, it explains how it is that persons who are mentally deranged are able to produce precisely the same images and associations that are known to us from the study of old manuscripts. . . .

Carl Jung seems to think of his collective unconscious in terms somewhat similar to Noam Chomsky's Deep Structure of language: as ancient inherited Mind or brain structure. And indeed Newton, investigating the mythological symbols of alchemy, took them as seriously as his astronomy, as keys to understanding (Westfall, 1965).

Neurological research is in an extremely active stage, with results from electro-physiology and biochemistry pouring out of laboratories all over the world. Currently, it seems that there are many special processing systems, with sensory channels tuned to particular features. The outputs from these analysers are no doubt combined by more general-purpose cross-relating systems. Increasing knowledge of chemical transmitters and the endorphins controlling pain give new conceptual insights – and a physical basis – to psychosomatic disturbances, and the sometimes remarkable effects of psychological suggestion. These findings blur the classical distinction between 'organic' and 'functional' disease. They also produce a curious situation vis-à-vis placebos. If, as seems demonstrated, psychological suggestion works, such as that red pain-killing pills are more effective than white pills chemically identical, then the logic of placebos used in experiments needs careful consideration. Placebos may be chemically neutral; but if they induce production of, say, analgesic endorphins, then they are effectively active chemically. And it may turn out that therapy by suggestion and by drugs work, at least sometimes, by the same mechanisms.

Much of psychotherapy is, however, no doubt a matter of 're-programming' to overcome loops and blocks, best described in information-processing terms. Will the brain ever be understood in these terms? If one were to analyse books, written in a completely unknown language, by statistical or any other means, one could easily judge that each book is very similar to others. But one might be on cookery, another on geology, a third a novel, and a fourth on philosophy. They would however look extremely different once the language was known and they could be read – though almost indistinguishable to purely physical analysis or tests. And indeed such difference as may appear – of colour, of typeface, of paper or hard cover, and even the number of proportions of identified letters may hardly matter. Thus: 'This is a silly argument' and 'Is this a fitting response?' for those of us who read are very different; but to those who do not they would hardly be distinguished as essentially different either from each other, or from what we would call a random set of letters, such as 'Hsit si a tintigtf psepnees'. Here, the meaning is entirely lost; but this is clear only for those of us who know the language. True, if we know the statistical restraints of the language though without being able to read it, we may recognize that 'Zwep', or 'Guass', or 'Einstein' are unlikely English words; but for the physical basis for supposed brain language, we should have to know in detail where to look and what to describe even to get that far. To read a language we need far more, for we need a common knowledge base. But given that we understand each other's spoken and written language from the base of our shared knowledge, it should be possible to read brain language if only the technical problems can be overcome, because we could bring our common knowledge to bear as we do for verbal communication – including philosophy and science.

The amazing thing about the successful Physical Sciences is how they have learned what to observe, and what to regard as significant. It is quite remarkable that careful observations of the positions of the planet Mars turned out to provide a vital key to the dynamics of the Universe – relating every body in free orbit, to the shapes one gets by cutting a cone obliquely. To discover such a connection, between conic sections and the paths of the planets, is magic – indeed better than magic, for it works. To predict the return dates or the non-return of a comet is surely no less than miraculous. But for Mind, and its dependence on physical brain structures, there are as yet no such keys in our hands as conic sections for the stars, which Newton used to open the Universe.

Throughout the history of science there have been analogies for Mind in technology – physical models from water tanks, and pipes, of fires in boilers, of magnetic fields, telephone exchanges, and now computers. It is however useful as a warning to think of examples such as Aristotle's account of the after-effect of movement (the 'waterfall effect') as being, he said, like the way a thrown missile leaves the hand, to *continue* its flight as an after-effect of motion. He seems to have been so taken with this physical analogy that he ignored the *reversed* direction of the visual after-effect, which is totally unlike inertia, even as he saw inertia. One must assume that the power of Aristotle's highly inappropriate model seduced him to this elementary error of observation.

So we may return by considering models, to background assumptions or paradigms and how, though essential for observing and explaining, they can be highly misleading – to distort percepts and concepts out of recognition. Like all tools, the Mind tools of models are dangerous. Tool-using needs background knowledge, but this unfortunately is curiously lacking for Mind. Perhaps there are no appropriate physical models for Mind. Why indeed should there be? Granted that Mind is entirely based on physical brain function (as I suppose to be the case), it is still very far from clear that physical descriptions of brain structure and function will describe Mind – at least before we learn the brain's language. If this is so we have a kind of cryptographic problem. But this would be significantly different from problems of physics, for the structures of sentences are not significantly related to the physical states representing them except by the rules of the language, which are not part of any physics. If brains work by representing, as we believe that perceptions are representations, then we need to know the code – the rules of representation. On this account, physics is almost entirely unimportant, just as the physics of chess is unimportant. If Aristotle is right, it is structures and symbolic processes that matter and not so much physics or even physiology; though these are necessary for carrying out the rules of Mind. This means that, in principle, it should be possible to write a Mind on another tablet: to transfer a Self to another brain, or to a computer. Is this science fiction? Of course; but what science is free of fiction? And what science has not benefited by untestable fantasies? They are at the least mutations for new varieties, and may initiate new species of ideas.

What this possibility of writing Minds into computers does suggest, is that although we have biological origins, these may not be much more relevant than, say, soil for flowers. Although flowers depend upon soil, and evolved from earlier soil-living organisms, yet soil chemistry tells us remarkably little about orchids. It is perhaps more important to know that insects respond to certain shapes and colours of flowers – that flowers are 'read' by insects.

The point is that our biological origins may set some restraints on Mind; and no doubt ancient restraints or limited degrees of freedom are important for survival. But since we can experience situations and tackle problems unique to us, with virtually no precedence from our biological past, obviously our origins cannot explain in detail our present beliefs, or set clear limits to Mind. On the computer analogy, the more general-purpose the brain is, the less it is determined by the past. My microprocessor computer can play games with me, do logic and mathematics, list things and control experiments never dreamed of by its designers. It may even make its own discoveries. It can do virtually anything that can be stated to it clearly, or what it has learned clearly for itself. Insofar as it is general-purpose, its functional processes cannot be guessed from its design. Since there are vast individual differences of human abilities and knowledge, given by experience, so we are general-purpose, and are not limited by biological 'read only' programs.

We do, however, have limitations. My computer is much better than I at many tasks – and from this I know that I am not like it in these respects. Perhaps our differences, at least from present-day computers, tell us as much about ourselves as do our similarities to computers. However this may be, my belief is that computers do at last provide a technology appropriate for studying and describing at least some aspects of Mind; for we can now study functional structures of information – logical relations and procedures isolated from the substance which bugged Aristotle, and us his heirs. Computers are test-beds for psychological theories. They may though turn out to be Procrustean test-beds torturing us to see ourselves in the mould of their limitations. So, whether or not Mind ever fully enters computers, we should mind out for ourselves and teach them to look after us. This may sound like science fiction; but fiction is the look-ahead of Mind that has created the Science in which we find ourselves.

Notes

Part 1 Forging science from myth

Chapter 1 Mind in myth

1 William Blake, *Songs of Experience:* 'The Tyger'.

2 William James (1907), p.170.

3 H.Frankfort *et al.* (1949), p.12.

4 Ibid., p.21.

5 G.S.Kirk and J.E.Raven (1960), p.24.

6 Mythical origins of the world and the Gods by union and separation of earth and sky are described with great subtlety by the Cambridge classical scholar Francis Cornford, in an essay, 'A Ritual Basis for Hesiod's Theogony', in his *The Unwritten Philosophy and Other Essays* (1967). He describes a gradual development from pure myth where the events are based on the personalities of the various Gods involved, to a degree of rationalization to account for the separation of the land, the sea, the sky and living things when it became philosophy or the beginning of speculative science. For a verse translation of *Theogony,* see D.Wender (transl.) (1973).

7 Whether these longstanding differences in the social roles of the sexes have a marked genetic basis is of course a most important question, though it will not be discussed in detail here. See David P.Barash (1973), Eleanor E.Maccoby and Carol N.Jacklin (1974), and Edward O. Wilson (1978).

8 For details of the origin of Eve see S.N.Kramer (1963), pp.147–9.

9 Ibid., p.115.

10 O.Neugebauer (1969), pp.42–3.

11 Babylonian Creation stories are collected in A.Heidel (1942). A general account of Middle Eastern mythology is S.H.Hooke (1963). A collection of Sumerian poetry is S.N.Kramer (1979). For a superb collection of Babylonian 'wisdom literature', see W.G.Lambert (1960).

12 A.E.Wallis Budge (transl.) (1895), p.ixix.

13 The most important papyrus (there are several versions) giving first-hand accounts of Egyptian notions of Mind and soul is *The Egyptian Book of the Dead (The Papyrus of Ani),* translated by A.E.Wallis Budge (1895). Egyptian mathematics is given most fully in T.E.Peet (1923). For a collection of Egyptian literature, see W.K.Simpson (ed.) (1972).

14 R.F.Gombrich (1975), p.111.

15 Ibid., p.112.

16 Ibid., p.116.

17 M.I.Finley (1970), p.130.

18 Ibid., p.132.

19 G.S.Kirk and J.E.Raven (1960), p.372.

20   Bertrand Russell, *History of Western Philosophy* (1946), p.51. Russell is very good on Pythagoras, as is J.A.Philip (1966).

21   For a history of mathematics, see C.B.Boyer (1968).

22   J.Burnet (1908), p.108.

23   This passage is part of a rather unsatisfactory account of respiration. Some authorities, such as G.S.Kirk and J.E.Raven (1960), pp.341–3, do not accept this as an experiment, in the modern sense. Others, however, such as J.Burnet (1908), hold that these early analogies and proto-experiments were the start of experimental science, and profoundly influenced philosophy. I accept this view.

24   G.S.Kirk and J.E.Raven (1960), p.344.

25   Ibid.

26   Ibid., p.360.

27   Aristotle (1957), A. 5 411 a 7, p.61.

28   J.Burnet (1908), p.52.

29   Hippocrates (1923), pp.139–83.

30   The physicist John Ziman has important things to say on the place of science in society (1968, 1976). There are insights here, also, of the practice of science and into the characters and achievements of many scientists: see John Ziman (1978).

31   See E.Durkheim (1915), B.Malinowski (1926), and Claude Levi-Strauss (1966).

32   J.Piaget and B.Inhelder (1956). Jean Piaget (1896–1980) was born at Neuchâtel in Switzerland. He was the pioneer in investigating the development of children's conceptions of the world; so Piaget's psychology is essentially experimental epistemology. This includes, especially, development of concepts of number, of space, of objects (including the famous experiments on liquids in tilted bottles), and of language. Among the most recent of his many books to be translated into English is *The Grasp of Consciousness: Action and Concept in the Young Child* (1974).

    Piaget was Professor of Psychology at Geneva University, and Director of the Centre d'Epistémologie Génétique. He was also a Director of the Institut des Sciences de L'Education. As his work is so well known I have not attempted a detailed account or appraisal here. I hope, rather, that the many students of Piaget may find some of the discussions given useful for interpreting his work and seeing its implications, which extend from individual discovery and education to the basis of all knowledge.

    Apart from his many books on child development there are also books on the development of arithmetic, geometry and language. His epistemology is, rather briefly, given in his *Psychology and Epistemology: Towards a Theory of Knowledge* (1972). His epistemology is based on sensations being symbols of the world of objects – symbols which are indirectly related to reality and which are learned by active experience of handling things. This allows behavioural observation of children's development to be evidence for development of perception and knowing. For a clear account of Piaget's work see Margaret A.Boden (1979).

33   There is surprisingly little known about the relation between touch and vision in the development of perception in children. This lack of knowledge is largely due to experimental difficulties, for only extremely handicapped children can serve as controls for investigating vision in the absence of active explorative touching. Such children almost invariably suffer from brain damage, and when this is not so we can hardly expect their motivations and so on to be sufficiently normal for absence of touch *per se* to be distinguishable from other problems. See E.Abravanel (1972) and A.V.Zaporozhets (1961). For the best review of visual learning and adaptation, see I.P.Howard and W.B.Templeton (1966). For evolutionary development of touch in various species, see D.R.Kenshalo (1978); see also Peter Bryant (1974).

34   For experimental data and theories on how children draw, for example, see N.H.Freeman (1980).

35   The American psychologist Jerome Bruner has contributed immeasurably to the understanding of infant and child development as well as perception. See especially his collected papers (1974). This includes a bibliography of his writings up to 1973. See especially 'On Perceptual Readiness', pp. 7–42, for the point discussed here.

Chapter 2   Tools that challenged myth

1   Alfred North Whitehead, 'Technical Education and its Relation to Science and Literature', in A. N. Whitehead (ed.) (1932).
2   Kenneth P. Oakley (1961), pp. 1–2.
3   Wilfred Le Gros Clark (1967), p. 3.
4   Kenneth P. Oakley (1961), p. 3.
5   Richard E. Leakey (1977), p. 98.
6   A useful general reference to early technology is Charles Singer *et al.* (eds.) (1954), especially Volume 1.
7   The main written early source is Theophilus (1980).
8   A highly imaginative, indeed poetic, discussion of form and function is given by R. M. Pirsig (1974).
9   V. Gordon Childe (1954), p. 199.
10   Mary Douglas (1966). For the habits and customs of present-day hunting and gathering societies which may provide evidence of ancient social organizations, and how work was shared, see John E. Pfeiffer (1969 and 1977). The most fully studied hunter–gatherer society, which provides interesting evidence on the roles of the sexes, is that of the !Kung bushmen of Botswana, studied by Richard B. Lee (1972). The effects of technology on primitive societies are discussed, with practical advice, by Margaret Mead (ed.) (1955).
11   Ancient prime movers (or rather, the almost complete lack of prime movers) and the details of many mechanisms are described by Robert S. Brumbaugh (1966). An excellent discussion of Aristotle's concepts of motion and the initiation of movement is given by A. C. Crombie (1961), pp. 65–98. This includes the effects of the classical notions on much later thinking. We discuss this further in Chapter 5. For a general account of Greek science, see Benjamin Farrington (1961).
12   I shall not embark on the development of the brain and nervous system, although no doubt this has a great deal to do with Mind in Science. For a recent account of the great deal that is known about the evolution of the human brain, see S. J. Gould (1977). For a general account of brain structure and function, see C. U. M. Smith (1970). For a detailed account of nerve, muscle and synapse function, see Bernard Katz (1966). For what is surely the best general account of the anatomy and physiology of mammals, together with the effects of environment, see J. Z. Young (1975); and, for a good introduction to the study of man, see J. Z. Young (1966). On the vexed question of when speech first developed, see R. L. Holloway (1974), which discusses some evidence for the speech areas in early hominids.
13   N. Chomsky (1965), p. 58.
14   Edward Sapir (1931), p. 578.
15   Paul Henle (1966), p. 5.
16   Ibid., p. 7.
17   For an important discussion of Navaho language see Clyde Kluckhohn and Dorothea Leighton (1948), especially p. 204.
18   Paul Henle (1966), p. 9.
19   Bertrand Russell, *History of Western Philosophy* (1946), p. 212.

20 Paul Henle (1966), pp. 13–14.

21 For an account of earliest writing, believed to be from monetary tokens, see Denise Shmandt-Besserat (1978).

22 David Diringer (1962), p. 15. For details of determinatives, see David Diringer (1962), Chapter 2, and David Diringer (1968), Chapters 1 and 2. Full lists of Egyptian hieroglyphic determinatives are given by Alan Gardiner (1957), especially p. 31.

23 David Diringer (1962), p. 19.

24 For early counting systems, see D. E. Smith and J. Ginsburg (1956). For the ability of animals to count, see Otto Koehler (1956).

25 The writings of Charles Babbage are collected, with most interesting related writings by his contemporaries, by Philip and Emily Morrison: see Charles Babbage (1961). For his own life, see Charles Babbage (1969). A recent rather 'popular' biography is M. Moseley (1964).

Babbage's dream was to build entirely automatic calculating engines to avoid human error, especially for calculating mathematical tables. He built the (uncompleted) Analytical Engine, which is still to be seen in the Science Museum, South Kensington, London. This is remarkable as embodying the principal features of electronic digital computers developed a hundred years later. In building his calculating machines Babbage pioneered precision mass production. He also suffered the current problems of getting money out of a government for novel technical development; having spent most of his own money he failed to find adequate support. His other inventions include the cow-catcher for railways, and the ophthalmoscope (anticipating Helmholtz by a couple of years); and he invented a symbolic system for describing functional relationships of machines, which has never been fully developed.

He became Lucasian Professor of Mathematics at Cambridge, although he never gave a lecture!

26 See J. M. Pullan (1969).

27 Chaucer wrote a remarkably complete account of the design and use of astrolabes: see Geoffrey Chaucer (1974), pp. 544–63. For a brief history, with fine photographs, see Seyyed Hossein Nasr (1976). For a source-book of Islamic science, see Edward Grant (1975).

28 These passages and some others are cited by Derek de Solla Price (1975), especially pp. 56–60. They can also be found in *Tusculan Disputations* (1945), I, xxv, 63:62–3, pp. 73–5; and *De Re Publica* (1928), I, xiii, 20, xiv, 21–2, pp. 41–3.

29 These passages in the *Timaeus* are from I, vii, 37–9, pp. 50–3 of Desmond Lee's translation (1974). There is much else of interest here, such as the question of the composition of the soul.

30 See Robert S. Brumbaugh (1964).

31 Claudius Claudianus, b. ?–404 A D, was the last Latin poet, although an Alexandrian by birth, and Greek-speaking by upbringing.

32 Claudian, no. LXVIII, *Shorter Poems* (1922), Vol. II, p. 279.

33 Derek de Solla Price (1975), p. 58.

34 The Chinese clock preceding Western mechanical clocks is described in detail by Joseph Needham, Ling Wang and Derek de Solla Price (1960). It is described non-technically by Derek de Solla Price (1975). For the fullest treatment of Chinese technology see Joseph Needham (1965), and his one-volume resumé (1970), which includes a description of this key link from the 'analogue' clepsydra to the remarkable 'digital' clock, pp. 203–38. The history of clocks is extremely important, for they were seen both as lawful models of the Universe and later as mechanisms approximating perpetual motion – so they suggested and provided mechanical models for Newtonian astronomy. A good short history of clocks is Alan H. Lloyd (1958).

35 Alighieri Dante (1265–1321) was a philosophical poet, whose greatest work, the *Divina Comedia,* has been said to have created the Italian language.

36 The effects of technology on art are described and discussed in Francis Klingender (ed.)(1972). Renaissance art with interesting references to science is discussed by Ernst Gombrich (1966). Gombrich (1979) also discusses the psychology of pattern and design.

Chapter 3   Lessons from machines

1 William Wordsworth, 'Phantom of Delight'.
2 Euripides (*c*. 480–406 BC) was among the three most highly regarded Greek dramatists. Eighteen complete plays from the eighty or so he is known to have written survive. He is later than Aeschylus and Sophocles, and wrote more than all of their combined works. Euripides has been criticized for over-use of the *deus ex machina*, which it is said he employed to cut the knot of dramatic situations rather than to unravel them.
3 Desmond Lee (transl.) (1974), p.42.
4 Ibid.
5 The Universe is described as a living being, of spherical shape, which (*Timaeus*, I, 5:33, p.45).

had no need of eyes, as there remained nothing visible outside it, nor of hearing, as there remained nothing audible; there was no surrounding air which it needed to breathe in. . . . He did not think there was any purpose in providing it with hands as it had no need to grasp anything or defend itself, nor with feet or any other means of support. For of the seven physical motions [uniform circular motion; up and down; backwards and forwards; right and left] he allotted to it the one which most properly belongs to intelligence and reason, and made it move with a uniform circular motion on the same spot. . . .

The description in the *Timaeus* of how the immortal soul is embodied in the machine Universe details constructional methods (IX, 2:42, p.58). The children of the Framer of the Universe

remembered and obeyed their father's orders . . . [and] borrowed from the world portions of fire and earth, water, and air – loans to be eventually repaid – and welded together what they had borrowed; the bonding they used was not indissoluble, like that by which they were themselves held together, but consisted of a multitude of rivets too small to be seen. . . . And into this body, subject to the flow of growth and decay, they fastened the orbits of the immortal soul. Plunged into this strong stream, the orbits were unable to control it, nor were they controlled by it, and because of the consequent violent conflict the motions of the whole creature were irregular, fortuitous, and irrational. . . . The motions caused by all these were transmitted through the body and impinged on the soul. For this reason they were later called, as they still are, 'sensations'. [This is from αισθησισ, supposed to be derived from ασσω, a verb meaning 'rapid movement': see the footnote on p.59 of Desmond Lee's translation of *Timaeus*.]

6 There is more of interest in *Mechanica;* for example the discussions on the circle and wheels (847b):

The original cause of all such phenomena – levers etc. – is the circle. It is quite natural that this should be so; for there is nothing strange in a lesser marvel being caused by a greater marvel, and it is a very great marvel that contraries should be present together, and the circle is made up of contraries. For to begin with, it is formed of motion and rest, things which are by nature opposed to one another.

7 *Mechanica*, 856a.
8 Norbert Weiner (1894–1964) was a prodigy, entering Tufts University at the age of eleven and completing his PhD. at nineteen. He applied servo-theory to the nervous system, and ended his highly distinguished mathematical career warning of the coming problems of automation.

9   These kinds of issues are well discussed by Richard Rorty (1972).
10   Karel Čapek (1890–1938) was born Karel Schadonitz. Also in 1921 he and his brother Josef wrote the *Insect Play*, which is in the genre of George Orwell. The word 'Robot' is the Czech for 'worker'.
11   Gilbert Ryle (1949), pp. 116–17.
12   The many histories of clocks include Alan H. Lloyd (1958).
13   For mechanical toys of the Greeks see R. S. Brumbaugh (1966).

Chapter 4   Exorcisms of Mind

1   William Shakespeare, *Cymbeline*, IV, ii, 258.
2   A useful history of alchemy is John M. Stillman (1960), which gives a good bibliography.
3   This is a short extract from a long account of alchemy in Chaucer's *Chanouns Yemannes Tale* (Canon's Yeoman's Tale) (1974), p. 216.
4   Paracelsus (1493–1541) was born at Einsiedeln, studied at Basel and practised in many towns. He was among the first to use chemistry for the practice of medicine, introducing laudanum, mercury, sulphur and lead into practical medicine. He was also among the first to study an occupational disease – fibroid phthisis, common in miners. Together with his sound practical sense, he held that there is a life force, which he called *archaeus*, and which he supposed to influence even dead matter. For him there were three principles of matter: 'salt', 'sulphur' and 'mercury' (see Charles Singer and E. Ashworth Underwood, 1962, pp. 100–1). In the *Coelum Philosophorum* Paracelsus sets out seven Canons of Metals – spiritual properties of mercury, tin, iron, copper, lead, silver and gold, together with their associated planetary astrological virtues.

Paracelsus was frank that most alchemists failed, but he did not see this as contrary evidence for the truth of alchemy, but rather thought that:

All the fault and cause of difficulty in Alchemy, whereby very many persons are reduced to poverty, and others labour in vain, is wholly and solely lack of skill in the operator, and the defect or excess of materials. . . . If the true process shall have been found, the substance itself while transmuting approaches daily more and more towards perfection. The straight road is easy, but is found by very few.

This important textbook of alchemy was translated by A. E. Waite (1894).
5   Robert Boyle (1627–91) was a founder member of the 'invisible College' of Oxford intellectuals who opposed contemporary scholasticism, and which became the Royal Society of London. With Robert Hooke as his assistant, he experimented on air, vacuum and respiration, publishing the *Sceptical Chymist* in 1661, which criticized current theories of matter, and defined the chemical elements as the practical limit of chemical analysis – so giving an operational criterion, in the absence of an adequate theoretical model. It is interesting that Boyle accepted alchemy together with his atomism, by arguing that if substances are merely various arrangements of a few kinds of elementary particles, transmutation of the elements should be possible. The development of atomic fission and fusion has shown that he was correct. There are several lives of Boyle, including R. Pilkington (1959), which has a good discussion of Boyle's alchemy.
6   An intriguing – and indeed profound – question is: 'Why is magic not "scientific"?' Exactly how is magic 'refuted' by science? Is it lack of predictive power? Actually the majority of mankind do accept magic, and believe for example in the Evil Eye: see C. L. Maloney (ed.) (1976). For a history of magic see T. Lynn (1923). Magic still has immense social power: see Mary Douglas (1966). For an important account of Newton's alchemy see R. Westfall (1975), pp. 189–238.
7   Friedrich Wöhler (1800–82) studied medicine at Heidelberg and chemistry at Stockholm, under

the Swedish chemist Johan Berzelius (1779–1848), who made accurate measurements of atomic weights – which was essential for testing and applying the atomic theory.

8   Justus Freiherr von Liebig (1803–73) studied at Bonn, and established the first important chemical laboratory, at Giessen. Among many achievements of first importance he was a discoverer of chloroform.

9   The Scandinavian poems of the *Edda* are in part pre-Christian. They are Homeric in scale, and important for later Germanic culture. The equivalent to the Biblical Flood is the attacking of the ice giant Ymir by the sons of Bor, who kill him, and whose descendants all drown in the sea of his blood except for one son and his wife, who escape with a boat and continue the line of giants. The brains of the dead giant scatter to form the clouds. On-off events often serve as explanations of laws of Nature, of the tides and so on. Thus the tides are caused by Thor's unwittingly sucking up the sea with his drinking-horn. See the quote from Poincaré on p.519.

10   Campbell Thomson Luzac (1900), no. 30.

11   Ibid., no. 82.

12   Ibid., no. 70.

13   Ibid., no. 49.

14   Ibid., no. 250.

15   Ibid., no. 253.

16   Ibid., no. 170.

17   Erastosthenes was an 'all-rounder' who founded systematic geography, inventing longitude and latitude lines and drawing the first map of the world, and he applied mathematics to measure the earth and the distance of the moon and the sun. He was also a poet, and he wrote a history of philosophy, but only fragments of his writings survive. His measurement of the size of the earth, from measurements of shadows from vertical gnomons at Syene (near Aswan) and Alexandria, is well described by G.E.R.Lloyd (1973), pp.49–50.

18   I have taken Aristarchus' hypotheses and propositions from W.C.Whetham and M.Dampier (eds.) (1924).

19   Galileo found that although the planets are seen as much larger in a telescope, this is not so for stars – which observation he correctly attributed to optical properties of images. So even here, he had to interpret his observations, and sort out which features of observations were significant and which artefacts. This remains a problem for microscopy even today.

20   R.J.Seeger (1966), Chapter 18.

21   This quotation is from Stillman Drake (transl.) (1962), p.341.

Chapter 5   Concepts in science

1   Charles Lamb, letter to Thomas Manning, 2 January 1810.

2   G.S.Kirk and J.E.Raven (1957), pp.188–9.

3   Aristotle, *Metaphysics* (1928), A.3, 983b, 8ff.

4   Bertrand Russell, *History of Western Philosophy* (1946), gives an excellent account and criticisms of Aristotle's probably unfortunate notion of substance and essence: see Chapter 22. Russell also discusses substance and relation to matter in physics and meaning, and categories and classes in language (1921): see Chapter 10. These are important statements.

5   Bertrand Russell, *History of Western Philosophy* (1946), p.61.

6   G.S.Kirk and J.E.Raven (1957), pp.196–7.

7   J.Burnet (1908), p.364.

8   This account of Zeno's paradoxes of motion I have based on J.Burnet (1908), pp.279–81.

9   A good technical discussion of Aristotle's concept of force is H.Carteron (1975). It is pointed out on p.167 that Aristotle regards force as an emanation of substance – and so is too like our concept

of mass. He did not however see force as graded, for he did not distinguish as we do between force and resistance: for him force was only demonstrated by observed movement. He does not explain the transmission of motive force, but thinks of this rather like biological stimuli, eliciting movement according to the natures of the moving objects. One might indeed say that Animism permeates the details of Aristotle's mechanics. Carteron suggests that his concept of force is analogous to chemical reactions; but I think reference to living organisms may be more appropriate here, as these would be far more familiar to Aristotle.

Geoffrey Lloyd (1970) points out (p. 114) that Aristotle could hardly be blamed for saying that heavier objects fall faster than light objects – for they do, when falling through a medium such as water. His failure was in not abstracting from common experience to the case of frictionless motion – which is nearly the case for air, but difficult to observe with sufficient accuracy. Galileo was forced to slow movements by rolling balls down inclined planes to establish laws of free fall (the leaning tower experiment may be apocryphal, and in any case it is a small part of the story of Galileo's experiments).

Aristotle's notion of force, and its gradual modifications up to Newton's account, are discussed by R. S. Westfall (1971), pp. 13–16. More generally, Aristotle distinguishes between caused and uncaused motion; and he allows that chance can play its part. He accepts a teleological account of Nature, such that Natural things (when not constrained by Unnatural forces, as when we make things such as beds out of 'Natural' wood) move towards states corresponding to their natures. He approves of the example of the wood of a bed flowering: the flower is a Natural development of the wood, not of the bed (see the *Physics*, Book 2). There are clear parallels to how he – and we – see 'Natural' and 'Unnatural' like free and constrained human and animal behaviour.

10   These quotations are taken from Aristotle, *De Caelo* (1939), III, ii and iii, 301b, pp. 279–85.

11   Johannes Kepler (1571–1630) studied at Tubingen, and became Professor of Mathematics at Graz. He corresponded, most fruitfully, with Tycho Brahe (1546–1601), who, from his island observatory of Hveen – the Castle of the Heavens – provided the star positions which founded modern astronomy. From these tables of star positions, Kepler finally derived the orbit of Mars and showed that the planets move in conic sections: an idea he had for long rejected, largely for aesthetic reasons, as the ellipses have two foci but only one is occupied, by the 'parent' object.

12   For a full description of the bucket experiment and its implications, see R. S. Westfall (1971), especially pp. 443–8.

13   Ernst Mach studied at Vienna, to become Professor of Mathematics at Graz in 1864, and Professor of Physics at Prague in 1867. His work on the flow of gases is basic for aeronautics, and his name was given to the units, Mach numbers. His writings on epistemology largely founded Logical Positivism, and they clearly influenced Einstein. *The Analysis of Sensations* (1897) describes original perceptual phenomena, and remains important. *The Science of Mechanics* (1838) was a source of modern philosophical physics, and remains well worth reading; and his *Popular Scientific Lectures* (1894) are also interesting, especially the lecture on symmetry. *Space and Geometry* (1906) perhaps especially influenced Einstein. See also R. H. Dicke (1964).

14   Empedocles' fame as a physician and soothsayer was so great that he was offered the sovereignty of his state; but he introduced a democracy. There is a story that he threw himself into Mount Etna, which rejected his sandals, but accepted Empedocles as its own.

15   Democritus (*fl.* fifth century BC) was the most learned man of his time. He had the supremely important idea that all the variety of physical properties of matter could result from various combinations of a very few (or even one) kind of atoms. This is the central notion of any mechanistic account. It is not now possible to separate his views or contributions from those of Leucippus, as all his seventy or so books were lost in antiquity. He seems to have been a consummate genius, but badly treated by his heirs. His views were however espoused and

protected by Lucretius. A working atomic theory of matter, in which observed properties of matter are attributed to a few kinds of insensible particles, was not developed before John Dalton at the end of the eighteenth century. Before this insight Leucippus and Democritus seemed far less important than they do now – there are many early disparaging comments.

16 Lucretius, *On the Nature of Things*, Book 1, p. 17 *et seq*. Titus Lucretius Carus probably lived from 94 to 55 BC. Almost nothing is known of his life, although he was probably a member of an aristocratic Roman family. *De Rerum Natura (On the Nature of Things)* is his only work. For a full, though somewhat dated, treatment of Greek Atomism, see Cyril Bailey (1928). This verse translation is by W. E. Leonard (1921).

17 The prose version is by Cyril Bailey (1947), Vol. I, p. 193.

18 The distinction between space as 'active' and space as 'passive' in relation to objects, and especially their motion, remains a matter for discussion even today; for some theorists of Relativity Theory hold that, following Mach, the mass of matter bends space, or curves space, while others hold that space is structured to constrain object motion – and that space would have this curved structure in the absence of matter. This is as difficult to resolve as the hen-and-egg question: which came first? But the structured space notion abandons Relativity in a pure thoroughgoing form. We discuss Relativity, with relevance to the observer, in Chapter 18.

19 An interesting popular account of concepts of fundamental particle physics is given by Nigel Calder (1977).

20 There is a very large body of literature on time. Especially interesting and useful are papers in J. J. C. Smart (ed.) (1964), and Richard M. Gale (ed.) (1968). These both contain full bibliographies. See also G. J. Whitrow (1975) and P. Fraisse (1964). The psychology of time is discussed by William James (1890) Chapter 15; and M. F. Cleugh (1937), Chapter 1, has a short discussion. The interesting issue of whether 'psychological' time is continuous or intermittent has most recently been investigated by Allan Allport (1968). He concludes it is continuous.

21 Aristotle, *Physics* (1957), IV, xiv, 223b, pp. 423–4.

22 Aristotle, *De Anima* (1957), I, iii, 407a, pp. 37–8.

23 Ibid., 407b, p. 41.

24 Aristotle, *De Caelo* (1939), I, ii, pp. 9–17.

25 Isaac Barrow is quoted by G. J. Whitrow (1975), pp. 83–4. Isaac Barrow, who arranged for Wren to build the library of Trinity College, Cambridge, where he was the first Lucasian Professor of Mathematics, devised geometrical methods which developed in the hands of Newton and Leibniz into the differential calculus. He gave up his chair in 1669, in favour of Newton. Barrow gradually turned from mathematics and science (mainly optics) to divinity, his sermons becoming famous for their profundity and length. Preaching in Westminster Abbey, the organist finally played 'Till they Blowed Him Down' to get him to stop. Barrow was a saintly man, and the originator of several key ideas underlying Newton's achievements. He was a principal founder member of the Royal Society. See P. H. Osmond (1943), which gives Barrow's account of time in full (Chapter 6), though with a different translation from the Latin; and his Life.

26 John Locke (1690), Chapter XIV, para. 21, p. 150.

27 G. J. Whitrow (1975), p. 85.

28 This problem of what could lie outside space was a reason for supposing the Universe to be spherical; or at least not having corners – for projecting corners must move into nothingness (?) as the Universe rotates, as it was supposed by Aristotle and many others to do.

29 See R. S. Westfall (1971), p. 397.

30 The later life of John Toland was dramatized by the Prime Minister Benjamin Disraeli's father, Isaac D'Israeli, in a book entitled *Calamities of Authors* which appeared in 1812.

31 John Toland (1704), p.188.

32 There is a large body of literature of experiments on perception of motion of seen objects; for example, from the flow of the visual array (J.J.Gibson,1950), and perceptual processes and rules of perception by which objects are seen to have unchanging form with rotation, though the retinal pattern changes in complex ways (G.Johanson, 1975, Johanson *et al.*, 1980). Thus a coin changes from a circle to ellipses of ever-increasing eccentricity at the retina as it is viewed from changing angles, though it is *seen* as a disk even when the image is highly eccentric. It is remarkable that this is achieved in perception even for highly complex objects. It is interesting also that when objects (such as billiard balls) collide causation is seen – as though there is pre-conceptual attribution of causes of events. Perceptual rules for seeing cause have been extensively investigated by the Belgian psychologist, of the Gestalt tradition, Albert Michotte (1881–1965), who worked at, and became Professor at, Louvain from 1903–56. Michotte's *La perception de la Causalité* (1954) is available in English in Peter Heath's scientifically excellent translation, as he is an experimental psychologist, with an introduction by R.C.Oldfield who was Professor of Psychology at Oxford (1963). Of importance here also is Jean Piaget (1955), Jean Piaget (1929) and Jean Piaget and B.Inhelder (1956).

33 F.C.Bakewell (1853), p.11. F.C.Bakewell's book is a rare book in my possession (published by Ingram, Cooke) and perhaps almost impossible to obtain; but these contemporary books are marvellous sources, uncontaminated by later scholarship. This holds for early editions of the *Encyclopaedia Britannica*, which are generally far more interesting than the current issue.

34 Benjamin Franklin was outstanding as a statesman and scientist. He lived mainly in Philadelphia, and became President of Pennsylvania. He was a principal negotiator for the Declaration of Independence signed on 4 July 1776. The *Autobiography*, and several of the many Lives, are interesting for the period, and for a man of remarkably various genius. His scientific work occupied only ten years. The definitive edition of his works is Leonard W.Labarel *et al.* (eds.) (1959–). For a readily available collection see L.Jesse Lewis (ed.) (1961), including the *Autobiography*.

35 F.C.Bakewell (1853), p.21.

36 Thomas Kuhn (1970), p.61.

37 F.C.Bakewell (1853), p.71.

38 Ibid., pp.71–2.

39 A 'popular' life of Mesmer is given by Vincent Buranelli (1975). There should be a full study of Mesmer and Mesmerism.

40 Anthony Clare (1976), p.223.

41 Nicolas Leonard Sadi Carnot was the son of Lazare Carnot (1753–1823), who was the 'organizer of victory' in the French Revolution, and was engineer, general and statesman. Nicolas Carnot was also a military engineer. He founded the new science of thermodynamics as a Captain of Engineers in the French army. He died in a cholera epidemic at the age of thirty-six, before he could generalize his concept – which was achieved in the next generation, largely by the German theoretical physicist Rudolf Clausius (1822–88). Clausius coined the term 'Entropy' in 1865. Hermann von Helmholtz, and also Lord Kelvin (William Thomson), made important contributions to the insight of the Second Law, and heat death.

42 Erwin Schrödinger (1958), pp.3–5.

43 Jacques Monod (1972), p.28.

44 Richard Feynman (1972), Vol. 1, Chapter 46.

45 An important though different treatment is L.Brillouin (1956); and also F.T.S.Yu (1976).

46 F.T.S.Yu (1976), pp.96–111.

47   The first mathematical treatment of increase of structure (reduction of Entropy) by Natural Selection was given by Ronald Fisher (1930). For Beyesian strategy in human decision-taking, see D. E. Broadbent (1973, especially Chapter 4; and 1965).

48   Denis Gabor invented holography, some sixteen years before the invention of the laser which made it practicable. He was awarded the Nobel Prize for physics in 1971.

49   Claude Shannon did this work at the Bell Telephone Laboratories in the USA.

50   The classical exposition of Information Theory is C. E. Shannon and W. Weaver (1949).

51   Useful accounts of Information Theory in psychology are Colin Cherry (1957); Fred Attneave (1959); and P. M. Fitts (1954). For signal detection theory applied to sensory discrimination see D. M. Green and J. A. Swets (1966) and Paul Fitts and Michael Posner (1973). For classical theoretical discussions see D. E. Broadbent (1958) and Alan Welford (1968).

52   Sophisticated experiments on visual information rate and short-term memory storage (or span of apprehension) have been carried out by George Sperling and his collaborators, at the Bell Telephone Laboratories. E. Averbach and G. Sperling (1961) found, with most ingenious experimental techniques, that the span of apprehension (span of immediate memory) limits the number of items that can be recalled from a single glance to only 4–5 (20–26 bits); but most interesting, these items can be correctly remembered from a much larger set of items presented in the single glance or flash presentation. This set can represent at least 70 bits of information which are 'seen' though they cannot be recalled.

     A buffer store is postulated to explain the high information capacity of vision though the number of items that can be recalled is small. This two-stage model is supported by many distinguished investigators, including Norman Mackworth, S. Sternberg, R. Conrad, N. Moray, Ann Taylor (later Ann Treisman), and Peter Ladefoged and Donald Broadbent, who extended the notion to hearing. References to papers may be found in Ralph Haber (1968), which is a most useful collection of papers on visual perception and information processing.

53   Donald MacKay (1969), p. 24.

54   William Coleman (1971), p. 17.

55   There is of course an enormous literature on Evolution and on Darwin and his work. An account which is still useful is by Julian Huxley (1942). Readily available and with more recent ideas is John Maynard-Smith (1958). An interesting account of the background and social and other consequences is by Gertrude Himmelfarb (1962). There are many biographies: the one by Howard E. Gruber (1974) is particularly interesting, and contains previously unpublished notebooks transcribed by Paul H. Barnett. The Life and Letters edited by his son Francis Darwin (1887) is an essential source, as also is Nora Barlow (1945). A delightful joint biography of Darwin and T. H. Huxley is by William Irvine (1956). The celebrated essay of Thomas Malthus which in part inspired both Darwin and Alfred Russell Wallace to the survival of the fittest concept is available (1959). Alfred Russell Wallace's autobiography (1905) and the Life and Letters of John Romanes (1896) are important for the historical background and in their own right.

56   Jean Baptiste Lamarck was a French army officer when he became interested in Mediterranean flora. He left the army after an injury and became Keeper of the Royal Gardens, where he lectured for over twenty years on invertebrate zoology. In about 1801 he considered an evolutionary view of the origin of species based on inheritance of acquired characteristics.

57   William Coleman (1971), pp. 68–9.

58   In Time and Free Will (1889) and Creative Evolution (1907), Henri Bergson compares the ever-changing and yet unitary character of consciousness with the world of spatial objects, which may be grasped by intellect while duration is seen only by immediate experience, or intuition enriched by memory. This thought is, actually, well expressed by Plato in the Timaeus, Section 17:

. . . in general we should never speak as if any of the things we suppose we can indicate by pointing and using the expression 'this thing' or 'that thing' have any permanent reality: for they have no stability and elude the designation of 'this' or 'that' or any that expresses permanence. We should not use these expressions of them, but in each and every case speak of a continually recurrent similar quality.

The creative urge is central for Bergson: it directs Evolution. There are no guiding principles for conduct, though there is free will. He does not give reasons or arguments; but relies on striking analogies. He influenced William James and Whitehead. He received the Nobel Prize in 1927. Perhaps his most interesting book for us now is his work on humour, *Laughter* (1911). This is surely an under-researched and fascinating aspect of Mind.

59   The matter of the available time for Darwinian Evolution is extremely important. Pressure from the physicists of his day effectively put Darwin off his stroke, so that in late editions of the *Origin of Species* he was forced to accept in large measure Lamarckian inheritance of acquired characteristics, in order to account for Evolution on the short time-scale allowed (on the assumption, then held by physicists, that the sun was essentially a coal fire with a short life so that geological time was restricted to a few million years). Darwin started by saying, in 1859, that 'we have unlimited time', but the outstanding physicist Lord Kelvin (William Thomson), with all the weight of the prestige of mathematical physics and its reputation for reliable knowledge, declared the life of the sun to be certainly less than ten million years. Today we put the origin of life about three billion years into a past of roughly constant solar activity. The point is of course that the discovery of atomic energy gives an entirely different estimate. It is interesting that the biologists were right and the physicists wrong; but it is also upsetting to find that Darwin was driven into error and confusion by the physicists.

The assumed lack of sufficient time was used to 'refute' Evolution by Natural Selection until well into the twentieth century. At that time it seemed physically impossible. Who indeed could have guessed that a source of heat from disintegrating atoms was round the corner? This is a dramatic example of the unreliability of empirical refutation; and it shows how dangerous is a rigid prestige hierarchy of the sciences, for surprises can come from anywhere to send ripples and even convulsions through the body of accepted knowledge. Observations and inferences from varieties of finches, on an island few had heard of, by an amateur naturalist of non-proven ability, did just this.

60   Erasmus Darwin put forward an evolutionary account for the origin of species in his *Zoonomia* (1794–6), in Section 39 entitled 'Of Generation'. It seems that he believed in some kind of evolution by 1771. He clearly rejected the theory at that time popular, of preformation – that embryos were miniature adults in form; and he held, as stressed by Charles Darwin in the *Origin of Species,* that species had been greatly changed by breeding and artificial selection. He takes several examples and concludes that 'All animals undergo perpetual transformations, which are in part produced by their own exertions.' And 'many of these acquired forms or propensities are transmitted to their posterity' (Vol. 1, p.503). This may be the first statement that variations must be inherited for evolution to occur. He stresses sexual rivalry: 'The final cause of this contest [for females] amongst the males seems to be, that the strongest and most active animal should propagate the species, which should therefore be improved.' Erasmus proceeds to show that competition for food, and for security, including avoiding more powerful animals, should produce inherited changes.

It is difficult not to be impressed by some of Erasmus Darwin's ideas – though they are presented without the care and immense body of evidence of the *Origin of Species*. For Erasmus's life see Desmond King-Hele (1963), and the curious volume by Ernst Krause (1879) with a note by Charles Darwin on his grandfather.

61   Thomas Robert Malthus (1766–1834) was an economist with a strong Cambridge mathematical background. He argued that the optimism of Rousseau was doomed because populations tend to outgrow their means of subsistence. This was the stimulus for both Darwin and Wallace to adopt the concept of Evolution by Selection of the fittest to survive.

62   Alfred Russell Wallace collected plants and animal specimens in the Amazon Basin and the Malay Archipelago – though he lost much of his collection and his notes by shipwreck, due to fire at sea, from which he was fortunate to escape. He independently arrived at Darwin's position, and in 1858 sent Darwin a paper which prompted the latter to complete the *Origin of Species*, on which he had spent twenty years assessing the evidence for Natural Selection. The theory was made public with a joint paper by Darwin and Wallace, at a meeting of the Linnaean Society on 2 July 1858. Darwin had already written extensive notes for his own use (1842 and 1844), and these have since been published, edited by Francis Darwin (1909). Wallace always accepted Darwin's priority, and altogether this awkward situation was handled well by both. Wallace went on to develop the theory, until, late in life, he turned to spiritualism. He received the Order of Merit in 1910. (It happens that my own father, Christopher Clive Langton Gregory, who became an astronomer and Director of the University of London Observatory, used when a boy to visit Wallace at his Parkstone house near Poole in Dorset, where my father was brought up. I believe that Wallace made a deep impression on him, as well he might.)

63   Thomas Henry Huxley studied medicine in London and, as surgeon on HMS *Rattlesnake*, collected marine animals off the Great Barrier Reef, which led him to a highly distinguished career in biology, becoming President of the Royal Society (1881–5). His autobiography is edited by Gavin de Beer (1974). Huxley coined the word 'agnostic'. Having at first rejected Evolution, he was persuaded of its truth by the *Origin of Species*.

64   J.T. Bonner (1980), pp. 25–6. This is a beautifully written account of pre-human origins of culture, and the transmission of culture. For a study of human culture see Leslie A. White (1949). For a highly suggestive account of science, symbols and civilization, see J.Z. Young (1951).

65   A uniquely interesting account of the relations between forms of different species, and how forms change, largely following topological principles, is D'Arcy Thompson (1961). Although universally admired by biologists, it seems that D'Arcy Thompson's ideas have not had much effect on later thinking. We should refer also to the great statistician, Ronald Fisher (1930).

66   The Neo-Darwinian account of Natural Selection is criticized in R.C. Lewontin (1974) on the grounds that there is no evidence from individual examples to show how selection occurs or how it is related to gene characteristics. Lewontin comments (p. 189):

For many years population genetics was an immensely rich and powerful theory with virtually no suitable facts on which to operate. . . . Quite suddenly the situation has changed. The mother-lode has been tapped and facts in profusion have been poured into the hoppers of this theory machine. And from the other end has issued – nothing. It is not as if the machinery does not work, for a great clashing of gears is clearly audible, if not deafening, but it somehow cannot transform into a finished product the great volume of raw material that has been provided. The entire relationship between the theory and the facts needs to be reconstructed.

Lewontin finds that the number of gene characteristics and their interactions are so many that description of the mechanisms at the microgene level is impossible. He concludes that here, as for thermodynamics, only statistical accounts are possible, and to describe gene characteristics in isolation is 'about as relevant to real problems of evolutionary genetics as the study of the psychology of individuals isolated from their social context is to an understanding of man's sociopolitical evolution. In both cases context and interaction are not simply second order effects. . . . Context and interaction are of the essence.'

The question of the circularity, and so apparent untestability of Darwinian Evolution is discussed in a more optimistic way (with important comments by C.H.Waddington and David Bohm) in a paper by John Maynard-Smith (1969). Maynard-Smith suggests that Neo-Darwinism would be refuted by (as a fanciful example) finding exact patterns of the constellations of the stars, on the tail of a fish; or, more seriously, if – which however is not the case – Evolution was found to occur too rapidly in laboratory conditions with fast-breeding species such as fruit-flies (actually Darwin was worried about the lack of time the physicists of his day allowed for Evolution: see note 59). See also P.P.Weiner (1949).

For C.H.Waddington's theoretical position see C.H.Waddington (1957).

67 Aristotle, *De Generatione Animalium* (1912), III, Chapter 10.
68 Ibid., 762 b, pp. 28 *et seq*.
69 Anton van Leeuwenhoek was a clerk at an Amsterdam warehouse until 1654, after which his fame spread as the most distinguished microscopist, with a quite remarkable series of discoveries. He made his own (simple) microscopes – one for each specimen. Many survive.
70 Louis Pasteur became Professor of Chemistry at the Sorbonne. He started working on fermentation, and applied his discoveries with great success to many diseases of plants; and to diseases of animals and men by vaccination.
71 Charles Darwin (1871), pp. 526–7.
72 Ibid., p. 566.
73 Ibid., pp. 563–6.
74 Fraser Harrison (1977), p. 95.
75 Charles Darwin (1871), p. 619.
76 J.Z.Young (1978), pp. 5–6.

Part 2   Links to Mind
Chapter 6   Links of light
1 Emily Brontë, 'The Prisoner'.
2 Joseph Blanco White, 'To Night'.
3 Democritus' works are not available for direct citation. The fullest account of his work I know is Cyril Bayley (1928), in which he is discussed through the whole of Part 3, pp. 109–213.
4 Plato, *Timaeus* (1974), Section 13, pp. 61–2.
5 Ibid., Section 14, p. 64.
6 Ibid., Section 36, p. 92.
7 Ibid., p. 93.
8 The remarkable twelfth-century Archbishop of Canterbury, John Peckham, has extremely interesting things to say on optics, just as the Islamic knowledge reached Europe. He argues that light-rays are like rods or rulers and can give distance directly as they can. See David C. Lindberg (transl.) (1970). For the major source-book of Islamic science in English, see Edward Grant (1974); and for a beautifully illustrated treatment, see Seyyed Hossein Nasr (1976).
9 Isaac Newton entered Trinity College Cambridge in 1661, graduated in 1665, and a year later wrote his first account of fluxions (the calculus) and in the next two years formulated his theory of gravity (quite possibly suggested by the falling apple), and started his work on optics, though this was not published until much later.

Newton was elected President of the Royal Society for life, and held the office for twenty-four years. He was elected Fellow of Trinity College in 1667, and Lucasian Professor of Mathematics in 1669, when Isaac Barrow made way for him. His knighthood (scientists were not knighted at that time) was given by Queen Anne in 1705 for political reasons when he stood for Parliament, though nothing came of this. He was however an extremely efficient administrator at the Mint, entirely reorganizing the coinage. Newton was a compulsive writer and note-taker, leaving some four million words. The manuscripts are housed in the Cambridge University Library. It is well known that he had something of a breakdown when he left Cambridge to live in London: possibly this was due to mercury poisoning, contracted from his extensive alchemical experiments.

Newton did not separate religion from science. He wrote a million words on the prophecies of Daniel, the Apocalypse, the Biblical account of Creation, and so on. He possibly attached as much importance to this as to unlocking the secrets of the stars and light by science. Given that he regarded the laws of Nature as characteristics of God's Mind, he had no need to make the distinction that his mechanical account did so much to create, and which today seems self-evident.

There are several Lives, notably David Brewster (1855), Louis Trenchard More (1934), and Frank Manuel (1968). The correspondence, which is extremely interesting, is edited by H. W. Turnbull and J. F. Scott (1959).

10 This long letter is not in Newton's handwriting, but is a transcript written by Wicken, who shared rooms with him at Trinity College Cambridge; there are corrections in Newton's hand. It appears in Vol. 1 of H. W. Turnbull and J. F. Scott (eds.) (1959), pp. 92–102.

11 Robert Hooke became Curator of Experiments to the Royal Society in 1662, showing remarkable ingenuity, inventing several instruments still in use. In 1677 he became Secretary to the Royal Society. After the Great Fire of London of 1666 he was appointed City Surveyor, and so worked closely with astronomer-turned-architect Christopher Wren (1632–1723). In any other age Hooke would have dominated the science of his generation, but he was upstaged throughout his life by Newton, whose attitude towards Hooke was ambivalent; for example Newton postponed publication of his *Opticks* until Hooke's death. Hooke is best known for his law of compression of springs, and his invention of the anchor escapement for clocks. There is also Hooke's joint for drive shafts which are not in line; the quadrant; meteorological instruments; and important developments of the telescope and microscope. His *Micrographia* (1665) is well worth reading today.

There are several Lives, and he wrote a diary.

12 The letter to Lord Brouncker is in H. W. Turnbull and J. F. Scott (eds.) (1959), Vol. 1, pp. 198–203.

13 H. W. Turnbull and J. F. Scott (eds.) (1959), Vol. 1, p. 204.

14 Robert Hooke (1665), p. 54. The reference to Francis Bacon's demand for crucial experiments as guides or landmarks is especially interesting in the light of discussions of Inductive enquiry (Chapter 8).

15 James Gregory was of a family of mathematicians, astronomers and Professors of Medicine at Edinburgh and Aberdeen. He became Professor of Mathematics at St Andrews in 1688 and at Edinburgh in 1674. His nephew, David Gregory (1661–1708), became Professor of Mathematics at Edinburgh in 1683, and moved south to become Savilian Professor at Oxford in 1691. He suggested the use of combinations of lenses to achromatize telescopes. For a history of the family see A. G. Stewart, *The Academic Gregories* (1901).

16 Plato, *The Republic* (1941), VII, 514A–521B, p. 222.

17 Vasco Ronchi (1970), p. 53.

18 Giovanni Battista della Porta was a Neapolitan physicist and philosopher who wrote on physiognomy and gardening as well as optics and painting.

19 Danielo Barbaro (1946), p. 18.

20 Leonardo da Vinci was not only among the most superb painters and sculptors of all time but was also a consummate inventor and a man of the widest scholarship in the arts and the sciences.

21 Filippo Brunelleschi was an equally superb architect and engineer: he alone dared and succeeded in completing the great cathedral of Florence, which was founded in 1296, by adding a dome of unprecedented and still unsurpassed size. The cathedral was built for the city, by the city, and its setting among other buildings was important. It seems that the impetus to devise and use the camera obscura notion came from the need to produce plans within perspective drawings of the surrounding buildings. This was an achievement of engineer-architects. It was later used by painters, such as Canaletto.

22 Leonardo da Vinci (1938), p. 215. It is interesting how confused he was over the optics of the eye, which occupies a large part of *The Notebooks*, and also over reversals experienced with mirrors. In addition he did not clearly appreciate that the two eyes give depth by stereopsis – though he does say (p. 238): 'Things seen with two eyes will seem rounder than those seen with one eye;' and although he understood the geometrical basis of perspective as seen in the camera obscura – and of course he was a superb exponent of perspective in his drawing – he still had the odd idea that (p. 225) 'Why objects as they come upon the small surface of the eye appear large arises from the fact that the pupil is a concave mirror, and so one sees for example with a glass ball filled with water that anything placed at the side either inside or outside appears larger [I take it that he is referring to one's own eye and not to reflections in another's eye].' It seems possible that he was unsatisfied and puzzled over the fact that things do not seem as they should by perspective because of what we call size constancy; for Leonardo says (p. 222): 'The flea and the man can approach the eye and enter into it at equal angles. For this reason does not the judgement deceive itself in that the man does not seem larger than this flea? Enquire as to the cause.'

23 Leonardo da Vinci (1938), p. 246.

24 This first demonstration of photography is described in a paper published in the *Journal of the Royal Institution of Great Britain*, Vol. 1, pp. 170–4, 1802. It was entitled 'An Account of a method of copying Paintings upon glass, and of making Profiles, by the agency of light upon Nitrate of Silver.' There are observations by Humphry Davy.

25 The reference for this is Aristotle, *De Sensu* (1936), II, vii, 418b, p. 105.

26 Descartes had extensive correspondence with Marin Mersenne, who was a Franciscan friar with considerable achievements in mathematics and physics. They were close friends from 1622. Mersenne's main works are *Cogitata Physico-Mathematica* (1644), and *L'Harmonia Universelle* (1636–7). Mersenne's correspondence was edited by Paul Tannery in 1932.

A useful though somewhat dated account of Descartes' scientific writings and ideas is J. F. Scott (1952).

27 James Bradley became Savilian Professor of Astronomy at Oxford in 1721, Regius Professor at Cambridge in 1742, and Astronomer Royal. His aberration-of-light discovery was published in 1729. It was the first experimental prediction, after two hundred years, to confirm the Copernican hypothesis.

Chapter 7   Links of nerve

1 René Descartes, *Principia Philosophiae* (1967), Principle CXCVI, p. 293.

2 Galen, *On The Natural Faculties* (1916), Vol. III, vi.

3 Keith Lucas was probably the first to use the electronic amplifier on nerve. He was killed, while young, in a flying accident in 1914. He demonstrated the refractory nature of nerve, which was the basis of all later work on neural conduction.

4 Edgar Douglas Adrian (1889–1977) was Professor of Physiology at Cambridge from 1937 to 1951. He received the Order of Merit, was President of the Royal Society, and Master of Trinity College, Cambridge. He was created Baron Adrian in 1955. His work on recording action potentials, on sensory adaptation and on the EEG with Sir Brian Matthews is fundamentally

important for neurophysiology. See W.Grey Walter in D.Hill and G.Parr (eds.) (1950) who made the principal contributions to the EEG following Adrian and Matthews.

5 E.D.Adrian (1928), p.28. It is interesting that Lord Adrian found it puzzling that sensation is not intermittent, as nerve signals are intermittent. Are we no longer surprised by this because, now, we are familiar with integration and smoothing of signals in electronic circuits? It is worth noting that Hans Berger's discovery of the alpha wave of the cortex was made by his apparently noticing a high-frequency (10/sec) intermittency or fluctuation of brightness. He recorded a 10-Hertz electrical activity from the brain, with scalp electrodes, using a string galvanometer without amplification (1928–35). This would just detect 0.1 millivolt. The only activity that could be recorded was the alpha wave. This was shown in 1935 by Lord Adrian and Sir Brian Matthews, using the Matthews amplifier and Adrian's head, to be a brain activity localized in the occipital cortex, and it was observed that the rhythm tended to stop when the subject opened his eyes or concentrated on a problem: see W.Grey Walter in D.Hill and G.Parr (eds.) (1950).

6 E.D.Adrian (1928), p.99.

7 Neville Maskelyne (1799), Section for 1795, pp.3, 319, and 339 *et seq.*

8 Johannes Müller (1801–58) became Professor of Physiology at Bonn and, in 1833, at Berlin. The highly influential *Handbuch der Physiologie des Menschen* was published in the years 1833–40. The (in modern terms) absurd estimate of the velocity of neural progation was derived from the highly inappropriate model that the pneuma is subject to Bernoulli's principle of velocity of fluid flow increasing as it is more restricted. But of course this assumes that fluid flow is involved, which is not the case. Müller's wild estimate is a dramatic example of how misleading an inappropriate reference to a well understood physical process can be.

9 J.Müller (1833–40).

10 This letter is given in Koenigsberger's life of Helmholtz (1906), p.67.

11 A recent study of the eye–ear method has been carried out by M.Hammerton and D.D.Stretch, as described in the *Journal of the British Astronomical Association* (1980). They find that for some observers it is a good method.

12 It is very generally assumed that events must precede causes, but there are philosophical discussions, especially M.Dummett and A.Flew (1954), and M.Black (1956). Further references will be found in R.M.Gale (ed.) (1968), p.510.

13 There is an extensive treatment on the Personal Equation in E.G.Boring (1950), Chapter 8.

Chapter 8   Links of logic

1 W.M.Pread, 'The Talented Man'.

2 David Hume (1758), ix, 'Of the Reasoning of Animals'.

3 Aristotle, *Metaphysics* (1956), Book A, ii, 982aff., p.54.

4 George Friedrich Bernhard Riemann (1826–66) wrote *On the Hypotheses Underlying Geometry* in 1854, and it was published in full in 1868.

5 The history of the subject–object distinction is touched upon by M.J.Morgan (1977), p.156. 'Subject' can mean (and originally meant) the underlying substance or medium which has properties which may be causal and may be observed. It was thus the inner essence of objective things. It now means individual experience – which may be of objects; but just how subject and object are supposed to be related depends upon the accepted theory of perception. Aristotelian logic is still known as 'subject–predicate' logic, although it refers to what we call 'objects' of the external world.

6 Gottlob Frege is the acknowledged originator of modern symbolic logic. The outstanding study is by Michael Dummett (1973).

7 George Boole (1847), pp.3–5.

8   Bertrand Russell (1919), p. 169.

9   Willard van Orman Quine (1953), p. 15.

10 See Willard van Orman Quine (1953 and 1960).

11 For a remarkably exciting account of Gödel's theorem and paradoxes in general, see Douglas R. Hofstadter (1979).

12 Aristotle, *Analytica Posteriora* (1964), Book A, 87b, p. 222.

13 Ibid., Book Z.

14 Ibid., Book A.

15 Francis Bacon (1620), *Aphorisms*, Book i, no. 50.

16 Ibid., no. 62.

17 Ibid., no. 63.

18 Ibid., no. 64.

19 William Gilbert became physician to Queen Elizabeth. He coined the word 'electricity'. His *De Magnete* (1600) demonstrated that the earth is a magnet, and he supposed that magnetism and electricity were ultimately one. Not only were his ideas and experimental results important for navigation and for physics, but this is the first truly scientific book, exhibiting scientific method as we understand it today.

20 Francis Bacon (1620), *Aphorisms*, Book i, no. 98.

21 Ibid., Book ii, no. 10.

22 Ibid., no. 11.

23 Ibid., no. 12.

24 Ibid., no. 13.

25 Ibid., no. 17.

26 Ibid., no. 19.

27 Ibid., no. 36.

28 Francis Bacon (1960), p. 206.

29 John Stuart Mill was the son of James Mill, who wrote *Analysis of the Human Mind* (1835). The latter took a leading part with Jeremy Bentham (1748–1832) in founding the University of London. He brought his son John up to be a genius. He was taught Greek at the age of three, Latin at eight and logic at twelve, with no recreation except a daily walk with his father, which was combined with an oral examination. J. S. Mill's *Autobiography* (1873) is a fascinating record of this unique upbringing, which nearly ended in disaster but in fact was highly successful. He was a champion of women's rights, and profoundly affected the political as well as the philosophical climate. He was Bertrand Russell's 'secular' godfather; and Russell was in many ways his political and philosophical heir.

30 John Stuart Mill (1843), Book 3, Chapter 3.

31 Pierre Laplace (1951), p. 180.

32 The following quotations are taken from John Stuart Mill (1843), Book 3, Chapter 3.

33 Jacob Bronowski (1973), pp. 323 *et seq*. The details are in Mendeleev's lecture to the Chemical Society, delivered at the Royal Institution, London, on Tuesday 4 June 1889. This is given in W. C. Whetham and M. Dampier (eds.) (1924), pp. 112–17.

34 Bertrand Russell, *History of Western Philosophy* (1946), p. 645.

35 Bertrand Russell (1912), p. 35.

36 Ibid., p. 35.

37 Bertrand Russell (1948), p. 439.

38 Ibid., p. 449.

39 It might be added that Bertrand Russell gave me a signed copy of *Human Knowledge* on that occasion, which I greatly treasure.

40   Bertrand Russell (1948), p.477.

41   Francis Bacon (1620), *Aphorisms*, Book ɪ, no.105.

42   Hans Reichenbach (1953), p.49.

43   There are very many discussions of whether Induction should be trusted. See R.B.Braithwaite (1953): I attended Richard Braithwaite's most stimulating lectures on logic and scientific method at Cambridge. To be recommended as a not-too-technical exposition of these and other problems of the philosophy of science is Norman Campbell (1921). For a useful collection of papers, see Richard Swinburne (ed.) (1974).

44   Auguste Comte became the germinal French philosopher and sociologist who founded Positivism. Apart from political writings, his main work is *Cours de philosophie positive* (1830–42). This is translated in condensed form by Harriet Martineau (1853). Comte's aim was to organize, relate and combine knowledge of the world, of man, and of society into a consistent whole. Human conceptions and theories are supposed to pass through a theological, and then an explicitly formulated, metaphysical stage, to end in a positive or testable stage – while society moves from militarism to industrialism. The ideal and aim is to gain and live by knowledge. There is surprisingly little reference to Comte in current British or American philosophy, perhaps because he is regarded as being too conservative and 'finalistic' in his views of society and science. He was less interested in discovering truth than in classifying the current knowledge of his time, so he has dated; but his Positivism was the principal inspiration of the Logical Positivism of the Vienna circle. It is odd that neither William James in *Pragmatism* (1907) nor Sir Freddy Ayer in *The Origins of Pragmatism* (1968) refer to Comte. There are however several discussions of Comte and Positivism in M.Farber (1950). Comte is now perhaps more important in sociology.

45   Karl Popper (1972), p.6.

46   An important logical discussion of the basis of language learning, in terms of considerations of analogies so that words can be applied in different situations, and with reference to Popper's position, is given by Mary Hesse (1974), pp.9–24. There is much more of direct relevance to issues we discuss here, including whether probabilities are objective (as Popper holds) or subjective (essentially Beyesian), and the relation between observations and theories. Mary Hesse argues cogently that there is no fundamental distinction between observational and theoretical statements. Allowing 'theoretical' to extend to myth, I share this view. A danger (which she avoids) is to assume or infer from this that animals (or ourselves) would need *formulated* theories in order to be able to observe or perceive.

   Conversely, perceptions do not have to be formulated in shared symbols or language to be implicit theories, or, as I would rather say, hypotheses. It is an interesting question just how far scientific theories (complex 'molecules' of hypotheses?) must be 'seeable' as potential perceptions. In various forms this question permeates philosophy.

47   Sir Peter Medawar (b. 1915) has been Professor of Zoology at Birmingham, Professor of Comparative Anatomy at London, and Director of the National Institute for Medical Research at Mill Hill, London. He is an outstanding scientist: in 1960 he shared the Nobel Prize with F.M.Burnet for their work on acquired immunological tolerance. He is also a philosopher of humanism. He became a Companion of Honour in 1971 and received the Order of Merit in 1981. His books include the Reith Lectures of 1959, *The Future of Man; Induction and Intuition in Scientific Thought* (1969); and *The Art of the Soluble* (1967).

48   Hermann Bondi (1967), p.9.

49   Sir Alexander Fleming (1881–1955), when Professor of Bacteriology at St Mary's Hospital London, noted that colonies of a common coccus on a culture plate which had become contaminated with a mould, instead of growing as expected, showed signs of dissolution. As this was unusual, the mould was placed in a pure culture, and later identified as *Penicillium notatum*.

Fleming showed that this strongly inhibits the growth of organisms responsible for many infectious diseases. He named it Penicillin. It was later concentrated for medical use.

50   Karl Popper (1959), p.50.
51   Thomas Kuhn (1970), p.14.
52   For an excellent treatment of the subjective, including Beyesian statistics, see Bruno de Finetti (1970). For applications to experimental design, see R.A.Fisher (1935).
53   Thomas Kuhn (1970), p.12. Imre Lakatos and Alan Musgrave (eds.) (1970) is an excellent collection of papers referring to Popper and Kuhn, in which they both contribute.
54   Karl Popper (1972), p.7.
55   Karl Popper (1959), p.251.
56   Karl Popper (1972), p.19.
57   R.C.Lewontin (1974), p.189.
58   Donald Campbell (1974), p.41.

Part 3   Links in Mind
Chapter 9   Links of memory

1   Alexander Pope, *An Essay on Criticism* II.
2   See Richard Sorabji (1972), Chapter 1.
3   Cicero, *De Oratore*, II, cccvi, 351–4.
4   For an interesting appraisal of Bartlett's work on memory see O.L.Zangwill (1972).
5   Cicero, *De Inventione* (1949), Book II, liii, p.160.
6   Quintilian, *Institutio oratoria*, III, iii, 4.
7   This quotation is given by Frances Yates in her superb book, *The Art of Memory* (1976). This discussion of early ways of learning is largely based on her research, and on her fascinating account, which should be read in full.
8   Wilhelm Max Wundt's laboratory was set up in an old building, which has not survived. It was influential not only for its experiments but also for training many distinguished psychologists, from Germany, America and Russia.

   It may be noted that William James set up a psychological laboratory at Harvard in 1875, in the Lawrence Scientific School, where he held a class: 'The Relation between Physiology and Psychology'. In 1877 James added space in Harvard's Museum of Comparative Zoology.

   The first psychological laboratory in Britain was set up by the anthropologist W.H.Rivers (1864–1922); but this did not become effective until its direction by Charles Myers (1843–1901), and its full flowering under Sir Frederic Bartlett, from 1931 to 1969, and Oliver Zangwill, from 1969 to 1981. For a bibliography of Bartlett's writings, see A. D. Harris and O. L. Zangwill (1973).

   In Russia, Vladimir Bekhterev, who studied under Wundt, founded a laboratory at Kazan to investigate learning, psychopathology and alcoholism. There was no officially recognized Institute of Psychology until 1911, at Moscow.
9   F.C.Bartlett (1932), p.3.
10   Ibid., p.4.
11   Ibid., p.5.
12   A.R.Luria (1975), p.17.
13   Ibid., p.26.
14   Ibid., p.27.
15   Ibid., p.33.
16   Ibid., pp.36–7.
17   Ibid., p.56.

18   A.J.Ayer (1972), Chapter 3. Bertrand Russell argues that memory is acquaintance with the past in *The Problems of Philosophy* (1912), where he says (p. 26):

The first extension beyond sense-data to be considered is acquaintance by *memory*. It is obvious that we often remember what we have seen or heard or had otherwise present to our senses, and that in such cases we are still immediately aware of what we remember, in spite of the fact that it appears as past and not present. This immediate knowledge by memory is the source of all our knowledge concerning the past: without it, there could be no knowledge of the past by inference, since we should never know that there was anything past to be inferred.

A few pages later Russell adds: *'Every proposition which we can understand must be composed entirely of constituents with which we are acquainted'* (his italics). This view he modified later on, as he came to deny that there is a Self – following David Hume – in *The Analysis of Mind* (1921). Here he argues that the sense of familiarity is our evidence that we are remembering the past. This again is similar to Hume's position. For a detailed discussion see David Pears, 'Russell's Theory of Memory 1912–1921' (1974), pp. 117 *et seq.*

19   For an excellent recent treatment of experiments on human memory, see A.D. Baddeley (1976).

20   Ivan Petrovich Pavlov studied medicine at St Petersburg and carried out research at Breslau and Leipzig. He became Professor at St Petersburg in 1891, and Director of the Institute of Experimental Medicine in 1913. He was awarded the Nobel Prize in 1904.

21   Vladimir Mikhailovich Bekhterev, as Professor of Neuropsychology at Kazan, investigated neural conduction and mechanisms of reflexes. He founded the Psychoneural Institute at Leningrad.

22   R.A. Hinde and J. Stevenson-Hinde (1973), p. 15.

23   Data on limits of learning are given by M.E.P. Seligman and Joan L. Hager (eds.) (1972). For details on the rapid learning of imprinting, see P.P.G. Bateson (1966).

24   Edward Lee Thorndike studied under William James at Harvard. As Professor of Psychology at Columbia he devised intelligence tests, as well as doing his highly original work on animal learning.

25   J.B. Watson's brilliant career as a psychologist was destroyed while he was still a young man at Johns Hopkins University, due to a love affair with a graduate student, whom he later married. Having founded Behaviourism, shown that experiments on infants could be highly rewarding, and effectively invented behaviour therapy, he left psychology for advertising, in which he was highly successful. Bertrand Russell described Watson as comparable to Aristotle. For his life see David Cohen (1979).

26   Conwy Lloyd Morgan (1852–1936) was a founder, perhaps the founder, of experimental animal behaviour. He was born in London. After extensive travel in North and South America, he became a student of T.H. Huxley, who greatly influenced him, at the Royal School of Mines in London. He finally settled in Bristol, as Professor of Geology and Zoology (1884), and later (1910) Ethics and Psychology. He was the first Vice-Chancellor of the University of Bristol, when it gained full University status, from its foundation as University College. He held this inaugural post for only a year, returning to his Department and research. His Gifford Lectures given at St Andrews University in 1922 and 1923, were published as *Emergent Evolution* (1923) and *Life, Mind and Spirit* (1926). Lloyd Morgan was the first, working in psychology, to be elected, in 1899, into the Royal Society.

He criticized even his hero Darwin, and especially Romanes, for accepting anecdotal evidence of animal behaviour rather than demanding controlled experiments. This was his strength. Later in life he became more interested in philosophy, and especially reconciling a mechanistic view of emergence of novelty with a theistic Universe.

27 C. Lloyd Morgan (1894), p. 53. Morgan's Canon is an interesting special case of Ockham's razor – that we should not generate theoretical postulates beyond necessity.

28 E. R. Hilgard (1958), p. 191.

29 Ibid., p. 470.

30 W. Grey Walter was a genuine pioneer, who made the first impressive life-like machines, with very few active components. He developed circuits of model nerves, and his mechanical tortoise M. Speculatrix, which had a sweeping photocell 'eye' and touch sensors, and was capable of simple learning, was a remarkable achievement, if only because it showed the power of simple models of behaviour – and Grey Walter stressed the simpler the better. This, together with a summary of the work up to that time on EEG in which he was also a major figure, is described (with working circuits) in his still suggestive book, *The Living Brain* (1953).

31 A. M. Uttley (1959), Vol. 1, p. 121.

32 Ibid., p. 142.

33 For an interesting general account of brain changes related to experience, see M. R. Rozenweig, E. L. Bennett and M. C. Diamond (1970).

34 See R. Mark (1974), Chapter 4.

35 Ibid., p. 59.

36 For further accounts of John Young's important work on memory, largely based on experiments on octopus learning, see *The Evolution of Memory* (1976), 'The Organization of a Memory System' (1965), and many experimental papers, in *Proc. Roy. Soc. 'B'*, the *Journal of Experimental Biology,* and elsewhere.

Chapter 10 The nature and nurture of intelligence

1 Francis Galton, *Hereditary Genius* (1869), p. 325. Francis Galton (1822–1911) was cousin of Charles Darwin, Erasmus Darwin being grandfather, with different wives, to them both. Francis Galton was born at Birmingham (the home of the important Lunar Society, of which Erasmus was a prominent member: see R. E. Schofield, 1963). After graduating from Trinity College, Cambridge, Galton explored unknown regions of South Africa. He went on to study colourblindness and mental imagery, and founded eugenics. *Hereditary Genius* is still well worth reading. For Galton's life see his autobiography (1908), the Life by Karl Pearson (1914–30), and the most recent Life, D. W. Forrest (1974).

2 I shall not present a review of the vast literature on intelligence and intelligence-testing, but will, rather, attempt a somewhat novel account. Standard works are J. P. Guilford (1967), L. M. Terman and M. Merrill (1960), P. E. Vernon (1950), D. Weschler (1958), C. E. Spearman (1927) and L. L. Thurstone (1924).

3 Herbert Spencer is today seldom read; but his influence in the nineteenth century was considerable as a voluminous writer on scientific and social matters. He coined the term 'Natural Selection' and he held a systematic evolutionary philosophy, which he extended to Ethics. His arguments were however apriorist; and Darwin with justice said that he could not understand them. His *Principles of Psychology* (1855), the writing of which impaired his health, has a thoroughgoing evolutionary approach, and might repay study.

4 Karl Pearson was Professor of Applied Mathematics and Mechanics at University College London (1884) and first Galton Professor of Eugenics (1911), also at University College. In *The Grammar of Science* (1892), Pearson likens the brain to a telephone exchange. He has interesting discussions on the physics of life. He introduced the notion of statistical normality, which was developed by R. A. Fisher in the 1930s into modern statistical method.

5 Alfred Binet became Director of Physiological Psychology at the Sorbonne in 1892. There is a

Life by R.Martin (1925). See A.Binet and T.Simon (1908), C.Spearman (1937), F.Galton (1883), and H.Spencer (1855).

6 An interesting approach, due to W.H.Hick (1951), is to apply Information Theory (see p. 160) to estimate what I call Kinetic Intelligence. The less the store of Potential Intelligence available, the more processing, decisions and surprise involved; in short, the greater the information transfer required. So far as I know this idea has not been followed up.

7 Social implications, especially concerning alleged race differences, I shall not discuss here, but see Gerald Dworkin (1977).

8 An account stressing the importance of analogies (part of 'Potential Intelligence') is Robert Sternberg (1977); and there are many studies on analogy in problem-solving, for example P.N.Johnson-Laird and P.C.Wason (1977) and J.P.Guilford (1967). For an excellent review of older work on thinking, see G.Humphrey (1951). George Humphrey was the first Professor of Psychology at Oxford.

9 Cognitive perceptual illusions – of distortion, fiction-generating, ambiguity and paradox – have parallels in thinking and belief. Considering paradoxes, there are dramatic examples of perceptual paradoxes such as 'impossible figures' and 'impossible objects': see chapter 12. And there is also cognitive dissonance, as investigated by Jerome Bruner and Leo Postman in their fascinating paper, 'On Perceptual Incongruity: A Paradigm' (1949), and L.Festinger, 'A Theory of Cognitive Dissonance' (1957).

10 For studies of inventive creation see J.Hadamard (1949), and for case studies of practical inventions, J.Jewkes, D.Sawers and R.Stillerman (1958). For psychological studies, see the pioneer Francis Galton (1869 and 1883). Interesting recent work on comparing tests with later success, and special achievement, have been carried out by the British psychologist Liam Hudson, including correlating scientists' degree classes with gaining Fellowship to the Royal Society, finding surprisingly low correlation (1958), and tests for science/art aptitude (1960). In *Contrary Imaginings* (1966), Hudson compares what he calls 'convergers' and 'divergers' at school.

11 Henry Fuseli (real name Johann Heinrich Füssli) was an art critic and a painter of horrific allegories. He introduced the somewhat mystical notions of the Swiss physiognomist Johann Kaspar Lavater (1741–1801) to William Blake. Fuseli's writings were collected by Knowles (1831).

12 Genius may be said to apply beyond extreme powers of thinking, but thinking is certainly central. It could be argued that until we know at least in principle how to program machines for creative thinking we can say little about it; and our present lack of success in this is an indication of the inadequacy of psychological studies and accounts of thinking. This does of course imply that thinking is machine-like; but a fundamental tenet of Artificial Intelligence is that anything that can be stated clearly can be programmed into a computer. For studies of human thinking, see George Humphrey (1951), Allen Newell and Herbert Simon (1972), Jean and George Mandler (eds.) (1964), and P.N.Johnson-Laird and P.C.Wason (eds.) (1977). Wolfgang Kohler's classical study of problem-solving in apes is *The Mentality of Apes* (1925). This describes the 'genius' ape Sultan joining sticks together to get out-of-reach bananas, and social problem-solving in apes.

13 Charles Darwin (1880), p.471.

14 For a collection of key papers on Artificial Intelligence see E.A.Feigenbaum and J.Feldman (eds.) (1963). This includes Alan Turing's prescient paper, 'Computing machinery and intelligence' (1950). A lively and 'personal' history of AI is Pamela McCorduck (1980). The British contribution does not get much of a look-in; but there are interesting quotations from the leading Americans in the field.

15 See James Watson (1968).

16 Sir Laurence Bragg (1890–1971) shared the Nobel Prize in 1915 with his father Sir William Bragg (1862–1942). Sir Laurence was Director of the Royal Institution until 1966.

17 For an excellent general account of current work on Artificial Intelligence see M. Boden (1977); and, for more technical detail of programming, see H. P. Winston (1977).

Chapter 11 Links of Mind

1 William Wordsworth, 'Lines composed a few miles above Tintern Abbey.'

2 Spinoza's *The Treatise on the Correction of the Understanding* was published anonymously in 1670. For accounts of Spinoza's difficult ideas, Bertrand Russell (1946) is very good, and indeed is responsible for the modern interest in Spinoza; and Stuart Hampshire (1962) gives a clear account. He is translated by Andrew Boyle (1959).

3 Immanuel Kant spent his entire life – at Königsberg – so systematically that his neighbours would set their clocks as he passed by on his daily walk. His most important works are *Critique of Pure Reason* (1781), *Prolegomena to any Future Metaphysics* (1783) and *Critique of Practical Reason* (1788). An excellent book on his philosophy is by Stephan Körner (1955). Kant's *Universal Natural History of the Heavens* (1755) is translated by W. Hastie – *Kant's Cosmogony* (1968). This was developed by Laplace (1825).

4 Stephan Körner (1955), pp. 34–5. See also Stephan Körner (1970).

5 John Locke (1690), 'Epistle to the Reader', p. 4.

6 Bertrand Russell, *History of Western Philosophy* (1946), p. 585.

7 John Locke (1690), Book II, Chapter 1, Section 2.

8 Ibid., Chapter 8, Sections 22–6.

9 Ibid., Chapter 9, Section 8.

10 This does not now look so simple. Early cases of recovery from blindness (see M. von Senden, 1960) seemed to show that development of perception is very slow after operations. These operations involved the removal of the lens for cataract; but this leaves the eye optically deficient for a long period, especially for the early primitive operations, which go back to the tenth century. On the other hand cases of restored vision by corneal graft give adequate retinal images almost immediately; and some of these show rapid, and indeed almost immediate, perceptual capacity.

 This was so for S. B., studied by Jean Wallace and myself (1963). We found evidence of transfer to vision of previous knowledge of the world from touch and other senses, and from what had been learned from other people, which was immediately available for his new visual perception. For example, S. B. could tell the time without having to be taught, as he could read the time when blind by touch from the hands of his large pocket watch.

 Cross-modal transfer has been demonstrated in young children by A. D. Milner and P. E. Bryant (1968). So cases of adults born blind (or, for S. B., blind from a few months of age), who are given sight when adult by operation, are much more difficult to use as evidence for innate *v.* learned perception than Locke supposed.

11 John Locke (1690), Book II, Chapter 2.

12 John Locke (1690), Book II, Chapter 1.

13 Ibid., Book IV, Chapter 5, Sections 4 and 5.

14 David Hume (1758), Book I, Section V.

15 Ibid., Book I, Section VI.

16 E. G. Boring (1950), p. 196.

17 David Hartley (1749), p. 521.

18 Isaac Newton (1730), Query 15.

19 Ibid., Query 14.

20 There is a biography of Thomas Reid by Dugald Stewart (1803). For well-chosen extracts from Reid's writings, see A. J. Ayer and Raymond Winch (eds.) (1952).

21 Thomas Reid (1785), Chapter XIV.

22   Ibid., Chapter xvi.
23   Bertrand Russell was son of Viscount Amberly and brother of the second Earl Russell, from
whom he succeeded to the earldom in 1931, upon his brother's death. His father was the eldest
son of the first Earl, John Russell (1792–1878), who was a younger son of the sixth Duke of
Bedford, who was twice Prime Minister. So Bertrand Russell was brought up in one of the leading
families of the country. His protected childhood was spent at Cumberland Lodge, Windsor Great
Park; and he developed as an eighteenth-century, rather than a nineteenth-century, aristocrat.
This perhaps makes his later extreme and often controversial political commitments even more
remarkable.
    Bertrand Russell graduated from Trinity College Cambridge, in mathematics in 1893 and in
Moral Sciences (essentially philosophy) the next year. He was elected a Fellow of his College in
1895, where he gave up an early infatuation with Hegelian Idealism through his logical studies of
Peano and Frege, as well as his researches into Leibniz. His main concern in the first decade was
to derive mathematics from logic, which culminated in *Principia Mathematica* (1910–13) with
A.N. Whitehead, following his *Principles of Mathematics* (1903). Seeing a contradiction in
Frege's work, he developed and used to great effect his Theory of Descriptions, which freed
logicians from supposing that there must be objects in the world corresponding to fictional
propositions.
    Russell's *The Problems of Philosophy* (1912) reflects his early Idealist position, though he
continued to like this book and it has remarkable clarity.
    One cannot consider Bertrand Russell in isolation from Ludwig Wittgenstein, who was his
student. Russell's short Introduction to the *Tractatus* remains important. I have the feeling that
Wittgenstein somewhat put Russell off his stroke, and perhaps this was partly why he gradually
gave up philosophy for politics, late in his life.
    For Russell's life see the *Autobiography* (1967–9), *My Philosophical Development* (1959),
and 'Reply to My Critics' (1946). A full biography is Ronald Clark (1975). Russell was elected
a Fellow of the Royal Society, and awarded the Order of Merit in 1949, and the Nobel Prize for
Literature in 1950.
24   Bertrand Russell (1914), p.75.
25   Bertrand Russell (1967), p.28.
26   A.J. Ayer (1972), p.72.
27   Bertrand Russell (1940), p.126. An invaluable collection of Russell's technical papers is Robert
C. Marsh (ed.) (1956). See also Bertrand Russell (1959 and 1967–9).
28   This letter is quoted by D. Shakow and D. Rapaport (1968), p.286.
29   John Hughlings Jackson became physician at the London Hospital, and later at the National
Hospital for the Paralysed. He established relations between brain regions and limb movements,
and developed the fundamental notion of a pyramidal hierarchy of brain centres, only the highest
centres being associated with consciousness and voluntary control. His work on speech and
aphasia was also of first importance. He was elected a Fellow of the Royal Society in 1878.

Chapter 12   Links of brain

1   Plato, *Theaitetos* (1961), p.78, expressing Protagoras' view; and Aristotle, *Analytica Posteriora*
(1974), 876, p.222, expressing his own view.
2   The retinal image is just one of many 'cross-sections' of the visual pathway. There is no more
reason to see the retinal image as a picture (upside-down and laterally inverted) than there is to
see action potentials of the optic nerve. It is believed that all species have elaborate anatomical
arrangements for cortical mapping, such that all 'primary' maps in the central nervous system are
upside-down but not laterally inverted with respect to the world as imaged on the retina: see John

Scholes (1981). This applies to compound eyes, whose images are not inverted; to the octopus eye, in which the optic nerve fibres come straightforwardly from the back of the retina as the receptors lie on its front surface; and to vertebrates, including humans, in which the receptors of the eye are 'irrationally' at the back, so that their fibres have to cross the retina from the periphery to the 'blind spot' where the fibres leave the retina. The reason for upside-down and laterally inverted brain maps is not clear, but may serve to make relating visual to touch information simpler. For a thoughtful discussion see Alan Cowey (1979).

3   Unconscious Inference and judgements were unpalatable to philosophers concerned with Ethics, for it was (and no doubt still is) accepted that *awareness* of reasons for action is necessary for morality. Traditionally philosophers combine an interest in epistemology with Ethics, so there is a professional conflict here, which does not seem to be adequately discussed.

4   Hermann von Helmholtz (first pub. 1867) (1962), section 26.

5   Richard and Roselyn Warren (1968), p. 18.

6   Ivan Pavlov (1928), pp. 126–7.

7   Hermann von Helmholtz (1962), p. 152.

8   Ibid., pp. 152–3.

9   Ibid., pp. 153–4.

10   Ernst Mach (1959), pp. 214–15.

11   The notion of 'channels' in the visual system – channels for orientation, movement, position, texture, spatial frequencies, and colour – is however highly important in current neuro-physiological accounts of visual perception. These 'channels' (which may not have anatomical but only functional criteria) can be separately adapted by maintained stimuli. They may represent dedicated analysing systems.

12   For the key Gestalt papers translated into English see W. D. Ellis (ed.) (1938).

13   Ibid., p. 550.

14   K. Koffka (1935), p. 56.

15   Ibid., pp. 32–3.

16   In my view there is a deep confusion here through failure to distinguish between energy and information stability. One might say that reaching a conclusion is attaining an informational stability; but this has almost nothing to do with the energy states of the computing mechanisms responsible for the decision. There may be occasional exceptions for special kinds of analogue computer dedicated to a highly restricted class of problems. But clearly vision allows and requires an enormous range of 'solutions'; and in general biasing by energy considerations of brain states would be irrelevant, except for very special cases. The more 'general-purpose' the computer, the more it must be free of energy considerations – except indeed for giving stability to its (digital) steps for computing, and the states of its memory stores. Gestalt tendencies towards solutions through reduction of potential energy of isomorphic brain traces implies that very general tendencies following energy paths will be usually appropriate for finding solutions; but clearly the more general-purpose (the greater the range of situations they are applied to), the less appropriate they tend to become – unless, indeed, they reflect almost universal structures of the problem-space, as for general theories in physics; but this is hardly applicable here, for formulated theories are essentially 'digital' and not analogue or 'holistic': they show how logical steps and calculations can be used for a variety of situations and they can readily be adapted for each particular situation, which is not so for holistic models.

Generalizations such that converging lines are often receding perspective, and generalizations of objects such that they are usually simple and closed in form and their parts move together, are, however, useful. Generalizations of the world as sensed can be used as powerful aids to object perception because they set restraints for what is likely; but this is quite different from the Gestalt properties of brain traces which are internal restraints. It is however true that many useful generalizations of object characteristics such as closure of object forms, and 'Common Fate' of parts of moving objects were suggested by the very Gestalt psychologists who held holistic accounts of brain traces.

17 Kenneth J.W.Craik (1943), p.51. Kenneth Craik was a pioneer in giving machine analogy explanations of neural function, both for visual perception and limb control, and was among the first to emphasize the importance of stored knowledge for giving the richness of perception and predictive control of behaviour– from what he called 'Internal Models'. Much of his experimental work was devoted to human visual processes, especially dark adaptation. It is entirely appropriate that a laboratory newly founded in Cambridge is named the Craik Laboratory. For Craik's life and an assessment of his work, see O.L.Zangwill (1980).

18 D.O.Hebb (1949), p.60.

19 Gibson uses the word 'active' in this sense especially in Chapter 7 of *The Perception of the Visual World* (1950).

20 J.J.Gibson (1966), p.5.

21 Ibid., p.226.

22 Ibid., p.258.

23 Gibson was forced to retract somewhat on his rejection of retinal images, largely through pressure from R.M.Boynton (1974) – as we see in Gibson (1974). I have attacked this notion, and compared the merits and snags of theories of perception, by putting up 'paradigm tests': see R.L.Gregory (1974).

24 A research explosion has followed the discovery by David Hubel and Torstin Wiesel (1962 and 1968) of 'feature detectors' in the striate cortex of the cat, and later of the monkey. Several electrophysiologists, including Colin Blakemore, have reared kittens in restricted visual environments such as environments of only horizontal, or only vertical, stripes, and have found that the receptive fields are mainly absent when there has been an early lack of appropriate stimulation (there is an absence of horizontal detectors when only vertical stripes have been visible, and so on). It is a matter for controversy whether these differences are due to a lack of development of feature detectors in the absence of stimulation, or whether they do develop normally but then atrophy. Reviews of these important issues include H.B.Barlow (1975) and R.M.Boynton (1974).

25 Terry Winograd (1972), p.1.

26 Terry Winograd did this outstandingly successful work as a PhD student at MIT. The quotations all come from Terry Winograd (1972), pp.1 and 3.

27 N.R.Hanson (1958), p.71.

28 Ibid., pp.71–2.

29 Ibid., p.72.

30 Ibid., p.5.

31 Ibid., pp.18–19.

32 Ibid., p.6. I do wish that Hanson did not refer here to the retinal receptors (rods and cones) as 'selenium cells'!

33 Ibid., p.6.

34 Ibid., p.7–8.

35 Ibid., p.8. Spontaneous reversals of visual depth were first described by a Swiss crystallographer, L.A.Necker, in 1832. Necker was drawing rhomboid crystals using his microscope, and noticed that the drawings suddenly looked different from the crystals seen in the microscope. He realized with a shock that it 'was his perception, and not the crystals or the drawings, that had changed.

36 Ibid., p.9.

37   Ibid., p.11.
38   Ibid., pp.12–13.
39   Ibid., p.13.
40   Ibid., p.19.
41   The quotations which follow are taken from Ludwig Wittgenstein (1954), pp.193–214. See the bibliography for a full list of Wittgenstein's works. A fascinating short life is Norman Malcolm (1958). For the intellectual and social background, see Allan Janik and Steven Toulmin (1973). Among the enormous number of books on his philosophy are David Pears (1971), Anthony Kenny (1973) and G.E.M.Anscombe (1959). A collection of papers on particularly relevant questions is edited by Harold Morick (1967).
42   The outstanding classical work on visual ambiguity of figures is due to the Danish psychologist Edgar Rubin, who worked in the Phenomenalist tradition. He emphasized the importance of selecting the 'figure' from the 'ground', and he devised many now famous examples of figure–ground ambiguity. His writings are not available in English except for selections in D.C.Beardslee and D.C.Westheimer (eds.) (1958). There is a discussion with examples in R.L.Gregory (1970), pp.15–18, where I quote from Rubin, showing his liveliness and humour:

If a figure looks like a beloved and admired professor from his homeland, this may remind the subject of the pleasure in having met him again as he stopped on his way to Gottingen. If a figure looks like a beautiful female torso, this indubitably calls forth certain feelings. When one succeeds in experiencing as figure areas which are intended as ground, one can sometimes see that they constitute aesthetically displeasing forms. If one has the misfortune in pictures of the Sistine Madonna to see the background as figure, one will see a remarkable lobster claw grasping Saint Barbara, and another odd pincer-like instrument seizing the holy sexton.

It is an important question how far picture and object ambiguities are normally resolved by (1) perceptual hypotheses accepted without challenge, or (2) by general, high-level object knowledge (of Madonnas' torsos, lobsters, or whatever), or by characteristic shapes of objects such as the closure and 'Common Fate' criteria stressed by the Gestalt psychologists (Kohler, 1920). This is important for AI scene analysis and object recognition (Guzman, 1971).

43   It may be added that, when studying the case of S.B., who gained his sight by operation when middle-aged after effective blindness since infancy, we found that he did not experience perceptual ambiguities for Necker cubes; but unfortunately we did not test for figure–ground ambiguity: see R.L.Gregory and J.Wallace (1963).
44   Where I say 'the spontaneous switches of perception must be neurally centrally determined' rather than set by sensory signals or stimuli, I am assuming a complete neurological basis for perception; but this is questioned by William James (1890), Vol.1, Chapter 9, and by Popper and Eccles (1977), pp.522–6, who suggest that mental forces have some control over neurological processes of perception. We discuss this later, on pp.468–72.
45   Thomas Kuhn (1970), p.114.
46   J.J.Gibson (1966), pp.247–8.
47   I shall not go into details of current experiments or cognitive theories of perception, if only because I have written at considerable length on these topics elsewhere (1966, 1970 and 1974). To be recommended are Ulrich Neisser (1966), H.W.Leibowitz (1965), J.E.Hochberg (1964), and R.N.Haber (ed.) (1968). The view of perception as essentially like hypotheses is published, in similar form, in R.L.Gregory (1980).
48   Object recognition by computers is discussed on pp. 321–32, 337, 377–82, and 402.
49   The impossible triangle drawing, and the impossible staircase, were invented and drawn by L.S. and R.Penrose (1958). It is unfortunate that they described these figures as 'impossible objects',

for, as we showed (R.L.Gregory, 1970), it is entirely possible to make simple three-dimensional objects which appear impossible from restricted points of view – including the impossible triangle. The fact that objects, and not only drawings, can appear impossible is significant, for the observer is selecting all the visual features from objects, while an artist is free to create paradoxes by combining, as he will, incompatible features. Fascinating use has been made of these effects by the Swiss artist Maurits Escher, as discussed with examples by Marianne L.Teuber (1974). Implications of impossible figures for computer scene analysis has been worked out by D.A.Huffman (1971), who compares them with nonsense sentences.

50   Gaetano Kanizsa is Professor of Psychology at Trieste, and is also a painter. He approaches perception more in the spirit of central European than American or British psychology: like Edgar Rubin he is concerned with the experience of perception rather than its use, or its practical limitations, or its place in a theory of knowledge. Kanizsa is cautious over explanations, and makes little use of controlled experimental situations or statistics. His strength is examining his own experience. This has an honourable tradition and has produced important discoveries. Indeed there is a danger of being too concerned with behaviourally measurable features of perception, which can reject phenomena whose physiology or behavioural significance is not clear; and it can make the study of perception opaque to painters, who are highly concerned with visual phenomena *per se*. In *Organization in Vision* (1979) Kanizsa presents his view of perception with superb original illustrations, and discussions which cast a different illumination from British Empiricism; it is important and delightful.

51   If we suppose that perceptions are hypotheses – and that perceptual hypotheses may be selected according to more-or-less random jumps – then we have lost the clear 'directness' of Deductive inference, and the 'directness' of Induction by extrapolation and analogy following the Methods of Induction. So the hypothesis account of perception which we are advocating is an extreme version of indirect perception.

52   The famous Ames room is non-rectangular, but appears as a normal rectangular room. This is so from a critical viewing position. Logically, this is like a picture of a room; though the room is not flat, but is some intermediate shape between a rectangular room and its plane picture. There can be an infinite set of such 'rooms', all giving the same retinal image, and so all appearing alike – like a rectangular room. Actually the principle was clearly stated by Helmholtz in 1866 (Vol. III, section 26):

But even when we look around a room . . . flooded with sunlight, a little reflection shows us that under these conditions too large a part of our perceptual image may be due to factors of memory and experience. The fact that we are accustomed to the perspective distortions of pictures of parallelopipeds and to the form of the shadows they cast has much to do with the estimation of the shape and dimensions of the room, as we shall see hereafter. Looking at the room with one eye shut, we think we see it distinctly and definitely as with both eyes. And yet we should get exactly the same view in case every point in the room were shifted arbitrarily to a different distance from the eye, provided they all remained on the same line of sight.

This is exactly the Ames room (W.H.Ittelson, 1952), built over eighty years later. Whether Helmholtz actually constructed such a room I do not know. In this passage, he puts his finger on a central problem of vision: why do we usually settle for just one out of the infinity of objects that could produce a given retinal image? (Is this similar to, why do we usually settle for one out of an infinity of possible hypotheses in science? I think it is. Helmholtz also puts his finger here on a vast class of illusions.)

For anthropological data on perception in people not familiar with rectangular rooms, see M.H.Segall et al. (1966).

53 I refer to these suggestions by Popper and Eccles, and by James, in a somewhat different context on p. 470 *et seq.*, from which the full references may be found in these notes.
54 For a further discussion of consciousness see Part 4.

Chapter 13 Meaning to say

1 James Boswell, *The Life of Samuel Johnson LL.D.*, April 1776.
2 A comparative study of ape and child development, when they are brought up together, is W.N. and L.A. Kellog (1933), showing superiority in the ape up to language in the child, when it shoots ahead. There is however evidence of evolution of culture in animals: see J.T. Bonner (1980). See also E.H. Lenneberg (1967).
3 It is possible and perhaps wise to be even more cautious about claims of animal language ability. There are so many experimental difficulties and possibilities of the animals picking up clues from the experimenters, given unwittingly, that extreme caution is essential. Also, the inevitable investment of time and patience does tend to some exaggerated claims, which unfortunately may have occurred in this research: it is all too seductive in its promise of our being able to communicate to ancestral species.
4 J.Z. Young discussed language and communication in his Reith lectures (1947) and recently in *Programs of the Brain* (1978), p. 176. He here defines language, biologically, as 'any species specific system of intentional communication between individuals'. He describes language as involving the encoding of desired messages by selecting appropriate items from a mutually known set of signs. This definition covers all species, but as Young points out, what is special about human speech is its ability to transmit a seemingly infinite set of messages on any subject. He stresses the importance of intention in language, and how confusing a problem it is. Margaret Boden (1972a) is useful, and Dan Dennett (1978).
   Useful books and collections of papers are Roger Brown (1958 and 1970); Jerry Fodor (1968 and 1975); J.J. Katz (1966 and 1972); B.F. Skinner (1957), who gives a Behaviourist account of language which is now hotly attacked; John Lyons (1970) for a clear account of Chomsky's highly influential theory of language; Shirley Weitz (ed.) (1974) on non-verbal communication; G.A. Miller and P.N. Johnson-Laird (1976); David Diringer (1966) for the best history of written languages; and G.A. Miller *et al.* (1960) on programmed language acquisition.
5 Charles Saunders Peirce is surely among the top half dozen modern philosophers. He is thoroughly worth reading. His original papers in which the bases of Pragmatism were set out were published in the highly obscure American journal, *Popular Science Monthly*, in 1878. They are reprinted in Philip Weiner (ed.) (1958). The historical and philosophical basis of Pragmatism is fully discussed by A.J. Ayer (1968). For the full collected works, see C. Hartshorne and P. Weiss (eds.) (1931–58). For commentaries, see W.B. Gallie (1952) and J. Feibleman (1946).
6 William James's famous squirrel example is on pp. 43–5 of *Pragmatism* (1907).
7 There is an excellent critical study of the Oxford Idealist philosopher Francis Herbert Bradley (1846–1924) by the London philosopher Richard Wolheim (1959).
8 These quotations are taken from A.J. Ayer, *Language, Truth and Logic* (1946 ed.); here, p. 47.
9 Ibid., p. 48.
10 Ibid., pp. 48–9.
11 Ibid., p. 50.
12 Ibid., p. 13.
13 Ibid., p. 13.
14 Ibid., p. 52.
15 Ibid., p. 17.

16   Ludwig Wittgenstein (1953), p. 196.
17   Ibid.
18   This is discussed by G.E.M.Anscombe (1959), especially in Chapter 4.
19   Ludwig Wittgenstein (1969), p. 17.

Chapter 14   Meaning to dream

1   William Shakespeare, *The Tempest,* IV, i, 156.
2   *Epic of Gilgamesh* (1972), pp. 92–3.
3   Much of interest on the history of dreams will be found in R.L. van de Castle (1971); an extract of this, with useful references, is in S.G.M.Lee and A.R.Mayes (1973), pp. 17–32.
4   Plato, *The Republic* (1941), Book IX, p. 290.
5   These particularly interesting discussions on dreams are in three sections of Aristotle's *Parva Naturalia* – 'On Sleep and Waking', 'On Dreams' and 'On Prophecy in Sleep'.
6   Aristotle, *Parva Naturalia* (1957), Vol. VIII, 'On Prophecy in Sleep', pp. 377–9.
7   The works of Artemidorus were translated by Wood in 1644. I have not consulted this – only secondary sources.
8   Bergson's theory is given in *Revue Scientifique*, 8 June 1901. The English translation is by J.A.Hadfield. For a present-day philosopher on dreams see Norman Malcolm (1962).
9   Wilder Penfield and L.Roberts (1959), Chapter 3.
10   Ernest Jones (1953–7), p. 225.
11   Sigmund Freud (1900), Chapter VII, B.
12   Thomas Hobbes, *Leviathan* (1651), Part 1, Chapter 2.
13   Sigmund Freud (1900), Chapter VII, B.
14   An especially interesting collection of comments on Freud is Richard Wolheim (ed.) (1977), which includes the remarkable 'Conversations on Freud' by Wittgenstein. Wittgenstein suggests that we may feel psychology to be unsatisfactory because we take physics as our ideal science, with its formulated laws, but for psychology we cannot use the same sort of 'metric' – and cause hardly applies to Mind because there are not corresponding mental experiments. Wittgenstein points out that Freud's notions of causes in dreams seem odd because 'Freud never knows how we know when to stop – or indicate where is the right solution. Sometimes he says that the right solution, or the right analysis, is the one which satisfied the patient. Sometimes he says that the doctor knows what the right solution or analysis of the dream is whereas the patient doesn't: "the doctor can say that the patient is wrong".' For Wittgenstein the decision is not on evidence, but is a speculation only. So, for him, Freudian psychology (and perhaps all psychology) is myth rather than science.

 Wittgenstein goes on to consider whether dreams are a language, or in what ways they are like a language – concluding that dreams are somewhat like doodles and equally hard to understand or read. Wittgenstein suggests that Freud sought as explanations simple *essences* of things – including essences of dreams – and that this (Aristotelian) kind of explanation lacks causes, as in functional accounts. For Wittgenstein, Freudian psychology is not science, though it may be interesting and important. He does not however seem to have studied Freud in depth.
15   Carl Jung (1964), p. 67. Jung developed word association as a probing tool for exploring mental structures and revealing what Freud called 'complexes' or blocks to recall. Jung's *Studies in Word Association* (1918) is impressive.
16   Sigmund Freud (1900), p. 215.
17   Ibid.
18   See F.Snyder (1966).

Part 4   Regarding consciousness

Chapter 15   What is consciousness?

1   Julian Jaynes (1976), p.21.

2   William Shakespeare, *Much Ado About Nothing*, v, i, 35.

3   This is the most difficult set of issues for discussing Mind in Science, for we do not normally consider science as conscious; and human consciousness (or rather our own consciousness) appears unique and extremely difficult to discuss in the familiar terms of science. On the other hand, though, and most curiously, almost all, if not quite all, accounts of consciousness are in terms of analogies to physical objects and states, such as William James's reference (1890), in Chapter 9, to the 'stream' of consciousness and thought, and the moving light analogy used in these pages.

There are many attempts to find close relationships between conscious states of awareness and physical brain activity, including E.G.Boring (1933) and P.A.Buser and A.R.Rougeul-Buser (eds.) (1977). These are in many ways highly suggestive but of course leave open the nature of the link (or identity) between brain states and consciousness. Then there are accounts of consciousness as being Emergent in some ways, such as C.D.Broad (1923) discusses philosophically; and the brain surgeon and neurophysiologist Wilder Penfield (1975) came to believe. Sir John Eccles holds a Dualistic position. A particularly useful collection of papers is S.Hook (ed.) (1960).

Emphasis on the two hemispheres as representing two kinds of consciousness has been most fully expressed by Robert Ornstein (1972), and by several writers in a useful collection of Ornstein's (1973). G.G.Globus *et al.* (1976) and the biologist Steven Rose (1973) also consider the physical basis of consciousness; and Godfrey Vesey (1964) has produced a useful collection of classical writings. A recent symposium is B.D.Josephson and V.S.Ramachandran (eds.) (1980).

4   How to speak about consciousness is discussed in a long series of papers originally published in the journal *Mind*, and collected in book form by John Wisdom (1952). John Wisdom is a disciple of Wittgenstein, and became Professor of Philosophy at Cambridge, where I attended his truly memorable lectures for two years.

5   See L. Weiskrantz *et al.* (1974), pp.707–8. See also A.Cowey (1979). For technical neurological accounts of cortical blindness, see I.Bodis-Wollner (1977) and I.Bodis-Wollner *et al.* (1977).

Chapter 16   How are Mind and brain linked?

1   From a postscript to a letter to Basnage de Beauval, written in January 1696, in L.F.Loemker (ed.), 1957. It is interesting that Leibniz tried to invent a clock. He also became somewhat involved with an unfortunate priority dispute over the invention of the hair spring for balances. Leibniz's notion was to transfer energy through a succession of springs; but it was never made (see J.E.Hoffman, 1974, Chapter 9). He was of course interested in mechanisms, and designed a calculating machine, which he demonstrated with moderate success at the Royal Society. The Leibniz cylinder was a feature of nineteenth-century calculating machines. He was a pioneer in regarding mathematics and logic as amenable to mechanization. This is an example of the close connection between technology and the most abstract concepts (see fig. 8, p.59).

2   The dreams are described by A.Baillet in *La Vie de Monsieur Des Cartes* (1691), and have been reconstructed from fragments and translated by L.A.Lafleur (1961).

3   'Rules for the Direction of the Mind' are in E.S.Haldane and G.R.T.Ross (transl.) (1967), Vol. I, pp.1–77.

4   René Descartes, *Meditations on First Philosophy* (1967), Meditation ii, p.149.

5   René Descartes, *Principia Philosophiae* (1967), Part 4, Principle 179.

6   Anthony Kenny (1968), p.218.

7   René Descartes, *The Passions of the Soul* (1967), Part i, Article 31, p.345.

8   Ibid., Article 32.

9   René Descartes, *Discourse on Method* (1967), Part 5.
10 William Harvey was physician to King James I and Charles I. As Lumleian Lecturer in the College of Physicians, he announced his radical conclusions in public. He studied at Caius College Cambridge, and at Padua, and became physician at St Bartholomew's Hospital. It is remarkable that one of his professional standing should risk his position by challenging a basic tenet of medicine and physiology. This he did, though blood vessels linking the venous and arterial systems, without which circulation would be impossible, were not to be seen. Their apparent absence was strong evidence against – indeed almost a refutation of – his theory, and yet he pushed it, and won the day. *De Motu Cordis* and *De Circulatione Sanguinis* (1628) are available in English (1963).
11 William James (1890), Vol. I, p.434.
12 Ibid., p.447.
13 Ibid., p.453.
14 Ibid., p.454.
15 Karl Popper and John Eccles (1977), p.514.
16 Wilder Penfield and L. Roberts (1959), p.51.
17 Ibid., p.52.
18 Ibid., p.53.
19 George John Romanes (1885), from G. Vesey (ed.) (1963), p.183. Romanes was a highly cultivated, charming man, and a neighbour and good friend to Darwin. He devoted his life to his estate and to science, being especially interested in the evolution of Mind. His Life and Letters are edited by his wife (1896).
20 Essential papers on the identity theory are in the collection edited by C. V. Borst (1970). See also D. M. Armstrong (1968), and J. J. C. Smart (1963).

Part 5 Mind in us
Chapter 17 The Self

1   Bertrand Russell (1921), p.41.
2   Lewis Carroll, *Alice in Wonderland*.
3   The best discussion, surely, of functional accounts of Mind are those of the American philosopher Hilary Putnam (1975). Here are also important discussions of Self, and of Minds and machines.
4   The quotations which follow are taken from Sigmund Freud (1900), Chapter 7, B, pp.684–92.
5   John Locke (1690), Chapter 27.
6   Norman Kemp Smith (1941), p.499.
7   David Hume (1740), Book II, Part III, Section I.
8   Bernard Williams, Professor of Philosophy at Cambridge, develops this kind of argument in detail and with interesting variations in *Problems of the Self* (1973). For a discussion of arguments of this kind by a neurophysiologist, see V. S. Ramachandran (1980). Further arguments are in Amélie Rorty (ed.) (1976).
9   See note 8. The 1978 reprint of Morton Prince (1905) has an illuminating introduction by the psychiatrist Charles Rycroft, who relates phenomena of dissociation of personality to hysteria in his *Anxiety and Neurosis* (1968), p.123.
10 William McDougall (1926), p.548.
11 The major collection of reported experiences of telepathy and communications from the deceased remains the classical work by F. W. H. Myers (1915).
12 William McDougall (1926), p.550.
13 Ibid., p.552.
14 Margaret Boden (1972), p.256.

15   Michael Gazzaniga (1967) is reprinted in a useful collection edited by R.C.Atkinson and
     W.H.Freeman (1971), pp.167–72.
16   Donald MacKay (1980), p.110.
17   Horace Barlow (1980), p.84.
18   See the first three chapters of Martin Gardner (1964). Kant discusses right–left reversal in
     Prolegomena to any Future Metaphysics (1783), Part 1, section 13. He uses the argument that a
     hand and its mirror image are not congruent, to try to deny that 'space and time are actual qualities
     inherent in things in themselves', and to suggest that they are rather 'mere forms of our sensuous
     intuition', and unfounded. This argument is surely incorrect: see J.Bennett (1970). Plato does not
     appreciate mirror reversal either: see *Timaeus,* section 14.

Chapter 18   Science in Mind

1   Norman Campbell (1921), p.110.
2   This question of illusory changes of sensed weight, brightness, or whatever, does seem to me
    both remarkably interesting and surprisingly under-considered. The experiments reported by
    Helen Ross and myself in 1964 and 1970 were carried out by Helen Ross for her Ph.D. thesis
    under my supervision at Cambridge, and she went on to further experiments (see also Gregory,
    1974, pp.228–52). Half-way through, we changed our minds from the notion that discrimina-
    tion may be a monotonic function of apparent rather than physical stimulus magnitudes. To test
    this, we separated scale weight from apparent weight of objects to be lifted, by using the
    size–weight illusion in which smaller objects are sensed and judged heavier than larger objects of
    the same scale weight. We expected discrimination to be worse for small than for larger weights
    of the same scale weight, as they feel heavier. At first this seemed true; but then a very light
    material became available – expanded polystyrene. With this, we made large cuboids which felt
    amazingly light when lifted; we matched these for weight with sets of much smaller tins. All the
    weights were fitted with identical wire loops for lifting them; so the weight signals to the hands
    were the same for all, but there were large differences in the apparent weights. With these large
    differences we found that weight discrimination, surprisingly, became worse for heavier or lighter
    objects and was best for a density of about unity (the density of water, or of the arm, and roughly
    the average density of objects). This suggested that even simple signals such as signals of weight
    are matched against internal standards based on expectation by comparators somewhat like
    Wheatstone bridges, which give best discrimination when their internal standards (or the internal
    arm of the bridge) are set nearly correctly to the sensed value.

    Kenneth Craik had several years before found that visual intensity discrimination is best when
    the eye is adapted to the background intensity, from which it is made. This may be a special
    feature of the eye, or it may be another example of matching against a standard with a bridge-type
    circuit, which may be a very general principle for the nervous system. It implies, however, that
    we must know the internal references in order to appreciate fully the effects of even simple
    stimuli. I do just wonder whether intractable pain might sometimes be due to mis-settings or
    failure to match such internal standards.
3   Ernst Mach held that this is so for any measurement in science, for he held that all 'objective'
    measures must ultimately be reducible to experience. See Mach, *The Science of Mechanics*
    (1883). Mach saw physics and psychology (or at least sensory and perceptual psychology) as
    ultimately not to be distinguished from physics. He calls this 'the complete parallelism of the
    psychical and physical' with 'no gulf between the two provinces': see Ernst Mach (1960), p.60.
    He is explicit in rejecting Fechner's Dualist conception of the physical and psychical as 'two
    aspects of the one and same reality', for which Mach claims that his view, unlike Fechner's, has
    no 'metaphysical background', and that the 'elements given in experience are always the same
    . . . of one nature', though they 'appear, according to the nature of the connection, at one moment

as physical and at another as psychical elements'. However, why this is not metaphysical seems far from clear.

Mach's emphasis on sensation as the basis of science is a precursor of Operationalism and Logical Positivism, and he quite directly affected the thinking of scientists, including Einstein, although he never made his position sufficiently clear. A useful critical discussion is Peter Alexander (1963), especially Chapters 1 and 2. Also relevant is a recent paper by Roger Shepard (1979). It might be argued that Mach's attempt to identify perception with what is perceived in the physical world is a precursor of Heisenberg's view (1930, 1958 and 1959); but this is not a close connection.

4   Ernst H.Weber (1978), p.132.
5   Ibid., p.132.
6   G.S.Brindley (1970), p.132.
7   Ibid., p.133.
8   For excellent discussions on sensation see J.L.Austin (1962) and D.O.Dennett (1978), Part 3.
9   C.Wade Savage (1970), p.197.
10   Ibid., p.197. An interesting technical paper on the effects of context is M.H.Birnbaum (1974).
11   S.S.Stevens's work on scaling is brought together in Geraldine Stevens (ed.) (1975).
12   For data based on many kinds of psychophysical measurements see S.S.Stevens (1951). For a clear general account of problems of measurement, see Brian Ellis (1968); and for a collection of papers, C.W.Churchman and P.Ratoosch (eds.) (1959).

The issues raised and the experimental difficulties in perceptual scaling are very great, and there is no consensus of opinion of their significance or status. For technical papers on these problems see E.C.Poulton (1968 and 1979), M.H.Birnbaum (1974), M.Treisman (1964), and R.M. and R.P.Warren (1963).

Part 6   Mind in Science
Chapter 19   The nurture of physics
1   Henry David Thoreau, *Walden*, 1854.
2   Jules Henri Poincaré (1914), pp.285–6.
3   Pierre Laplace (1951), p.4. See also Laplace (1823–39), Vol. III, p.xiii.
4   Einstein's papers on Relativity, together with related papers by H.A.Lorentz, M.Hinkowski and H.Weyl are in *The Principle of Relativity* (1923). Superb papers of exposition and comment are in P.A.Schilp (1949). There are countless more or less technical expositions, including Ernst Cassirer (1953), D.W.Sciama (1969), and, for a more popular and highly readable account, Martin Gardner (1976). Three Lives are Philip Frank (1947), Ronald W.Clark (1971), and Jeremy Bernstein (1973).
5   Poincaré delivered this paper in St Louis. It is quoted by Jeremy Bernstein (1973), p.75.
6   D.W.Sciama (1969), p.4.
7   Sir Arthur Eddington gives a fascinating argument in the form of a discussion between an experimental physicist, a pure mathematician, and a Relativist, on the geometry of space (1920), pp.1–16. On p.10 we find this exchange:

Phys. Apart from measures, we have a general perception of space, and the space we perceive is at least approximately Euclidean.
Rel. Our perceptions are crude measures. It is true that our perception of space is very largely a matter of optical measures with the eyes. If in a strong gravitation field optical and mechanical measures diverged, we should have to make up our minds which was the preferable standard. . . . So far as we can ascertain, however, they agree in all circumstances, and no such difficulty

arises. So, if physical measures give us a non-Euclidean space, the space of perception will be non-Euclidean. If you were transplanted into an extremely intense gravitational field, you would directly perceive the non-Euclidean properties of space.

Phys. Non-Euclidean space seems contrary to reason.

Math. It is not contrary to reason, but contrary to common experience, which is a very different thing, since experience is very limited.

Phys. I cannot imagine myself perceiving non-Euclidean space!

Math. Look at the reflection of the room in a polished door knob, and imagine yourself one of the actors in what you see going on there.

Later, on p. 15, Eddington's characters discuss the relation between the physics of Relativity, space and perception, in these terms:

Rel. The term dimension seems to be associated with relations of *order*. I believe that the order of events in nature is one indissoluble four-dimensional order. We may split it arbitrarily into space and time, just as we can split the order of space into length, breadth and thickness. But space without time is as incomplete as a surface without thickness.

Math. Do you argue that the real world behind the phenomena is four-dimensional?

Rel. I think that in the real world there must be a set of entities related to one another in a four-dimensional order, and that these are the basis of the perceptual world so far as it is yet explored by physics. But it is possible to pick out as four-dimensional set of entities from a basal world of five dimensions, or even of three dimensions. The straight lines in a three-dimensional space form a four-dimensional set of entities, i.e. have a fourfold order. So one cannot predict the ultimate number of dimensions in the world – if indeed the expression *dimensions* is applicable.

The philosopher concerned with psychology is advised to view Relativity in this way:

Rel. In so far as he is a psychologist our results must concern him. Perception is a kind of crude physical measurement; and perceptual space and time is the same as the measured space and time, which is the subject-matter of natural geometry.

8   Richard Feynman (1963), Vol. 1, pp. 5–15.
9   J. H. Poincaré (1914), p. 115.
10   For a sophisticated discussion of the 'clock paradoxes', see Alfred Schild (1959).
11   Arthur Eddington (1959), Chapter 4.
12   Hecht, Schlaer and Pirenne's experiment is based on the steepness of 'frequency of seeing' curves derived from reporting the occurrence of small dim flashes of light. The curves follow a Poisson distribution (integral of the Gaussian normal distribution), which indicates, here, the numbers of effective photons in each flash, since the photon distribution is Gaussian, and the steeper the integral function, the greater the number of photons required for detection. The experiment is such a beautiful example of physics applied to vision that T. N. Cornsweet (1970) takes it as the model that other experiments in perception should follow – roughly that perception should be reduced to physics.
13   Heisenberg's philosophy of physics is examined usefully by Patrick Heelan (1965). The most recent discussion, with a novel approach, is by the British theoretical physicist David Bohm (1980). David Bohm's basic idea is to extend the familiar experience of the very different viewpoints to saying that experiments and general theories such as Relativity and Quantum Theory reveal different aspects of the same underlying and not directly observable reality, is not, and perhaps cannot ever be verified in detail. So it does not seem clear how consistency tests can be applied in theoretical physics. Perhaps, ultimately, physics is unverifiable.

Some theoretical physicists have become interested in alleged psychokinetic and other 'paranormal' phenomena, considering these from the standpoint of Quantum Theory. A series of papers along these lines is A. Puharich (ed.) (1979). Perhaps of these R. D. Mattuck is the most relevant. Whatever the evidence, it is surely interesting to see how apparent violations of physics could be fitted into present or some imaginable future physics. What, after all, is empirically impossible? A great deal of what we now believe true would have been rejected as impossible, absurd and unimaginable only a few years ago.

The Heisenberg quotations are taken from J. R. Newman (1956), pp. 1051–2.

14   Bernard d'Espagnat (1976), p. 264.
15   John Archibald Wheeler (1980), p. 148.

Chapter 20   The nature of knowledge

1   Plato, *Theaitetos*.
2   Ludwig Wittgenstein (1969), para. 383.
3   Ibid., para. 115.
4   Ibid., para. 105.

Postscript

1   Boscoe Pertwee, eighteenth-century wit.
2   William James (1907), p. 12.
3   William James (1907), p. 6.
4   Francis Cornford (1967), p. viii.
5   A. N. Whitehead (1925), p. 71.
6   Carl Jung (1920), p. 140.

Bibliography

Eugene Abravanel   'How children combine vision and touch when perceiving the shape of objects',
Percept and Psychophysics, 12, 1972
E.D. Adrian   *The Basis of Sensation: The Action of the Sense Organs*, Cambridge, 1928
Peter Alexander   *Sensationalism and Scientific Explanation*, London, 1963
Francesco Algarotti   *Essay on Painting*, London, 1764
D.A. Allport   'Phenomenal Simultaneity and the Perceptual Moment Hypothesis', *Br. J. Psychol.*,
59, pp. 395–406, 1968
Adelbert Ames   *The Morning Notes of Adelbert Ames*, New Jersey, 1960
G.E.M. Anscombe   *An Introduction to Wittgenstein's Tractatus*, London, 1959
Aristotle   *Analytica Posteriora* (John Warrington transl.), London, 1964
    De Anima (Parts of Animals, Movement of Animals and Progression of Animals) (A.L. Peck and
E.S. Forster transl.), London, 1957
    De Caelo (W.K.C. Guthrie transl.), London, 1939
    De Generatione Animalium (A. Platt transl., W.D. Ross ed.), Oxford, 1912
    De Sensu, in *Parva Naturalia*, q.v.
    Metaphysica (W.D. Ross transl. and ed.), Oxford, 1908
    Metaphysics (J. Warrington transl.), London, 1956
    Parva Naturalia (W.D. Ross ed.), London, 1957
    Physica (R.P. Hardie and R.K. Gaye transl., W.D. Ross ed.), Oxford, 1930
    Physics (P.H. Wicksteed transl.), 2 vols., 1957
——(pupil of)   *Mechanica*, in *The Works of Aristotle* (Sir D. Ross ed.), Vol. vi, Oxford, 1913
D.M. Armstrong   *A Materialist Theory of the Mind*, London, 1968
W.R. Ashby   *Design for a Brain*, London 1952
R.H. Atkin   'Time as a pattern on a multi-dimensional structure', *Journal of Social Biological
Structure*, 1, pp. 281–95, 1978
Fred Attneave   *Applications of Information Theory to Psychology*, New York, 1959
    'Multi-stability in perception', *Scient. Amer.*, 540, p. 62, 1971
J.L. Austin   *Sense and Sensibilia*, Oxford, 1962
E. Averbach and G. Sperling   'Short-term storage of information in vision', in *Symposium on
Information Theory* (Colin Cherry ed.), London, 1961
A.J. Ayer   *The Foundations of Empirical Knowledge*, London, 1940
    Language, Truth and Logic, London, 1936
    The Origins of Pragmatism, London, 1968
    The Problem of Knowledge, Harmondsworth, 1956

Russell, London, 1972

―――― and Raymond Winch (eds.)   *British Empirical Philosophers*, London, 1952

Charles Babbage   *Charles Babbage and his Calculating Engines* (Philip and Emily Morrison eds.), New York, 1961

Passages from the Life of a Philosopher (first pub. 1864), Farnborough, 1969

Francis Bacon   *The Advancement of Learning* (first pub. 1605), London, 1973

The Great Restauration, in *Novum Organum, q.v.*

Instauration II, in *Works of Francis Bacon, q.v.*

New Atlantis (first pub. 1627), Oxford, 1915

Novum Organum (first pub. 1620) (F.H. Anderson ed.), Indianapolis, 1960

Works of Francis Bacon (J. Spedding, Robert L. Ellis and Douglas D. Heath eds.), 7 vols., New York, 1869

Alan D. Baddeley   *The Psychology of Memory*, New York, 1976

C. Bailey   *The Greek Atomists and Epicurus*, New York, 1928

F. C. Bakewell   *Electric Science: Its History, Phenomena and Applications*, London, 1853

David P. Barash   *Patriarchy*, New York, 1973

Danielo Barbaro   'Practica della Perspettiva', *Bulletin of the Metropolitan Museum of Art* (A. H. Mayor transl.), Summer 1946

H. B. Barlow   'Nature's Joke: A Conjecture on the Biological Role of Consciousness', in *Consciousness and the Physical World* (B. D. Josephson and V. S. Ramachandran eds.), Oxford, 1980
'Visual Experience and Cortical Development', *Nature*, 258, pp. 199–204, Nov. 1975

Nora Barlow   *Charles Darwin and the Voyage of the Beagle*, London, 1945

Cyril Barnett (ed.)   *Wittgenstein: Lectures and Conversations*, Oxford, 1966

F. C. Bartlett   *Remembering*, Cambridge, 1932

P. P. G. Bateson   'The Characteristics and Context of Imprinting' *Biol. Rev.*, 41, pp. 177–220, 1966

Cyril Bayley   *The Greek Atomists and Epicurus*, New York, 1928

D. C. Beardslee and D. C. Westheimer (eds.)   *Readings in Perception*, New York, 1958

Vladimir M. Bekhterev   *Objective Psychology*, n.p., 1907–12, especially p. 399

Jonathan Bennett   'The difference between right and left', in *American Philosophical Quarterly*, 7:3, pp. 175–91, July 1970

Henri Bergson   *L'Evolution Créatrice* (first pub. 1907), translated as *Creative Evolution* (A. Mitchell transl.), London, 1911

Laughter: An Essay on the Meaning of the Comic, London, 1911

Time and Free Will, London, 1910

George Berkeley   *A New Theory of Vision* (first pub. 1907), London, 1910

Three Dialogues Between Hylas and Philonous, in *A Treatise Concerning the Principles of Human Knowledge, q.v.*

A Treatise Concerning the Principles of Human Knowledge, London, 1962

Jeremy Bernstein   *Einstein*, London, 1973

A. Binet and T. Simon   'Le développement de l'intelligence des enfants', *L'Année Psychologique*, 14, pp. 1–94, 1908

M. H. Birnbaum   'Using contextual effects to derive psychological scales', *Perception and Psychophysics*, 15:1, pp. 89–96, 1974

M. Black   'Why cannot an effect precede its cause?', *Analysis*, 16, 1956

C. Blakemore   'The Confounded Brain', in *Illusion in Nature and Art* (R. L. Gregory and E. H. Gombrich eds.), London, 1973

——— and D.E.Mitchell   'Environmental modification of the visual cortex and the neural basis of learning and memory', *Nature*, 241, pp.467–8, 1973

——— and Richard C.van Sluythers   'Innate and environmental factors in the development of the kitten's visual cortex', *J. Physiol.*, 248, pp.663–716, 1975

M.A.Boden   *Artificial Intelligence and Natural Man*, Sussex, 1977

'Intention and Mechanism', *Science*, 177:4045, 1972

Piaget, London, 1979a

Purposive Explanation in Psychology, Cambridge, Mass., 1972b

Ivan Bodis-Wollner 'Recovery from cerebral blindness: evoked potential and psychophysical measurements', *Electroencephalography and Clinical Neurophysiology*, 42, pp.178–84, 1977

——— and Adam Atkin, Edward Raab and Murray Wolkstein   'Visual association cortex and vision in man: pattern-evoked occipital potentials in a blind boy', *Science*, 198, pp.629–31, 1977

David Bohm   *Wholeness and the Implicate Order*, New York, 1958

Niels Bohr   *Atomic Physics and Human Knowledge*, New York, 1958

Sissela Bok   *Lying: Moral choice in public and private life*, New York, 1978

Sir Hermann Bondi   *Assumption and Myth in Physical Theory*, Cambridge, 1967

J.T.Bonner   *The Evolution of Culture in Animals*, Princeton, 1980

George Boole   *The Mathematical Analysis of Logic* (first pub. 1847), Oxford, 1948

E.G.Boring   *A History of Experimental Psychology*, New York, 1950

The Physical Dimensions of Consciousness, New York, 1933

Sensation and Perception in the History of Experimental Psychology, New York, 1942

C.V.Borst (ed.)   *The Mind/Brain Identity Theory*, London, 1970

Brian Bowers   *Sir Charles Wheatstone*, London, 1975

John Bowlby   *Attachment*, New York, 1969

C.B.Boyer   *A History of Mathematics*, New York, 1968

R.M.Boynton   'The visual system: environmental information', in *Handbook of Perception, Vol. I: Historical Roots of Perception* (E.C.Carterette and M.P.Friedman eds.), London, 1974

Francis Herbert Bradley   *Appearance and Reality: A Metaphysical Essay*, London, 1899

R.B.Braithwaite   *Scientific Explanation*, Cambridge, 1953

David Brewster   *The Life of Sir Isaac Newton*, London, 1855

L.Brillouin   *Science and Information Theory*, London, 1956

G.S.Brindley   *Physiology of the Visual Pathway* (2nd ed.), London, 1970

———'Nerve net models of plausible size that perform many simple learning tasks', *Proc. R. Soc. Lond. 'B'*, 174, pp.173–91, 1969

C.D.Broad   *Mind and Its Place in Nature*, London, 1929

D.E.Broadbent   'Applications of information theory and decision theory to human perception and reaction', in *Cybernetics of the Nervous System* (N.Wiener and J.P.Schade eds.), Amsterdam, 1965

In Defence of Empirical Psychology, London, 1973

Perception and Communication, Oxford, 1958

J.Bronowski   *The Ascent of Man*, London, 1973

Roger Brown   *A First Language: The Early Stages*, Cambridge, Mass., 1973

Psycholinguistics: Selected Papers by Roger Brown, London, 1958

Words and Things: An Introduction to Language, London, 1958

Robert S.Brumbaugh   *Ancient Greek Gadgets and Machines*, Westport, Conn., 1966

The Philosophers of Greece, New York, 1964

Jerome S.Bruner   *Beyond the Information Given*, London, 1974

'On perceptual readiness', *Psychol. Rev.*, 64, pp.123–52, 1957

——— and L.Postman   'On the perception of incongruity: a paradigm', *Journal of Personality*, 18, pp.206–23, 1949

P. Bryant   *Perception and Understanding in Young Children: An Experimental Approach*, London, 1974

A. E. Wallis Budge   *The Egyptian Book of the Dead (The Papyrus of Ani)* (First pub. 1895), New York, 1967

V. Buranelli   *The Wizard from Vienna: Franz Anton Mesmer*, London, 1975

J. Burnet   *Early Greek Philosophy*, London, 1908

P. A. Buser and A. Rougeul-Buser (eds.)   *Cerebral correlates of conscious experience*, Insern Symposium No., 6, Institut de la Santé et de la Recherche Médicale, 1978

Nigel Calder   *The Key to the Universe*, London, 1977

Norman Campbell   *What is Science?*, London, 1921

T. D. Campbell   'Evolutionary Epistemology', in *The Philosophy of Karl Popper* (P. A. Schilpp ed.), Illinois, 1974

W. B. Cannon   *The Wisdom of the Body*, New York, 1932

Karel Čapek   *R. U. R. Rossum's Universal Robot*, London, 1923

Fritjof Capra   *The Tao of Physics*, London, 1976

Rudolph Carnap   'Replies and Systematic Expositions', in *The Philosophy of Rudolph Carnap*, q.v.
The Philosophy of Rudolph Carnap (P. A. Schilpped.), Illinois, 1964

H. Carteron   'Does Aristotle have a Mechanics?', in *Articles on Aristotle* (Jonathan Barnes, M. S. Schofield and R. Sorabji eds.), London, 1975

E. Cassirer   *Einstein's Theory of Relativity* (first pub. 1923), New York, 1953
Substance and Function (first pub. 1923), New York, 1953

R. L. van de Castle   'The Psychology of Dreaming', in *Dreams and Dreaming* (S. G. M. Lee and A. R. Mayes eds.), London, 1973

Geoffrey Chaucer   *The Complete Works of Geoffrey Chaucer* (F. N. Robinson ed.), London, 1974

Colin Cherry   *On Human Communication*, London, 1957

V. Gordon Childe   'Rotary Motion', in *A History of Technology* (Charles Singer, E. J. Holmyard and A. R. Hall eds.), Vol. 1, Oxford, 1954

Noam Chomsky   *Aspects of the Theory of Syntax*, MIT Press, 1965

C. W. Churchman and P. Ratoosch   *Measurement: Definitions and Theories*, New York, 1959

Cicero   *De Inventione* (H. Hubbell transl.), London, 1949
————*De Re Publica* (C. E. Keyes transl.), London, 1928
————*Tusculan Disputations* (J. E. King transl.), London, 1945

Anthony Clare   *Psychiatry in Dissent*, London, 1976

Ronald W. Clark   *Einstein: The Life and Times*, New York, 1971
The Life of Bertrand Russell, London, 1975

E. Clarke and K. Dewhurst   *An Illustrated History of Brain Function*, California, 1972

Claudian   *Shorter Poems* (M. Platnauer transl.), London, 1922

J. F. Clauser and A. Shimony   'Bell's Theorem: Experimental tests and implications', *Rep. Prog. Phys.*, 41, pp. 1881–1927, 1978

James Clerk-Maxwell   *Matter and Motion*, New York, 1952

M. F. Cleugh   *Time, and its Importance in Modern Thought*, London, 1937

D. Cohen   *J. B. Watson: The Founder of Behaviourism*, London, 1979

William Coleman   *Biology in the Nineteenth Century: Problems of Form, Function and Transformation*, New York, 1971

Auguste Comte   *Cours de philosophie positive* (first pub. 1830–42), 6 vols. (Harriet Martineau transl.), 2 vols., New York, 1853

F. M. Cornford   *From Religion to Philosophy*, London, 1912
　The Laws of Motion in Ancient Thought, Inaugural Lecture, Cambridge, 1931
　Plato's Cosmology, New York, 1957
　(transl.) *The Republic of Plato*, Oxford, 1941
　The Unwritten Philosophy and Other Essays (W. K. C. Guthrie ed.), Cambridge, 1967
T. N. Cornsweet   *Visual perception*, London, 1970
A. Cowey   'Cortical Maps and Visual Perception', *Q. Jl exp. Psychol.*, 31, pp. 1–17, 1979
Kenneth Craik   *The Nature of Explanation*, Cambridge, 1943
A. C. Crombie   *Augustine to Galileo, Vol. 1: Science in the Middle Ages: V–XIII Centuries*, London, 1961
E. R. F. W. Crossman   'The information-capacity of the human motor-system in pursuit tracking', *Q. Jl exp. Psychol.*, 12:1, pp. 1–16, 1960

A. D'Abro   *The Evolution of Scientific Thought* (first pub. 1927) (2nd ed.), New York, 1950
Charles Darwin   *Descent of Man*, London, 1871
　Expression of the Emotions in Man and Animals, London, 1872
　The Foundations of the Origin of Species, essays written 1842 and 1844 (Francis Darwin ed.), London, 1909
　Journal of Researches into the Natural History and Geology of the Countries visited during the Voyage of HMS *Beagle Round the World* (2nd ed.), London, 1845
　Origin of Species, London, 1859
　The Power of Movement in Plants, London, 1880
Erasmus Darwin   *Zoonomia; or, the Laws of Organic Life*, London, 1794–6
Francis Darwin (ed.)   *The Life and Letters of Charles Darwin*, 3 vols., London, 1887
Richard Dawkins   *The Selfish Gene*, Oxford, 1976
Julien Offray de la Mettrie   *Man a Machine* (first pub. 1748), Illinois, 1912
William Dement   'The Biological Basis of REM Sleep (circa 1968)', in *Dreams and Dreaming* (S. G. M. Lee and A. R. Mayes eds.), Harmondsworth, 1973
W. Dement and N. Kleitman   'The relation of eye movements during sleep to dream activity: an operational method for studying dreaming', *J. exp. Psychol.*, 53, pp. 339–46, 1957
Daniel O. Dennett   *Brainstorms: Philosophical Essays on Mind and Psychology*, Bradford, 1978
René Descartes   *De la Lumière* (*Optics*) in *Discourse on Method, Optics, Geometry and Meteorology, q.v. Discourse on Method, Optics, Geometry, and Meteorology* (first pub. 1637), Indianapolis, 1965
　Meditationes de Prima Philosophia, in *The Philosophical Works of Descartes, q.v.*
　The Passions of the Soul, in *The Philosophical Works of Descartes, q.v.*
　The Philosophical Works of Descartes (E. S. Haldane and G. R. T. Ross transl.), 2 vols., Cambridge, 1967
　Principia Philosophiae, in *The Philosophical Works of Descartes, q.v.*
　'Rules for the Direction of Mind', in *The Philosophical Works of Descartes, q.v.*
Adrian Desmond   *The Ape's Reflection*, London, 1979
Bernard d'Espagnat   *Conceptual Foundations of Quantum Mechanics*, Boston, Mass., 1976
　The Quantum Theory and Reality, Reading, Mass., 1971
R. H. Dicke   'The Many Faces of Mach', in *Gravitation and Relativity* (Hong-Yee Chiu and William F. Hoffman eds.), New York, 1964
D. Diringer   *The Alphabet: a Key to the History of Mankind* (3rd ed.), 2 vols., London, 1968
　Writing, London, 1962
Isaac D'Israeli   *Calamities of Authors*, London, 1812
Margaret Donaldson   *Children's Minds*, London, 1978

Mary Douglas   *Purity and Danger: An Analysis of Concepts of Pollution and Taboo*, London, 1966

Stillman Drake (transl.)   *Discoveries and Opinions of Galileo*, London, 1957

——— (transl.)   *Galileo: Dialogue Concerning Two Chief World Systems*, California, 1962

J.L.E.Dreyer   *A History of Astronomy from Thales to Kepler*, New York, 1953

   History of the Planetary System from Thales to Kepler, Cambridge, 1906

Michael Dummett   *Frege: Philosophy of Language*, London, 1973

——— and A.Flew   'Can an effect Precede its Cause?', *Proc. Aristotelian Soc.*, Supp. Vol. 28, 1954

E.Durkheim   *Les Formes Elémentaires de la vie Religieuse*, Paris, 1912, translated as *The Elementary Forms of the Religious Life*, London, 1915

E. du Bois-Reymond   *Animal Electricity*, 2 vols., Paris, 1848

Gerald Dworkin (ed.)   *The IQ Controversy*, London, 1977

Hermann Ebbinghaus   *Uber das Gedächtniss*, Leipzig, 1885

J.C.Eccles   *The Neurological Basis of Mind*, Oxford, 1953

Sir Arthur Eddington   *The Philosophy of Physical Science*, Cambridge, 1939

   Space Time and Gravitation: An Outline of General Relativity Theory, Cambridge, 1920

A.Einstein   *Relativity: The Special and General Theory. A Popular Exposition*, London, 1920

Brian Ellis   *Basic Concepts of Measurement*, Cambridge, 1968

Havelock Ellis   *Man and Woman*, London, 1926

W.D.Ellis (ed.)   *A Source Book of Gestalt Psychology*, London, 1938

S.T.Emlen   'The Stellar-Orientation System of a Migrating Bird', *Scient. Am.*, 233:2, pp. 102–11, 1975

E.S.Eriksson   *A Cognitive Theory of Three-Dimensional Motion Perception*, Uppsala, 1970

Euclid   *Optics and Catoptrics*, in *A Source Book of Greek Science* (M.R.Cohen and I.E.Drabkin eds.), Cambridge, Mass., 1966

   Elements (T.L.Heath ed.), 3 vols., New York, 1926

   H.J.Eysenck   'The Effects of Psychopathology: an Evaluation', *J. Consult. Psychol.*, 16, pp.319–24, 1952

   Experiments in Behaviour Therapy, Oxford, 1964

M.Farber (ed.)   *Philosophical Thought in France and the United States: Essays representing major trends in contemporary French and American philosophy*, New York, 1950

Benjamin Farrington   *Greek Science*, Harmondsworth, 1961

Gustav Fechner   *Elemente der Psychophysic*, 1860

J.K.Feibleman   *An Introduction to Peirce's Philosophy, interpreted as a System*, New York, 1946

E.A.Feigenbaum and J.Feldman (eds.)   *Computers and Thought*, New York, 1963

L.Festinger   *A Theory of Cognitive Dissonance*, Evanston, 1957

P.Feyerabend   'Explanation, reduction, and empiricism,' in *Minnesota Studies in the Philosophy of Science* (H.Feigl and G.Maxwell eds.), 3, pp.28–97, 1962

Richard Feynman   *Lectures on Physics*, Reading, Mass., 1972

Bruno de Finetti   *The Theory of Probability*, New York, 1970

M.I.Finley   *Early Greece: The Bronze and Archaic Ages*, London, 1970

R.A.Fisher   *The Design of Experiments*, London, 1935

   The Genetical Theory of Natural Selection, Oxford, 1930

P.M.Fitts   'The information capacity of the human motor system in controlling the amplitude of movement', *J. exp. Psychol.*, 47, pp.381–91, 1954

—— and M.I.Posner   *Human Performance*, New Jersey, 1973

K.A.Flowers   'Visual "closed loop" and "open loop" characteristics of voluntary movement in patients with parkinsonian and intention tremor', *Brain*, 99, pp.261–310, 1976

J.A.Fodor   *The Language of Thought*, Sussex, 1976

Psychological Explanation, New York, 1968

—— and J.D.Fodor and M.Garrett   *The Psychology of Language: An Introduction to Psycholinguistics and Generative Grammar*, New York, 1974

E.B.Ford   *Evolution studied by observation and experiment*, Oxford, 1973

D.W.Forrest   *Francis Galton: the Life and Work of a Victorian Genius*, London, 1974

S.W.Fox   'Simulated natural experiments in spontaneous organisation of morphological units from protenoid', in *The Origins of Prebiological Systems* (S.W.Fox ed.), London, 1965

'A theory of macromolecular and cellular origins', *Nature*, 205, pp. 328–40, 1965

P.Fraisse   *The Psychology of Time* (J.Leith transl.), London, 1964

Philip Frank   *Einstein: His Life and Times*, New York, 1947

H.Frankfort, H.A.Frankfort, J.A.Wilson and T.Jacobsen   *Before Philosophy*, Harmondsworth, 1949

Benjamin Franklin   *Autobiography*, London, 1903

Benjamin Franklin: the Autobiography and other Writings (L.Jesse Lemisch ed.), London, 1961

The Papers of Benjamin Franklin (Leonard W.Labarel *et al.* eds.), 40 vols., New Haven, Conn., 1959

Works, Boston, Mass., 1905–7

Norman H.Freeman   *Strategies of Representation in Young Children: Analysis of Spatial Skills and Drawing Processes*, London, 1980

Sigmund Freud   *The Interpretation of Dreams* (first pub. 1900) (James Strachey transl.), Harmondsworth, 1979

The Introductory Lectures on Psychoanalysis (James Strachey and Angela Richards eds., James Strachey transl.), Harmondsworth, 1973

Jokes and their Relation to the Unconscious (James Strachey transl.), Harmondsworth, 1976

—— and Carl G.Jung   *The Freud/Jung Letters* (W.McGuire ed.), London, 1974

R.M.Gale (ed.)   *The Philosophy of Time*, Sussex, 1968

Galen   *On the Natural Faculties* (A.J.Brock transl.), London, 1916

Galileo Galilei   *Dialogue Concerning the Two Chief World Systems* (first pub. 1632) (S.Drake transl.), California, 1962

Discoveries and Opinions of Galileo (S.Drake transl.), New York, 1957

W.B.Gallie   *Peirce and Pragmatism*, Harmondsworth, 1952

Gordon G.Gallup   'Self Recognition in Primates: A Comparative Approach to the Bidirectional Properties of Consciousness', *American Psychologist*, 32:5, pp.329–38, 1977

F.Galton   *Hereditary Genius*, London, 1869

Inquiries into Human Faculty, London, 1883

Memories of My Life, London, 1908

Alan Gardiner   *Egyptian Grammar* (3rd ed.), Oxford, 1957

B.T.Gardiner and R.A.Gardiner   'Teaching Sign Language to a Chimpanzee', *Science*, 165, pp.664–72, 1969

Martin Gardner   *The Ambidextrous Universe: Left, Right, and the Fall of Parity*, Harmondsworth, 1964

The Relativity Explosion, New York, 1976

Michel Gauquelin   *The Cosmic Clocks*, New York, 1967

———, F. Gauquelin and S. B. G. Eysenck   'Personality and the position of the planets at birth: An empirical study, *Br. J. soc. and clin. Psychol.*, 18, pp. 71–5, 1979

Johan Karl Friedrich Gauss   *Theoria Motus Corporum Coelestium*, Hamburg, 1809

R. M. Gaze   *The Formation of Nerve Connections*, London, 1970

Michael S. Gazzaniga   'The split brain in man', *Scient. Amer.*, August 1967

Patrick Geddes   *Sex*, London, 1911

——— and J. A. Thompson   *Evolution of Sex*, London, 1889

N. Geschwind   *The Development of the Brain and the Evolution of Language*, Monograph Series on Language and Linguistics No. 17, Report on the 15th Annual R.T.M. on Linguistic and Language Studies, or *Monograph Series on Language and Linguistics*, 17 April 1964

James J. Gibson   'A Note on Ecological Optics', in *Handbook of Perception, Vol. 1: Historical Roots of Perception* (E. C. Carterette and M. P. Friedman eds.), London, 1974
    The Perception of the Visual World, Boston, Mass., 1950
    The Senses Considered as Perceptual Systems, Boston, Mass., 1966

William Gilbert   *De Magnete* (first pub. 1600) (P. S. Mottelay transl.), New York, 1958

Gilgamesh   *The Epic of Gilgamesh* (N. K. Saunders transl.), Harmondsworth, 1972

G. G. Globus, G. Maxwell and I. Savodnik   *Consciousness and the Brain: A Scientific and Philosophical Enquiry*, London, 1976

C. Glymour   *Philosophers on Freud: New Evaluations* (Richard Wolheim ed.), New York, 1974

Kurt Gödel   *On formerly undecidable propositions of Principia Mathematica and related systems* (B. Meltze transl.), New York, 1962

Ernst Gombrich   *Art and Illusion*, Oxford, 1960
    Norm and Form, Oxford, 1966
    The Sense of Form, Oxford, 1979

R. F. Gombrich   'Ancient Indian Cosmology', in *Ancient Cosmologies* (C. Blacker and M. Loewe eds.), London, 1975

S. J. Gould   *Ontogeny and Philogeny*, Cambridge, Mass., 1977

Edward Grant (ed.)   *A Source Book in Medieval Science*, Cambridge, Mass., 1975

D. M. Green and J. A. Swets   *Signal Detection Theory and Psychophysics*, New York, 1966

R. L. Gregory   'The Brain as an Engineering Problem', in *Current Problems in Animal Behaviour* (W. H. Thorpe and O. L. Zangwill eds.), Cambridge, 1961
    'Choosing a paradigm for perception', in *Handbook of Perception, Vol. 1: Historical Roots of Perception* (E. C. Carterette and M. P. Friedman eds.), London, 1974
    'Cognitive Contours', *Nature*, 238, pp. 51–2, 1972
    Concepts and Mechanisms of Perception, London, 1974
    'Consciousness', in *The Encyclopaedia of Ignorance, Vol. 2* (R. Duncan and M. Weston-Smith eds.), London, 1977
    'Distortion of visual space as inappropriate constancy scaling', *Nature*, 199:678, 1963
    Eye and Brain, London, 1966
    'The Grammar of Vision', *The Listener*, 83:242, 1971
    The Intelligent Eye, London, 1970
    'On how so little information controls so much behaviour', in *Towards a Theoretical Biology* (C. H. Waddington ed.), Edinburgh, 1968
    'Perceptions as Hypotheses', *Phil. Trans. R. Soc. Lond. 'B'*, 290 , pp. 181–97, 1980
    'Perceptual Illusions and Brain Models', *Proc. R. Soc. Lond. 'B'*, 171, pp. 278–96, 1968

——— and Jean Wallace   *Recovery from Infant Blindness: a Case Study*, Experimental Psychological Society Mongr., Cambridge, 1963

W.Grey Walter   *Electroencephalography: a Symposium on its Various Aspects* (Denis Hill and Geoffrey Parr eds.), London, 1950
The Living Brain, London, 1953
H.E.Gruber   *Darwin on Man: A Psychological Study of Creativity,* and *Darwin's Unpublished Notebooks,* New York, 1974
Adolf Grünbaum   'Logical and Philosophical Foundations of the Special Theory of Relativity', *Am. J. Phys.,* 23, pp.450–64, 1955
J.P.Guilford   *The Nature of Human Intelligence,* New York, 1967
A.Guzman   'Analysis of curved line drawings using content and global information', in *Machine Intelligence* (B.Meltzer and D.Michie eds.), 6, Edinburgh, 1971
'Computer recognition of three-dimensional objects in a visual scene', PhD thesis: available as Project Report M.A.C. TR 59, AD 692 200

Ralph N.Haber   *Contemporary Theory and Research in Visual Perception,* New York, 1968
J.Hadamard   *Psychology of Invention in the Mathematical Field,* Princeton, 1949
F.Hallam and S.Weed   'A Study of Dream Consciousness' *Am. J. Psychol.,* 7:405, 1896
W.D.Hamilton   'The Genetical Theory of Social Behaviour', *J. theor. Biol.,* 7, pp.1–52, 1964
M.Hammerton and D.D.Stretch   'The Timing of Occultations', *J. Br. Astron. Assn.,* 1980
Stuart Hampshire   *Spinoza,* Harmondsworth, 1962
N.R.Hanson   *Observation and Explanation: A Guide to Philosophy of Science,* London, 1972
Patterns of Discovery, Cambridge, 1958
A.D.Harris and O.L.Zangwill   'The Writings of Sir Frederic Bartlett', *Br. J.Psychol.,* 64:4, pp.493–510, 1973
Fraser Harrison   *Dark Angel: Aspects of Victorian Sexuality,* London, 1977
David Hartley   *Observations on Man,* London, 1749
C.Hartshorne and P.Weiss   *Collected Papers of Charles Saunders Peirce,* Cambridge, Mass., 1931–58
William Harvey   *The Circulation of the Blood and other writings* (K.J.Franklin transl.), London, 1963
An Anatomical Disputation Concerning the Movement of the Blood in Living Creatures (G.Whitteridge transl.), Oxford, 1976
Donald O.Hebb   *Organisation of Behaviour,* London, 1949
S.Hecht, S.Schlaer and M.H.Pirenne   'Energy quanta and vision', *J. gen. Physiol.,* 25:819, 1942
Patrick Heelan   *Quantum Mechanics and Objectivity: A Study of the Physical Philosophy of Werner Heisenberg,* The Hague, 1965
A.Heidel   *The Babylonian Genesis,* Chicago, 1942
Werner Heisenberg   *The Physical Principles of Quantum Theory,* Chicago, 1930
Physics and Philosophy, London, 1958
Hermann von Helmholtz   *On the sensations of tone . . . ,* New York, 1954
Popular Scientific Lectures (Morris Kline ed.), New York, 1962
Treatise on Physiological Optics (J.P.C.Southall ed.), New York, 1862
Paul Henle   *Language, Thought and Culture,* Ann Arbor, 1966
[Hero of Alexandria]   *The Pneumatics of Hero of Alexandria* (Bennett Woodcroft transl.), London, 1851
W.Heron, B.K.Doane and T.H.Scott   'Visual disturbances after prolonged perceptual isolation', *Can. J. Psychol.,* 10, pp.13–18, 1956
Hesiod   *Theogony, Works and Days,* and *Theognis* (D.Wender transl.), London, 1973
E.H.Hess   'Development of the chick's responses to light and shade cues of depth', *J. comp. Physiol.,* 43, pp.112–22, 1950

Mary Hesse   *The Structure of Scientific Inference*, London, 1977

W. E. Hick   'Information theory and intelligence tests', *Br. J. Psychol.* (Stats Section), 4:3, pp. 157–64, 1951

   'On the rate of gain of information', *Q. Jl exp. Psychol.*, 4, pp. 11–26, 1952

D. Hilbert and S. Cohn-Vossen   *Geometry and the Imagination*, New York, 1952

E. R. Hilgard   *Theories of Learning* (2nd ed.), New York, 1958

D. Hill and G. Parr (eds.)   *Electroencephalography*, London, 1950

Gertrude Himmelfarb   *Darwin and the Darwinian Revolution*, New York, 1962

Robert A. Hinde   *Animal Behaviour* (2nd ed.), New York, 1970

——— and J. Stevenson Hinde (eds.)   *Constraints on Learning: Limitations and Predispositions*, London, 1973

Hippocrates   *The Sacred Disease* (E. Capps, T. E. Page and W. H. D. Rouse eds.), Vol. II, London, 1923

J. E. Hochberg   *Perception*, New Jersey, 1964

J. E. Hofmann   *Leibniz in Paris 1672–1676*, Cambridge, 1974

D. R. Hofstadter   *Gödel, Escher, Bach: A Metaphorical Fugue of Minds and Machines*, Sussex, 1979

R. L. Holloway   'The casts of fossil hominid brains', *Scient. Am.*, 231, pp. 106–15, 1974

Homer   *The Odyssey* (S. O. Andrew transl.), London, 1953

Robert Hooke   *Micrographia*, New York, 1965

S. Hooke (ed.)   *Dimensions of Mind*, New York, 1960

-   *Middle Eastern Mythology*, Harmondsworth, 1963

I. P. Howard and W. B. Templeton   *Human Spatial Orientation*, New York, 1966

Fred Hoyle   *Astronomy and Cosmology*, Reading, 1975

D. H. Hubel and T. N. Wiesel   'Receptive fields, binocular interaction and functional architecture in the cat's visual cortex', *J. Physiol.*, 165, pp. 559–68, 1962

   'Receptive fields and functional architecture of monkey striate cortex', *J. Physiol.*, 195, pp. 215–43, 1968

L. Hudson   *Contrary Imaginings*, London, 1966

   'A differential test of Arts/Science aptitude', *Nature*, 186:413, 1960

   'Undergraduate academic record of Fellows of the Royal Society', *Nature*, 182:380, 1958

D. A. Huffman   'Impossible pictures as nonsense sentences', in *Machine Intelligence* (B. Meltzer and D. Michie eds.), 6, Edinburgh, 1971

J. Hughlings Jackson   'Hughlings Jackson on aphasia and kindred affections of speech, together with a complete bibliography of his publications on speech and a reprint of some of the more important papers', *Brain*, 38, pp. 1–190, 1915

David Hume   *An Enquiry concerning Human Understanding* (first pub. 1758), London, 1938

   The Life of David Hume, Written by Himself (first pub. 1777), London, 1965

   A Treatise of Human Nature (first pub. 1739–40), London, 1934–40

G. Humphrey   *Thinking*, London, 1951

Julian Huxley   *Evolution: The Modern Synthesis*, London, 1942

T. H. Huxley   *Autobiography* (Gavin de Beer ed.), London, 1974

R. Hyman   'Stimulus information as a determinant of reaction time', *J. exp. Psychol.*, 45, pp. 188–96, 1953

B. Inhelder and J. Piaget   *The Child's Conception of Space*, London, 1956

   The Growth of Logical Thinking from Childhood Through Adolescence, New York, 1958

W.Irvine   *Apes, Angels and Victorians: Joint Biography of Darwin and Huxley*, London, 1956

W.H.Ittelson   *The Ames Demonstrations in Perception*, Princeton, 1952

Jahnke-Emde-Lösch   *Tables of Higher Functions* (6th ed.), New York, 1960

William James   *Pragmatism*, New York, 1907

  Principles of Psychology, New York, 1890

  The Varieties of Religious Experience, New York, 1902

Allan Janik and Steven Toulmin   *Wittgenstein's Vienna*, New York, 1973

Julian Jaynes   *The Origin of Consciousness and the Breakdown of the Bicameral Mind*, Boston, Mass., 1976

J.Jewkes, D.Sawers and R.Stillerman   *The Sources of Invention*, London, 1958

G.Johanson   'Visual motion perception', in *Scient. Am.*, 232:6, pp.76–88, 1975

—— and C. von G.Hofsten and G.Jansson   'Event perception', in *A. Rev. Psychol.*, 31, pp.27–38, 1980

P.N.Johnson-Laird and P.C.Wason (eds.)   *Thinking: Readings in Cognitive Science*, Cambridge, 1977

Ernest Jones   *The Life and Work of Sigmund Freud*, 3 vols., New York, 1953–7

B.D.Josephson and V.S.Ramachandran (eds.)   *Consciousness and the Physical World*, Oxford, 1980

C.G.Jung   *Analytical Psychology* (2nd ed.), London, 1920

  Man and his Symbols, London, 1964

  Psychologische Abhandlungen, Vol. v, Zurich, 1944

  Studies in Word Association (first pub. 1918), London, 1969

  Two Essays of Analytical Psychology, London, 1953

Gaetano Kanizsa   'Margini quasi-percettivi in campi con stimolazione omogenea', *Rivista di Psicologia*, 49, pp.7–30, 1955

  Organization of Vision: Essays on Gestalt Perception, New York, 1979

Immanuel Kant   *Cosmogony* (first pub. 1755) (W.Hastie transl., W.Ley ed.), New York, 1968

  Critique of Practical Reason, Riga, 1788

  Critique of Pure Reason, London, 1781

  Prolegomena to any Future Metaphysics (first pub. 1783), Indianapolis, 1950

Bernard Katz   *Nerve, Muscle and Synapse*, New York, 1966

J.J.Katz   *The Philosophy of Language*, New York, 1966

W.N.Kellog and L.A.Kellog   *The Ape and the Child*, New York, 1933

C.R.Kelly   *Manual and Automatic Control*, New York, 1968

G.A.Kelly   *The Psychology of Personal Constructs*, New York, 1955

N.Kemp-Smith   *The Philosophy of David Hume*, London, 1941

Anthony Kenny   *Descartes: A Study of his Philosophy*, Harmondsworth, 1968

  Wittgenstein, Harmondsworth, 1973

D.R.Kenshalo   'Phylogenetic development of feeling', in *Handbook of Perception, Vol. VI: Feeling and Hurting* (E.C.Carterette and M.P.Friedman eds.), London, 1978

Johannes Kepler   *Mysterium*, 1596

D.King-Hele   *Erasmus Darwin*, London, 1963

G.S.Kirk   *Myth: Its meanings and Functions in Ancient and other Cultures*, Cambridge, 1970

—— and J.E.Raven   *The Pre-Socratic Philosophers* (with corrections), Cambridge, 1960

Sheila Kitsinger   *Mothering*, London, 1977–8

Francis D.Klingender (ed.)   *Art and the Industrial Revolution*, London, 1972

Clyde Kluckhohn and Dorothea Leighton   *The Navaho*, Cambridge, Mass., 1948

H. Klüver   'An experimental study of the Eidetic type', *Genet. Psychol. Mongr.*, 1, pp.77–230, 1926

J. Knowles (ed.)   *Life and Writings of Henry Fuseli*, 3 vols., London, 1831

O. Koehler   'The Ability of Birds to Count', in *The World of Mathematics, Vol. 1* (James R. Newman ed.), New York, 1956

L. Koenigsberger   *Hermann von Helmholtz* (F. A. Welby transl.), New York, 1906

Arthur Koestler   *The Sleep Walkers*, London, 1959
    The Ghost in the Machine, London, 1967

K. Koffka   *Principles of Gestalt Psychology*, New York, 1935

Wolfgang Kohler   *The Mentality of Apes*, London, 1925
    'Physical Gestalten', in *A Source Book of Gestalt Psychology* (W. D. Ellis ed.), London, 1920

Stephan Körner   *Categorical Frameworks*, Oxford, 1970
    Kant, Harmondsworth, 1955

Samuel Noah Kramer   *From the Poetry of Sumer: Creation, Glorification, Adoration*, California, 1979
    The Sumerians: Their History, Culture, and Character, Chicago, 1963

E. Krause   *Erasmus Darwin* (W. S. Dallas transl.), London, 1879

I. Krechevsky   'The hereditary nature of "hypotheses" ', *J. comp. Psychol.*, 16, pp.99–116, 1933
    ' "Hypotheses" versus "chance" in the pre-solution period in sensory discrimination learning', *Univ. Calif. Publ. Psychol.*, 6:3, 1932

Thomas Kuhn   'Logic of Discovery or Psychology of Research?', in *Criticism and the Growth of Knowledge* (Imre Lakatos and Alan Musgrave eds.), Cambridge, 1970
    The Structure of Scientific Revolutions, Chicago, 1962

Imre Lakatos   'Falsification and the Methodology of Science', in *Criticism and the Growth of Knowledge* (Imre Lakatos and Alan Musgrave eds.), Cambridge, 1970

Antoine Lamarck   *Histoire des Animaux sans Vertèbres*, 7 vols., Paris, 1815–22
    Philosophie Zoologique, 2 vols., Paris, 1809

W. G. Lambert   *Babylonian Wisdom Literature*, Oxford, 1960

Pierre Laplace   *Mécanique Céleste* (Nathaniel Boditch transl.), 4 vols., Boston, 1823–39
    A Philosophical Essay on Probabilities (first pub. 1820) (F. W. Truscott and F. L. Emory transl.), New York, 1951

K. S. Lashley   *Brain Mechanisms and Intelligence*, Chicago, 1929
    'In search of the emgram', *Symposia of the Society for Experimental Biology*, 4, pp.454–82, 1950

Richard E. Leakey   *Origins*, London, 1977

Richard B. Lee   'The !Kung Bushmen of Botswana', in *Hunters and Gatherers Today* (M. G. Bicchieri ed.), New Jersey, 1972
    —— and I. De Vore (eds.)   *Man the Hunter*, Chicago, 1968

S. G. M. Lee and A. R. Mayes (eds.)   *Dreams and Dreaming*, Harmondsworth, 1973

W. E. LeGros Clark   *Man-Apes or Ape Man? The Story of Discoveries in Africa*, New York, 1967

Gottfried Leibniz   *The Leibniz–Clarke Correspondence* (H. G. Alexander ed.), 1956
    G. W. Leibniz: Philosophical Letters and Papers (L. F. Loemker transl. and ed.), Chicago, 1957
    Leibniz: Philosophical Writings, London, 1934
    Monodologie (first pub. 1714) (R. Latta transl.), 1898
    The Principles of Nature and Grace, 1714
    Theodicée (first pub. 1710) (R. Latta transl.), 1898

H. W. Leibowitz   *Visual Perception*, London, 1965

E. H. Lenneberg   *Biological Foundations of Language*, New York, 1967

Leonardo da Vinci   *The Notebooks of Leonardo da Vinci* (Edward MaCurdy ed.), 2 vols., London, 1938

Claude Levi-Strauss   *La Pensée Sauvage* (first pub. 1962), translated as *The Savage Mind*, London, 1966–8

R.C.Lewontin   *The Genetic Basis of Evolutionary Change*, Columbia, 1974

Willy Ley (ed.)   *Kant's Cosmogony* (first pub. 1755) (W.Hastie transl.), New York, 1968

Benjamin Libet   'Cortical activation in conscious and unconscious experience', *Perspectives in Biology and Medicine*, 9, pp.77–86, 1965

'Neuronal vs. subjective timing, for a conscious sensory experience', in *Cerebral Correlates of Conscious Experience* (P.Buser and A.Rougeul-Buser eds.), Amsterdam, 1965

David C.Lindberg (transl.)   *John Peckham and the Science of Optics: Perspectiva Communis*, Wisconsin, 1970

Alan H.Lloyd   *Old Clocks*, London, 1951

Some Outstanding Clocks over Seven Hundred Years 1250–1950, London, 1958

G.E.R.Lloyd   *Early Greek Science: From Thales to Aristotle*, London, 1970

'Greek Cosmologies', in *Ancient Cosmologies* (C.Blacker and M.Loewe eds.), London, 1975

Greek Science after Aristotle, London, 1973

John Locke   *Correspondence* (E.S.de Beer ed.), 2 vols., Oxford, 1978–9

Correspondence with Clarke (B.Rand ed.), Oxford, 1927

Essay Concerning Human Understanding (first pub. 1690) (P.H.Nidditch ed.), Oxford, 1975

Lucretius   *De Rerum Natura* (Cyril Bailey transl. and ed.), 3 vols., Oxford, 1947

On the Nature of Things (W.E.Leonard transl.), London, 1921

A.R.Luria   *The Mind of a Mnemonist*, New York, 1968

Campbell Thomson Luzac   *The Reports of the Magicians and Astrologers of Nineveh and Babylon*, London, 1900

Thorndike Lynn   *A History of Magic and Experimental Science*, New York, 1923

John Lyons   *Chomsky*, London, 1970

Eleanor E.Maccoby and Carol N.Jacklin   *The Psychology of Sex Differences*, Stanford, California, 1974

Celleste McCollough   'Colour adaptation of the edge-detectors in the human visual system', *Science*, 149, pp.1115–16, 1965

Pamela McCorduck   *Machines Who Think: A Personal Inquiry into the History and Prospects of Artificial Intelligence*, Reading, 1980

W.S.McCulloch and W.H.Pitts   'A Logical calculus of the ideas imminent in nervous activity', *Bull. Math. Phys.*, 5, pp.115–33, 1943

William McDougall   *An Outline of Abnormal Psychology*, London, 1926

Ernst Mach   *The Analysis of Sensations* (5th ed.), New York, 1959

Knowledge and Error: Sketches in the Psychology of Enquiry, Dordrecht, 1976

Popular Scientific Lectures (first pub. 1898) (Thomas J.McCormack transl.), Illinois, 1943

The Science of Mechanics: A Critical and Historical Account (first pub. 1883), Illinois, 1960

Space and Geometry, Illinois, 1906

Donald MacKay   *Information, Mechanism and Meaning*, MIT Press, 1969

'Conscious Agency with Unsplit and Split Brains', in *Consciousness and the Physical World* (B.D.Josephson and V.S.Ramachandran eds.), Oxford, 1980

Norman Malcolm   *Dreaming*, New York, 1962

Ludwig Wittgenstein: A Memoir, Oxford, 1958

B. Malinowski   *Crime and Custom in Savage Society*, London, 1926

C. L. Maloney (ed.)   *The Evil Eye*, Columbia, 1976

Thomas Robert Malthus   *Essay on the Principles of Population* (first pub. 1798), Holland, 1973

Jean and George Mandler   *Thinking: From Association to Gestalt*, New York, 1964

Frank Manuel   *A Portrait of Isaac Newton*, n.p., 1968

Richard Mark   *Memory and Nerve Connections*, Oxford, 1974

E. A. Marland   *Early Electrical Communication*, New York, 1964

D. Marr   *Early Processing of Visual Information* (Artificial Intelligence Lab., MIT Report), 1975
'Representing and Computing Visual Information', in *Artifical Intelligence: An MIT Perspective, Vol. II: Understanding Vision, Manipulation, Computer Design, Symbol Manipulation* (P. H. Winston and R. H. Brown eds.), MIT Press, 1979

——— and T. Poggio   'From understanding computation to understanding neural circuitry', *Neurosciences Res. Prog. Bull.*, 51:3, pp. 470–88, 1977

N. Maskelyne   *Astronomical Observations at Greenwich* (section for 1795), 1799

Margaret Masterman   'The Nature of a Paradigm', in *Criticism and the Growth of Knowledge* (Imre Lakatos and Alan Musgrave eds.), Cambridge, 1970

Richard D. Mattuck   'A Quantum Mechanical Theory of Psychokinesis', in *The Iceland Papers* (A. Puharich ed.), Wisconsin, 1979

J. Maynard-Smith   'The Status of Neo-Darwinism', in *Towards a Theoretical Biology* (C. H. Waddington ed.), 2, 1969
The Evolution of Sex, Cambridge, 1978
The Theory of Evolution (3rd ed.), Harmondsworth, 1975

J. Mayo, O. White and H. J. Eysenck   'An empirical study of the relation between astrological factors and personality', *J. soc. Psychol.*, 105, pp. 229–36, 1978

Otto Mayr   *The Origins of Feedback Control*, MIT Press, 1970

Margaret Mead (ed.)   *Cultural Patterns and Technical Change*, n.p., 1955

Sir Peter Medawar   *The Art of the Soluble*, London, 1967
Induction and Intuition in Scientific Thought, London, 1969
The Uniqueness of the Individual, London, 1957
'Unnatural Science', in *New York Review of Books*, 24, No. 1, pp. 13–18, 3 February 1977

——— and Julian H. Shelly (eds.)   *Structure in Science and Art*, Proceedings of the 3rd C. H. Boehringer Sohn Symposium, Kronberg, Taunus, 1979, 1980

Franz Anton Mesmer   *De Planetarium Influx*, n.p., 1766

C. Meyer and I. Allen (eds.)   *Source Materials in Chinese History*, London, 1970

H. W. Meyer   *A History of Electricity and Magnetism*. MIT Press, 1971

A. Michotte   *The Perception of Causality* (P. Heath transl.), London, 1963

James Mill   *Analysis of the Phenomena of the Human Mind*, 2 vols., New York, 1967

J. S. Mill   *Autobiography*, London, 1873
A System of Logic, in *John Stuart Mill: Philosophy of Scientific Method* (first pub. 1843) (Ernest Nagel ed.), New York, 1950

G. A. Miller   'The Magic Number seven plus or minus two: some limits on our capacity for processing information', *Psychol. Rev.*, 63, pp. 81–97, 1956

——— and E. Galanter and K. H. Pribram   *Plans and the Structure of Behaviour*, New York, 1960

——— and P. N. Johnson-Laird   *Language and Perception*, Cambridge, Mass., 1976

Jonathan Miller   *The Body in Question*, Cape, 1978

S. L. Miller   'Formation of Organic Compounds on the Primitive Earth', in *The Origin of Life on Earth* (A. I. Oparin ed.), Oxford, 1959

A. D. Milner and P. E. Bryant   'Cross-modal matching by young children', *J. comp. Physiol. Psychol.*, 71, pp. 453–8, 1968

Marvin Minsky   'Steps towards Artificial Intelligence', *Proc. Inst. Radio Engrs.*, 49, pp.3–30, 1961

Jacques Monod   *Chance and Necessity*, London, 1972

M. Montessori   *Dr Montessori's Own Handbook* (first pub. 1914), New York, 1965

E. A. Moody   'Laws of motion in medieval physics', in *Towards Modern Science Vol. I* (R.M.Palmer ed.), New York, 1961

Louis Trenchard More   *Isaac Newton*, New York, 1962

Lloyd C. Morgan   *Emergent Evolution*, London, 1923

An Introduction to Comparative Psychology, London, 1894

Life, Mind and Spirit, London, 1926

Harold Morick (ed.)   *Wittgenstein and the Problems of Other Minds*, New York, 1967

M. Moseley   *Irascible Genius: A Life of Charles Babbage, Inventor*, London, 1964

Frederic F. W. H. Myers   *Human Personality and its Survival After Death*, 2 vols., London, 1915

Jayant Narlikar   *The Structure of the Universe*, Oxford, 1977

Seyyed Hossein Nasr   *Islamic Science: An Illustrated Study*, Westerham, 1976

L. A. Necker   'Observations of some remarkable phenomena seen in Switzerland: and an optical phenomenon which occurs on viewing of a crystal or geometrical solid', *Phil. Mag.*, 1 (3 ser.) 329, 1832

Joseph Needham   *Clerks and Craftsmen in China and the West*, Cambridge, 1970

'The Cosmology of Early China', in *Ancient Cosmologies* (C.Blacker and M.Loewe eds.), London, 1975

Science and Civilization in China, Cambridge, 1965

——— and Ling Wang and Derek de Solla Price   *Heavenly Clockwork, the Great Astronomical Clocks of Medieval China*, Cambridge, 1960

Ulrich Neisser   *Cognitive Psychology*, New York, 1966

O. Neugebauer   *The Exact Sciences in Antiquity*, New York, 1969

John von Neumann   *The Computer and the Brain*, Yale, 1958

Beaumont Newball   *The History of Photography from 1839 to the Present Day*, New York, 1964

Allen Newell and H. A. Simon   'GPS.: A Program that simulates human thought', in *Lernende Automaten*, (K.G.Oldenbourgh ed.), Munich, 1961

Human Problem Solving, New Jersey, 1972

James R. Newman (ed.)   *The World of Mathematics*, 4 vols., New York, 1956

Isaac Newton   *The Correspondence of Isaac Newton* (H.W.Turnbull and F.J.Scott eds.), Cambridge, 1959

Opticks (4th ed. 1730), New York, 1952

Principia (first pub. 1687) (Andrew Motte transl., 1729), California, 1934

Kenneth P. Oakley   *Man the Tool Maker*, London, 1961

William of Ockham   *Ockham: Philosophical Writings* (P.Boemer ed.), London, 1926

C. K. Ogden and I. A. Richards   *The Meaning of Meaning*, London, 1944

J. Olds   'Self-Stimulation Experiments and Differentiated Reward Systems', in *Reticular Formation of the Brain* (H.H.Jasper, L.D.Proctor, R.S.Knighton, W.C.Noshay and T.R.Costello eds.), Boston, Mass., 1958

——— and P. Milner   'Positive reinforcement produced by electrical stimulation of septal area and other regions of the rat's brain', *J. comp. Physiol. Psychol.*, 47, pp.419–27, 1954

A. I. Oparin   *Origin of Life* (first pub. 1938), New York, 1953

Robert E. Ornstein (ed.)   *The Nature of Human Consciousness: A Book of Readings*, Reading, 1973

620   Bibliography

The Psychology of the Unconscious, Reading, 1972
C.E.Osgood, G.J.Suci and M.Tannenbaum   *The Measurement of Meaning*, Urbana, 1957
P.H.Osmond   *Isaac Barrow: His Life Work and Times*, London, 1943
Ian Oswald   'Human Brain Protein, Drugs and Dreams', *Nature*, 223, pp.893–7, 1969

Paracelsus   *Hermetic and Alchemical Writings of Paracelsus* (A.E.White transl.), n.p., 1894
I.P.Pavlov   *Lectures on Condiitioned Reflexes* (W.H.Grant transl.), New York, 1928
David Pears   'Russell's Theory of Memory 1912–1921', in *Bertrand Russell's Philosophy* (G.Nakhnikian ed.), London, 1974
Wittgenstein, London, 1971
Karl Pearson   *The Grammar of Science*, London, 1892
Life, Letters and Labours of Francis Galton, 2 vols., Cambridge, 1914–30
T.E.Peet   *The Rhind Mathematical Papyrus*, Liverpool, 1923
Charles Saunders Peirce   *Collected Papers of Charles Saunders Peirce* (C.Hartshorne and P.Weiss eds.), Cambridge, Mass., 1931–5
Essays in the Philosophy of Science (V.Tomos ed.), New York, 1957
'How to Make Our Ideas Clear', *Popular Science Monthly*, pp.286–302, Jan. 1878
Values in a Universe of Chance: Selected Writings of Charles S.Peirce (1839–1914) (Philip P.Weiner ed.), New York, 1958
Wilder Penfield   *The Mystery of Mind*, Princeton, 1975
——— and H.H.Jasper   *Epilepsy and the Functional Anatomy of the Human Brain*, Boston, Mass., 1954
———H.H.Jasper and T.Rasmussen   *The Cerebral Cortex of Man*, New York, 1950
——— and L.Roberts   *Speech and Brain Mechanisms*, Princeton, 1959
L.S. and R.Penrose   'Impossible Objects: A Spacial Type of Illusion', *Br. J. Psychol.*, 49: 31, 1958
John E.Pfeiffer   *The Emergence of Man*, New York, 1969
The Emergence of Society, New York, 1977
J.A.Philip   *Pythagoras and Early Pythagoreanism*, Toronto, 1966
J.Piaget   *The Child's Conception of the World* (J. and A.Tomlinson transl.), London, 1929
The Child's Construction of Reality (Margaret Cook transl.), London, 1955
The Grasp of Consciousness: Action and Concept in the Young Child (S.Wedgwood transl.), London, 1974
The Origins of Intelligence in Children, New York, 1952
Psychology and Epistemology: Toward a Theory of Knowledge (P.A.Wells transl.), London, 1972
The Psychology of Intelligence, New York, 1951
——— and B.Inhelder   *The Child's Conception of Space*, London, 1956
R.Pilkington   *Robert Boyle: Father of Chemistry*, London, 1959
Robert M.Pirsig   *Zen and the Art of Motorcycle Maintenance: an Enquiry into Values*, London, 1974
Plato   *Apology* (J.Burnet transl.), Oxford, 1924
Crito (J.Burnet transl.), Oxford, 1924
Phaedo (J.Burnet transl.), Oxford, 1911
Theaitetos (John Warrington transl.), London, 1961
The Republic (F.M.Cornford transl.), Oxford, 1941
Timaeus and Critias (2nd ed.) (Desmond Lee transl.), Harmondsworth, 1974
Jules Henri Poincaré   'L'Espace et la géometrie', *Revue de métaphysique et de morale*, 3, pp.631–46, 1895

Science and Method (first pub. 1908) (F. Maitland transl.), London, 1914

Science and Hypothesis (first pub. 1902) (G. B. Halsted transl.), New York, 1905

Karl Popper   *The Logic of Scientific Discovery*, London, 1959

'Logic of Discovery or Psychology of Research?' in *Criticism and the Growth of Knowledge* (I. Lakatos and A. Musgrave eds.), Cambridge, 1970

Objective Knowledge: An Evolutionary Approach, Oxford, 1972

———— and John Eccles   *The Self and Its Brain*, New York, 1977

Giovanni Battista della Porta   *Natural Magic* (first pub. 1668), n.p.

E. C. Poulton   'Models for biases in judging sensory magnitude', *Psychol. Bull.*, 86, pp. 777–803, 1979

'The New Psychophysics: Six Models for Magnitude Estimation', *Psychol. Bull.*, 69, pp. 1–19, 1968

David A. Premack   'A Functional Analysis of Languages', *Journal of Experimental Analysis of Behaviour*, 14, pp. 107–25, 1970

'Language and Intelligence in Ape and Man', *Am. Scient.*, 64:6, pp. 674–83, 1976

Karl Pribram   'Some dimensions of remembering: steps toward a neuropsychological model of memory', in *Macromolecules and Behaviour* (J. Gaito ed.), London, 1966

Derek de Solla Price   *Gears from the Greeks: the Antikythera Mechanism. A Calender Computer from ca. 80 BC*, New York, 1974

Science Since Babylon (2nd ed.), Connecticut, 1975

H. H. Price   *Perception*, London, 1932

Ilya Prigogine   *From Being to Becoming*, London, 1980

Morton Prince   *The Dissociation of a Personality: The Hunt for the Real Miss Beauchamp*, London, 1905

Andrija Puharich (ed.)   *The Iceland Papers*, Wisconsin, 1979

J. M. Pullan   *The History of the Abacus*, New York, 1969

Hilary Putnam   *Mind Language and Reality: Philosophical Papers*, 2 vols., Cambridge, 1975

Willard van Orman Quine   *From a Logical Point of View*, Cambridge, Mass., 1953

Word and Object, MIT Press, 1960

V. S. Ramachandran   'Twins, split brains and personal identity', in *Consciousness and the Physical World* (B. D. Josephson and V. S. Ramachandran eds.), Oxford, 1980

J. W. Reeves   *Body and Mind in Western Thought*, Harmondsworth, 1958

Hans Reichenbach   'Bertrand Russell's Logic', in *The Philosophy of Bertrand Russell* (P. A. Schilpp ed.), Illinois, 1946

The Rise of Scientific Philosophy, California, 1953

Thomas Reid   *An Inquiry into the Human Mind* (first pub. 1764) (J. Duggan ed.), Chicago, 1803

Essays on the Intellectual Powers of Man (A. D. Woozley ed.), MIT Press, 1969

Philosophy of the Intellectual Powers, 1785

Colin Renfrew   *Before Civilization: The Radio Carbon Revolution and Prehistoric Europe*, London, 1973

J. O. Robinson   *The Psychology of Visual Illusion*, London, 1972

George John Romanes   *Life and Letters of George John Romanes* (Mrs Romanes ed.), London, 1896

'Mind and Motion', in *Mind and Motion and Monism*, London, 1896

Vasco Ronchi   *The Nature of Light* (V. Barocas transl.), London, 1970

Amelie Oksenberg Rorty (ed.)   *The Identity of Persons*, California, 1976

Richard Rorty   'Dennett on Awareness', *Philosophical Studies*, 23, n.d.

'Functionalism, machines and incorrigibility', *The Journal of Philosophy*, 69:8, 20 April 1972

Steven Rose   *The Conscious Brain*, London, 1973

H.E.Ross and R.L.Gregory   'Weight illusion and weight discrimination – a revised hypothesis', *Q. Jl exp. Psychol.*, 22:318, 1970

H.E.Ross and R.L.Gregory   'Is the Weber fraction a function of physical or perceived input?', *Q. Jl exp. Psychol.*, 16:2, p.116, 1964

H.E.Ross and G.M.Ross   'Did Ptolemy understand the moon illusion?', *Perception*, 5:4, pp.377–86, 1976

M.R.Rozenweig, E.L.Bennett and M.C.Diamond   'Brain Changes in Response to Experience', *Scient. Am.*, 226, pp.22–9, 1970

E.Rubin   *Synsoplevede Figurer* (first pub. 1915) (D.C.Beardsley and M.Westheimer transl.), in *Readings in Perception*, Princeton, 1958

Bertrand Russell   *The Analysis of Mind*, London, 1921

The Autobiography of Bertrand Russell, 3 vols., London, 1967–9

An Enquiry into Meaning and Truth, London, 1940

History of Western Philosophy, London, 1946

Human Knowledge: Its Scope and Limits, London, 1948

Introduction to Mathematical Philosophy (first pub. 1919), London, 1967

'Knowledge by Acquaintance and Knowledge by Description', *Proceedings Aristotelian Society*, new series, v, xi, pp.108–28, 1910–11

Logic and Knowledge: Essays (Robert C.Marsh ed.), London, 1956

My Philosophical Development, London, 1959

'On Denoting', *Mind*, new series, v, xiv, pp.479–93, 1905

Our Knowledge of the External World, London, 1914

The Problems of Philosophy (first pub. 1912), Oxford, 1967

'Reply to My Critics', in *The Philosophy of Bertrand Russell* (P.A.Schilpp ed.), Cambridge, 1946

——— and A.N.Whitehead   *Principia Mathematica*, Cambridge, 1910–13

M.G.Rutten   *The Origin of Life by Natural Causes*, Amsterdam, 1971

Charles Rycroft   *Anxiety and Neurosis*, Harmondsworth, 1968

Gilbert Ryle   *The Concept of Mind* (first pub. 1949), Harmondsworth, 1963

M.D.Sanders, E.K.Warrington, K.Marshall and L.Weiskrantz   *Blindsight: Vision in a field defect*, *The Lancet*, pp.707–8, 20 April 1974

N.K.Sanders   *The Epic of Gilgamesh*, Harmondsworth, 1960

E.Sapir   'Conceptual Categories of Primitive Language', *Science*, 74: 578, 1931

R.J.Seeger   *Galileo Galilei, His Life and Works*, Oxford, 1966

C.Wade Savage   *The Measurement of Sensation: A Critique of Perceptual Psychophysics*, California, 1970

Alfred Schild   'Clock paradoxes in relativity theory', *Am. Math. Monthly*, Jan. 1959

P.A.Schilpp   *Albert Einstein: Philosopher-Scientist*, Cambridge, 1949

(ed.) *The Philosophy of Bertrand Russell*, Cambridge, 1946

Denise Schmandt-Besserat   'The earliest precursor of writing', *Scient. Amer.*, June 1978

R.E.Schofield   *The Lunar Society of Birmingham: A Social History of Provincial Science in Eighteenth Century England*, Oxford, 1963

J.H.Scholes   'Retinal fibre projection patterns in the primary visual pathways to the brain', in *Sense Organs* (M.S.Laverack and D.J.Cosens eds.), Glasgow, 1981

Erwin Schrödinger   *What is Life?*, Cambridge, 1958

F.Schumann   *Zeitschrift für Psychologie*, i, 23, pp.1–32, 1904

R.M.Schwartz and Margaret O.Dayhoff   'Origins of Procaryotes, Eucaryotes, Itochondria and Chloplasts', *Science*, 199, pp.395–403, 1978

D.W.Sciama   *Modern Cosmology*, Cambridge, 1971
The Physical Foundations of General Relativity, London, 1969

J.F.Scott   *The Scientific Work of René Descartes*, London, 1952

R.J.Seeger   *Galileo Galilei, His Life and Works*, Oxford, 1966

M.H.Segall, D.T.Campbell and M.J.Herskovits   *The Influence of Culture on Visual Perception*, New York, 1966

M.E.P.Seligman and Joan L.Hager (eds.)   *Biological Boundaries of Learning*, New Jersey, 1972

M.von Senden   *Space and Sight: The Perception of Space and Shape in the Congenitally Blind before and after Operation* (first pub. 1932) (Peter Heath transl.), 1960

D.Shakow and D.Rapaport   *The Influence of Freud on American Psychology*, Ohio, 1968

Claude E.Shannon   'Prediction and entropy in printed English', *Bell System Tech. Journ.*, 30, pp.59–64, 1951

—— and Warren Weaver   *The Mathematical Theory of Communication*, Illinois, 1949

Roger N. Shepard   'Psychophysical Complementarity,' in *Perceptual Organization* (M.Kubovy and J.R.Pomerantz eds.), New Jersey, 1979

—— and J.Metzler   'Mental Rotation of Three-Dimensional Objects', *Science*, 171, pp.701–3, 1971

C.S.Sherrington   *The Integrative Action of the Nervous System* (first pub. 1905), Cambridge, 1947

F.Sherwood Taylor   *The Alchemists*, London, 1952

Abner Shimony   *Is observation theory-laden? A problem in naturalistic epistemology*, Pittsburgh, 1969
'Perception from an evolutionary point of view', *J. of Philosophy*, 68:19, October 1971
'Quantum Physics and the Philosophy of Whitehead', *Boston Studies in the Philosophy of Science*, Vol. II (R.S.Cohen and M.W.Wartofsky eds.), New York, 1965
'Role of the Observer in Quantum Theory', *Am. J. Phys.*, 31:10, pp.755–73, 1963
'The status of hidden variable theories', in *Methodology and Philosophy of Science* (P.Suppes et al. eds.), London, 1973

W.K.Simpson (ed.)   *Literature of Ancient Egypt* (R.O.Faulkner transl.), Yale, 1972

Charles Singer and E. Ashworth Underwood   *A Short History of Medicine*, Oxford, 1962

Charles Singer, E.J.Holmyard and A.R.Hall (eds.)   *A History of Technology, Vol. 1: From Early Times to the Fall of Ancient Empires c. 500 BC*, Oxford, 1954

B.F.Skinner   *Behaviour of Organisms: An Experimental Analysis*, London, 1938
Cumulative Record: A Selection of Papers (3rd ed.), New York, 1972
Science and Human Behaviour, London, 1953
Verbal Behaviour, New York, 1957

Lawrence Sklar   *Space, Time and Spacetime*, California, 1974

J.J.C.Smart   *Philosophy of Scientific Realism*, London, 1963
(ed.) *Problems of Space and Time*, London, 1964
'Sensations and Brain Processes', in *The Mind/Brain Identity Theory* (C.V.Borst ed.), London, 1970

C.U.M.Smith   *The Brain: Towards an Understanding*, London, 1970

D.E.Smith and J.Ginsburg   'From numbers to numerals and from numerals to computation', in *The World of Mathematics Vol. 1* (James R.Newman ed.), New York, 1956

Frederick Snyder   'Toward an Evolutionary Theory of Dreaming', *Am. J. Psychiat.*, 123, pp.121–42, 1966

Richard Sorabji   *Aristotle on Memory*, London, 1972

C.E.Spearman   *The Abilities of Man*, London, 1927

Herbert Spencer   'Brain bisection and mechanisms of consciousness', in *Brain and Conscious Experience* (J.C.Eccles ed.), New York, 1966
   Principles of Psychology, 1855
R.W.Sperry   The Growth of Nerve Circuits', *Scient. Am.*, 201:5, pp.68–75
   'Restoration of vision after crossing of optic nerves and after contralateral transplantation of eye', *J. Neurophysiol.*, 8, pp.15–28
Benedict Spinoza   *Treatise on the Correction of the Understanding* (first pub. 1670) (A.Boyle transl.), London, 1959
   Ethics, (first pub. 1677) (A.Boyle transl.), London, 1959
R.J.Sternberg   *Intelligence, Information Processing and Analogical Reasoning. The Componential Analysis of Human Abilities*, New York, 1977
Geraldine Stevens (ed.)   *Psychophysics: An Introduction to its Perceptual, Neural and Social Prospects*, New York, 1975
S.S.Stevens (ed.)   *Handbook of Experimental Psychology*, New York, 1951
A.G.Stewart   *The Academic Gregories*, n.p., 1901
Dugald Stewart   *Reid's Works*, Edinburgh, 1803
John M.Stillman   *The Story of Alchemy and Early Chemistry*, New York, 1960
G.M.Stratton   'Vision without inversion of the retinal image', *Psychol. Rev.*, 4:341, 463, 1897
J.M.Stroud   'The fine structure of psychological time', in *Information Theory in Psychology* (H.Quastler ed.), Illinois, 1955
   The Justification of Induction (Richard Swinburne ed.), Oxford, 1974

L.M.Terman and M.Merrill   *Measuring Intelligence: A Guide to the Administration of the New Revised Stanford Binet Tests of Intelligence*, Boston, Mass., 1960
L.Marianne Teuber   'Sources of ambiguity in the prints of Maurits C.Escher', *Scient. Am.*, July 1974
Theophilus   *On Divers Arts* (John G.Hawthorne and Cyril S.Smith transl.), New York, 1980
D'Arcy Thompson   *On Growth and Form* (first pub. 1917), Cambridge, 1961
Edward L.Thorndike   *Animal Intelligence*, London, 1898
L.Thorndike   *A History of Magic and Experimental Science*, 2 vols., London, 1923
L.L.Thurstone   *The Nature of Intelligence*, London, 1924
O.L.Tinklepaugh   'An experimental study of representative factors in monkeys', *J. comp. Psychol.*, 8, pp.197–236, 1928
John Toland   *Letters to Serena*, London, 1704
S.E.Toulmin   *Human Understanding*, Princeton, 1972
M.Treisman   'Sensory Scaling and the Psychophysical Law', *Q. Jl exp. Psychol.*, 16, pp.11–12, 1964
R.L.Trivers 'The evolution of reciprocal altruism', *Q. Rev. Biol.*, 46, pp.35–75, 1971
A.M.Turing   'Computing machinery and intelligence', *Mind*, 59, pp.433–60, 1950

A.M.Uttley   'Conditional Computing in a Nervous System', in *Mechanization of Thought Processes*, London, 1959

A. Valvo   *Sight restoration after long-term blindness: the problems and behavior patterns of visual rehabilitation*, New York, 1971
P.E.Vernon   *The Structure of Human Abilities*, New York, 1950
G. Vesey (ed.)   *Body and Mind*, London, 1964
L.S. Vygotsky   *Thought and Language*, MIT Press, 1962

C.H. Waddington   *The Strategy of Genes*, London, 1957

C. Wade Savage   *The Measurement of Sensation*, California, 1970

Alfred Russell Wallace   *My Life: A Record of Events and Opinions*, 2 vols., London, 1905

R.M. Warren and R.P. Warren   'A Critique of Steven's new Psychophysics', *Perceptual and Motor Skills*, 16, pp. 797–810, 1963

Helmholtz on Perception: Its Physiology and Development, New York, 1968

J.B. Watson   *Psychology from the Standpoint of a Behaviourist* (2nd ed.), Philadelphia, 1924

James D. Watson   *The Double Helix*, London, 1968

Ernst Heinrich Weber   *De Tactu* (first pub. 1834), translated as *The Sense of Touch* (H.E. Ross and D.J. Murray transl.), London, 1978

Steven Weinberg   *The First Three Minutes*, London, 1977

L. Weiskrantz, E.K. Warrington, M.D. Sanders and J. Marshall   'Visual capacity in the hemianopic field following a restricted occipital ablation', *Brain*, 97, pp. 709–28, 1974

'Blindsight: Vision in a field defect', *The Lancet*, pp. 707–8, 20 April 1974

August Weismann   *Studies in the Theory of Descent*, London, 1882

Shirley Weitz   *Nonverbal Communication*, Oxford, 1974

A.T. Welford   *Fundamentals of Skill*, London, 1968

Max Wertheimer   'Laws of Organization in Perceptual Forms', in *A Source Book of Gestalt Psychology* (W.D. Ellis ed.), London, 1923

D. Weschler   *The Weschler Intelligence Scale for Children*, New York, 1958

R.S. Westfall   *Force in Newton's Physics*, London, 1971

'The role of alchemy in Newton's career', in *Reason, Experiment, and Mysticism in the Scientific Revolution* (M.L.R. Bonelli and W.R. Shea eds.), London, 1975

John A. Wheeler   'Beyond the Black Hole', in *Princeton Einstein Centenary Lectures* (H. Woolf ed.), Reading, Mass., 1980

Frontiers of Time, Amsterdam, 1979

'Include the observer in the wave function?', in *Quantum Mechanics a Half Century Later* (J. Leiter Lopes and M. Paty eds.), Reidal, 1977

'Law without Law,' in *Structure in Science and Art* (Peter Medawar and Julian Shelley eds.), Oxford, 1980

'The Universe as a home for Man', in *The Nature of Scientific Discovery* (O. Gingerich ed.), Washington, 1975

W.C. Whetham and M. Dampier (eds.)   *Cambridge Readings in the Literature of Science*, Cambridge, 1924

Leslie A. White   *The Science of Culture: A Study of Man and Civilization*, New York, 1949

Alfred North Whitehead   *Adventures of Ideas*, Harmondsworth, 1933

(ed.) *The Aims of Education*, London, 1932

Concept of Nature, Cambridge, 1920

Science and the Modern World, Cambridge, 1925

J.G. Whitrow   *The Nature of Time*, Harmondsworth, 1975

C.W.M. Whitty and O.L. Zangwill (eds.)   *Amnesia*, London, 1966

N. Wiener   *Cybernetics*, New York, 1948

P. Wiener   *Evolution and the Founders of Pragmatism*, Cambridge, Mass., 1949

(ed.) *Values in a Universe of Chance: Selective Writings of Charles S. Peirce 1839–1914*, New York, 1958

Eugene Wigner   *Symmetries and Reflections*, Westpoint, Conn., 1967

Bernard Williams   *Problems of the Self*, Cambridge, 1973

Moyra Williams   *Brain Damage and the Mind*, London, 1966

A.T. Wilson   'Synthesis of Macromolecules', *Nature*, 188, pp. 1007–9, 1960

Edward O.Wilson   *On Human Nature*, Cambridge, Mass., 1978
  Sociobiology: The New Synthesis, Cambridge, Mass., 1975
Terry Winograd   *Understanding Natural Language*, Edinburgh, 1972
Henry P.Winston   *Artificial Intelligence*, Reading, Mass., 1977
John Wisdom   *Other Minds*, Oxford, 1952
Ludwig Wittgenstein   *The Blue and Brown Books* (G.E.M.Anscombe ed.), Oxford, 1972
  On Certainty (G.E.M.Anscombe and G.H.von Wright eds., Denis Paul and G.E.M.Anscombe
  transl.), Oxford, 1969
  Philosophical Investigations (G.E.M.Anscombe ed.), Oxford, 1953
  Remarks on Colour (G.E.M.Anscombe ed.), Oxford, 1977
  Remarks on the Foundations of Mathematics (G.H.von Wright, R.Rhees and G.E.M.Anscombe
  eds., G.E.M.Anscombe transl.), Oxford, 1956
  Tractatus Logico-Philosophicus (D.F.Pears and B.F.McGuiness transl.), London, 1961
  Zettel (G.E.M.Anscombe and G.H.von Wright eds., G.E.M.Anscombe transl.), Oxford, 1967
  Wittgenstein: Lectures and Conversations (Cyril Barrett ed.), Oxford, 1966
Friedrich Wöhler   1828 paper on synthesis of urea, in W.C.Whetham and M.Dampier, *Cambridge
  Readings in the Literature of Science*, Cambridge, 1924
Richard Wolheim   *Bradley*, Harmondsworth, 1959
  (ed.) *Philosophers on Freud: New Evaluations*, New York, 1974
Bennett Woodcroft (transl.)   *The Pneumatics of Hero of Alexandria*, London, 1851
R.S.Woodworth   *Experimental Psychology*, New York, 1938
D.E.Wooldridge   *The Machinery of the Brain*, New York, 1963
  Mechanical Man: The Physical Basis of Intelligent Life, New York, 1968
Leonard Wooley   'Excavations at Ur', *Antiq. J.*, 10, p.332, 1930
B.L.Whorf   *Language, Thought and Reality*, New York, 1956

F.A.Yates   *The Art of Memory*, London, 1966
J.Z.Young   *Doubt and Certainty in Science*, Oxford, 1951
  The Evolution of Memory (J.J.Head ed.), Carolina, 1976
  An Introduction to the Study of Man, Oxford, 1971
  The Life of Mammals (2nd ed.), Oxford, 1975
  The Memory System of the Brain, Oxford, 1966
  The Organization of a memory system', *Proc. Roy. Soc. 'B'*, 163, pp.285–320, 1965
  Programs of the Brain, Oxford, 1978
F.T.S.Yu   *Optics and Information Theory*, New York, 1976

O.L.Zangwill   'Kenneth Craik: The Man and his Work,' *Br. J. Psychol.*, 71, pp.1–16, 1980
  'Remembering revisited', *Q. Jl exp. Psychol.*, 24, pp.123–38, 1972
A.V.Zaporozhets   'The Origin and Development of the Conscious Control of Movements in man',
  in *Recent Soviet Psychology* (N.O'Connor ed.), New York, 1961
Christopher Zeeman and Peter Buneman   'Tolerance Spaces and the Brain', in *Towards a Theo-
  retical Biology* (C.H.Waddington ed.), Edinburgh, 1968
John M.Ziman   *The Force of Knowledge: the Scientific Dimension of Society*, Cambridge, 1976
  Public Knowledge, Cambridge, 1968
  Reliable Knowledge, Cambridge, 1978

Index

Page numbers in *italic* refer to the illustrations